PLANETARY NEBULAE: MULTI-WAVELENGTH PROBES OF STELLAR AND GALACTIC EVOLUTION

IAU SYMPOSIUM 323

COVER ILLUSTRATION:

The image of planetary nebula NGC 2346 shows unprecedented resolution of the molecular hydrogen gas. The image is about 1 arc minute on a side: north is up, east is to the left. Image from GeMS/GSAOI Multi-Conjugat Adaptive Optics System, Gemini Observatory/AURA.

IAU SYMPOSIUM PROCEEDINGS SERIES

INTERNATIONAL ASTRONOMICAL UNION
UNION ASTRONOMIQUE INTERNATIONALE

PLANETARY NEBULAE: MULTI-WAVELENGTH PROBES OF STELLAR AND GALACTIC EVOLUTION

PROCEEDINGS OF THE 323RD SYMPOSIUM OF THE INTERNATIONAL ASTRONOMICAL UNION HELD IN BEIJING, CHINA OCTOBER 10–14, 2016

Edited by

XIAOWEI LIU
Peking University Kavli Institute for Astronomy and Astrophyics

LETIZIA STANGHELLINI
National Optical Astronomy Observatory

and

AMANDA KARAKAS
Monash Centre for Astrophysics

CAMBRIDGE
UNIVERSITY PRESS

CAMBRIDGE UNIVERSITY PRESS
University Printing House, Cambridge CB2 8BS, United Kingdom
1 Liberty Plaza, Floor 20, New York, NY 10006, USA
10 Stamford Road, Oakleigh, Melbourne 3166, Australia

First published 2017

Printed in the UK by Bell & Bain, Glasgow, UK

Typeset in System LaTeX 2ε

A catalogue record for this book is available from the British Library Library of Congress Cataloguing in Publication data

ISBN 9781107169913 hardback
ISSN 1743-9213

Table of Contents

Part 2. Poster Presentation

Preface

Planetary nebulae are the spectacular products of low- and intermediate-mass stellar evolution. They are key laboratories for the understanding of atomic, molecular, dust and plasma processes in different astrophysical environments, they probe fundamental physics of single and binary star evolution, including nucleosynthesis, rotation, mass transfer and loss, and magnetic fields, and they trace stellar populations, kinematics and chemistry of galaxies including our own galaxy the Milky Way.

The planetary nebula community, now represented by IAU Commission H3 of Division H, has had a history of very successful IAU symposia, beginning in 1967 with IAU Symp. 34 held in Tatranska Lomnica (Czechoslovakia), followed in 1977 by IAU Symp. 76 in Ithaca, NY (USA), IAU Symp. 103 in London (England) in 1982, IAU Symp. 131 in Mexico City (Mexico) in 1987, IAU Symp. 155 in Innsbruck (Austria) in 1992, IAU Symp. 180 in Groningen (Holland) in 1996, IAU Symp. 209 in Canberra (Australia) in 2001, IAU Symp. 234 in Hawaii (USA) in 2006, and IAU Symp. 283 in Puerto de la Cruz (Spain) in 2011.

Following the endorsement and sponsorship by IAU Divisions and approved by the IAU Executive Committee in April 2015, the latest of this series of IAU Symposia of planetary nebulae, IAU Symposium 323: Planetary nebulae: Multi-wavelength probes of stellar and galactic evolution, was held in Beijing on October 10-14, 2016. This was the first time an IAU symposium on this topic was held in Asia. Over one hundred registered participants from seventeen countries attended the symposium, and interacted and discussed the wide range of topics related to the field. In total 17 invited reviews, 44 contributed talks and 36 posters were presented.

The meeting covers broad themes including new challenges to surveys of planetary nebula, AGB and post-AGB populations in the Milky Way, in the Local Group of galaxies and beyond, planetary nebulae as versatile laboratories for atomic, molecular, dust and plasma physics, stellar nucleosynthesis and chemical enrichment of the interstellar medium, hydrodynamic and radiative modeling of planetary nebulae, single and binary stellar evolution from the asymptotic giant branch to white dwarf stellar phases, magnetic fields, mass loss and mass transfer in binaries and connection to the planetary nebula phenomena, planetary nebulae in nearby galaxies as tracers of galactic dynamics and chemical evolution. In addition, two discussion sessions, each of one hour duration, were held on the topics "Modelling and data" and "Planetary nebulae and forefront instrumentation", moderated by specific SOC members and other scientists who are experts in these areas.

The programme was carefully put together by the Scientific Organizing Committee, consisting of Magda Arnaboldi (Germany), You-Hua Chu (China Taipei), Romano Corradi (Spain), Orsola De Marco (Australia), Adam Frank (U.S.A.), Amanda Karakas (co-chair, Australia), Karen Kwitter (U.S.A.), Sun Kwok (China Nanjing), Xiaowei Liu (co-chair, China Nanjing), Alberto Lopez (Mexico), Arturo Manchado (Spain), Roberto Mendez (U.S.A.), Quentin Parker (Australia), Miriam Pena (Mexico), Letizia Stanghellini (co-chair, U.S.A.), Albert Zijlstra (U.K.). The Committee made an early decision that, in order to help bring in early career speakers, no Committee members were invited to give a review talk. During the symposium, three of the Committee members, Orsola De Marco, Letizia Stanghellini, and Albert Zijlstra volunteered to run a best student poster competition. One first class, two second class and a number of encouragement awards were awarded, with generous prizes provided by the Local Organizing Committee. The activity was well received by all student participants.

The Local Organizing Committee, consisting of Bingqiu Chen, Jianning Fu, Biwei Jiang (co-chair), Stephen Justham (co-chair), Xiaowei Liu (co-chair), Juanjuan Ren, Jieling Yu, Juanjuan Ren and Haibo Yuan provided the local support. The symposium website was built and maintained by Jian Li. Ning Zhang provided indispensable logistic assistance. The symposium was operated by the Overseas Exchange Center of Peking University. The smooth and efficient operation of the symposium was widely and highly praised by the participants. As one of the participants commented afterwards, "Everything was outstanding. From the scientific program to the logistics assistance, the social activities and everything. The meeting was perfect."

Finally, it is a great pleasure to acknowledge the financial support from the International Astronomical Union, National Natural Science Foundation of China, National Astronomical Observatories of Chinese Academy of Sciences, and Peking University. The active involvement of members of the Scientific Organizing Committee, crucial for the success of the meeting, is also acknowledged.

Xiaowei Liu, Letizia Stanghellini, and Amanda Karakas
Co-chairs of Scientific Organizing Committee Beijing,
Rome, and Melbourne, February 2017

The Organizing Committees

Scientific

Magda Arnaboldi (Germany)
Romano Corradi (Spain)
Adam Frank (U.S.A.)
Karen Kwitter (U.S.A.)
Xiaowei Liu (co-chair, China Nanjing)
Arturo Manchado (Spain)
Quentin Parker (Australia)
Letizia Stanghellini (co-chair, U.S.A.)
You-Hua Chu (China Taipei)
Orsola De Marco (Australia)
Amanda Karakas (co-chair, Australia)
Sun Kwok (China Nanjing)
Alberto Lopez (Mexico)
Roberto Mendez (U.S.A.)
Miriam Pena (Mexico)
Albert Zijlstra (U.K.)

Local

Juanjuan Ren
Jianning Fu
Biwei Jiang (co-chair)
Xiaowei Liu (co-chair)
Haibo Yuan
Bingqiu Chen
Stephen Justham (co-chair)
Jifeng Liu
Jieling Yu

Acknowledgements

The symposium was sponsored and supported by IAU Division H (Interstellar Matter and Local Universe), Division G (Stars and Stellar Physics), Division J (Galaxies and Cosmology), as well as by IAU Commission H3 (Planetary Nebulae).

The Local Organizing Committee operated under the auspices of School of Physics, Peking University.

Funding by the
International Astronomical Union,
National Natural Science Foundation of China,
National Astronomical Observatories of Chinese Academy of Sciences,
Peking University
is gratefully acknowledged.

CONFERENCE PHOTOGRAPH

Part 1. Invited Reviews and Contributed Presentations

1. SOC co-Chair Xiaowei Liu gives the welcome address

2. IAU President, and Planetary Nebula scholar, Silvia Torres-Peimbert delivers a very interesting introductory talk

3. Mike Barlow, SOC Member Arturo Manchado, and others are attentively following an interesting review

4. Participants discussing science during a break

5. SOC co-Chair Letizia Stanghellini enjoys coffee while discussing science with Invited Speaker Maria Lugaro

6. Astronomers connect with old and new friends and drink coffee

7. SOC Member Magda Arnaboldi and Devika Kamath at the coffee break

8. Invited speaker Flavia dell'Agli and Jorge Gardia-Rojas enjoying coffee break

Part 2. Poster Presentation

9. Invited speaker Gloria Delgado-Inglada views the posters

10. Poster viewing and science discussion

Participants

First Name	Last Name	Nationality	Gender	Institute	Country	City	Email
Xiaowei	Liu	China	M	PKU	China	Beijing	x.liu@ pku.edu.cn
Xuan	Fang	Chinese	M	The University of Hong Kong	China	HongKong	fangx@ hku.hk
Letizia	Stanghellini	Italy	F	NOAO	United States	Tucson	lstanghellini@ noao.edu
Orsola	De Marco	Australian	F	Macquarie University	Australia	Sydney	orsola.demarco@ mq.edu.au
Quentin	Parker	Australian	M	The University of Hong Kong	China	HongKong	quentinp@ hku.hk
Lizette	Guzman	Mexican	F	ESO	Mexico	Saltillo	guzmanl@ eso.org
Nicholas	Sterling	USA	M	University of West Georgia	United States	Atlanta	nsterlin@ westga.edu
José Jairo	Díaz-Luis	Spanish	M	IAC	Spain	Santa Cruz de Tenerife	jdiaz@ iac.es
Mercedes	Mollá	Spain	F	CIEMAT	Spain	Madrid	mercedes.molla@ ciemat.es
Hans	Van Winckel	Belgian	M	Institute of Astronomy, KU Leuven	Belgium	Leuven	hans.vanwinckel@ ster.kuleuven.be
Mike	Barlow	UK	M	UCL	United Kingdom	London	mjb@ star.ucl.ac.uk
Gloria	Delgado Inglada	Spanish	F	Instituto de Astronomía, UNAM	Mexico	Mexico City	gdelgado@ astro.unam.mx
Jorge	Garcia-Rojas	Spanish	M	Instituto de Astrofisica de Canarias	Spain	La Laguna	astroyorch@ gmail.com
Marcelo	L. Leal Ferreira	Brazilian	M	Leiden Observatory	Netherlands	Leiden	ferreira@ strw.leidenuniv.nl
Nazım	Aksaker	Turkish	M	Çukurova Üniversity	Turkey	Adana	naksaker@ cu.edu.tr
Suryashree	Aniyan	Indian	F	Australian National University	Australia	Canberra	suryashree.aniyan@ anu.edu.au
Yong	Zhang	Chinese	M	University of Hong Kong	China	HongKong	zhangy96@ hku.hk
christophe	morisset	french	M	instituto de astronomia - UNAM	Mexico	Mexico City	chris.morisset@ gmail.com
Sun	Kwok	Canada	M	University of Hong Kong	China	Anshan	sunkwok@ hku.hk
Jackeline	Rechy-García	Mexican	F	Instituto de Astronomia UNAM	Mexico	Mexico City	jrechy@ astro.unam.mx
Muthumariappan	Chinnathambi	Indian	M	Indian Institute of Astrophysics	India	Bengaluru	muthu@ iiap.res.in
Sheila Nathanya	Flores Duran	Mexican	F	Instituto de Astronomia UNAM	Mexico	Mexico City	nathanyaster@ gmail.com
Arturo	Manchado	Spanish	M	IAC	Spain		amt@ iac.es
Isabella	Soeldner-Rembold	German	F	Max Planck Institute for Extraterrestrial Physics	Germany	Muenchen	ism@ mpe.mpg.de
Zhuo	Chen	China	M	University of Rochester	United States	Rochester	zchen25@ ur.rocchester.edu
Domingo Anibal	Garcia-Hernandez	Spanish	M	Instituto de Astrofisica de Canarias (IAC)	Spain	Santa Cruz de Tenerife	agarcia@ iac.es
Lodovico	Coccato	Italian	M	ESO	Germany	Muenchen	lcoccato@ eso.org
Vasiliki	Fragkou	Greek	F	HKU	China	Anshan	vfrag@ hku.hk
Hyosun	Kim	Korea	F	ASIAA	Korea, South	Seoul	hkim@ asiaa.sinica.edu.tw
Ivan	Bojicic	Serbian	M	The University of Hong Kong	China	HongKong	ibojicic@ hku.hk
Mohammad	Akashi	Israeli	M	Technion, Israel	Israel	Nazareth	akashi@physics. technion.ac.il

First Name	Last Name	Nationality	Gender	Institute	Country	City	Email
Liliana	Hernández-Martínez	Mexicana	F	ICN-UNAM	Mexico	Mexico City	lilihe.mtz@gmail.com
Helen	Barker	British	F	University of Manchester	United Kingdom	Manchester	helen.barker@postgrad.manchester.ac.uk
Emma Margarita	Pereyra Talamantes	Mexican	F	University of Southampton	United Kingdom	Southampton	mally@astrosen.unam.mx
Jason	Nordhaus	US	M	Rochester Institute of Technology	United States	Rochester	nordhaus@astro.rit.edu
Laura	Boyle	Irish	F	NUI Galway	Ireland	Galway	l.boyle2@nuigalway.ie
Miriam	Pena Cardenas	chilean	F	Instituto de Astronomia, UNAM	Mexico	Mexico City	miriam@astro.unam.mx
Kyung-Won	Suh	South Korea	M	Chungbuk National University	Korea, South	Seoul	kwsuh@chungbuk.ac.kr
REBECA	GALERA	SPANISH	F	INSTITUTO DE ASTROFÍSICA DE CANARIAS	Spain	La Laguna	rgr@ iac.es
Helge	Todt	German	M	Institute of Physics and Astronomy, University of Potsdam	Germany	Potsdam	htodt@astro.physik.uni-potsdam.de
Flavia	Dell'Agli	Italian	F	INAF - Observatory of Rome	Italy	Rome	flaviadellagli@gmail.com
Romano Luigi Maria	Corradi	italian	M	GRANTECAN	Spain	Santa Cruz de Tenerife	romano.corradi@gtc.iac.es
Magda	Arnaboldi	Italian	F	ESO	Germany	Muenchen	marnabol@eso.org
Carolyn	Doherty	Australian	F		Hungary	Budapest	carolyn.doherty@csfk.mta.hu
Alessia	Longobardi	Italian	F	Max Planck for extraterrestrial Physics (MPE)	Germany	Muenchen	alongobardi@mpe.mpg.de
Nicholas	Walton	Great Britain	M	University of Cambridge	United Kingdom	Cambridge	naw@ast.cam.ac.uk
Johanna	Hartke	German	F	ESO	Germany	Muenchen	jhartke@ eso.org
Nicole	Reindl	German	F	University of Leicester	United Kingdom	Leicester	nr@ le.ac.uk
Jose Alberto	Lopez	Mexico	M	Instituto de Astronomia, UNAM	Mexico	Ensenada	jal@astrosen.unam.mx
Bruce	Hrivnak	USA	M	Valparaiso University	United States	Chicago	bruce.hrivnak@valpo.edu
Rodolfo	Montez	US	M	Smithsonian Astrophysical Observatory	United States	Boston	rmontez@cfa.harvard.edu
Leticia	Juan de Dios	Mexican	F	INAOE	Mexico	Puebla de Zaragoza	leticiajuandedios@gmail.com
Roberto	Mendez	Argentine	M	Institute for Astronomy	United States	Honolulu	mendez@ifa.hawaii.edu
Shane	Vickers	Australian	M	Macquarie University	Australia	Sydney	shane.vickers@students.mq.edu.au
Albert	Zijlstra	Dutch	M	University of Manchester	United Kingdom	Manchester	a.zijlstra@ manchester.ac.uk
Masaaki	OTSUKA	JAPAN	M	ASIAA	Japan	Soka	otsuka@asiaa.sinica.edu.tw
Greg	Sloan	USA	M	Cornell	United States	New York City	sloan@isc.astro.cornell.edu
Devika	Kamath	Indian	F	Institute of Astronomy, KU Leuven	Belgium	Leuven	devika@ster.kuleuven.be
Marcin	Gladkowski	polish	M	Nicolaus Copernicus Astronomical Center, Polish Academy of Sciences	Poland	Warsaw	seyfert@ncac.torun.pl
Marcelo	Miller Bertolami	Argentine	M	CONICET-IALP	Argentina	La Plata	m3b-muadib@hotmail.com
Ryszard	Szczerba	Polish	M	NCAC	Poland	Torun	szczerba@ncac.torun.pl

First Name	Last Name	Nationality	Gender	Institute	Country	City	Email
Jan	Cami	Belgium	M	The University of Western Ontario / SETI	Canada	London	jcami@ uwo.ca
Fabio	Bresolin	Italy	M	University of Hawaii	United States	Honolulu	bresolin@ ifa.hawaii.edu
Kentaro	Asano	JAPAN	M	ISAS/JAXA	Japan	Sagamihara	kenasano@ ir.isas.jaxa.jp
Diana	Estrella	Mexican	F	UNAM	Mexico	Mexico City	diana.trujillo@ correo. nucleares. unam.mx
Bruce	Balick	Balick	M	Univ. Washington	United States	Seattle	balick@ uw.edu
Griet	Van de Steene	Belgian	F	Royal Observatory of Belgium	Belgium	Brussels	G.Vandesteene@ oma.be
Detlef	Schoenberner	German	M	Leibniz-Instiute for Astrophysics Potsdam	Germany	Potsdam	deschoenberner@ aip.de
Tim	Gledhill	UK	M	University of Hertfordshire	United Kingdom	Saint Albans	t.gledhill@ herts.ac.uk
Martin A	Guerrero	Spain	M	IAA	Spain	Granada	mar@ iaa.es
Alba	Aller	Spanish	F	Intituto de Física y Astronomía, Universidad de Valparaíso	Chile	Valparaiso	alba.aller@ ifa.uv.cl
Roger	Wesson	British	M	University College London	United Kingdom	London	rw@ nebulousre- search.org
Monica Wendolyn	Blanco Cardenas	Mexican	F	Instituto de Astronomia UNAM	Mexico	Ensenada	mwblanco@ astrosen. unam.mx
George	Pagomenos	British	M	The Open University	United Kingdom	Milton Keynes	george. pagomenos@ open.ac.uk
YU	WANG	China	M	Fudan University	China	Shanghai	wang_yu@ fudan.edu.cn
Vladimir	Escalante	Mexican	M	IRyA	Mexico	Mexico City	v.escalante@ crya.unam.mx
Minia	Manteiga Outeiro	Spain	F	Universidad de A Coruna	Spain	A Coruna	minia.manteiga@ udc.es
David	Jones	British	M	IAC	Spain	Santa Cruz de Tenerife	djones@ iac.es
Harriet	Dinerstein	U.S.A.	F	U. Texas, Austin	United States	Austin	hdinerstein@ gmail.com
Paul	Woods	British	M	Nature Astronomy	United Kingdom	London	paul.woods@ nature.com
Eva	Villaver	Spanish	F	UAM	Spain	Madrid	villaver.eva@ gmail.com
Maria	Lugaro	Italian	F	MTA CSFK	Hungary	Budapest	maria.lugaro@ csfk.mta.hu
Jeffrey	Cummings	USA	M	Johns Hopkins University	United States	Baltimore	jcummi19@ jhu.edu
Noam	Soker	Israel	M	Technion	Israel	Haifa	soker@physics. technion.ac.il
SILVIA	TORRES-PEIMBERT	MEXICAN	F	IASTRONOMIA, UNIV NACIONAL AUTONOMA MEXICO	Mexico	Mexico City	silvia@astro. unam.mx
Gerardo	Ramos	Mexican	M	Universidad de Guadalajara	Mexico	Guadalajara	gerardo@astro. iam.udg.mx
Yu	Zhang	China	M	Xinjiang Astronomical Observatory	China	Urumqi	zhy@ xao.ac.cn
Tao	Chen	Chinese	M	Leiden Observatory	Netherlands	Leiden	chen@strw. leidenuniv.nl
Pedro	Garcia-Lario	Spanish	M	ESA	Spain	Madrid	pedro.garcia. lario@esa.int
Haibo	Yuan	China	M	Beijing Normal University	China	Beijing	yuanhb@ bnu.edu.cn
Yang	Huang	China	M	Peking University	China	Beijing	yanghuang@ pku.edu.cn
Zhiying	Huo	China	F	NAOC	China	Beijing	zhiyinghuo@ bao.ac.cn
Maosheng	Xiang	Chinese	M	NAOC	China	Beijing	msxiang@ nao.cas.cn
Lang	Cui	Chinese	M	Xinjiang Astronomical Observatory, CAS	China	Urumqi	cuilang@xao. ac.cn
Hongli	Ma	China	M	Xinjiang Astronomical Observatory, UCAS	China	Urumqi	mahongli@ xao.ac.cn
Haifeng	Yang	China	M	Taiyuan University of Science and Technology	China	Taiyuan	hfyang@nao. cas.cn

First Name	Last Name	Nationality	Gender	Institute	Country	City	Email
Zhijia	Tian	Chinese	M	Peking University	China	Beijing	tianzhijia@ pku.edu.cn
chun	wang	China	F	Department of Astronomy, Peking University	China	Beijing	wchun@ pku.edu.cn
yong	yang	Chinese	M	Beijing Normal University	China	Beijing	yang_yong@ mail.bnu.edu.cn
Bingqiu	Chen	Chinese	M	Peking University	China	Beijing	bchen@ pku.edu.cn
Huawei	Zhang	Chinese	M	Department of Astronomy, Peking University	China	Beijing	zhanghw@ pku.edu.cn
Gaochao	Liu	Chinese	F	Three Gorges university	China	Wuhan	gcliu@ nao.cas.cn
Jiawen	Yao	Chinese	M	Beijing Normal University	China	Beijing	201621160012@ mail.bnu.edu.cn
David	Frew	Australian	M	HKU	Hongkong	Hongkong	djfrew@ hku.hk
Eric	Lagadec	France	M	Observatoire Cote d'Azur	France	Nice	elagadec@ oca.eu
Kris On Yan	Chui	Chinese	F	HKU	China	Hongkong	koychui@ hku.hk
Meng	Zhang	Chinese	F	Hebei Normal University	China	Shijiazhuang	2479995140@ qq.com

Part 1. Invited Reviews and Contributed Presentations

Session 1: A census of the PostAGB stars and PN populations

Chair: Arturo Manchado

Planetary Nebulae: Multi-Wavelength Probes of Stellar and Galactic Evolution
Proceedings IAU Symposium No. 323, 2016
X. Liu, L. Stanghellini & A. Karakas, eds.

doi:10.1017/S1743921317000916

Some Historical Notes

Silvia Torres-Peimbert

Instituto de Astronomía, Universidad Nacional Autónoma de México,
Apartado Postal 70-264, México, D. F., México
email: silvia@astro.unam.mx

Abstract. The study of planetary nebulae started more than a century ago. Since then the understanding of these exciting objects has advanced extraordinarily. I present a personal selection of topics to reflect on some developments of PNe research.

Keywords. planetary nebulae: general

1. More than a century ago

Firstly I will present some historical events that took place more than a hundred years ago, for us to ponder on the changes that we have attained on this very exciting topic.

First identifications and images. We are all aware that the first PNe (M27, M57, M76 and M97) were catalogued in 1764 by Charles Messier. I also found of interest to comment on the first drawings of what we now know as PNe by William Parsons, 3rd Earl of Rosse (Rosse 1844). From what was at the time the world largest telescope, he observed and sketched by hand the appearance of M57, M97, NGC 2438, NGC 7662 and NGC 7009. It required more than 70 years for Curtis (1918) to present photographic plates of a large number of PNe obtained with the 0.9 m Crossley Telescope at Lick Observatory at a scale of 30″/mm.

Optical Spectroscopy and radial velocities. One of the first spectra of planetary nebulae reported is from the photographic plate of NGC 6543 by Huggins & Miller (1864) who recorded 3 nebular emission lines (Hβ, 4959 and 5007 Å) with no continuum; the last two lines are from [O III], of which their origin was not known at the time. Following this work, the first radial velocities of Orion and a dozen PNe, previous to the identification of the [O III] 5007 feature, were obtained by Keeler (1894) at Lick Observatory with the 36-in equatorial telescope. He determined very accurately at 5007.07 Åthe rest wavelength of the strong emission line; from this measurement he could derive the velocities of a dozen objects, with uncertanties of a few miles/sec and concluded that "...The observations show that the nebulae are moving in space with velocities comparable with those of the stars...". At the time it was of great importance to determine whether the nebulous objects had the same kinematical properties as the stars to decide whether they were Galactic or extragalactic.

2. PNe Symposia

As we all know this meeting is part of a series of 11 symposia that have taken place, at intervals of approximately five years where we have witnessed the development of the field. Ten of these symposia have been held as IAU activities (the exception was the symposium held in Liege in 1972). These meetings have not been the only scientific events dedicated to PNe; since 1984 a series of conferences on Asymmetrical Planetary Nebulae was started, which have attracted an important number of participants in the field. In addition, there have been several other conferences dedicated to specific aspects of PNe. The proceedings of these meetings allow us to glimpse at a snapshot of our knowledge on this topic at a given time.

From IAU Symposium 34. In particular, I would like to mention some highlights of the meeting held in Czechoslovakia in 1967. The program included the different topics that we continue to address: space distributions, observational data (some of the first photoelectric data, and infrared emission from bolometers that were able to record up to 15 μm), model predictions for UV emission lines, free-free radio data, updates on physical processes, determinations of chemical compositions, data on central stars and studies on the origin and evolution of these objects. It was clear that the recently published Catalogue of Planetary Nebulae by Perek & Kohoutek (1967) gave a new perspective to PNe research: there were 1067 galactic objects uniformly catalogued, with as many characteristics as were available, as well as overall properties like Galactic distribution, kinematics, etc. For his part, Terzian (1968) presented observations at radio wavelengths of about 80 PNe which were then compared to the Hβ flux. These results confirmed the recombination theory of emission and thus established PNe as thermal sources. Osterbrock (1968) considered that "...we have the theoretical tools we need to investigate planetary nebulae...collision strengths, transition probabilities, recombination coefficients and photo-ionization cross-sections..."; according to Feast (1968) the "...major problem in discussion on the distribution and kinematics of the planetaries is the problem of distance determination...". Gurzadian (1968) listed several questions of particular interest: (a) internal motions, (b) two-envelope nebulae, (c) magnetic fields, (d) gas dynamics, and (d) condensations of NGC 7293. Savedoff (1968) reflected that the ancestor for the planetary nebulae are the Red giants; but at that meeting there was also the concern about the large amount of energy required to lift the envelope material from a degenerate core. In the field of extragalactic nebulae, Westerlund (1968) indicated that there were 30 true planetary nebulae identified in the SMC and about 45 in the LMC to the limiting magnitude $M_{pg} = -3.0$.

3. From Early Discoveries to the Present

Now returning to more modern studies, I would now like to call attention to the great leaps achieved on several topics.

Radio observations of PNe. Heckathorn (1971) in a very interesting review noted that in the first radio survey of PNe by Lynds (1961) at NRAO with the 85 ft dish, only 5 PNe were detected. Later, Menon & Terzian (1965) measured the flux density of 10 PNe with the 300 ft transit dish at NRAO; these observations were used to check the recombination theory of emission by comparison with Hβ fluxes. In Heckathorn catalogue there were about 150 PNe with radio observations at 2 wavelengths (750 & 1410 MHz) and the main concern was the difference between recombination theory and the validity of the models. We can compare these early efforts with the recent work by Chhetri *et al.* (2015) who provided radio properties at high frequencies up to 20 GHz, for 65 PNe which comprise a complete radio flux limited sample of planetary nebulae in the Galactic plane at $|b| < 1.5°$ in the southern hemisphere.

Ultraviolet observations. The pioneer rocket observation of NGC 7027 by Bohlin *et al.* (1975) was the precursor of the PNe ultraviolet observations. They were followed by Copernicus data and very shortly by the IUE mission which provided extensive studies of a large sample of PNe yielding the very important C abundance in the nebulae and mass loss determinations from central stars. Furthermore the IUE database set up a precedent regarding uniformly secured spectra that could be retrieved and analyzed. This facility was used widely by a large number of investigators. HST allowed new opportunities for observations to observe UV spectra. The FUSE mission again opened up the access for extensive studies of PNe (see Guerrero *et al.* 2010) who analized P-Cygni profiles of 60 central stars.

X-ray observations. Among the first data on PNe were the results by Tarafdar & Apparao (1988) who searched the archival observations of the Einstein Observatory for 19 central stars of PNe for X-ray emission. They found detections in X-rays in four of them (NGC 246, NGC 1360, NGC 6853 and NGC 7293); NGC 246 had also been observed by the high-resolution imaging instrument, from which they concluded that this object is a point source.

More sensitive observations from the Chandra X-ray Observatory and the XMMNewton missions revealed the presence of difuse X-ray emission within PNe, lending strong support to the interacting stellar winds model of PN formation (e.g., Kastner *et al.* 2000). Chandra has been particularly successful in the detection of point-sources at CSPNe (Ruiz *et al.* 2013).

The review by Guerrero (2015) emphasize that two types of X-ray sources are mostly found in planetary nebulae (PNe): point sources at their central stars and diffuse emission inside hot bubbles. The available data on these hot bubbles have been displayed in conjunction with the optical observations and the match for the inner regions is outstanding.

At present there are very complete surveys at different energies (see Kastner *et al.* 2012). Building over multiple programs for X-ray observation of individual sources, the PN community has joined efforts into the Chandra Planetary Nebulae Survey, ChanPlaNS, aiming at the X-ray study of a volume-limited sample of PNe within 1.5 kpc of the Sun. In this survey, they find an overall X-ray detection rate of 70% for a sample of 35 objects, and 50% of the PNe observed by Chandra present X-ray-luminous CSPNe, while soft, diffuse X-ray emission tracing shocks (in most cases hot bubbles, formed by energetic wind collisions) is detected in 30% of them; five objects display both diffuse and point-like emission components. Further results of this same project have been presented by Montez *et al.* 2015 who carried out X-ray spectral analysis of 20 point-like X-ray sources detected in the solar neighborhood. Most of these detections are associated with luminous central stars within relatively young, compact nebulae.

Proto-planetary nebulae. A new class of stellar objects, called proto-planetary nebulae, PPNe, were discovered on the data acquired by the IRAS satellite. They are objects in transition between the asymptotic giant branch, AGB, and planetary nebula phases in the evolution of intermediate- and low- mass stars. As such, they are characterized by a central star of spectral type B - G surrounded by an expanding envelope of gas and dust. Selecting PPNe candidates is difficult using only visible light observations, but the excess mid-infrared emission produced by their cool dust affords a good means to identify PPN candidates. PPNe have been studied widely in the IR and now there are extensive identifications and a significant number of them have been imaged in the IR.

Thus, following the IRAS mission, it became possible to identify a number of candidates and to confirm them with ground-based observations, followed up with Hubble Space Telescope imaging and ISO and Spitzer mid- infrared spectroscopy. It was seen that most PPNe could be characterized as oxygen-rich ($C/O < 1$) or carbon-rich ($C/O > 1$) based on mid-infrared spectral features. Hu *et al.* (1993) from a sample of PPNe candidates based on the IRAS colors, were able to secure photometry, spectroscopy and molecular line observations which led to the confirmation of their nature as PPNe.

At the same time theoretical calculations through the AGB phase with the inclusion of mass loss, and also through the following evolutionary stages down to the white dwarf stage have become available (e. g. Vassiliadis & Wood 1994; Bloecker 1995) to compare with observations. These calculations predict for a typical remnant of 0.6 M$\odot$, evolutionary lifetimes of several 1000 yr in the transition region between the tip of the AGB and the planetary-nebulae region. Thus it became possible to compare theory and observations in the very early phase of the post-AGB evolution. However, recent post-AGB evolutionary sequences computed by Miller Bertolami (2016) for the same stellar mass are at least three to ten times faster than those previously published. These results will allow a better comparison between studies of post-AGB stars and planetary nebulae.

Extragalactic Planetary Nebulae. The number of catalogued extragalactic PNe has increased significantly from the early works on point source emission line objects in the Magellanic Clouds which comprised less than a hundred. From high sensitivity, high resolution narrow-band surveys very deep catalogues have been produced to study planetary nebulae in nearby galaxies. These studies have been coupled with access to complimentary, deep, multi-wavelength surveys across near-IR, mid-IR and radio regimes in particular from both ground-based and spaceborne telescopes. Planetary nebulae have been identified not only in Local Group Galaxies, but also at larger distances including the nearby Virgo and Fornax galaxy clusters and even there are planetary candidates in the distant Coma Cluster (Gerhard 2007). They are being used not only as standard candles to determine distance scales to galaxies (starting with the seminal work by Jacoby 1989), but also to determine kinematics and metallicity gradients in local universe galaxies (e. g. Magrini *et al.* 2016).

Abundance determinations. The study of abundance determinations has advanced considerably, not only from forbidden lines from but also from permitted lines. Altogether the abundance determination of 17 elements had been obtained by 1990. Later neutron capture elements in Galactic planetary nebulae, undertaken to study enrichments from s-process nucleosynthesis in their progenitor stars were identified for the first time in NGC 7027 by Pequignot & Baluteau (1994).

These results have been achieved due to the increasing observational capabilities, the far UV and far IR observations and to the possibility of line identifications and atomic parameter determinations. Even in the optical and near infrared observations it is possible to derive abundances of a significant number of elements. For example, in the near infrared Sterling, *et al.* (2016) discovered Rb, Ca & Ge in the near infrared spectra and García-Rojas *et al.* (2015) based on deep, high-resolution spectrophotometric data obtained with a high resolution echelle spectrograph, from 3100 - 10500 Å, identified and measured more than 750 emission lines of NGC 3918 and from them determined the abundances of 16 elements (He, C, N, O, Ne, Na, S, Cl, Ar, K, Ca, Fe, Se, Kr, Rb, Xe). Delgado-Inglada (2017) indicate that also P, F, and Br have been determined in several PNe.

The ionization structure model determinations that started in late 1960s (e. g. Flower 1968; Williams 1968) have acquired more complete representations and options through codes such as CLOUDY (Ferland *et al.* 2013) and with codes by other authors that allow a better understanding of the ionization structure within the nebulae.

Nevertheless there is a standing problem due to the difference in abundance determinations between the forbidden lines and the recombination lines. The recombination lines systematically yield higher ionic abundances than the forbidden line determinations. This difference has been called the Abundance Discrepancy Factor, ADF, and great efforts have been devoted to explain this discrepancy. An explanation of this problem has been proposed being due to temperature inhomogeneities in the nebula (Peimbert 1967). Recently Peimbert *et al.* (2014) found that in a group of 20 well observed PNe, at least 16 of them are chemically homogeneous and present large temperature fluctuations. On the other hand, Corradi *et al.* (2015) and Jones *et al.* (2016) have found several chemically inhomogeneous PN with binary central stars.

4. Conclusions

I have presented just a few samples of the great advances that have taken place on the study of planetary nebulae. Many of the topics of interest have not been included, and will be discussed in this meeting and in several meetings to come.

The increase in our understanding of the origin, development and status of PNe has been remarkable. This has been possible through the concerted efforts of more extended and deeper observations coupled to great efforts in theoretical and computational progress. There are still many pending issues that we have not yet been able to solve, and we are looking forward for many more future meetings on PNe.

References

Bloecker, T. 1995, *A&A*, 299, 755

Bohlin, R. C., Stecher, T. P., & Marionni, P. A. 1975, *ApJ*, 202, 415

Chhetri, R., Ekers, R. D., Kimball, A., *et al.* 2015, *MNRAS*, 451, 3228

Corradi, R. L. M., García-Rojas, J., Jones, D., & Rodríguez-Gil, P. 2015, *ApJ*, 803, 99

Curtis, H. D. 1918, *Pub. Lick Obs.*, 13, 9

Delgado-Inglada, G. 2017, in IAU Symp. 323: Planetary nebulae: Multi-wavelength probes of stellar and galactic evolution, eds. X.-W. Liu, L. Stanghellini & A. Karakas

Dinerstein, H. & Stirling, E. 1993, *ApJ*, 28, 490

Feast, M. W. 1968, in IAU Symp. 34, Planetary Nebulae, ed. D. E. Osterbrock & C. R. O'Dell (New York, Springer-Verlag), 445

Ferland, G. J., Porter, R. L., van Hoof, P. A. M., *et al.* 2013, *RMxAA* 49, 137

Flower, D. R. 1968, in IAU Symp. 34, Planetary Nebulae, ed. D. E. Osterbrock & C. R. O'Dell (New York, Springer-Verlag), 205

García-Rojas, J., Madonna, S., Luridiana, V., *et al.* 2015, *MNRAS*, 452, 2606

Gerhard, O., Arnaboldi, M., Freeman, K. C., Okamura, S., Kashikawa, N., & Yasuda, N. 2007, *A&A*, 468, 815

Guerrero, M. A., Ramos-Larios, G., & Massa, D. 2010, *PASA*, 27, 2, 210

Guerrero, M. A. 2015, in *Wolf-Rayet Stars* eds. W-R Hamann, A. Sander, & H. Todt (Potsdam: Univ-Verlag), 263

Gurzadian, G. A. 1968, in IAU Symp. 34: Planetary Nebulae: ed. D. E. Osterbrock & C. R. O'Dell (New York, Springer-Verlag), 448

Heckathorn, J. 1971, *Ap&SS*, 11, 309

Hu, J. Y., Jiang B. W., de Jong, T., & Slijkuis, S. 1993, in IAU Symp 155: Planetary nebulae: ed. R. Weinberger & A. Acker (Dordrecht, Kluwer), 333

Huggins, W. & Miller, W. A. 1864, *Proc. R. Soc. Lond.*, 13, 491

Jacoby G. H. 1989, *ApJ*, 339, 39

Jones, D., Wesson, R., García-Rojas, J., Corradi, R. L. M., & Boffin, H. M. J. 2016, *MNRAS*, 455, 3263

Kastner, J. H., Soker, N., Vrtilek, Saeqa D., & Dgani, R. 2000, *ApJ*, 545, L5

Kastner, J. H., Montez, R., Jr, Balick, B., *et al.* 2012, *AJ*, 144, 58

Keeler, J. E. 1894, *Pub. Lick Obs*, III, 161

Lynds, C. R. 1961, *Pub. NRAO*, 1. No. 5

Magrini, L., Coccato, L., Stanghellini, L., Casasola, V., & Galli, D. 2016, *A&A*, 588, A91

Menon T K. & Terzian Y. 1965, *ApJ*, 141, 745

Miller Bertolami, M. M. 2016, *A&A*, 588, 25

Montez, R., Jr., Kastner, J. H., Balick, B., *et al.* 2015, *ApJ*, 800, 8

Osterbrock, D. E. 1968, in IAU Symp. 34, Planetary Nebulae, ed. D. E. Osterbrock & C. R. O'Dell (New York, Springer-Verlag), 441

Peimbert, M. 1967, *ApJ*, 150, 825

Peimbert, A., Peimbert, M., Delgado-Inglada, G.,García-Rojas, J., & Peña, M. 2014, *RMxAA*, 50, 329

Pequignot, D. & Baluteau, J.-P. 1994, *A&A*, 283, 593

Perek, L. & Kohoutek, L. 1967, Catalogue of Galactic Planetary Nebulae (Prague: Academia)

Rosse, Earl of 1844, *Phil. Trans. R. Soc. London*, 134, 321

Ruiz, N., Chu, Y.-H., Gruendl, R. A., *et al.* 2013, *ApJ*, 767, 35

Savedoff, M. P. 1968, in IAU Symp. 34, Planetary Nebulae, ed. D. E. Osterbrock & C. R. O'Dell (New York, Springer-Verlag), 463

Sterling, N. C., Dinerstein, H. L., Kaplan, K. F., & Bautista, M. A. 2016, *ApJ* (Letters), 819, L9

Tarafdar, S. P. & Apparao, K. M. V. 1988, *ApJ*, 327, 342

Terzian, Y. 1968, in IAU Symp. 34, Planetary Nebulae, ed. D. E. Osterbrock & C. R. O'Dell (New York, Springer-Verlag), 87

Vassiliadis, E. & Wood, P. R. 1994, *ApJS*, 92, 125

Westerlund, B. E. 1968, in IAU Symp. 34, Planetary Nebulae, ed. D. E. Osterbrock & C. R. O'Dell (New York, Springer-Verlag), 23

Williams, R. E. 1968, in IAU Symp. 34, Planetary Nebulae, ed. D. E. Osterbrock & C. R. O'Dell (New York, Springer-Verlag), 190

Discussion

ARNABOLDI: Important most developments is to establish the properties of PNe with respect to the late phase of stellar evolutions and relation with AGB and metallicity of the parent stellar population.

Planetary Nebulae: Multi-Wavelength Probes of Stellar and Galactic Evolution
Proceedings IAU Symposium No. 323, 2016
X. Liu, L. Stanghellini & A. Karakas, eds.

doi:10.1017/S1743921317003507

Challenges faced by current Galactic planetary nebula surveys

David J. Frew[1,2]

[1]The University of Hong Kong, Department of Physics, Hong Kong SAR, China
[2]Laboratory for Space Research, The University of Hong Kong, Hong Kong SAR, China
email: djfrew@hku.hk

Abstract. Determining the demographics of the Galactic planetary nebula (PN) population is an important goal to further our understanding of this intriguing phase of stellar evolution. The Galactic population has more than doubled in number over the last 15 years, particularly from narrowband Hα surveys along the plane. In this review I will summarise these results, with emphasis on the time interval since the last IAU Symposium. These primarily optical surveys are not without their limitations and new surveys for PNe in the infrared similarly face a number of challenges. I will discuss the need for multi-wavelength approaches to discovery and analysis. The desire to have accurate volume-limited samples of Galactic PNe at our disposal is emphasised, which will be impacted with new data from the Gaia satellite mission. We need robust surveys of PNe and their central stars, especially volume-limited surveys, in order to clarify and quantify their evolutionary pathways.

Keywords. planetary nebulae: general, stars: evolution, circumstellar matter, Galaxy: stellar content, catalogs, surveys, techniques: miscellaneous

---

1. Introduction

Planetary nebulae (PNe) are a common end-point of stellar evolution and are an important contributor to the evolution of the Galaxy, but there is still much to learn regarding their mass-loss and nucleosynthetic processes and the importance of binarity in their formation. Widely-held assumptions on their progenitor stars and visibility lifetimes have needed revision (Miller Bertolami 2016). Ultimately, determining the ages, masses, and chemical compositions of a complete census of PNe is necessary to understand how low- to intermediate-mass stars influence the evolution of our Galaxy. While close binary central stars have now been proven to have a shaping role for their PNe (Hillwig *et al.* 2016a), the common-envelope phenomenon is still poorly understood (De Marco, these proceedings). Moreover, a subset of close binaries are double-degenerate systems, and potential SN Ia progenitors, which are important cosmological probes.

Many new Galactic PNe have been found over the last two decades, especially from narrow-band surveys along the Galactic plane, although these optical surveys are not without their limitations and undertaking new surveys for PNe similarly face a number of challenges. In this review I will summarise these results, especially over the time interval since the last IAU Symposium. Here I will only concentrate on Milky Way surveys for PNe; for studies of their progenitors, the post-AGB stars and pre-PNe (e.g. Szczerba *et al.* 2012; Vickers *et al.* 2015) the reader is referred to Lagadec (these proceedings).

2. PN Surveys in the Milky Way

Current PN search techniques are imperfect in one way or another, with limitations depending on the spatial resolution and areal coverage of the survey, and on the size,

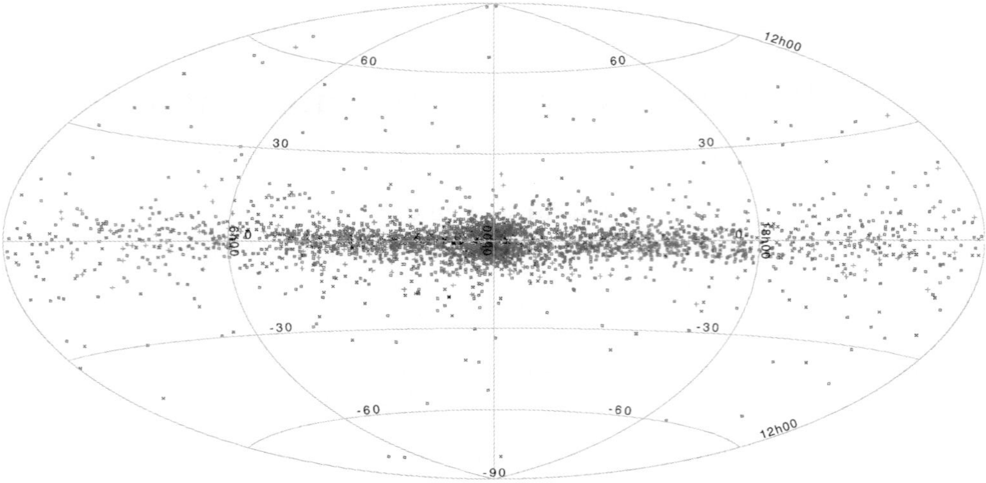

Figure 1. Aitoff projection showing the Galactic distribution of all 3540 PNe currently in HASH. True, likely, and possible PNe are the red, green, and black symbols respectively.

surface brightness and extinction of individual PNe. Using a range of complementary techniques should in principle recover most of the detectable PNe in a particular wavelength regime, assuming that each survey was systematically completed and covered the whole sky, which is clearly not the case. Since the Spitzer surveys only covered a region very close to the Galactic plane (Benjamin *et al.* 2003; Carey *et al.* 2009), mid-IR surveys for PNe are quite incomplete, Consequently, only a fraction of the true population of Galactic PNe has been recovered. Past compilations have been seriously incomplete and biased, and only now has an attempt been made to consolidate all extant data sets into one database in a systematic manner (Parker *et al.* 2016a,b; Bojičić *et al.* 2016). Currently, the Hong Kong/AAO/Strasbourg/Hα (HASH) database contains 3540 true, likely, and possible PNe (Parker *et al.* 2016b). This number is only about 15–30% of the estimated total of Galactic PNe (Frew 2008; Jacoby *et al.* 2010), showing that only a small fraction of Galactic PNe have been catalogued. The distribution of the current population of Galactic PNe is seen in Fig. 1.

Most of these undiscovered PNe are invisible at optical wavelengths and future confirmations and discoveries will require surveys in the IR and radio domains across wide swaths of the sky with enhanced sensitivity and resolution. Since the work of Bojičić *et al.* (2011), some progress has already been made in these areas (e.g. Hoare *et al.* 2012; Norris *et al.* 2011; Umana *et al.* 2015). Ultimately, IR imaging and spectroscopy will be required to validate future discoveries in a similar way to current classification protocols.

3. Detection Techniques, Past and Present

It is germane to briefly review the various detection techniques that have been used to discover Galactic PNe over the last two centuries (for a detailed review, see Frew & Parker 2010). Each technique is sensitive to PNe of differing size and brightness as illustrated in Fig. 2, and summarised below:

• Visual discoveries, typified by the NGC PNe (e.g. those found between 1780 and 1838 by W. and J. Herschel), which are relatively large *and* bright (Fig. 2).

• Broadband photographic imagery e.g. Abell (1966), and the recent work by the Deep-Sky Hunters (DSH) team, an amateur-professional consortium (Jacoby *et al.* 2010;

Kronberger *et al.* 2016, and references therein). Candidates were typically identified from direct inspection of images, or the online scanned data in the case of the DSH discoveries.

• Narrowband imagery, e.g. the MASH catalogues (Parker *et al.* 2006, Miszalski *et al.* 2008), IPHAS catalogue (Sabin *et al.* 2014), and some finds by the DSH team (Kronberger *et al.* 2014, 2016), and other smaller surveys (Fig. 2); and potentially with the J-PAS survey (Gonçalves *et al.* 2016). These PNe are usually of low surface brightness, being identified via direct inspection, image processing, or using color-color plots.

• Near-infrared (near-IR) imagery, particularly utilising narrowband images to find faint PNe; e.g. Jacoby & Van de Steene (2004) in the [S III] line, and discoveries from the UWISH2 survey (Froebrich *et al.* 2015; Gledhill, these proceedings) in the light of molecular hydrogen. Near-IR imaging of IRAS mid-IR candidates has also been useful for classifying new PNe and pre-PNe (Ramos-Larios *et al.* 2009, 2012)

• Mid-IR imagery or colors, with radio or near-IR confirmation in some cases (e.g. Preite-Martinez 1988; Van de Steene & Pottasch 1993; García-Lario *et al.* 1997; Kwok *et al.* 2008; Cohen *et al.* 2010; Mizuno *et al.* 2010; Ingallinera *et al.* 2016). WISE mid-IR imagery has also been used to find very faint PNe (Kronberger *et al.* 2016).

• Low-resolution objective-prism spectroscopic surveys by many authors, particularly before the 1970s, that revealed modestly bright but generally compact PNe.

• Higher-resolution spectroscopic discoveries using multiplexed fiber spectroscopy, e.g. Yuan & Liu (2013), and §7.

• Direct discoveries of low-mass [WR] central stars (Kanarek *et al.* 2015, 2016), that may be surrounded by currently undetected faint PNe. This approach has the potential to find many obscured H-deficient central stars.

Most discoveries in the last decade have utilised narrowband imaging or near- and mid-IR surveys. Since the last IAUS review in 2011 (Parker *et al.* 2012), over 750 PNe and PN candidates have been found, many still needing assessment in the HASH database. Sabin *et al.* (2014) announced the discovery of 159 true, likely and possible PNe, 284 PN candidates have been found from UWISH2 (Froebrich *et al.* 2015; Gledhill, these proceedings), and another 180 from the DSH team and related efforts (Kronberger *et al.* 2016; and personal communication; Acker *et al.* 2015). Fragkou *et al.* (2016) have found 70 candidates using radio and IR surveys, and other contributions in the infrared have been made by Flagey *et al.* (2014), Nowak *et al.* (2014), and Ingallinera *et al.* (2014a,b, 2016) who are systematically investigating the 300+ unidentified MIPSGAL nebulae found by Mizuno *et al.* (2010), though most remain un-assessed at the time of writing.

Small additional lists of new objects or one-off discoveries, not including our own unpublished objects, include those of Ramos-Larios *et al.* (2012), Parker *et al.* (2012b), Miszalski *et al.* (2013), Aller *et al.* (2013), Blanco Cárdenas *et al.* (2013), Frew *et al.* (2013), Munari *et al.* (2013, see Frew *et al.* 2014c), Yuan & Liu (2013), Miranda *et al.* (2014), Miszalski & Mikołajewska (2014), Chhetri *et al.* (2015) and Kanarek *et al.* (2016).

4. The Problem of PN Mimics

Contaminating objects in previous PN surveys have been the rule, and may never be reduced to zero, though Parker *et al.* (2016a,b) has explained the rationale for HASH. Both emission-line stars and extended objects of various types have been confused with PNe (for a review, see Frew & Parker 2010). Emission-line stars, historically found through objective prism surveys, and now from multi-band CCD surveys, can be subdivided into several classes: symbiotic stars, Be and B[e] stars, Wolf-Rayet stars, luminous blue variables (LBVs), pre-main sequence stars, cataclysmic variables of various types, and even late-type giant stars, with or without Balmer emission.

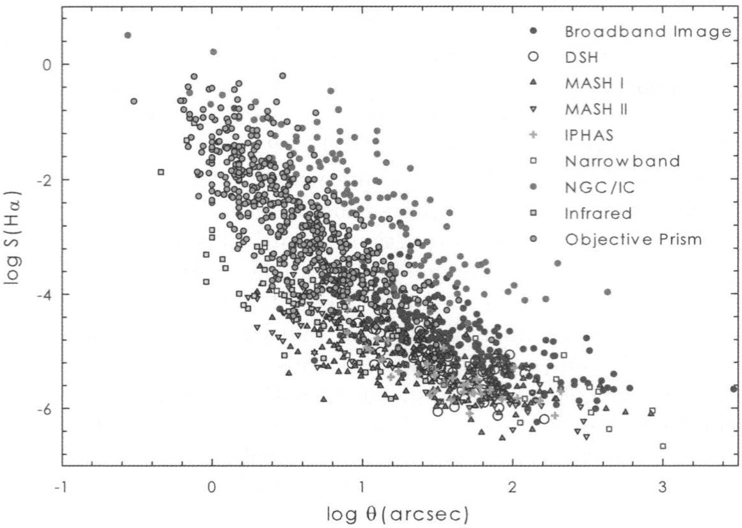

Figure 2. PNe with reliable Hα surface brightnesses from HASH. Here the surface brightness is plotted against nebular radius. Different discovery techniques are sensitive to different samples of PNe, as outlined in §3.

Extended nebulae, usually found via imagery, have as mimics included H II regions, reflection nebulae, galaxies, symbiotic outflows, Wolf-Rayet shells and LBV ejecta nebulae, supernova remnants, young stellar objects, Herbig-Haro objects, other miscellaneous nebulae and even plate defects. Indeed, several of the nearest candidates are likely to be Strömgren zones in the interstellar medium ionized by a hot pre-white dwarf or subdwarf, with any original PN having long since dispersed. If the ionizing star is hot, these H II regions can have [O III] emission, and emission-line ratios which mimic true PNe; e.g. Hewett 1, DeHt 5, Sh 2-68, and Sh 2-174 (Hewett *et al.* 2003; Chu *et al.* 2004; Frew 2008; cf. Ransom *et al.* 2010, 2015).

A suite of diagnostic diagrams are now available to help eliminate these mimics and to discriminate between them and bona fide PNe, transitional objects between the post-AGB and PN phases, and a number of peculiar PN-like nebulae (Anderson *et al.* 2010; Frew & Parker 2010; Sabin *et al.* (2013); Frew *et al.* (2014b, 2016a,b). Symbiotic stars and their resolved nebulae, if present, remain a thorn in the side of classification efforts. Corradi *et al.* (2010) have devised useful diagnostic plots for symbiotic stars, to complement the diagrams mentioned previously.

5. Volume-limited Surveys

Extant PN surveys in the Milky Way, and the studies utilising them, have until now been flux-limited, and biased against the most abundant kind of PN – evolved ones! It should be apparent that volume-limited surveys are needed for unbiased statistical analysis of PNe and their central stars (multiplicity, central star properties, nebular properties, nebular kinematics), and for robust comparison with population synthesis models and evolutionary theory. Specifically we need to clarify and quantify their evolutionary pathways, which may be several in number (Frew & Parker 2010; Reindl *et al.* 2014; Hillwig *et al.* 2016). Furthermore the discovery of Galactic PNe continues, including several nearby objects, discussed in §3 and §7.

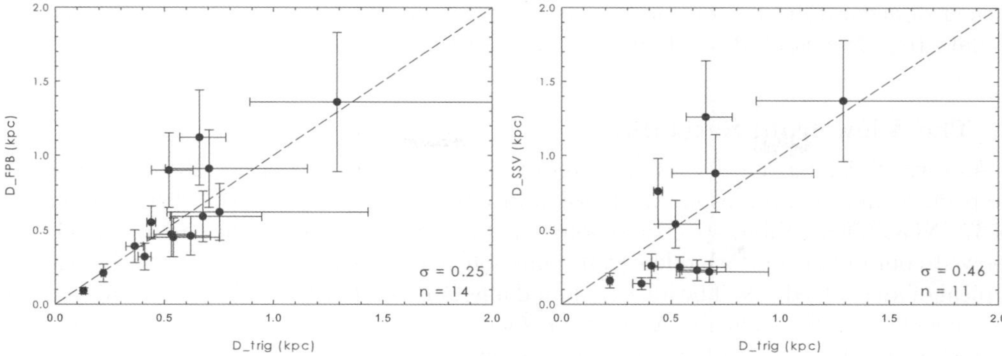

Figure 3. Comparison of trigonometric distances with published statistical distances. The comparison is with the Frew *et al.* (FPB) scale (left) and the Stanghellini *et al.* (SSV) scale (right).

Current developments on volume-limited surveys are being undertaken by Frew *et al.* (in prep.). At present, relatively few accurate primary distances are available (see §6), so we have relied on statistical distances; e.g. Stanghellini *et al.* (2008), and most recently, Frew *et al.* (2016a). To date, only a few studies have utilised volume-limited samples, but particularly noteworthy are the ChanPLaNS (Kastner *et al.* 2012; Freeman *et al.* 2014; Montez *et al.* 2015) and HerPlaNS (e.g. Ueta *et al.* 2014; Aleman *et al.* 2014) surveys in the X-ray and thermal-IR domains respectively.

6. A short look at Gaia Data Release 1

The local census of PNe will be greatly helped in future with data from the Gaia survey (Gaia Collaboration 2016). Gaia Data Release 1 (Lindegren *et al.* 2016) contains positions and G magnitudes (in the Gaia system) for some hundreds of central stars, but less than two dozen have proper motions and parallaxes (Stanghellini *et al.* 2016; Manteiga, these proceedings), and only four of these parallaxes are statistically meaningful, with errors of <50%). To make a better comparison with the latest statistical distance scales, the Gaia data are supplemented with the HST fine-guidance sensor parallaxes (Benedict *et al.* 2009), and high-quality USNO data from Harris *et al.* (2007). A comparison of the trigonometric distances against the statistical distances from Frew *et al.* (2016a) show good agreement (Fig. 3), but less so with the older Stanghellini *et al.* (2008) distance scale (see Smith 2015). Two strongly discrepant points, Cn 3-1 and M 1-77, are both young, compact objects (not shown in Fig. 3). Their Gaia distances are only 520 and 930 pc respectively, compared to estimates of 4 – 6 kpc from both statistical scales. Either the preliminary Gaia parallaxes for these two compact nebulae are seriously in error, or we are dealing with a new PN-like phenomenon at low luminosities.

Even after later Gaia data releases become available, statistical distances will remain complementary. In the optical, the Hα emission-line is the most useful calibrating waveband, as it is brighter than Hβ, is less affected by moderate reddening, and because several flux-calibrated Hα imaging surveys are now available (Gaustad *et al.* 2000; Drew *et al.* 2005, 2014; Frew *et al.* 2014a), from which large numbers of accurate integrated Hα fluxes can be determined (Frew *et al.* 2013, 2014a). For highly-reddened nebulae, Paschen-α is the brightest detectable hydrogen line but is only accessible from space (Wang *et al.* 2010), but Brackett-γ is also useful in the near-IR. Forbidden mid-IR fine-structure lines are almost immune to reddening, but the intensities are strongly influenced by the excitation of the PN, and thus are more problematic to calibrate. Ultimately, the advent of accurate radio-continuum fluxes for the many small, faint PNe current lacking in data (e.g. Norris *et al.* 2011) will warrant a recalibration of the radio distance scale.

7. The View from Serendip

As I was finishing this review, an interesting paper by Núñez *et al.* (2016) appeared in preprint form, on chromospheric emission in late-type members of the rich open cluster M 37 (NGC 2099). Fiber spectra of some of the stars in the northern part of the cluster showed contaminating forbidden [N II] emission lines which were strong relative to Hα, typical of an evolved PN. The authors found a nebula on IPHAS Hα imagery but made no firm conclusion on its nature. An evening spent with online databases revealed a 19-mag blue star near the centre of the box-shaped nebula on the DSS blue image. Applying the well-established cluster reddening to the available GALEX/SDSS photometry gives colors consistent with a hot star, and if a member of M 37 ($D = 1.6$ kpc), the absolute magnitude ($M_{\rm V} = +7.5$) is typical of an evolved ionizing star. Curiously, this star was already noted as a candidate white dwarf by Gentile Fusillo *et al.* (2015) and independently found by Chang *et al.* (2015) to be a periodic variable with a period of 0.445 days. This is undoubtedly a new bipolar PN, and potentially a physical member of M 37, so is of great interest, especially if the CSPN is a close binary (see Jones 2016). We are planning to follow-up on this object shortly. This serendipitous discovery shows that our census of even relatively nearby PNe is still incomplete.

8. Future Directions

Future discovery surveys in the Milky Way will continue to be useful, as part of efforts to characterise the entire Galactic PN population (Parker *et al.* 2016a,b). More powerful imaging and spectroscopic IR surveys of representative samples of PNe are necessary to better understand mass-loss processes and dust chemistry. The advent of the Large Synoptic Survey Telescope (LSST) will be a major breakthrough in time-domain studies of central stars in particular. For the nebulae, high-resolution spatial and kinematic and spatial surveys with ALMA will be rewarding, for both discovery, and importantly in the characterisation of post-AGB stars and younger PNe. The flagship James Webb Space Telescope (JWST), to be launched in 2018, will potentially have a major impact on Galactic PNe studies, though less so in the mid-IR owing to its faint saturation limits (Barlow 2012). The recently launched Astrosat mission should also provide useful complementary data at X-ray and UV wavelengths. Continued work on volume-limited surveys will constrain more tightly the relative statistics of different PN populations, and this is where where the final Gaia data release at the end of the decade will have the greatest impact. In addition, the common envelope phenomenon remains poorly understood, as is its relationship to the newly recognised abundance discrepancy problem (Corradi *et al.* 2015). We need deep spectra of all central stars in the local volume (cf. Weidmann & Gamen 2011) for comparison with theoretical models, and to also constrain the relative frequency of binary and ternary systems (see Akashi, these proceedings), and their potential role in forming and shaping PNe.

References

Abell, G. O. 1966, *ApJ*, 144, 259
Acker, A. 2015, *EAS Publications Series*, 71-72, 309
Aleman, I., *et al.* 2014, *A&A*, 566, A79

Aller, A., *et al.* 2013, *A&A*, 552, A25

Anderson, L. D., Zavagno, A., Barlow, M. J., García-Lario, P., & Noriega-Crespo, A. 2012, *A&A*, 537, A1

Barlow, M. 2012, in A. Manchado, L. Stanghellini & D. Schönberner (eds), *Proc. IAU Symp.*, 283, Planetary Nebulae: An Eye to the Future (Cambridge: CUP), p. 295

Benedict, G. F., *et al.* 2009 *AJ*, 138, 1969

Benjamin, R. A., *et al.* 2003, *PASP*, 115, 953

Blanco Cárdenas, M. W., *et al.* 2013, *A&A*, 551, A64

Bojičić, I. S., Parker, Q. A., Filipović, M. D., & Frew, D. J. 2011, *MNRAS*, 412, 223

Bojičić, I. S., Parker, Q. A., & Frew, D. J. 2016, these prceedings, `arXiv:1612.01521`

Carey S. J., *et al.* 2009, *PASP*, 121, 76

Chang, S.-W. Byun, Y.-I., & Hartman, J. D., 2015, *AJ*, 150, 27

Chhetri, R., Ekers, R. D., Kimball, A., Miszalski, B., Cohen, M., & Manick, R. 2015, *MNRAS*, 451, 3228

Chu, Y.-H., Gruendl, R. A., Williams, R. M., Gull, T. R., & Werner, K. 2004, *AJ*, 128, 2357

Cohen, M., Parker, Q. A., Green, A. J., Miszalski, B., Frew, D. J., & Murphy, T. 2011, *MNRAS*, 413, 514

Corradi, R. L. M., *et al.* 2010, *A&A*, 509, A41

Corradi, R. L. M., García-Rojas, J., Jones, D., & Rodríguez-Gil, P. 2015, *ApJ*, 803, 99

Drew, J. E., *et al.* 2005, *MNRAS*, 362, 753

Drew, J. E., *et al.* 2014, *MNRAS*, 440, 2036

Flagey, N., Noriega-Crespo, A., Petric, A., & Geballe, T. R. 2014, *AJ*, 148, 34

Fragkou, V., Bojičić, I., Frew, D. J., & Parker, Q. 2016, these proceedings, `arXiv:1611.08590`

Freeman, M., *et al.* 2014, *ApJ*, 794, 99

Frew, D. J. 2008, PhD thesis, Macquarie University, Sydney

Frew, D. J. & Parker, Q. A. 2010, *PASA*, 27, 129

Frew, D. J., Bojičić, I. S., & Parker, Q. A. 2013, *MNRAS*, 431, 2

Frew, D. J., Bojičić, I. S., & Parker, Q. A., Pierce M. J., Gunawardhana M. L., Reid W. A. 2014a, *MNRAS*, 440, 1080

Frew, D. J., *et al.* 2014b, *MNRAS*, 440, 1345

Frew, D. J., Bento, J., Bojičić, I. S., & Parker, Q. A. 2014c, *MNRAS*, 445, 1605

Frew, D. J., Parker, Q. A., & Bojičić, I. S. 2016a, *MNRAS*, 455, 1459 (FPB)

Frew, D. J., Bojičić, I. S., & Parker, Q. A. 2016b, *Journal of Physics: Conf. Series*, 728, 032015

Froebrich, D., *et al.* 2015, *MNRAS*, 454, 2586

Gaia Collaboration 2016, *A&A*, in press, `arXiv:1609.04153`

Garcia-Lario, P., Manchado, A., Pych, W., & Pottasch, S. R. 1997, *A&AS*, 126, 479

Gaustad, J. E., McCullough, P. R., Rosing, W., & Van Buren, D. 2001, *PASP*, 113, 1326

Gentile Fusillo, N. P., Gänsicke, B. T., & Greiss, S. 2015, *MNRAS*, 448, 2260

Gonçalves, D. R., *et al.* 2016, in *Proc. IAU Symp.*, 317, The General Assembly of Galaxy Halos: Structure, Origin and Evolution (Cambridge: CUP), p. 304

Harris, H. C., *et al.* 2007, *AJ*, 133, 631

Hillwig, T., Jones, D., De Marco, O., Bond, H., Margheim, S., & Frew, D. 2016a, *ApJ*, 832, 125

Hillwig, T. C., Frew, D. J., Reindl, N., Rotter, H., Webb, A., & Margheim, S. 2016b, *AJ*, in press, `arXiv:1612.01420`

Hoare, M. G., *et al.* 2012, *PASP*, 124, 939

Ingallinera, A., *et al.* 2014a, *MNRAS*, 437, 3626

Ingallinera, A., Trigilio, C., Umana, G., Leto, P., Agliozzo, C., & Buemi, C. 2014b, *MNRAS*, 445, 4507

Ingallinera, A., *et al.* 2016, *MNRAS*, 463, 723

Jacoby, G. H. & Van de Steene, G. 2004, *A&A*, 419, 563

Jacoby, G. H., *et al.* 2010, *PASA*, 27, 156

Jones, D. 2016, these proceedings, `arXiv:1611.08454`

Kanarek, G. C., Shara, M. M., Faherty, J. K., Zurek, D., & Moffat, A. 2015, *MNRAS*, 452, 2858

Kanarek, G. C., Shara, M. M., Faherty, J. K., Zurek, D., & Moffat, A. 2016, *MNRAS*, in press, `arXiv:1512.09306`

Kastner, J. H., *et al.* 2012, *AJ*, 144, 58
Kronberger, M., *et al.* 2014, in C. Morisset, G. Delgado-Inglada & S. Torres-Peimbert, (eds.), *Proc. APN VI Conf.*, see `http://www.astroscu.unam.mx/apn6/PROCEEDINGS/`, article 48
Kronberger, M., *et al.* 2016, *Journal of Physics: Conf. Series*, 728, 072012
Kwok, S., Zhang, Y., Koning, N., Huang, H.-H., & Churchwell, E. 2008, *ApJS*, 174, 426
Lindegren, L., *et al.* 2016, *A&A*, in press, `arXiv:1609.04303`
Miller Bertolami, M. M. 2016, these proceedings, `arXiv:1611.09801`
Miranda, L. F., Rodríguez, L. F., Pereira, C. B., & Vázquez, R. 2014, *MNRAS*, 442, 995
Miszalski, B., Parker, Q. A., Acker, A., Birkby, J. L., Frew, D. J., & Kovacevic, A. 2008, *MNRAS*, 384, 525
Miszalski, B. & Mikołajewska, J. 2014, *MNRAS*, 440, 1410
Miszalski, B., Mikołajewska, J., & Udalski, A. 2013, *MNRAS*, 432, 3186
Mizuno, D. R., *et al.* 2010, *AJ*, 139, 1542
Munari, U., Corradi, R. L. M., Siviero, A., Baldinelli, L., & Maitan, A. 2013, *A&A*, 558, A2
Montez, R., *et al.* 2015, *ApJ*, 800, 8
Norris, R. P., *et al.* 2011, *PASA*, 28, 215
Nowak, M., Flagey, N., Noriega-Crespo, A., Billot, N., Carey, S. J., Paladini, R., & Van Dyk, S. D. 2014, *ApJ*, 796, 116
Núñez, A., Agüeros, M. A., Covey, K. R., & López-Morales, M. 2016, *ApJ*, in press, `arXiv:1611.09313`
Parker, Q. A., *et al.* 2005, *MNRAS*, 362, 689
Parker, Q. A., *et al.* 2006, *MNRAS*, 373, 79
Parker, Q. A., Frew, D. J., Acker, A., & Miszalski, B. 2012a, in A. Manchado, L. Stanghellini & D. Schönberner (eds), *Proc. IAU Symp.*, 283, Planetary Nebulae: An Eye to the Future (Cambridge: CUP), p. 9
Parker, Q. A., *et al.* 2012b, *MNRAS*, 427, 3016
Parker, Q. A., Bojičić, I. S., & Frew, D. J. 2016a, *Journal of Physics: Conf. Series*, 728, 032008
Parker, Q. A., Bojičić, I. S., & Frew, D. J. 2016b, these proceedings, `arXiv:1612.00167`
Preite-Martinez, A. 1988, *A&AS*, 76, 317
Purcell, C. R., *et al.* 2013, *ApJS*, 205, 1
Ramos-Larios, G., Guerrero, M. A., Suárez, O., Miranda, L., & Gómez, J. 2009, *A&A*, 501, 1207
Ramos-Larios, G., Guerrero, M. A., Suárez, O., Miranda, L., & Gómez, J. 2012, *A&A*, 545, A20
Ransom, R. R., Kothes, R., Wolleben, M., & Landecker, T. L. 2010, *ApJ*, 724, 946
Ransom, R. R., Kothes, R., Geisbuesch, J., Reich, W., & Landecker, T. L. 2015, *ApJ*, 799, 198
Reindl, N., Rauch, T., Werner, K., Kruk, J. W., & Todt, H. 2014, *A&A*, 566, A116
Sabin, L., *et al.* 2013, *MNRAS*, 431, 279
Sabin, L., *et al.* 2014, *MNRAS*, 443, 3388
Smith, H., Jr. 2015, *MNRAS*, 449, 2980
Stanghellini, L., Shaw, R. A., & Villaver, E. 2008, *ApJ*, 689, 194 (SSV)
Stanghellini, L., Bucciarelli, B., Lattanzi, M. G., & Morbidelli, R. 2016, `arXiv:1609.08840`
Szczerba, R., Siódmiak, N., Stasińska, G., Borkowski, J., García-Lario, P., Suárez, O., Hajduk, M., & García-Hernández, D. A. 2012, in A. Manchado, L. Stanghellini & D. Schönberner (eds), *Proc. IAU Symp.*, 283, Planetary Nebulae: An Eye to the Future (Cambridge), 506
Ueta, T., *et al.* 2014, *A&A*, 565, A36
Umana, G., *et al.* 2015, *MNRAS*, 454, 902
Van de Steene, G. C. M. & Pottasch, S. R. 1993, *A&A*, 274, 895
Vickers, S. B., Frew, D. J., Parker, Q. A., & Bojičić, I. S. 2015, *MNRAS*, 447, 1673
Wang, Q. D., *et al.* 2010, *MNRAS*, 402, 895
Weidmann, W. A. & Gamen, R. 2011, *A&A*, 526, A6
Yuan, H. B. & Liu, X. W. 2013, *MNRAS*, 436, 718

Discussion

STANGHELLINI: I think your work on mimics is very important. How would you see it to extend to PNe outside the Local Group?

De Marco: How many post RGB stars are there in the post common envelope PN group? Population synthesis studies claim quite a large fraction of post RGB post common envelope PN.

Frew: One very likely (Hf 2-2) but other than that only may be approx. 4 could be post CE post RGB out of a sample of approx. 60 post common envelope PN. So too few to corroborate pop. synthesis models.

Planetary Nebulae: Multi-Wavelength Probes of Stellar and Galactic Evolution
Proceedings IAU Symposium No. 323, 2016
X. Liu, L. Stanghellini & A. Karakas, eds.

doi:10.1017/S1743921317000928

Post-AGB nebular studies

Eric Lagadec

Université Côte d'Azur, Observatoire de la Côte d'Azur, CNRS, Lagrange, France
email: elagadec@oca.eu

Abstract. This review presents the latest advances in the nebular studies of post-AGB objects. Post-AGB stars are great tools to test nucleosynthesis and evolution models for stars of low and intermediate masses, and the evolution of dust in harsh environment. I will present the newly discovered class of post-RGB stars, formed via binary interaction on the RGB. Binary systems can also lead to the formation of two class of aspherical post-AGB, the Proto-Planetary Nebulae and the naked post-AGBs (dusty RV Taus, a.k.a. Van Winckel's stars).

Keywords. post-AGB stars

1. Introduction

In this review, I present the most recent results based on the study of post-AGB stars. For a thorough review of the post-AGB objects, I refer the reader to the very nice work by Hans van Winckel (Van Winckel (2003)).

Post-AGB stars, as it can be guessed from their names, are evolved stars of low to intermediate initial masses, in the evolutionary phase after the Asymptotic Giant Branch. The beginning of the post-AGB phase is usually defined as when the envelope of the AGB star gets detached. The post-AGB phase ends when this envelope gets ionised by the central star, which has shrunk and became hotter. Observationally, post-AGB stars are often identified by a double-peaked spectral energy distribution, with a peak in the optical due to the contribution from stellar emission, and a peak at longer wavelengths, in the infrared, due to cool dust in the detached envelope. We will see in this review that this observational identification is changing.

The advances in instrumentation, with new instruments/telescopes such as ALMA, SPHERE/VLT and the VLTI enable us to probe deep in the core of post-AGB objects and study them intensively. We can now map the different components of these objects (discs/tori, jets, molecular envelope, shocks) with great details down to a few milliarcseconds using either narrowband filters or integral field spectroscopy. One of the nicest illustrations for this is the impressive work by Gledhill & Forde(2015), where they map a spherically symmetric ionised envelope (in Brγ) expanding in a bipolar neutral gas envelope (in H_2). I will review here how these advances are helping us to study these objects, which are great test benches for nucleosynthesis and the timing of stellar evolution, the evolution of gas and dust in a harsh radiation field and the understanding of the shaping of non-spherical planetary nebulae.

2. Finding post-AGB stars

As the central star quickly shrinks and get hotter during the pos-AGB phase, this phase of stellar evolution is fast, making post-AGB stars rather rare (less than 500 post-AGB stars are known). As the temperature of the central star quickly increase during this phase, post-AGB stars harbour spectral types from B to M. As they are embedded in the dusty envelope formed during the AGB phase, they are highly obscured. All this makes it very difficult to find post-AGB stars.

The success of the IRAS mission was a great leap forward for the field, as post-AGB stars were identified as luminous object with infrared excess. IRAS colour-colour diagrams (particularly the ([12]-[25] vs [25]-[60]) were thus widely used for follow-up studies to identify stars located between the AGB and the PN phases. Post-AGB objects were also found by correlating optical catalogues and infrared ones looking for bright optical objects with an infrared excess (Van de Steene *et al.* 2000 and references therein).

As the post-AGB objects have circumstellar environment very similar to other dusty objects, confusion can occur, mostly with Young Stellar Objects or massive evolved stars (post-Red SuperGiants). A very useful resource to study post-AGB objects is the online, evolving Torun catalogue of post-AGB objects (Szczerba *et al.* 2012).

3. Post-RGB stars

The success of IRAS in the 80s lead to the discovery of many Galactic post-AGBs. The *Spitzer Space Telescope* mission now enables the study of extragalactic post-AGB stars. Studying post-AGB stars in different galaxies can bring a lot of new insights, despite the distances to the objects. With well known distance to the host galaxies, it is straightforward to determine the luminosities of the studied objects. With nearby galaxies such as the Magellanic Clouds that an overall lower metallicity than the Milky Way, one can also study the effect of metallicity on the late stages of the evolution of low and intermediate mass stars.

A very nice study of post-AGB objects in the Large and the Small Magellanic clouds is presented by Kamath *et al.* (2014) and Kamath *et al.* (2015). Their work combines photometry from Spitzer (to look for IR excess due to dust) and ground-based optical spectroscopy (to suppress contaminants). This led to the identification of 63 post-AGB candidates in the SMC and 154 in the LMC.

As the distances to these objects are well constrained, which is not the case for most of the Galactic post-AGBs, they estimated their luminosities. Surprisingly, most of the objects appear to be emerging from the Red Giant Branch, rather than from the Asymptotic Giant Branch. It is very likely that these object have evolved blue-wards before reaching the AGB phase. Kamath *et al.* (2016) suggest that these objects are the results of binary interaction, which led to the ejection of the star's envelope before it could reach the AGB phase. They thus identified a new class of objects, the post-RGB stars. These objects certainly have a Spectral Energy Distribution (SED) consistent with the presence of a disc due to the presence of a binary.

The SEDs of these binary objects are different to the ones classically assumed for post-AGB objects with a double peak (a blue peak due to emission from the star and an infrared peak due to dust emission from the shell). If a disc is formed via the interaction with the companion, warm dust will form and be trapped in the disc, leading to a near-infrared excess, filling the gap between the two peak of the SED. Such a shape of the SED is observed for the dusty RV Tau objects (a.k.a. the Van Winckel's stars), which are binary systems on the post-AGB with a stable dusty disc and no visible nebulae (van Winckel *et al.* (2009)).

4. What can we learn from post-AGB stars studies?

As we just mentioned, post-AGB objects are not easy to identify but can be key objects to understand stellar evolution and a large variety of physical processes. Determining the chemical composition of their envelopes is easier than on the AGB phase (as molecules and dust make it very difficult to determine the continuum's position in the spectra of AGBs). As the envelope is ejected after the AGB, post-AGB envelopes' compositions reflect the final chemical enrichment of the gas due to nucleosynthesis. Post-AGB stars are thus ideal test-benches for models of nucleosynthesis. They can also be used to time stellar evolution.

After the AGB phase, the envelope gets detached and the stars shrinks and becomes hotter. Post-AGB stars are ideal to study the evolution of gas and dust in a harsh radiation field, and thus better understand the life cycle of matter.

AGB stars are also widely thought to be spherically symmetrical, while many Planetary Nebulae (PNe) are aspherical. Post-AGB stars can thus be key objects for the study of the onset of asymmetry and better understand the impact of binaries and magnetic field in the shaping of those nebulae.

4.1. *Test for nucleosynthesis and stellar evolution*

As stated before, post-AGBs are ideal to study the products of nucleosynthesis on the AGB (as they suffer from less molecular absorption). Studying post-AGB stars in the SMC and LMC can help understand the effect of metallicity on nucleosynthesis. For example, a discrepancy appears for Galactic objects between models and observations for the abundance of lead (one of the product of s-process nucleosynthesis on the AGB). De Smedt *et al.* (2014) have shown that this discrepancy is even larger at low metallicity. Neutron capture with neutron densities between s- and r- neutron capture processes may provide an explanation to this discrepancy (Lugaro 2015). A recent study of post AGB stars in the LMC also confirms a correlation between the third dredge-up efficiency and the neutron exposure (van Aarle *et al.* (2013)).

It has also been shown recently that some meteorites have isotopic ratios consistent with a post-AGB origin (Jadhav 2013) with a composition similar to the prediction for the hydrogen injection phase during a Very Late Thermal Pulse (a thermal pulse during the post-AGB phase).

Finally, and this came out to be one of the most important results from this symposium, the study of post-AGB stars have revealed that the timing of stellar evolution was certainly wrong. A study of PNe in the Galactic Bulge (Gesicki 2014) has shown that Bloecker's evolution models are too slow to explain the presence of PNe around such low mass stars as found in the Bulge (the models takes too much time to reach a temperature high enough for the gas to be ionised). The predicted final stellar masses are also higher than what is measured for White Dwarfs or measurements from asteroseismology. Bloecker's post-AGB evolution models need to be accelerated by a factor of 3. This helps produce new stellar evolutionnary models, that should replace Bloecker's models (Miller Bertolami(2016)).

4.2. *Dust evolution after the AGB*

At the end of the AGB, the envelope gets detached, so that the dust goes further away from the star and gets cooler. At the same time, the star becomes hotter and the dust is exposed to a harsher radiation field. The presence of a companion can also lead to the formation of a disc or a torus and dust processing. All these factors lead to a very complex chemistry on the post-AGB, with very complex components observed such as PAHs, fullerenes, and the unidentified 21 microns feature. The most recent knowledge about dust evolution during the post-AGB phase has been obtained thanks to the study of objects in the Magellanic Clouds. A very thorough work by Sloan *et al.* (2014) using infrared spectra from the *Spitzer Space Telescope* has identified 5 classes of objects clearly clustered in color-color diagrams. The difference of chemistry is thus likely linked to a difference in density that could be linked to the presence of discs/torii. A morphological study of these objects would be necessary to understand their dust properties.

These studies however enabled a better understanding of the dust/gas composition after the AGB. For example, the so-called 21 micron feature, an unidentified broad feature seen only in carbon-rich post-AGB stars seems to be very common at low metallicity (Volk *et al.* 2011) and to be always associated with objects harbouring PAHs in their spectra. Thus, Cerrigone *et al.* (2011) proposed that it was associated with hydrocarbons. Sloan *et al.* (2014) showed that it was always associated with two unidentified features at 15.8 and 17.1 microns. They showed that these two features were clearly linked to alkynes, confirming that the 21 micron feature is associated with hydrocarbons. Finally, stars harbouring the 21 micron feature have a special class of newly defined PAHs (Sloan *et al.*, 2014), and seem to have a line of sight to the central star is free of dust absorption. *Spitzer* spectroscopy of post-AGB stars in the Magellanic Clouds also revealed a new class of PAHs (Matsuura *et al.* 2014), with a peak at 7.7 microns. They also showed that the observed class of PAHs for a given object was linked to its evolutionary state rather than its metallicity, i.e. that the different kinds of PAHs observed is due to the dust density and central star's radiation field rather than to its initial composition.

Finally, an unexpected discovery from the *Spitzer* surveys of post-AGB stars in the Magellanic Clouds was the presence of very strong and broad emission features around 11.3 microns. A very common feature due to SiC is usually observed at this wavelength but AGB studies have shown that the strength of this feature tend to decrease with metallicity (Sloan *et al.* 2006; Lagadec *et al.* (2007)), as less Si is available in low Z environments to produce silicon carbide. One would thus expect post-AGBs and PNe in the Magellanic Clouds to have a rather weak SiC emission. But these lower metallicity objects harbour more prominent 11.3 microns features than in the Galaxy (where less than a handful of PNe with SiC are known). This 11.3 microns feature could be due to PAHs emission, but its shape is fully consistent with SiC (Sloan *et al.* (2014)). SiC can indeed be less abundant but a bright SiC feature can be observed if the condensation sequence is different at low Z, leading to grain coating, with SiC forming on top of amorphous carbon grains (Sloan *et al.* (2014); Lagadec *et al.* 2007; Leisenring *et al.* 2008).

5. Binary post-AGB stars and shaping

5.1. *Determining the morphology of post-AGB objects*

While most of the stars on the RGB and the AGB are thought to be more ore less spherical, Planetary Nebulae (PNe) can harbour a wide variety of shapes and be elliptical, bipolar or multipolar. It is also widely accepted that the shaping of these objects is likely to occur during the post-AGB phase and that binarity and magnetic field (sustained by a companion) are important shaping agents.

The observed morphology of the objects appears as projected on the sky, making it difficult to know their intrinsic shape. In a recent work, Koning *et al.* (2013) constructed a model for post-AGB nebulae based on a pair of hollow cavities in a spherical dust envelope. They were able to reproduce most of the observed shapes by only changing the orientation and dust densities. One thus has to keep that in mind when trying to study the morphology of an object.

Another difficulty arises when one wants to determine the morphology of an object is the wavelength-dependence induced by radiative transfer effects. This is very well illustrated by a study of the highly aspherical post-AGB star HD 161796 by Min *et al.* (2013). Using imaging polarimetry, they mapped the dust shell around the object. Polarimetry is a great tool to separate the light scattered by dust in the envelope and the unpolarised light from the central star. HD 161796 clearly hosts a central density enhancement, which is optically thick at short wavelengths and thin towards longer wavelengths. Thus, optical images reveal bright lobes due to scattered light from the torus, while in the infrared one sees emission from this equatorial ring. One thus also has to keep in mind while studying morphologies that this is something wavelength-dependent, and that to see direct emission from the dust shell one has to observe at mid- or far- infrared wavelengths (see e.g. Lagadec *et al.* (2011)).

5.2. *Binaries as shaping agents*

Hydrodynamical models explain many of the observed aspherical structures from a structure-magnification mechanism, where a fast wind from the smaller, hotter, central star ploughs into the earlier slow, dense, AGB wind (Kwok *et al.* 1978) and amplifies any density asymmetry already present (Balick *et al.* 1987): the Generalised Interacting Stellar Wind model or GISW. Fast, collimated and precessing jets could also explain the formation of multipolar nebulae , not explained by the GISW model (Sahai & Trauger 1998). The presence of jets was confirmed by a study of PPNe by Bujarrabal *et al.* (2001).They found that in about 80% of the PPNe, the momentum of the outflowing material is too high to be powered by radiation pressure only. An extra source of angular momentum is thus needed to explain the presence of these jets. To explain the formation of jets, two kinds of models have been proposed, implying either a magnetic field or a bianry companion, leading to a long debate in the PN community. Noam Soker and Jason Nordhaus settled the debate by showing that magnetic fields could play an important role, but a single star could not supply enough angular momentum to shape the nebulae: a companion is needed (Soker 2006; Nordhaus & Blackman 2006).

More and more binary systems are being discovered in PNe (see the contribution by David Jones to these proceedings (Jones 2016). The detection of binaries in post-AGB systems is made complex by the pulsation of the central stars and their dusty envelopes. Binaries have however been discovered in two emblematic bipolar post-AGBs, OH 231.8+4.2 (Gómez & Rodríguez 2001) and the Red Rectangle (Waelkens *et al.* 1996). A long lasting quest for binaries in post-AGBs has been performed by Bruce Hrivnak and his undergraduate students at Valparaiso University, with radial velocity monitoring since 1994. They might have detected a binary system with P>22 years (Hrivnak *et al.* 2011). Binaries are also being discovered by Hans van Winckel and his collaborators, mostly in dusty RV Tau stars, that certainly form a special class of post-AGB stars, as we will discuss now.

6. Two classes of post-AGB objects

It is indeed important to make the distinction between two kind of observed post-AGB objects. Proto-Planetary Nebulae are post-AGB objects that (as their name indicates) will from PNe, but not all post-AGB objects will form PNe.

6.1. *Proto-Planetary Nebulae*

PPNe have nebulae visible in reflection in the optical and emission in the infrared. The most bipolar PPNe are often associated with massive central torii (masses of the order of about a solar mass or more: Lagadec *et al.* 2006). They have a low expansion velocity (typically a few km/s: Peretto *et al.* 2007. They have a small angular momentum and expand radially (i.e. they have almost no rotation). If the mass loss that created them stops, they will vanish rapidly.

In the recent years, very interesting studies of such objects have been obtained, in particular thanks to new instruments like ALMA, that enable the study of gas kinematics at very high angular resolution, and thus a better understanding of the formation of torii/jets. Sahai *et al.* (2013) resolved a hourglass-shaped torus in the core of the Boomerang nebula using ALMA CO observations. They found this torus to be a hollow cavity with very thick walls of gas and dust. This torus is surrounded by a roughly round, but patchy, high velocity outflow. This outflow appears to be ultra-cool, cooler than the cosmic microwave background, so that they dubbed this object the coolest object in the universe (PPNe are very cool objects to study!). Using the SMA, Lee *et al.* (2013) spatially resolved the CO gas around the PPN CRL 618. The circumstellar environments consist of a dense equatorial torus and different fast molecular outflows roughly perpendicular to it. Two episodes of bullet-like ejection are needed to explain these outflows, that clearly are not long lived jets. Similar structures were found in the Water Fountain Nebula (Sahai *et al.* 2016). The magneto-rotational explosion model of Matt *et al.* (2006) seems to be a good explanation for these bullet like outflows.

That leads us to the study of magnetic fields at the surface of these objects. For a thorough review, I will refer the reader to the contribution to this proceedings by Laurence Sabin. More and more magnetic fields measurements have been lately obtained for post-AGB stars. One of the most impressive one is the detection of a magnetically collimated jet in the water fountain post-AGB IRAS 15445−5449 by Pérez-Sánchez *et al.* (2013). The direction and extent of the jet are clearly coincident with the extended envelope observed in the mid-infrared (Lagadec *et al.* 2011). The short-lived jet is collimated by a magnetic field of the order of mG, at $\sim$ 7000 AU from the central star. The surface magnetic fields of a couple of RV Tau post-AGB stars was also measured by Sabin *et al.* (2014). Their study suggest that the magnetic field varies with the pulsation phase of the star and monitoring programs are needed to study the impact of this weak magnetic field on the evolution of the stars. Finally, as we mentioned before, theoretical models predict that a companion is necessary to sustain a magnetic field strong enough to shape the nebulae. Studying a post-AGB system harbouring a known binary star is thus important to better understand the impact of binaries and magnetic fields on the shaping of nebulae. Leal-Ferreira *et al.* (2012) confirmed the presence of a magnetic field around OH231.8+4.2, and measured its strength within a few tens of AU of the binary system, but were not able to determine the morphology of the field. This was done by Sabin *et al.* (2015) with CARMA observation that showed polarisation maps indicating an overall organised magnetic field within the nebula, with a toroidal component aligned with the central equatorial over-density. This seems to indicate the presence of a magnetic launching mechanism in the Rotten Egg Nebula.

Recent works are thus showing that the shaping of PPNe is very likely due to the presence of a companion and a magnetic field, sustained by this companion.

6.2. *Naked post-AGB stars*

Many post-AGB stars are also of the dusty RV Tau type, also called naked post-AGB stars, discs post-AGB stars or the Van Winckel's objects. They harbour a near-infrared excess in their SED, due to the presence of dust near the sublimation temperature, in a compact (R$\sim$ 10 AU) disc. This was confirmed by their infrared interferometric measurements (Deroo *et al.* 2007). These discs are long lived, as indicated by crystalline dust, have a very small aperture angle and a Keplerian kinematics (see e.g. Bujarrabal *et al.* (2013)[Bujarrabal *et al.* 2013]; Bujarrabal *et al.* 2015. Their lifetimes are much larger than those of the torii in the core of PPNe. The photospheres of the central stars are also depleted in refractory elements, as a result of gas and dust separation in the disc, followed by re-accretion of the gas. This gas is poor in refractory elements, as these elements with a high condensation temperature are trapped in dust. Radial velocities monitoring indicates that close binary systems (0.5–3 AU) are present in the cores of those objects (van Winckel *et al.* 2009). A recent study of one of these objects, IRAS08544-4431, by Hillen *et al.* (2016) led to the first image of a post-AGB binary system with a circumbinary disc. Using PIONIER at the Very Large Telescope Interferometer (VLTI), they obtained an image with a resolution down to the milliarcsec scale and managed to map a circum-companion accretion disc. This opens a new window to study binary evolution and circumstellar disc evolution in detail over space and over time.

6.3. *Naked vs clothed post-AGBs*

As described in the previous sections, two very distinct classes of post-AGB stars are observed, the dusty RV Taus (Naked post-AGBs) and the PPNe. Naked post-AGBs all have dusty Keplerian discs (with almost no gas), crystalline dust, binaries but no long-lived nebulae. They are thus very unlikely to form PNe, as when the central stars will become hot enough for the ionisation to occur, no gas will be present around them. PPNe emerging from binary systems have slowly, radially, expanding torii (no Keplerian rotation), and contain amorphous dust and gas. They are short-lived. One of the explanation for the naked post-AGB stars is that the configuration of the binary system led to gas accretion from the central star. It then does not lose mass on the post-AGB and remains cold longer. There are no ionising photons while gas is present in the circumstellar environment and thus no PN (O. De Marco, private communication).

7. Conclusion and perpectives

Post-AGB stars are great tools to test nucleosynthesis and evolution models for stars of low and intermediate masses. The post-AGB phase starting by definition when the envelope of an AGB star gets detached and the star becomes hotter. They are also great test-benches for the study of the evolution of dust in harsh environments of different densities and harbour a very larger variety of complex molecules and dust features such as PAHs, crystaline and amorphous dust. They are also not easy to identify as the post-AGB phase is short and the stars are embedded in dust. If they are in close binary systems, the envelopes can be ejected before the AGB phase, leading to the formation of objects of a newly discovered class: the post-RGB stars. If binary systems manage to evolve till the end of the AGB without ejecting the envelope, they can lead to the formation of bipolar post-AGB stars. Two kinds of aspherical post-AGB object can be formed. PPNe harbour a dense torus and will become bipolar planetary nebulae. Naked post-AGBs harbour Keplerian discs and no visible nebulae and will not form PNe. Finally, the recent advances in observing techniques with high angular resolution now enable us to map circumbinary discs around post-AGB stars and study in real-time the evolution of the discs and the binaries. These high angular resolution techniques also show that, contrarily to what has been thought for decades, the shaping of aspherical nebulae does not necessarily start during the post-AGB phase, as more and more AGB stars appear to be aspherical. This was demonstrated very strikingly by the observations of the nearby AGB star L_2 Pup using the extreme adaptive optics system SPHERE on the VLT, which revealed the clear presence of a disc, outflows perpendicular to it and a companion to the central star (Kervella *et al.* (2015)). We might thus have to study stars on the AGB phase rather than on the post-AGB to understand the beautiful shapes of PNe.

References

Balick, B., Preston, H. L., & Icke, V. 1987, *AJ*, 94, 1641
Balick, B. & Frank, A. 2002, *ARA&A*, 40, 439
Bujarrabal, V., Castro-Carrizo, A., Alcolea, J., & Sánchez Contreras, C. 2001, *A&A*, 377, 868
Bujarrabal, V., Castro-Carrizo, A., Alcolea, J., *et al.* 2013, *A&A*, 557, L11
Bujarrabal, V., Castro-Carrizo, A., Alcolea, J., & Van Winckel, H. 2015, *A&A*, 575, L7
Cerrigone, L., Hora, J. L., Umana, G., *et al.* 2011, *ApJ*, 738, 121
Deroo, P., Acke, B., Verhoelst, T., *et al.* 2007, *A&A*, 474, L45
De Smedt, K., Van Winckel, H., Kamath, D., *et al.* 2014, *A&A*, 563, L5
Gesicki, K., Zijlstra, A. A., Hajduk, M., & Szyszka, C. 2014, *A&A*, 566, A48
Gledhill, T. M. & Forde, K. P. 2015, *MNRAS*, 447, 1080
Gómez, Y. & Rodríguez, L. F. 2001, *ApJL*, 557, L109
Hillen, M., Kluska, J., Le Bouquin, J.-B., *et al.* 2016, *A&A*, 588, L1
Hrivnak, B. J., Lu, W., Bohlender, D., *et al.* 2011, *ApJ*, 734, 25
Jadhav, M., Pignatari, M., Herwig, F., *et al.* 2013, *ApJL*, 777, L27
Jones, D. 2016, arXiv:1611.08454
Kamath, D., Wood, P. R., & Van Winckel, H. 2014, *MNRAS*, 439, 2211
Kamath, D., Wood, P. R., & Van Winckel, H. 2015, *MNRAS*, 454, 1468
Kamath, D., Wood, P. R., Van Winckel, H., & Nie, J. D. 2016, *A&A*, 586, L5
Kervella, P., Montargès, M., Lagadec, E., *et al.* 2015, *A&AL*, 578, A77
Koning, N., Kwok, S., & Steffen, W. 2013, *ApJ*, 765, 92
Kwok, S., Purton, C. R., & Fitzgerald, P. M. 1978, *ApJL*, 219, L125
Lagadec, E., Chesneau, O., Matsuura, M., *et al.* 2006, *A&A*, 448, 203
Lagadec, E., Zijlstra, A. A., Sloan, G. C., *et al.* 2007, *MNRAS*, 376, 1270
Lagadec, E., Verhoelst, T., Mékarnia, D., *et al.* 2011, *MNRAS*, 417, 32
Leal-Ferreira, M. L., Vlemmings, W. H. T., Diamond, P. J., *et al.* 2012, *A&A*, 540, A42
Lee, C.-F., Yang, C.-H., Sahai, R., & Sánchez Contreras, C. 2013, *ApJ*, 770, 153
Leisenring, J. M., Kemper, F., & Sloan, G. C. 2008, *ApJ*, 681, 1557-1573
Lugaro, M., Campbell, S. W., Van Winckel, H., *et al.* 2015, *A&A*, 583, A77
Matsuura, M., Bernard-Salas, J., Lloyd Evans, T., *et al.* 2014, *MNRAS*, 439, 1472
Matt, S., Frank, A., & Blackman, E. G. 2006, *ApJL*, 647, L45
Miller Bertolami, M. M. 2016, *A&A*, 588, A25
Min, M., Jeffers, S. V., Canovas, H., *et al.* 2013, *A&A*, 554, A15
Nordhaus, J. & Blackman, E. G. 2006, *MNRAS*, 370, 2004
Peretto, N., Fuller, G., Zijlstra, A., & Patel, N. 2007, *A&A*, 473, 207
Pérez-Sánchez, A. F., Vlemmings, W. H. T., Tafoya. *et al.*, 2013, *MNRAS*, 436, L79
Sabin, L., Zhang, Q., Zijlstra, A. A., *et al.* 2014, *MNRAS*, 438, 1794
Sabin, L., Hull, C. L. H., Plambeck, R. L., *et al.* 2015, *MNRAS*, 449, 2368
Sahai, R. & Trauger, J. T. 1998, *AJ*, 116, 1357
Sahai, R., Vlemmings, W. H. T., Huggins, P. J., Nyman, L.-Å., *et al.*, I. 2013, *ApJ*, 777, 92
Sahai, R., Vlemmings, W. H. T., Gledhill, T., *et al.* 2016, arXiv:1612.05616
Sloan, G. C., Lagadec, E., Zijlstra, A. A., *et al.* 2014, *ApJ*, 791, 28
Sloan, G. C., Kraemer, K. E., Matsuura, M., *et al.* 2006, *ApJ*, 645, 1118
Soker, N. 2006, *PASP*, 118, 260
Szczerba, R., Siódmiak, N., Stasińska, G., *et al.* 2012, *IAU Symposium*, 283, 506
van Aarle, E., Van Winckel, H., De Smedt, K. *et al.*, P. R. 2013, *A&A*, 554, A106
Van de Steene, G. C., van Hoof, P. A. M., & Wood, P. R. 2000, *A&A*, 362, 984
van Winckel, H. 2003, *ARA&A*, 41, 391
van Winckel, H., Lloyd Evans, T., Briquet, M., *et al.* 2009, *A&A*, 505, 1221
Volk, K., Hrivnak, B. J., Matsuura, M., *et al.* 2011, *ApJ*, 735, 127
Waelkens, C., Van Winckel, H., Waters, L. B. F. M., & Bakker, E. J. 1996, *A&A*, 314, L17

Planetary Nebulae: Multi-Wavelength Probes of Stellar and Galactic Evolution
Proceedings IAU Symposium No. 323, 2016
X. Liu, L. Stanghellini & A. Karakas, eds.

doi:10.1017/S1743921317002204

Infrared studies of post-AGB stars and PNe

Kyung-Won Suh

Dept. of Astronomy and Space Science, Chungbuk National University,
Cheongju, Chungbuk 28644, Republic of Korea
email: kwsuh@chungbuk.ac.kr

Abstract. We present infrared two color diagrams (2CDs) for a large sample of AGB stars, post-AGB stars, and planetary nebulae (PNe) to study their spectral evolution in infrared bands. We discuss possible evolutionary tracks from AGB stars to PNe on the 2CDs. We use catalogs from the available literature for the sample of AGB stars, post-AGB stars, and PNe in our Galaxy.

Keywords. stars: AGB and post-AGB, planetary nebulae: general, infrared: stars

1. Introduction

The intermediate phase between the end of the asymptotic giant branch (AGB) phase and the planetary nebula (PN) phase is called the post-AGB or pre-PN phase. During the post-AGB phase, the dust shell formed in the AGB phase detaches from the central star and becomes optically thin after a few hundred years (e.g., Hrivnak *et al.* 1989; van Hoof *et al.* 1997; Suh 2015).

The *Infrared Astronomical Satellite* (*IRAS*) Point Source Catalog (PSC) (version 2.1) provides useful photometric data in four bands (12, 25, 60, and 100 μm). In characterizing the circumstellar environments of AGB and post-AGB stars, the two-color diagram (2CD) in the *IRAS* PSC has been useful (e.g., Suh & Kwon 2011; Suh 2015). The Midcourse Space Experiment (*MSX*; Egan *et al.* 2003) surveyed the Galactic plane in four bands (8.28, 12.13, 14.65, and 21.34 μm). *AKARI* (Murakami *et al.* 2007) provides PSC data in two bands (9 and 18 μm) from an all-sky survey.

The data from Wide-field Infrared Survey Explorer (WISE), which completed an entire sky survey in the 3.4, 4.6, 11.6 and 22.6 μm bands (hereafter named W1, W2, W3, and W4, respectively) with better angular resolutions and higher sensitivities (Wright *et al.* 2010), would be useful especially for dim objects.

In this work, we present infrared two-color diagrams (2CDs) for AGB stars, post-AGB stars, and PNe using the MSX and WISE data. We cross-identify the *IRAS*, *AKARI*, *MSX*, and WISE counterparts for each object in the sample. We discuss possible evolutionary tracks on the 2CDs.

Table 1. Sample of AGB stars, post-AGB stars, and PNe

Class	Reference	Number	*IRAS* PSC	*AKRAI* PSC	*MSX* PSC	WISE
O-AGB	Kwon & Suh (2012)	3373	3373	3241	2248	3371
C-AGB	Suh & Kwon (2011)	1168	1168	1157	758	1168
post-AGB	Szczerba *et al.* (2007)	326	236	254	117	325
Pre-PNe	Kohoutek (2001)	334	326	253	177	253
PNe	Kohoutek (2001)	1510	927	808	414	808
PNe (MASH1)	Parker *et al.* (2006)	903	161	135	103	900
PNe (MASH2)	Miszalski *et al.* (2008)	335	89	99	35	334

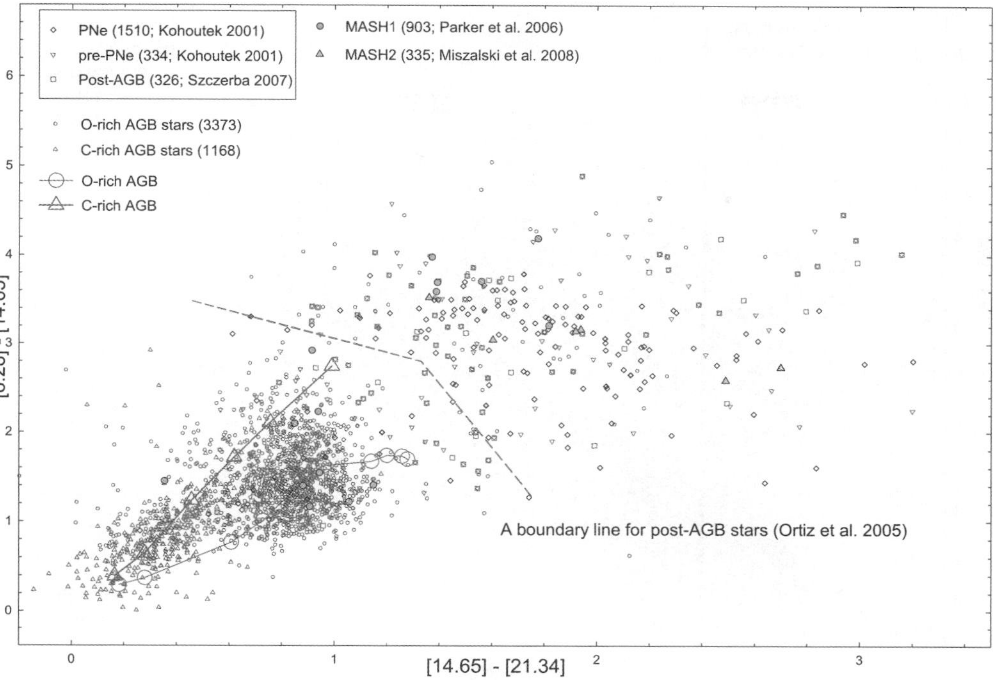

Figure 1. The MSX 2CD for AGB stars, post-AGB stars, and PNe.

2. Sample stars

We use catalogs from the available literature for the sample of 4541 AGB stars (3373 O-AGB; 1168 C-AGB), 660 post-AGB stars (326 post-AGB; 334 pre-PN), and 2748 PNe. For each object in the catalog, we cross-identify the *IRAS* PSC, *AKARI* PSC, *MSX* PSC, and WISE counterparts.

In Table 1, we list the reference, total number of objects, and numbers of the cross-identified *IRAS* PSC, *AKARI* PSC, *MSX* PSC, and WISE counterparts. Because *IRAS* has very low angular resolution, the WISE counterpart obtained from the position data from *AKARI* would be much more reliable. In this work, we improve the identification method presented by Suh (2015) using the more reliable scheme presented by Abrahamyana *et al.* (2015).

2.1. *AGB stars*

Suh & Kwon (2011) present a catalog of AGB stars consisting of 3003 O-rich, 1168 C-rich objects in our Galaxy. Kwon & Suh (2012) present a revised sample of 3373 O-rich AGB stars.

Using the *IRAS* PSC name, we cross-identify the *AKARI* and *WISE* counterparts of objects in the catalog of Abrahamyana *et al.* (2015). For *MSX*, we find the closest counterpart within 30″ of the best position from Abrahamyana *et al.* (2015). Note that the numbers of cross-identified sources are slightly larger than those presented in Suh (2015).

2.2. *Post-AGB stars*

In this work, we use the list of 326 objects from the catalog of post-AGB stars by Szczerba *et al.* (2007) and 334 pre-PNe from the catalogue of Galactic Planetary Nebulae (Updated Version 2000; Kohoutek 2001).

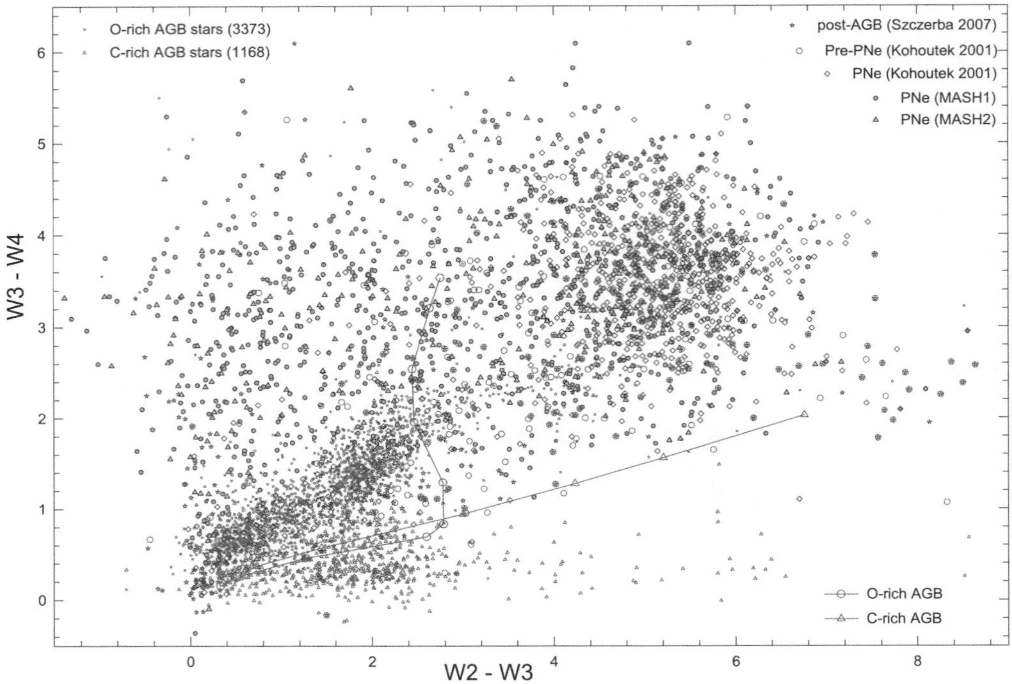

Figure 2. The WISE 2CD for AGB stars, post-AGB stars, and PNe.

For post-AGB stars, we cross-identify the *IRAS*, *AKARI*, *MSX*, and *WISE* counterparts by finding the nearest source within 30″ using the position information in Szczerba *et al.* (2007) for each object. For pre-PNe, we use the *IRAS* PSC counterparts listed in Kohoutek (2001). We cross-identify the *AKARI* and *MSX* counterparts by finding the nearest source within 30″ using the position information in Kohoutek (2001). For the WISE counterpart, we find the closest source within 15″ of the position of the cross-identified *AKARI* PSC source.

2.3. *Planetary Nebulae*

For this paper, we use the catalogue of Galactic Planetary Nebulae (Updated Version 2000; Kohoutek 2001). This catalogue contains 1510 objects classified as Galactic PNe up to the end of 1999. For PNe listed in Kohoutek (2001), we use *IRAS* PSC counterparts listed in the reference. We cross-identify the *AKARI* and *MSX* counterparts by finding the nearest source within 30″ using the position information in Kohoutek (2001) and Kerber *et al.* (2003). For the WISE counterpart, we find the closest source within 15″ of the position of the cross-identified *AKARI* PSC source.

We also use the Macquarie/AAO/Strasbourg Hα (MASH) catalogues (Parker *et al.* 2006; Miszalski *et al.* 2008). MASH PNe are typically more evolved, obscured, and of lower surface brightness than those found in most previous surveys. For MASH PNe, we cross-identify the *IRAS*, *AKARI*, *MSX*, and *WISE* counterparts by finding the nearest source within 30″ of the position given in Parker *et al.* (2006) and Miszalski *et al.* (2008).

3. Infrared Two-Color Diagrams and implications

Figures 1 and 2 show infrared 2CDs for the sample stars. We plot only those objects with good quality measurements at wavelengths. Suh (2015) presents various infrared 2CDs for AGB stars, post-AGB stars, and PNe and investigates theoretical evolutionary model tracks for AGB and post-AGB stars. The theoretical model tracks for AGB stars from Suh (2015) are plotted on the 2CDs.

Figure 1 shows the *MSX* 2CD using [8.28]−[14.65] versus [14.65]−[21.34]. Suh (2015) presents theoretical model tracks for the 2CD. MASH1 objects are scattered over wide regions but MASH2 objects are confined to the post-AGB region as suggested by Ortiz *et al.* (2005).

Figure 2 shows the *WISE* 2CD using W2−W3 versus W3−W4. Note that a major portion of the observed bright sources are severely saturated in the WISE photometry. So the theoretical model tracks for AGB stars show severe deviations from the observed points. Though MASH objects are scattered over wide regions in the 2CD, they are more densely populated in the upper-right region just like other PNe. Theoretically, the *WISE* two-color diagram would be very useful to investigate the evolutionary tracks from AGB to PNe because the *WISE* covers new wavelength bands (3.4, 4.6, 11.6 and 22.6 μm) which may contain some important dust features.

References

Abrahamyana, H. V., Mickaelian, A. M., & Knyazyanb, A. V. 2015, *Astronomy and computing*, 10, 99.

Egan, M. P., Price, S. D., Kraemer, K. E., *et al.* 2003, Air Force Research Laboratory Technical Report AFRL-VS-TR-2003-1589

Hrivnak, B. J., Kwok, S., & Volk, K. M. 1989, *ApJ*, 346, 265

Kerber, F., Mignani, R., Guglielmetti, F., & Wicenec, A. 2003, *A&A*, 408, 1029

Kohoutek, L. 2001, Catalogue of Galactic Planetary Nebulae (Updated Version 2000). Abhandlungen Hamburger Sternwarte, XII C

Kwon, Y.-J. & Suh, K.-W. 2012, *Journal of the Korean Astronomical Society*, 45, 139

Miszalski, B., Parker, Q. A., Acker,A., Birkby, J. L., Frew, D. J., & Kovacevic, A. 2008, *MNRAS*, 525, 534

Murakami, H., Baba, H., Barthel, P., *et al.* 2007, *PASJ*, 59, S369

Ortiz, R., Lorenz-Martins, S., Maciel, W. J., & Rangel, E. M. 2005, *A&A*, 431, 565

Parker, Q. A., Acker, A., & Frew, D. J., *et al.* 2006, *MNRAS*, 373, 79

Suh, K.-W. & Kwon, Y.-J. 2011, *MNRAS*, 417, 3047

Suh, K.-W. 2015, *ApJ*, 808, 165

Szczerba, R., Siódmiak, N., Stasińska, G., & Borkowski, J. 2007, *A&A*, 469, 799

van Hoof, P. A. M., Oudmaijer, R. D., & Waters, L. B. F. M. 1997, *MNRAS*, 289, 371

Wright, E. L., Eisenhardt, P. R. M., Mainzer, A. K., *et al.* 2010, *AJ*, 140, 1868

Planetary Nebulae: Multi-Wavelength Probes of Stellar and Galactic Evolution
Proceedings IAU Symposium No. 323, 2016
X. Liu, L. Stanghellini & A. Karakas, eds.

doi:10.1017/S1743921317000813

Planetary Nebulae in the UWISH2 Galactic Plane survey

T. M. Gledhill[1] and D. Froebrich[2]

[1]Centre for Astrophysics Research, University of Hertfordshire, College Lane, Hatfield, AL10 9AB, UK
email: t.gledhill@herts.ac.uk

[2]Centre for Astrophysics & Planetary Science, The University of Kent, Canterbury, CT2 7NH, UK

Abstract. UWISH2 is the first unbiassed imaging survey of v $= 1-0$ S(1) molecular hydrogen emission ($\lambda = 2.122$ μm) in the northern Galactic Plane. Here we discuss 284 extended emission line objects which we consider to be candidate planetary or pre-planetary nebulae. Some are clearly associated with known PN, but the majority (60%) have no previous detection. We have classified the objects according to morphology and find 53% are bipolar with half of these being new detections. The remaining objects are mostly elliptical/round (35%), fainter than the median sample flux (4.4×10^{-17} W m^{-2}) and previously undetected (74%).

Keywords. planetary nebulae: general, surveys, stars: AGB and post-AGB, stars: evolution

1. Introduction

The number of known planetary nebulae (PN) in the Galaxy is in excess of 3,400 (e.g. Frew *et al.* 2016 and these Proceedings). Most of these discoveries were made at optical wavelengths, with a large number coming from wide-field Galactic Plane (GP) surveys in Hα, such as IPHAS (Sabin *et al.* 2014) and SHS (Parker *et al.* 2006; Miszalski *et al.* 2008). However, the population of optically-obscured PN remains largely unexplored, and a significant number of objects may be expected to lie hidden behind dust clouds in the GP. Neglecting these optically-invisible objects will lead to an underestimate of the total number of PN as well as potential biassing of the population to the brighter and/or closer (and therefore possibly atypical) objects. The need to move the emphasis to detection at longer wavelengths has already been highlighted, for example by establishing mid-IR diagnostics based on Spitzer IRAC colours and mid-IR-to-radio flux ratios (Cohen *et al.* 2011; Parker *et al.* 2012).

In the near-IR K band (between 2 and 2.5 μm) molecular hydrogen (H_2), excited by UV photons or by collisions, can produce a characteristic ro-vibrational spectrum with the v$= 1-0$ S(1) line at 2.122 μm being most prominent. This line has often been used as a signpost of photodissociation regions and of gas heated in shocks, and is a particularly useful tracer of PN as well as their non-ionised post-AGB predecessors, pre-PN or PPN. A number of targeted H_2 observations of PN and PPN samples have been made over the years (e.g. Kastner *et al.* 1996; Kelly & Hrivnak 2004) but until now there has been no systematic H_2 survey of the GP from which to identify previously unknown PN and PPN candidates.

We have recently completed the first unbiassed imaging survey for H_2 v$= 1-0$ S(1) emission in the northern GP using the Wide Field Camera (WFCAM) at the 3.8-m UK Infrared Telescope. The UWISH2 survey (Froebrich *et al.* 2011) covers a contiguous area of the Plane, typically with latitudes between $\pm1.5°$, and extending from $l = 357°$

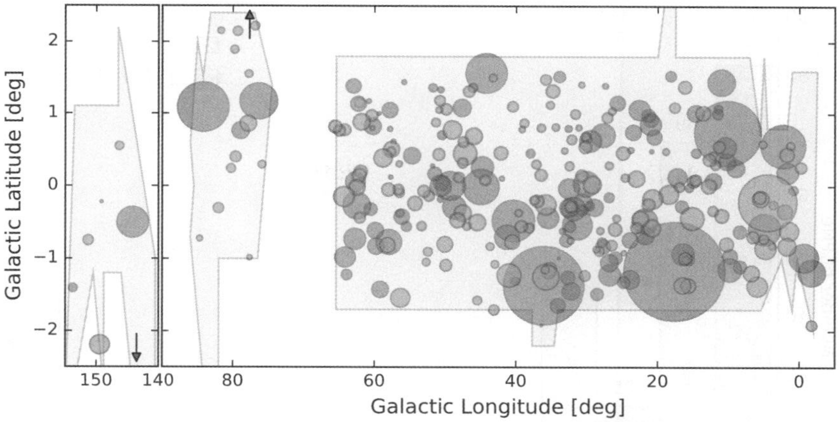

Figure 1. The distribution of H_2 emission line objects flagged as PN or PN candidates. Circle size is proportional to area on the sky. The two arrows point to objects lying off the diagram. See text for further details.

through the Galactic Centre to $l = 65°$. With the addition of smaller fields centred on the Cygnus and Cassiopeia/Auriga star-forming regions, the survey includes a total area of 287 square degrees. A large number (over 32,000) of individual extended H_2 features were identified, leading to 284 unique groups of features classified as candidate PN or PPN (Froebrich *et al.* 2015). Some of these detections are clearly associated with known PN whereas others have no previously-recorded affiliation. The number of candidate PN is likely to increase, as faint H_2 detections initially classed as likely HII regions or "unknown", are reassessed.

2. Distribution, morphology and Hα emission

The distribution of the 284 H_2 objects is shown in Fig. 1 with the polygonal outlines indicating the footprint formed by the survey tiles and including the two smaller areas in Cygnus ($74° < l < 86°$) and Auriga ($140° < l < 155°$). The circle sizes are scaled linearly according to the extent of emission on the sky, and we divide objects into previously known PN (purple) and those which appear to be new detections of candidate (P)PN (green). The latter number 170, or 60% of the total, with many seen to be small (and faint) relative to the known objects.

We have classified the sample according to H_2 morphology using the "ERBIAS" system (Parker *et al.* 2006) with results summarized in Fig. 2. The majority of objects (151 or 53%) appear bipolar with 75 of these being new detections. The remaining objects are mostly elliptical or round (aspect ratio less than 1.05) totalling 100 or (35%) and of these 74% are new detections.

The UWISH2 survey area is included within two Hα surveys: the SHS (Parker *et al.* 2006) covering the Galactic Centre and inner plane to longitudes of $l \approx 35°$ and the IPHAS (Drew *et al.* 2005) which extends outwards along the plane from $l \approx 29°$. Any point in UWISH2 therefore lies within one or both of these surveys which allows us to search for Hα emission in our H_2 targets and to compare morphologies. Fig. 2 also shows the proportion of objects present/absent from the Hα surveys. The majority of B (bipolar) PN with detectable H_2 emission also emit in Hα although 69/151 (46%) do not. In the remaining categories (ERAIS), objects without Hα outnumber those with,

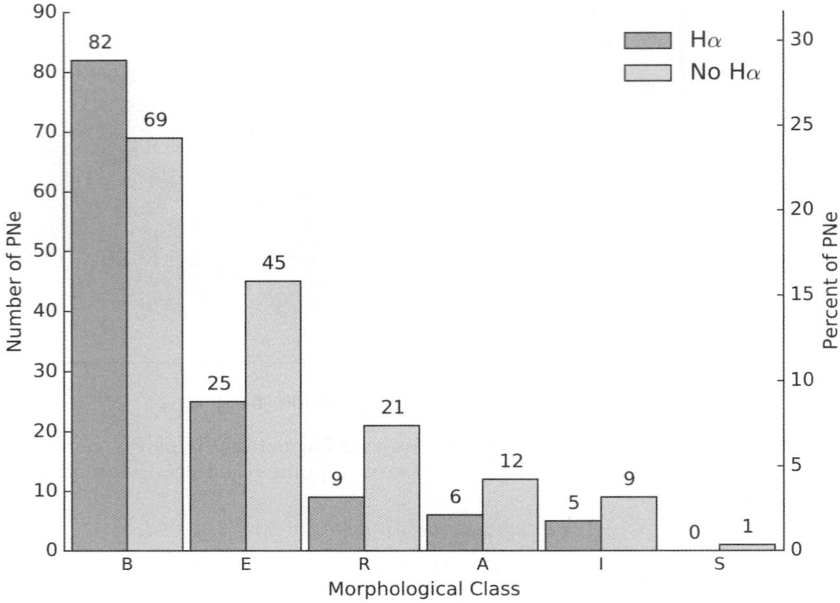

Figure 2. ERBIAS classification of UWISH2 objects and presence/absence of Hα emission.

by approximately 2:1. Of the 170 new candidate (P)PN in UWISH2, 142 (84%) are not seen in Hα.

3. Flux and surface brightness distribution

The flux in the v= 1 − 0 S(1) line is calculated for each object by summing the background-corrected emission within regions lying above the 5σ pixel noise. More than one region may contribute to an object and the surface area (in square arcseconds) is then the sum of the area of each region (see Froebrich *et al.* 2015 for details and values). In Fig. 3 we show the distribution of both flux and surface brightness (i.e. flux divided by area) for our sample, divided according to the presence/absence of detectable Hα emission. The median flux is 438×10^{-19} W m^{-2} with a significant proportion of faint objects and a long tail towards bright objects. This bright tail is made up mostly of known PN, with Hα emission (median 906×10^{-19} W m^{-2}), whereas the peak towards the faint end is almost exclusively due to objects without Hα emission (median 316×10^{-19} W m^{-2}). Hence UWISH2 is detecting a population of new and faint objects which are potentially optically-obscured PN. The surface brightness distribution, however, is very similar across the sample, regardless of Hα detection, and sharply peaked with a median value of 4×10^{-19} W m^{-2} arcsec^{-2}. In other words, the faint objects are small, which is consistent with the idea of probing longer sightlines to more distant and reddened objects in the K-band. If we split the sample according to morphology then we find that the majority of objects fainter than the median flux are E/R; the median fluxes for "E/R" and "B" types are 262 and 664×10^{-19} W m^{-2} respectively. To ascertain how many of these objects are obscured PN, PPN without ionized regions or other unrelated objects, will require further observations such as near-IR spectroscopy, which is in progress.

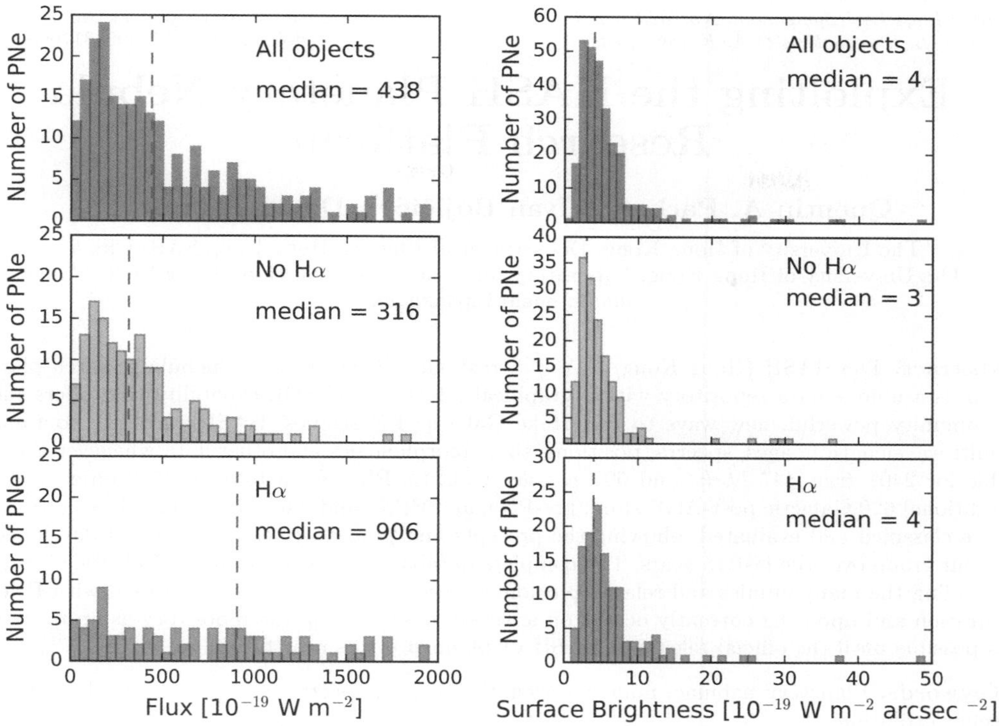

Figure 3. v= 1 − 0 S(1) emission line flux and surface brightness distributions.

References

Cohen, M., Parker, Q. A., Green A. J., Miszalski, B., Frew, D., & Murphy, T. 2011, *MNRAS*, 413, 514

Drew, J., *et al.* 2005, *MNRAS*, 362, 753

Frew, David J., Parker, Q. A., & Bojicic, I. S. 2016, *MNRAS*, 455, 1459

Froebrich, D., *et al.* 2011, *MNRAS*, 413, 480

Froebrich, D., *et al.* 2015, *MNRAS*, 454, 4679

Kastner, J. H., Weintraub, D. A., Gatley, I., Merrill, K. M., & Probst, R. G. 1996, *ApJ*, 462, 777

Kelly, D. M. & Hrivnak, B. J., 2005, *ApJ*, 629, 1040

Miszalski, B., Parker, Q. A., Acker, A., Birkby, J. L., Frew, D. J., & Kovacevic, A. 2008, *MNRAS*, 384, 525

Parker, Q. A., *et al.* 2006, *MNRAS*, 373, 79

Parker, Q. A., Cohen, M., Stupar, M., Frew, D. J., Green, A. J., Bojicic, I., Guzman-Ramirez, L., Sabin, L., & Vogt, F. 2012, *MNRAS*, 427, 3016

Sabin, L., *et al.* 2014, *MNRAS*, 443, 3388

Discussion

SCHOENBERNER: It would be very important to place all these objects in the HR diagram! So you have to look for the central stars, too.

DE MARCO: Why such a large fraction of bipolars?

GLEDHILL: Some objects are distant. The bipolar structure stands out in H2 while a possible elliptical structure may be too faint at larger distances.

Planetary Nebulae: Multi-Wavelength Probes of Stellar and Galactic Evolution
Proceedings IAU Symposium No. 323, 2016
X. Liu, L. Stanghellini & A. Karakas, eds.

doi:10.1017/S1743921317000904

Exploiting the HASH Planetary Nebula Research Platform

Quentin A. Parker[1,2], Ivan Bojičić[1,2], David J. Frew[1,2]

[1]The University of Hong Kong, Department of Physics, Hong Kong SAR, China
[2]The University of Hong Kong, Laboratory for Space Research, Hong Kong SAR, China
email: quentinp@hku.hk

Abstract. The HASH (Hong Kong/ AAO/ Strasbourg/ Hα) planetary nebula research platform is a unique data repository with a graphical interface and SQL capability that offers the community powerful, new ways to undertake Galactic PN studies. HASH currently contains multi-wavelength images, spectra, positions, sizes, morphologies and other data whenever available for 2401 *true*, 447 *likely*, and 692 *possible* Galactic PNe, for a total of 3540 objects. An additional 620 Galactic post-AGB stars, pre-PNe, and PPN candidates are included. All objects were classified and evaluated following the precepts and procedures established and developed by our group over the last 15 years. The complete database contains over 6,700 Galactic objects including the many mimics and related phenomena previously mistaken or confused with PNe. Curation and updating currently occurs on a weekly basis to keep the repository as up to date as possible until the official release of HASH v1 planned in the near future.

Keywords. Planetary nebulae, multi-wavelength imagery, spectroscopy, databases, late-stage stellar evolution

1. Introduction

Planetary Nebulae (PNe) are a useful tracer and touchstone population for studies of late-stage stellar evolution. Their strong emission lines permit the size, expansion velocity and age of the PN to be determined. We can use them to determine the chemical abundance pattern of the ejected gas while their varied forms provide clues to their formation, evolution, mass-loss processes, and possible shaping mechanisms. Such properties make PNe ideal age, abundance, and kinematic probes. Much of this power comes from statistical studies of PNe. Even some recent studies (e.g. Stanghellini & Haywood 2011) have relied on unrepresentative flux-limited samples taken from the early catalogues of Acker *et al.* (1992, 1996) and Kohoutek (2001). We now have a better understanding of the true underlying PN population within the Galaxy, with the discovery of many hundreds of PNe that are fainter, more obscured, and more evolved, as illustrated by the MASH survey (Parker *et al.* 2006; Miszalski *et al.* 2008). Earlier compilations were affected by significant numbers of contaminants that have degraded inferences and results from such studies (e.g. Cohen *et al.* 2011).

This situation has changed and a range of techniques have now been developed to effectively remove PN mimics (e.g. Frew & Parker 2010). There is a need to consolidate and re-evaluate the PN population from earlier catalogues in light of these tools, combined with the significant number of new discoveries (Frew 2016, these proceedings). We also undertook new measurements of known PNe from the latest available high-resolution, multi-wavelength imaging surveys. We now provide the community with a more robust master catalogue of Galactic PNe that offers the best possible data available to undertake statistical investigations on the Galactic PN population. This repository is known as the

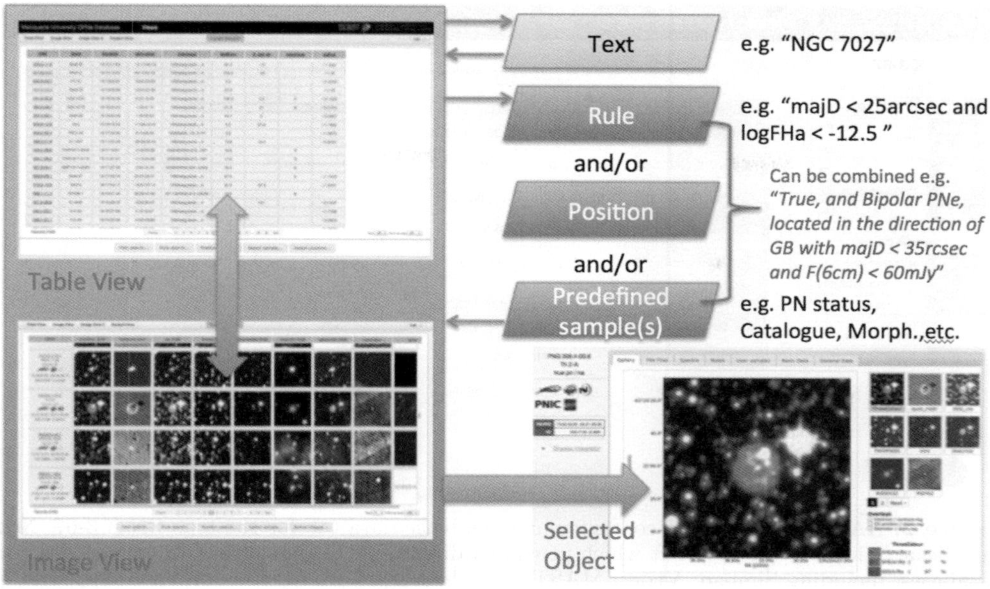

Figure 1. Example of the process by which the HASH catalogue can be visualised, first as a table of selected objects, then as a set of selected images, and then as an individual page for a single PN, presenting key data. It also shows how predefined samples selected according to user-applied rules can be extracted for further study.

HASH (Hong Kong/ AAO/ Strasbourg/ Hα) PN research platform (Parker, Bojičić & Frew 2016) and is recommended for use for all PNe studies.

2. Background

Over the last 10 years there has been a significant increase in the numbers of Galactic PNe known thanks in particular to the narrowband Hα Galactic plane surveys of Parker *et al.* (2005), and Drew *et al.* (2005, 2014). These offer exceptional discovery potential as reported in the MASH catalogues and IPHAS catalogues (e.g. Sabin *et al.* 2014). The amateur community, including the *Deep Sky Hunters* consortium, has also been actively trawling digitised broadband and narrowband surveys to reveal over a hundred new PN candidates (Jacoby *et al.* 2010; Kronberger *et al.* 2016), many of which have now been spectroscopically confirmed. Significant numbers of candidates have also been revealed by their multi-band characteristics in the mid-IR (Cohen *et al.* 2011) or radio (Hoare *et al.* 2012) domains. As most are heavily obscured, spectroscopic confirmation is difficult. In some cases PN candidates identified purely on their mid-IR and radio properties do have faint optical counterparts and have now been spectroscopically confirmed (e.g. Parker *et al.* 2012; Fragkou *et al.* 2016, these proceedings).

The HASH research platform brings all these different samples together into the same repository and attempts to apply the same rigorous set of parameter estimations for all. Full details of the scope, organisation, and utility of HASH will be presented in a refereed paper currently in train (Bojičić *et al.*, in preparation). In this brief conference paper only a little of the power and flexibility of this facility is described, as shown in Figure 1.

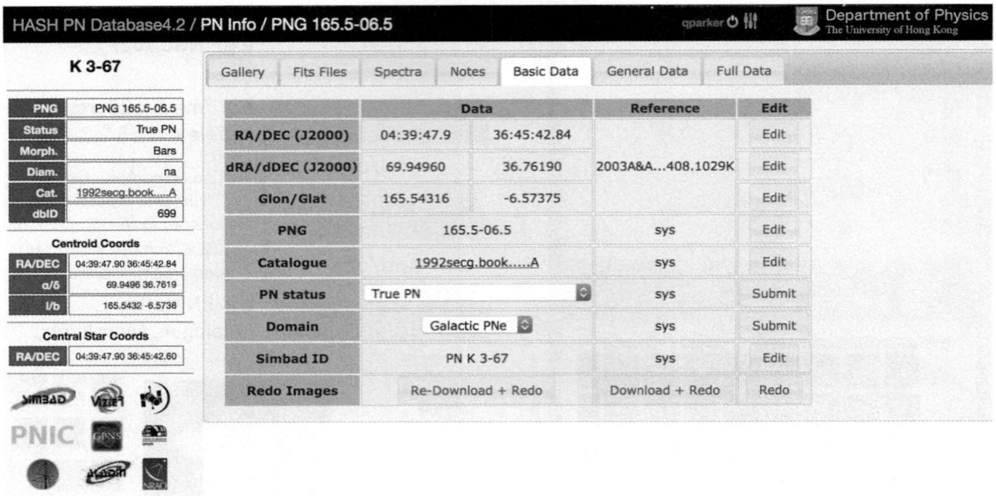

Figure 2. HASH page showing content under the basic data tab for the true PN K 3-67. Note the clickable icons at bottom left which provide a direct link to that object in various external databases, including Simbad, Vizier, MAST, and the San-Pedro Martir database (Lopez *et al.* 2012), etc. A highlighted icon indicates data are available from these external resources.

3. What can HASH do?

HASH (hashpn.space) is an integrated online system that is extremely flexible. It allows any registered user access not only to the best catalogue of Galactic PNe ever compiled, but also offers a powerful research platform to visualise and extract data for analysis from the database according to any valid SQL query. This can currently be done on more than 30 columns of data including positions (RA/Dec and l, b), usual names, sizes, morphologies, original catalogue origin, and many more. All entries have been checked, verified and in most cases re-measured from the latest imaging surveys in an attempt to provide the most robustly determined data possible. Many of these data columns refer to measured properties such as optical narrowband, mid-IR, or radio photometry whose values are being incorporated into HASH.

Additionally, wherever possible, optical (and some near-IR) spectra are included. Spectra are a pre-requisite for any PN to be classed as T (true). Thousands of these spectra come from our own follow-up observations on a series of mainly 2-metre class telescopes made over the last 15 years. A further 1050 low-resolution spectra of varying quality are also made available in HASH for the first time from the Stenholm-Acker survey (Stenholm & Acker 1987). Published emission-line fluxes from ELCAT, converted into line plots are also available for 1020 PNe (see Kaler, Shaw & Browning 1997). HASH includes a spectral visualisation tool, where the user can highlight various emission lines, choose and over-plot multiple spectra when available, and expand these spectra on screen; the user can also download all the 1-D fits spectra for more careful measurement and analysis. Furthermore, the user can download all the individual WCS fits files comprising the multi-wavelength color images, all put on a common astrometric grid as far as possible. An additional feature of HASH is the incorporation of an interactive flux integrator when suitable image data are available from the SHS, SHASSA, or VTSS Hα surveys (we plan to add IPHAS and VPHAS+ images in the near future). This tool offers users the capacity to undertake their own flux estimations following Frew, Bojičić & Parker (2013) and Frew *et al.* (2014), who estimated Hα fluxes for over 1300 PNe.

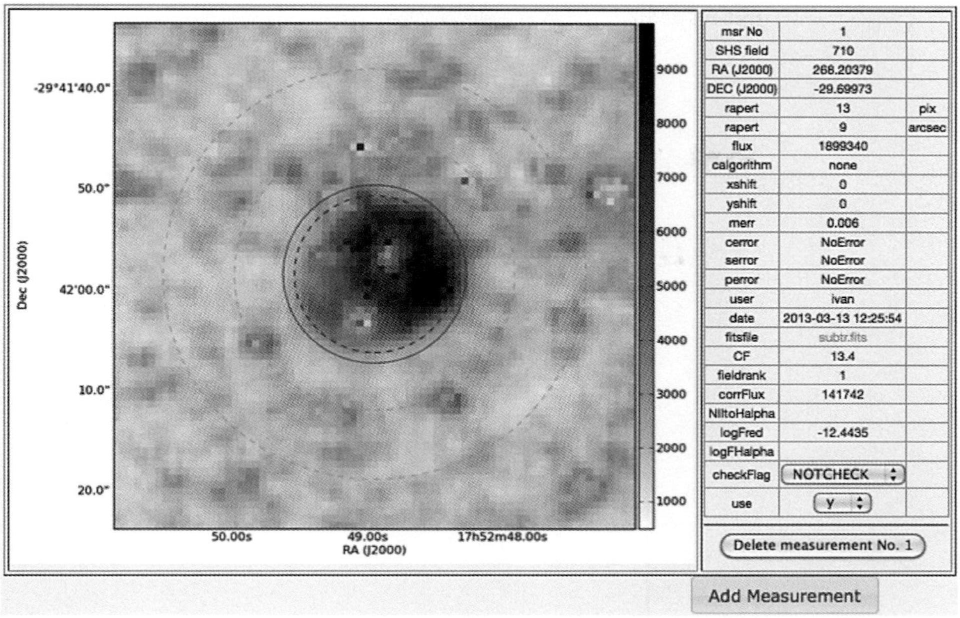

Figure 3. HASH page showing the interactive flux integrator that is available.

References

Acker, A., *et al.* 1992, *Strasbourg-ESO Catalogue of Galactic Planetary Nebulae.* Garching
Acker, A., Marcout, J., & Ochsenbein, F. 1996, *First Supplement to the SECGPN.* Strasbourg
Cohen, M., *et al.* 2011, *MNRAS*, 413, 514
Drew, J. E., *et al.* 2005, *MNRAS*, 362, 753
Drew, J. E., *et al.* 2014, *MNRAS*, 440, 2036
Fragkou, V., Bojičić, I., Frew, D., & Parker, Q. 2016, these proceedings (arXiv.1611.08590)
Frew, D. J. 2016, these proceedings (arXiv preprints)
Frew, D. J. & Parker, Q. A., 2010, *PASA*, 27, 129
Frew, D. J., Bojičić, I. S., & Parker, Q. A. 2013, *MNRAS*, 431, 2
Frew, D. J., *et al.* 2014, *MNRAS*, 440, 1080
Hoare M. G., *et al.* 2012, PASP, 124, 939
Jacoby, G. H., *et al.* 2010, *PASA*, 27, 156
Kaler, J. B., Shaw, R., & Browning, L. 1997, *PASP*, 109, 289
Kohoutek, L., 2001 *A&A*, 378, 843
Kronberger, M., *et al.* 2016, *JPhCS*, 728, 072012
López, J. A., *et al.* 2012, *Rev. Mex. Astron. Astrofís.*, 48, 3
Miszalski, B., *et al.* 2008, *MNRAS*, 384, 525
Parker, Q. A., *et al.* 2005, *MNRAS*, 362, 689
Parker, Q. A., *et al.* 2006, *MNRAS*, 373, 79
Parker, Q. A., *et al.* 2012, *MNRAS*, 427, 3016
Parker, Q. A. & Bojičić, I. S., Frew D. J. 2016, *JPhCS*, 728, 032008
Sabin, L., *et al.* 2014, *MNRAS*, 443, 3388
Stanghellini, L. & Haywood, M. 2010, *ApJ*, 714, 1096
Stenholm B. & Acker, A. 1987, *A&AS*, 68, 51

Discussion

STANGHELLINI: Please use the SIMBAD PNG nomenclature in your catalogs. What is the actual number of spectroscopy-confirmed Galactic planetary nebulae?

Session 2: PNe as a versatile laboratory I
Chair: Michael Barlow

Planetary Nebulae: Multi-Wavelength Probes of Stellar and Galactic Evolution
Proceedings IAU Symposium No. 323, 2016
X. Liu, L. Stanghellini & A. Karakas, eds.

doi:10.1017/S1743921317001004

Photoionization models of Planetary Nebulae

Christophe Morisset

Instituto de Astronomia
Universidad Nacional Autonoma de Mexico
Circuito Exterior, area de la Investigacion Cientifica
Ciudad Universitaria, C.P. 04510, CDMX, Mexico
email: chris.morisset@gmail.com

Abstract. The understanding of astronomical nebulae is based on observational data (images, spectra, 3D data-cubes) and theoretical models. In this review, I present my very biased view on photoionization modeling of planetary nebulae, focusing on 1D multi-component models, on 3D models and on big database of models.

Keywords. methods: numerical, planetary nebulae: general, ISM: abundances

1. Introduction

This talk could well be subtitled "Fifty Years of Nebular Modeling", inspired by the introduction of the review by Harrington (1989) at the IAU meeting on Planetary Nebulae (PNe) in Mexico in 1987. The first published models of PNe are indeed almost 50 years old: Goodson (1967, on theoretical study of the ionization structure of nebulae), Flower (1968, making models for IC418 and NGC7662), Harrington (1968, on ionization structure and thermal stability) and Williams (1968, presenting models at the IAU PNe Symposium 34). Impressively, most of those results were obtained using atmosphere models (from Böhm & Deinzer 1965, 1966) instead of simple Planck functions for the ionizing spectral energy distributions (SEDs).

Planetary nebulae (as well as any other components of the ionized interstellar medium, like H II regions, nebulae around AGNs, WR bubbles, novae and supernovae remnants) can be considered as active filters to the ionizing radiation: they transform the almost continuous incident stellar Lyman continuum radiation into a spread-out continuum spiked with emission lines produced following recombinations and collisional excitations (more on theory in Osterbrock & Ferland 2006). Photoionization codes solve the equations of ionization and thermal equilibria for gas and dust particles and compute the radiation transfer. This allows the determination of the ionic fractions, the electron temperature and the line and continuum emissivities in each cell of the model. One can view a photoionization code as a detailed quantitative summary of our knowledge of gas-photon interactions in the interstellar medium.

Today's most commonly used photoionization codes are: Cloudy (Ferland *et al.* 2013, and the recent conference to honor Gary Ferland, Mexico, 2016), MAPPINGs IV (Dopita *et al.* 2013, this code also deals with shocks) and MOCASSIN (Ercolano *et al.* 2003, 2008, this code being Monte Carlo full 3D). The python library pyCloudy (Morisset 2013, 2014) allows to manage the Cloudy code and to run pseudo-3D models (Morisset *et al.* 2005, and Section 4). Notice the impressive effort by Gary Ferland and collaborators to provide a very complete and useful manual to the users of Cloudy: this is certainly one of the most important ingredients to the success of this code (among the quality of the code itself, of course).

Pequignot (1997, at the IAU PNe Symposium 180) centered his review on the need for ultra deep spectra to identify and measure lines from the third row elements. This is still a challenge and recent efforts have led to observations (and sometimes first identifications) of emission lines from Se, Kr, Rb, Xe, Ca, K, Cr, Mn, Fe, Co, Ni, and Cu (García-Rojas *et al.* 2015), and Cd, Ge and Rb in the infra-red (Sterling *et al.* 2016). Ercolano (2006, at the IAU PNe Symposium 234) focused her review on the advances obtained in 3D modeling, models with dust and PDRs. All those comments remain valid, see the following Sections. Both point out the importance of atomic data in obtaining trustable models.

In the following pages I will first describe the way photoionization codes work, and what they are useful for. I will then focus on a few illustrations of some tailored models, big database of models and 3D models.

2. Photoionization models : ingredients and outputs

To run a photoionization model, one needs to give a description of the nebula to be modeled in terms of the ionizing radiation field and of the properties of the ionized gas. The first item involves the intensity and the shape of the ionizing stellar continuum. It can be a simple black body or a more complex result from a stellar atmosphere modeling code, for example the models by Rauch (1997). The shape is then determined by the effective temperature, the gravity, and the stellar metallicity. On the side of the ionized gas, one must specify: 1) the density, usually in form of the hydrogen density; 2) the elemental abundances; 3) the dust/gas ratio for each type of dust – all these parameters may vary with the position in the nebula; and 4) the total mass of the nebular gas, which leads to a matter- or radiation-bounded nebula. Distance to the nebula may be needed if one wants to compare the model with absolute fluxes. A velocity field must be provided to allow comparison with emission line profiles or Position-Velocity diagrams.

To this specific description of the object one must add a set of atomic data to describe quantitatively every process that occurs in the nebula. This includes photoionization cross sections, recombination coefficients including dielectronic processes, transition probabilities, collision strengths, charge exchange coefficients, molecular formation and dissociation rates as well as dust optical properties. In the case of Cloudy, the data directory weights 432 Mb, while the source directory is "only" 8.2 Mb (this still represents almost 200,000 lines of C++ code...). A great effort was made by Cloudy's team to define a new (universal?) format for atomic data, as described by Lykins *et al.* (2015). The models can be used to determine or constrain atomic data, as done for the charge transfer rates in a pioneering work by Pequignot *et al.* (1978). More recently Escalante *et al.* (2012) determine the contributions of fluorescence to the IC418 recombination spectra of C II, N I, N II, O I, and O II.

The results of a photoionization code applied to a PNe are (among others): the electron temperature, the ionic fractions, the continuum and line emissivities, and the dust grains temperatures. For all these parameters, the values are determined for each cell in the nebula†. It is important to realize that the electron temperature is a result of the code, not an input. If the line ratios of the temperature diagnostics are not reproduced, the user must change something in the model: this may be the SED hardness that modifies the

† A cell can be a spherical shell between R and R + ΔR in the case of a spherical 1D model, or an element of a Cartesian grid in the case of a 3D model.

heating, or the abundances that act on the cooling, or consider additional processes that may modify the heating or the cooling. Otherwise, the determination of the abundances from the model is meaningless.

2.1. *Limitations of static photoionization models*

The limitation of the photoionization models are basically the ones of their ingredients, for example: 1) the input SEDs are obtained from atmosphere models that have never been confronted to reality in the – unobservable – ionizing domain, 2) the atomic data are still incomplete and suffer inconsistencies between different authors, 3) the presence of additional physical process that are not always taken into account (κ-distribution, magnetic reconnection, shocks) that may explain some discrepancies.

3. Classical 1D models

3.1. *Tailored models*

They are used to determine the properties of a given object, e.g abundances. A converged model is supposed to reproduce all available observations.

As an example, one can consider the detailed model of IC418 by Morisset & Georgiev (2009). In this work, we combined a tailored stellar model with a tailored nebular model. This allowed us to determine the distance to the object, using its age determined from its size and expansion velocity, and the stellar age, from stellar evolutionary tracks and stellar properties. Without this information, one cannot resolve the degeneracy between the distance and filling factor (see Morisset & Georgiev 2009, for details). The nebula is the only witness of the ionizing SED. In this work, we demonstrated the importance of a correct description of the ionizing radiation of hν > 40 eV ([Ar III] and [Ne II] lines).

Another example is the detailed model of TS01 by Stasińska *et al.* (2010). Without any observational constraint on the electron temperature (i.e. no [O III]4363 nor [N II]5755 observed intensity), it has still been possible to obtain a rather acceptable determination of the abundances: this is possible only in the case of very strong constraints on almost every other parameter: very good stellar observations, nebular images and spectra (IR, optical and UV). And the use of a 3D model to take into account the morphology and the slit positions (see below).

Some models to study the properties of dust from its infrared (IR) emission are shown by e.g. Otsuka *et al.* (2015). A detailed model must include the dust emission coming from the PDR (if exists), which can be constrained by [O I] and [C II] IR lines.

3.2. *Grid of models*

They are are based on toy models and are used to study some specific effects or to determine some properties when varying parameters.

The most recent and complete grid of PNe models is the one included in the Mexican Million Models database (3MdB Morisset *et al.* 2015) under the references 'PNe_2014', 'PNe_2014_c13' (subset of the previous runs with Cloudy version 13.03) and 'PNe_2016' (refined mesh in effective temperature), with a total of 625,974 models, in which were varied the effective temperature, the stellar luminosity, the nebular inner radius, the gas density, the O/H abundance, the density law, the nebular mass and the presence or not of dust. An easily accessible criterion allows the user to select the most realistic models (92,513 models). The database was used to calculate PN ionization correction factors (ICFs) by Delgado-Inglada *et al.* (2014) – see below. It can also be used to look for models that fit some line ratio and thus determine the PN properties, like the stellar effective temperature, as done by García-Rojas *et al.* (2016).

3.3. *Abundance determinations*

The direct method is used to determine elemental abundances, in a 3-step way: 1) determination of the physical properties: electron temperature and density (multiple zones can be considered); 2) determination of the ionic abundances; and 3) determination of the elemental abundances by applying ICFs if needed.

One of the main source of uncertainties in this method is certainly linked to the ICFs. Perhaps the most cited PNe photoionization models are those used by Kingsburgh & Barlow (1994) to determine famous ICFs for N, O, C, Ne, Ar, and S. The 10 PNe models on which the ICFs are based should had been described in a never published paper! New ICFs have been recently published by Delgado-Inglada *et al.* (2014) where the uncertainties are also specified.

Another way to determine abundances is to compute photoionization models that fit all the observables: in this case one has to carefully define the input parameters, which is more complicated than applying the direct method. But there is no need for ICFs as the models compute the ionization state for all the elements at once. One can also try to find the set of models that fit the observables within a certain tolerance, which naturally gives us the uncertainties in parameter determination. This has been done for example by Esteban *et al.* (2016) for WR bubbles. Ad-hoc ICFs have been obtained for each object, by fitting [O III]/[O II], He I/H I and He II/H I line ratios with models from 3MdB (see previous Section).

4. 3D modeling

Current modern observational facilities yield 3D spectra (or IFUs for Integral Field Units), like VIMOS (Le Fèvre *et al.* 2002), MUSE (Bacon *et al.* 2010), MEGARA (Gil de Paz *et al.* 2014), or SITELLE (Brousseau *et al.* 2014). The spatial mapping allows a detailed study of the morphology and emission line profiles, leading in particular to the determination of the velocity fields of each region where the emitting ion is dominant. These improvements in observational capacities require corresponding theoretical tools needed to model the nebulae in 3D. The first 3D photoionization models are indeed from Kirkpatrick (1970), that included axially symmetric nebulae and optically thin inhomogeneities.

3D models are used for different purposes: 1) reproducing the morphology given by imaging the object; 2) taking into account the size and the position of the spectrometer aperture; and 3) constraining the velocity field by high spectral resolution observations. Examples of 3D models can be found in Zhang (2008), Morisset & Stasinska (2008), Wright *et al.* (2011), Monteiro *et al.* (2013), Danehkar *et al.* (2013), and Akras *et al.* (2016).

4.1. *Aperture effects*

3D models, even simple spherical ones, are useful if one wants to reproduce the observations obtained through slits that only partly overlap the nebula. This is especially true when one needs to use ICF to determine abundances, as the classical values have been determined for full nebulae. As an illustration, in Figure 1 I show a simple 3D ellipsoidal model of a nebula ionized by a 50 kK star, with a mean ionization parameter of log(U)$= -2.5$. The left panel shows the main location of the low ionization (N^+ in red) and the high ionization (O^{++} in green) regions. We can compute for each spaxel the ICF to apply to N^+/H^+ to obtain the N/H abundance, and compare it to the classical ICF ($N/N^+ = O/O^+$). This is shown in the right panel, where the color bar depicts any departure from 1.0 for this ratio. It is obvious that the ICF to be applied to N^+/H^+

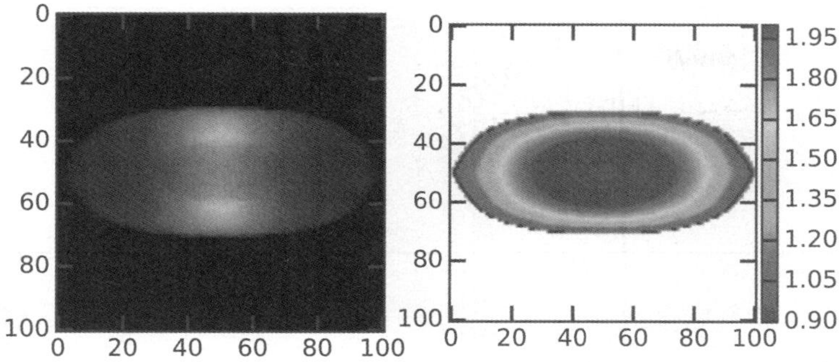

Figure 1. Left: RGB image of the 3D toy model (Red: [N II], Green: [O III], Blue: H I). Right: ICF: the color bar is following $(1+\mathrm{N}^{++}/\mathrm{N}^{+})/(1+\mathrm{O}^{++}/\mathrm{O}^{+})$.

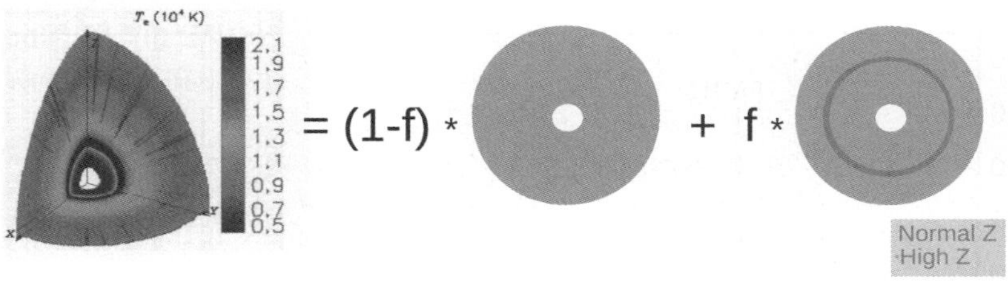

Figure 2. Illustration of topologically equivalent models. Left image from Yuan *et al.* (2011).

is strongly dependent on the position of the aperture (the mean value of the departure $(1+\mathrm{N}^{++}/\mathrm{N}^{+})/(1+\mathrm{O}^{++}/\mathrm{O}^{+})$ over the whole nebula is 1.26).

4.2. *Nebular kinematic*

Once a 3D model is obtained, one can add a velocity field and compute the emission line profiles as seen through any aperture. The effects of the geometry on the determination of the velocity fields from line profiles have been studied by Morisset & Stasinska (2008). By fitting imaging with 3D models and high resolution profiles with expansion velocities Gesicki *et al.* (2016) determine dynamical properties of 8 PNe. One must notice the long standing effort of the Potsdam group to develop hydrodynamical photoionization models: emission line profiles are used by Schönberner *et al.* (2014) to constrain 1D models of the expansion properties and the internal kinematics of PNe.

4.3. *Topologically equivalent models*

One may wonder if a very detailed 3D model of a PN is actually needed if the nebula is not fully spherically symmetric. One can mentally reorganize the density distribution to a much simpler morphology that would almost have the same properties. As an illustration, in Figure 2, I show that the 3D model of NGC 6153 by Yuan *et al.* (2011) [made of high metallicity inclusions embedded in a normal metallicity diffuse component] could have been equivalently obtained with two runs of a 1D code: the first one with only the normal background component and the second one including a thin hydrogen-poor shell. The result of the combined model is obtained as the weighted sum of the line intensities of each individual model (with f being the covering factor of the metallic clumps). This is exactly what has been done by Péquignot *et al.* (2002) in the first study of metallic inclusions in NGC 6153. One can elaborate a more complex model by changing the radius of the metallic shell and summing up more that two models. The extrapolation of this method combined with a reconstruction of a given morphology leads to what we term

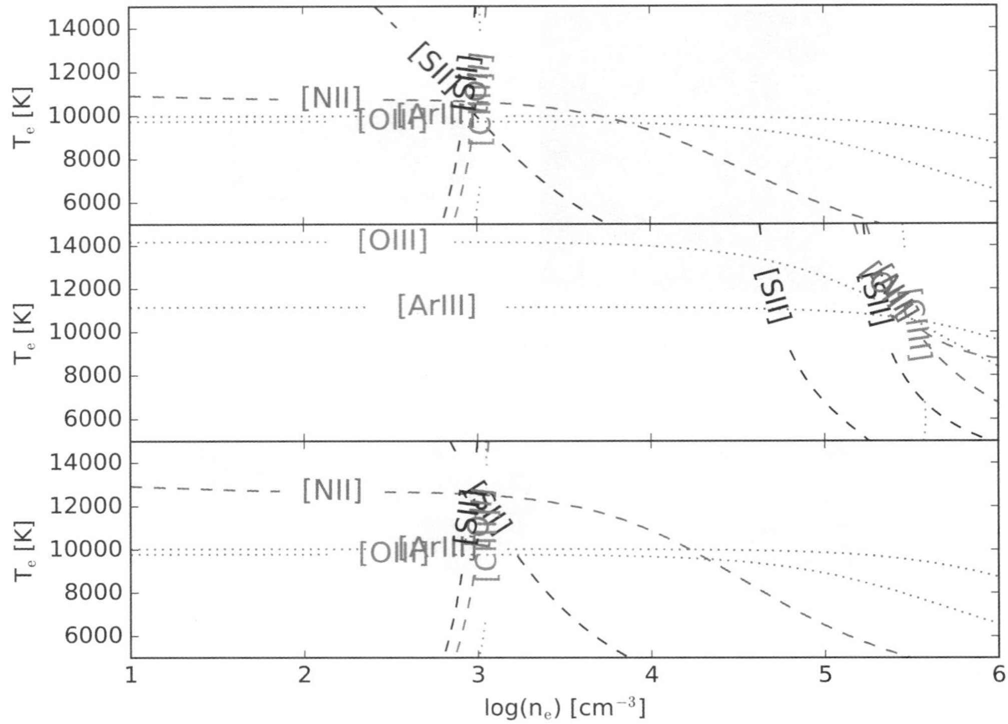

Figure 3. Example of a 2-component model. From upper to lower panel: diagnostic diagrams for the low density medium (n_H=10^3cm^{-3}, log(U) = -1.4, $T_e \sim 9,700$ K), high density medium (n_H=$10^{5.5}$cm^{-3}, log(U) = -3.2, $T_e \sim 10,500$ K) and the composite model (1% of the volume occupied by the high density medium).

pseudo-3D models (Morisset *et al.* 2005). The limitations of the pseudo-3D models are reached when non-radial radiation dominates an important process. This is the case in only two situations: shadows behind optically thick clouds† (Morisset & Ercolano 2004) and a region ionized by multiple and well separated sources (which is not relevant in the PNe case). Otherwise, pseudo-3D models are totally legitimate to model complex structures. The simplicity of this approach and the gain in CPU time execution (from hours/days for the full 3D case to minutes for the equivalent or pseudo-3D model) will give the user alternative options to explore more of the space parameter and converge to a solution in a faster and more economic way as well as address the question of multiple solutions.

4.4. *Hidden inhomogeneities*

The fact that on the same line of sight two or more different components of gas are mixed leads to wrong results if trying to interpret the observations as coming from a single component. This is illustrated in Figure 3, where Te-Ne-diagnostic diagrams are shown for two models representing regions ionized by the same star but differing in density (n_H=10^3 and $10^{5.5}$ cm^{-3} in the upper and middle panel respectively). The classical diagnostics line ratios are showing this difference in density, as well as the rather small difference in electron temperature (9,700 and 10,500 K respectively). But if one combines these two models into a single observation and repeat the same diagram (lower panel), it

† Yuan *et al.* (2011) use optically thin metallic clouds.

becomes impossible to detect the high density clumps and furthermore an apparent high temperature (13,000 K) is inferred from the low ionization [NII] emission lines, while the nebula is almost isothermal. This will in turn lead to wrong determinations of ionic and elemental abundances.

5. Conclusions

Photoionization codes are simply physical codes. They don't know about astrophysics. A set of input parameters that does not correspond to any realistic astrophysical situation will still lead a result in most situations, so be careful! It is the responsibility of the user to check the validity of the input parameters.

It is also the responsibility of the user to take care of the aperture effects and not compare a full model with observations obtained with a slit which is small compared to the size of the nebula. This is especially true when dealing with IFUs observations.

Topologically equivalent models can save a lot of CPU time. Finally, carefully exploring the space parameter is more important than fitting second-order parameters.

Acknowledgements

Participation of this conference has been possible due to grant PAPIIT (DGAPA-UNAM) 107215. Many thanks to Daniel Péquignot who taught me 20 years ago the basics (and a lot more) of photoionization models, especially the topologically equivalent models. Many thanks to Grażyna Stasińska who during the last 10 years helped me a lot in understanding in even greater details. Many thanks also to Grażyna Stasińska, Jorge García-Rojas, Luc Binette, and Xiaowei Liu for reviewing this paper.

References

Akras, S., Clyne, N., Boumis, P., *et al.* 2016, *MNRAS*, 457, 3409

Bacon, R., Accardo, M., Adjali, L., *et al.* 2010, in *Proc. SPIE*, Vol. 7735, Ground-based and Airborne Instrumentation for Astronomy III, 773508

Böhm, K. H. & Deinzer, W. 1965, *ZAp*, 61, 1

Böhm, K. H. & Deinzer, W. 1966, *ZAp*, 63, 177

Brousseau, D., Thibault, S., Fortin-Boivin, S., *et al.* 2014, in *Proc. SPIE*, Vol. 9147, Ground-based and Airborne Instrumentation for Astronomy V, 91473Z

Danehkar, A., Parker, Q. A., & Ercolano, B. 2013, *MNRAS*

Delgado-Inglada, G., Morisset, C., & Stasińska, G. 2014, *MNRAS*, 440, 536

Dopita, M. A., Sutherland, R. S., Nicholls, D. C., Kewley, L. J., & Vogt, F. P. A. 2013, *ApJS*, 208, 10

Ercolano, B. 2006, in IAU Symposium, Vol. 234, Planetary Nebulae in our Galaxy and Beyond, ed. M. J. Barlow & R. H. Méndez, 259–266

Ercolano, B., Barlow, M. J., Storey, P. J., & Liu, X. 2003, *MNRAS*, 340, 1136

Ercolano, B., Young, P. R., Drake, J. J., & Raymond, J. C. 2008, *ApJS*, 175, 534

Escalante, V., Morisset, C., & Georgiev, L. 2012, *MNRAS*, 426, 2318

Esteban, C., Mesa-Delgado, A., Morisset, C., & García-Rojas, J. 2016, *MNRAS*, 460, 4038

Ferland, G. J., Porter, R. L., van Hoof, P. A. M., *et al.* 2013, *Rev. Mexicana Astron. Astrofis.*, 49, 137

Flower, D. R. 1968, *Astrophys. Lett.*, 2, 205

García-Rojas, J., Madonna, S., Luridiana, V., *et al.* 2015, *MNRAS*, 452, 2606

García-Rojas, J., Peña, M., Flores-Durán, S., & Hernández-Martínez, L. 2016, *A&A*, 586, A59

Gesicki, K., Zijlstra, A. A., & Morisset, C. 2016, *A&A*, 585, A69

Gil de Paz, A., Gallego, J., Carrasco, E., *et al.* 2014, in *Proc. SPIE*, Vol. 9147, Ground-based and Airborne Instrumentation for Astronomy V, 91470O

Goodson, W. L. 1967, *ZAp*, 66, 118
Harrington, J. P. 1968, *ApJ*, 152, 943
Harrington, J. P. 1989, in IAU Symposium, Vol. 131, Planetary Nebulae, ed. S. Torres-Peimbert, 157–166
Kingsburgh, R. L. & Barlow, M. J. 1994, *MNRAS*, 271, 257
Kirkpatrick, R. C. 1970, *ApJ*, 162, 33
Le Fèvre, O., Mancini, D., Saisse, M., *et al.* 2002, The Messenger, 109, 21
Lykins, M. L., Ferland, G. J., Kisielius, R., *et al.* 2015, *ApJ*, 807, 118
Monteiro, H., Gonçalves, D. R., Leal-Ferreira, M. L., & Corradi, R. L. M. 2013, *A&A*, 560, A102
Morisset, C. 2013, pyCloudy: Tools to manage astronomical Cloudy photoionization code
Morisset, C. 2014, in Asymmetrical Planetary Nebulae VI conference, Proceedings of the conference held 4-8 November, 2013. Edited by C. Morisset, G. Delgado-Inglada and S. Torres-Peimbert.
Morisset, C., Delgado-Inglada, G., & Flores-Fajardo, N. 2015, *Rev. Mexicana Astron. Astrofis.*, 51, 103
Morisset, C. & Ercolano, B. 2004, in Astronomical Society of the Pacific Conference Series, Vol. 313, Asymmetrical Planetary Nebulae III: Winds, Structure and the Thunderbird, ed. M. Meixner, J. H. Kastner, B. Balick, & N. Soker, 240–+
Morisset, C. & Georgiev, L. 2009, *A&A*, 507, 1517
Morisset, C. & Stasinska, G. 2008, *Rev. Mexicana Astron. Astrofis.*, 44, 171
Morisset, C., Stasińska, G., & Peña, M. 2005, *MNRAS*, 360, 499
Osterbrock, D. E. & Ferland, G. J. 2006, Astrophysics of gaseous nebulae and active galactic nuclei
Otsuka, M., Hyung, S., & Tajitsu, A. 2015, *ApJS*, 217, 22
Pequignot, D. 1997, in IAU Symposium, Vol. 180, Planetary Nebulae, ed. H. J. Habing & H. J. G. L. M. Lamers, 167
Péquignot, D., Amara, M., Liu, X., *et al.* 2002, in Revista Mexicana de Astronomia y Astrofisica Conference Series, Vol. 12, Revista Mexicana de Astronomia y Astrofisica Conference Series, ed. W. J. Henney, J. Franco, & M. Martos, 142–143
Pequignot, D., Stasińska, G., & Aldrovandi, S. M. V. 1978, *A&A*, 63, 313
Rauch, T. 1997, *A&A*, 320, 237
Schönberner, D., Jacob, R., Lehmann, H., *et al.* 2014, Astronomische Nachrichten, 335, 378
Stasińska, G., Morisset, C., Tovmassian, G., *et al.* 2010, *A&A*, 511, A44
Sterling, N. C., Dinerstein, H. L., Kaplan, K. F., & Bautista, M. A. 2016, *ApJ*, 819, L9
Williams, R. E. 1968, in IAU Symposium, Vol. 34, Planetary Nebulae, ed. D. E. Osterbrock & C. R. O'dell, 190
Wright, N. J., Barlow, M. J., Ercolano, B., & Rauch, T. 2011, *MNRAS*, 418, 370
Yuan, H., Liu, X., Péquignot, D., *et al.* 2011, *MNRAS*, 411, 1035
Zhang, Y. 2008, *A&A*, 486, 221

Discussion

SLOAN: You noted the effect of grain shape on the MgS feature profile. The temperature of the grains also has an effect on position (lambda), as explored by Sacha Hony.

MORISSET: Yes, you are right. The distance to the central star is then an additional parameter to take into account.

SCHOENBERNER: In very metal-poor objects, like e.g. TS01, cooling by expansion may affect the electron temperature. Can this be considered in models?

Planetary Nebulae: Multi-Wavelength Probes of Stellar and Galactic Evolution
Proceedings IAU Symposium No. 323, 2016
X. Liu, L. Stanghellini & A. Karakas, eds.

doi:10.1017/S1743921317001053

Gaseous abundances in planetary nebulae: What have we learned in the past five years?

Gloria Delgado-Inglada

Instituto de Astronomía, Universidad Nacional Autónoma de México, Apdo. Postal 70264, Ciudad de México, 04510, Mexico
email: gdelgado@astro.unam.mx

Abstract. Nearly 50 years ago, in the proceedings of the first IAU symposium on planetary nebulae, Lawrence H. Aller and Stanley J. Czyzak said that *"the problem of determination of the chemical compositions of planetary and other gaseous nebulae constitutes one of the most exasperating problems in astrophysics"*. Although the situation has greatly improved over the years, many important problems are still open and new questions have arrived to the field, which still is an active field of study. Here I will review some of the main aspects related to the determination of gaseous abundances in PNe and some relevant results derived in the last five years, since the last IAU symposium on PNe.

Keywords. ISM: abundances, planetary nebulae: general, galaxies: abundances, stars: AGB and post-AGB, nuclear reactions, nucleosynthesis, abundances.

1. Introduction

Planetary nebulae (PNe) are the result of the evolution of many stars with masses between ~1 and ~8 $M_\odot$. During their lives, these stars may go through several nucleosynthesis and mixing events that change their initial surface composition and, at the end of their evolution, they eject their outer layers enriching the interstellar medium (ISM) with the newly synthesized products (see, e.g., Karakas & Lattanzio 2014). The PN phase emerges if the previously ejected material becomes ionized by the central star before being diluted in the ISM. The ionized gas of PNe contains the chemical footprints of the former stellar nucleosynthesis and, at the same time, stores information about the chemical composition at the time and place where the progenitor stars were formed.

PNe constitute important tools to understand the production and evolution of chemical elements in the Universe. The comparison between PN abundances and the predictions from stellar models is very useful to improve our knowledge on the efficiency of nucleosynthesis and transport mechanisms occurring in low- and intermediate-mass stars. The analysis of the element abundances that remain unaltered during the stellar life serves as constraints for galactic chemical evolution models. And the study of the gaseous abundances and depletions of refractory elements in PNe, and their elemental depletions, helps to understand dust formation and evolution in these objects.

I will review some aspects that should be considered when calculating chemical abundances in PNe and I will highlight some of the main results obtained in the last five years since the previous IAU symposium on PNe.

2. Comments on the abundance determination in PNe

2.1. *The abundance discrepancy factor*

This is one of the most intriguing problems in the field and it remains unsolved. We have known for almost 80 years (Wyse 1942) that the abundances derived from recombination lines (RLs) are systematically higher than those computed using collisionally excited lines (CELs). The so-called abundance discrepancy factor (ADF), ratio of ionic abundances obtained from RLs to corresponding values from CELs, ranges from 1.5 to 3 in most PNe but reaches $\sim$ 100 in a few PNe (Liu *et al.* 2006; Corradi *et al.* 2015a).

The abundance discrepancy (AD) may critically affect our knowledge of the evolution of metals in the universe because most of what we know about abundances, especially at large distances, comes from the bright forbidden lines of ionized nebulae. Various scenarios have been proposed to explain this discrepancy, the two preferred ones being the presence of spatial temperature fluctuations (Peimbert 1967; Torres-Peimbert *et al.* 1980) or chemical inhomogeneities in the nebulae (Liu *et al.* 2000). In both cases, the origin of the inhomogeneities needs to be explained.

Over the last years several works have been devoted to unravel the reason of this discrepancy and some new ideas have emerged. Nicholls *et al.* (2012) suggest that the presence of a significant number of non-thermal electrons in ionized nebulae could affect the determination of temperatures and abundances. The authors are able to explain the differences between the RL and CEL abundances by adopting a kappa distribution for the energies of electrons, instead of the Maxwell-Boltzmann one. However, Ferland *et al.* (2016) argue that non-thermal electrons do not survive long enough to affect the collisionally excited lines or the recombination lines.

Another important contribution to the AD problem has been made by Corradi *et al.* (2015a). These authors found that the PNe with the most extreme ADFs contain close binary systems. The sample of PNe that satisfies this condition keeps growing (Jones *et al.* 2016; García-Rojas *et al.* 2016a). This finding supports the idea of the existence of chemical inhomogeneities in those PNe, with two different components: one cold ($\sim$ 1000 K) and metal-rich and another hot ($\sim$ 10000 K) and with a near solar composition. More discussion on the presence of chemical inhomogeneities in PNe can be found elsewhere in this volume (see articles by M. Peña, J. García-Rojas, and R. Wesson).

Although the existence of chemical inhomogeneities in some PNe seems doubtless, Peimbert *et al.* (2014) claim that this should not be the case for most of the PNe. They find that in 16 PNe for which they can derive reliable electron densities from both CELs and RLs, the densities derived from CELs are similar to or higher than those derived from RLs. They conclude that there is no evidence of oxygen-rich high-density clumps in those PNe, and therefore, that they are chemically homogeneous.

2.2. *Atomic data*

Atomic data are a key ingredient to derive reliable chemical abundances. Over the past decades, a huge effort has been made by collaborations like the IRON Project† and CHIANTI‡ and also by individual scientists, leading to new and more accurate atomic data (a discussion on recent atomic calculations for neutron-capture elements can be found in the article by N. Sterling in this volume). Moreover, softwares, such as PyNeb (Luridiana *et al.* 2015), make very easy to explore the effect of different sets of atomic data in the calculations.

† http://www.usm.uni-muenchen.de/people/ip/iron-project.html
‡ http://www.chiantidatabase.org/

PNe have proven to be excellent laboratories to test atomic data. For example, Stasińska *et al.* (2013) compare the values of the densities derived from the observed intensity ratios of [S II] λ6716/6731 and [O II] λ3726/3729 for a sample of Galactic PNe with the ratios of the transition probabilities from two sources: Mendoza & Zeippen (1982) and Tayal & Zatsarinny (2010). Stasińska *et al.* find that the oldest atomic data (Mendoza & Zeippen 1982) reproduce better the observations and thus, they should be used to compute abundances. This kind of analysis can be easily done to test new atomic data.

The importance of atomic data in nebular analysis is illustrated by Juan de Dios & Rodríguez (2017, in prep.; and also in this volume). They find that the differences in the values of O/H and N/O derived using different atomic data can reach up to 0.6 and 1.0 dex, respectively, at densities above $\sim$10000 cm^{-3}. They point out that some atomic data should be avoided in the calculations of physical conditions and ionic abundances.

2.3. *Ionization correction factors*

The total chemical abundances can be computed by adding all the ionic abundances of each element. When some ions are not observed, one needs to take into account their contribution through ionization correction factors (ICFs). Originally, the proposed ICFs were based on similarities between the ionization potentials of different ions (e.g., Peimbert & Costero 1969; Peimbert & Torres-Peimbert 1971). Later, Kingsburgh & Barlow(1994) compiled a set of ICFs based on around ten photoionization models that were computed to fit the observations of some bright PNe. These ICFs are widely used in the literature but they are found to be inadequate for some elements, such as neon and carbon, since the proposed expressions (Ne/O = Ne^{++}/O^{++} and C/O = C^{++}/O^{++}) are too simplistic (Delgado-Inglada *et al.* 2014).

This is illustrated in Fig.1 for carbon. The ratio between ionic fractions of O^{++} and C^{++}, $x(O^{++})/x(C^{++})$, is plotted as a function of the degree of ionization for a large sample of photoionization models representative of PNe. The grid is available in the Million Model Database† (Morisset *et al.* 2015). Overplotted are the ICF proposed by Delgado-Inglada *et al.* (2014) (solid line) and ICF = 1 (i.e., adopting C/O = C^{++}/O^{++}; dashed line). It can be seen that whereas for PNe with relatively high degree of ionization, $O^{++}/(O^{+}+O^{++}) > 0.8$, both ICFs are similar, at lower degrees of ionization using C/O = C^{++}/O^{++} is clearly incorrect (the abundances may be overestimated by as much as 2 dex).

In the last few years, new ICFs from large grids of photoionization models have become available. Delgado-Inglada *et al.* (2014) provide analytical expressions for the ICFs of He, C, N, O, Ne, S, Cl, and Ar and also for the uncertainties associated to each ICF. Other ICFs recently published in the literature are the ones by Smith *et al.* (2014) for Zn, Sterling *et al.* (2015) for Se and Kr, and Delgado-Inglada *et al.* (2016) for Ni.

3. Results from the study of gaseous abundances in PNe

3.1. *The sulfur anomaly*

This is one open problem in the field discovered by Henry *et al.* (2004) more than ten years ago. It refers to the fact that, at the same metallicity, PNe have systematically lower sulfur abundances than H II regions. Nucleosynthesis models do not predict such a destruction of sulfur in the AGB stars (Shingles & Karakas 2013). Another option is blaming the ICF for not correcting properly the contribution of unseen ions in PNe (Henry *et al.* 2012), but the anomaly is still present when using more modern ICFs to compute sulfur

† https://sites.google.com/site/mexicanmillionmodels/

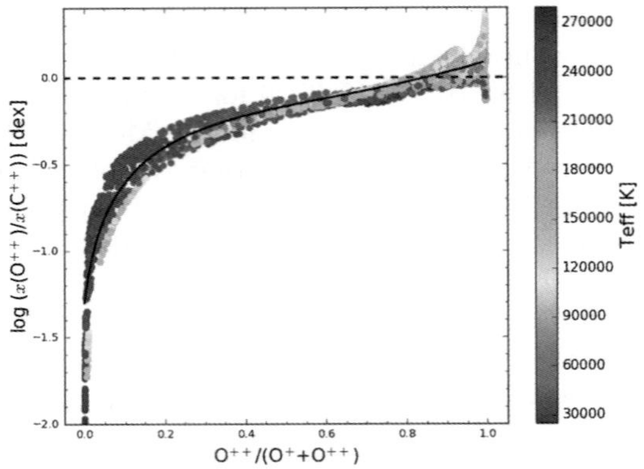

Figure 1. Values of $x(\mathrm{O}^{++})/x(\mathrm{C}^{++})$ as a function of $\mathrm{O}^{++}/(\mathrm{O}^{+}+\mathrm{O}^{++})$ for a sample of photoionization models available in the Million Model Database (Morisset *et al.* 2015). The color bar runs from low to high values of the effective temperature. The dashed line represents ICF = 1 while the solid line represents the ICF proposed by Delgado-Inglada *et al.* (2014).

abundances (Delgado-Inglada *et al.* 2014). Jacob *et al.* (2012) propose that considering a variable density in PNe (instead of a constant one) may solve the problem. This approach is explored by Delgado-Inglada *et al.* 2014 with a grid of photoionization models with a variable density distribution but does not solve the problem. Finally, uncertainties in the dielectronic recombination rates of S^{++} have also been explored without success (Badnell *et al.* 2015). In summary, this problem is far from being settled and deserves more study.

3.2. *Oxygen production by low mass stars in the Galaxy*

Oxygen is one of the elements for which more reliable abundances can be calculated in ionized nebulae and its abundance is generally used to trace the metallicity of the interstellar medium at different epochs. This is correct if this element is only produced in massive stars. Delgado-Inglada *et al.* (2015) find evidence of oxygen enrichment (of ~0.3 dex) in PNe of our Galaxy arising from stars of ~1.5 $M_{\odot}$ formed in a subsolar (around half solar) metallicity regime. At low metallicities, it is known that low-mass stars can produce and dredge up oxygen to the surface (Péquignot *et al.* 2000; Leisy & Dennefeld 2006) but this is the first observational indication of oxygen enrichment in Galactic disk PNe.

Fig. 2 shows the abundances derived for the PNe and H II regions studied in Delgado-Inglada *et al.* (2015). According to the dust features identified in infrared spectra, the PNe are divided into those with carbon-rich dust, CRD, and those with oxygen-rich dust, ORD. The theoretical predictions from Ventura *et al.* (2013) and Karakas *et al.* (2016) are also plotted in the figures. It can be seen that the models including a more efficient dredge-up, such as the ones by Ventura *et al.* (2013), reproduce better the derived abundances than those with no extra-mixing, such as those computed by Karakas *et al.* (2016).

One important question is how significant is the amount of oxygen produced by low mass stars for the overall enrichment of the Galaxy. The calculations by Carigi & Peimbert (2005), that use the old yields by Marigo *et al.* (1996), Marigo *et al.* (1998), and Portinari *et al.* (1998) for low and intermediate stars, predict that only around 1% of the present oxygen in the solar neighborhood is produced by low-mass stars. New calculations using the yields by Ventura *et al.* (2013) predict a somewhat higher production of oxygen (~2 %). All the results using different yields will be discussed in a forecoming paper (Delgado-Inglada *et al.* 2017, in prep.). Although oxygen production by low-mass

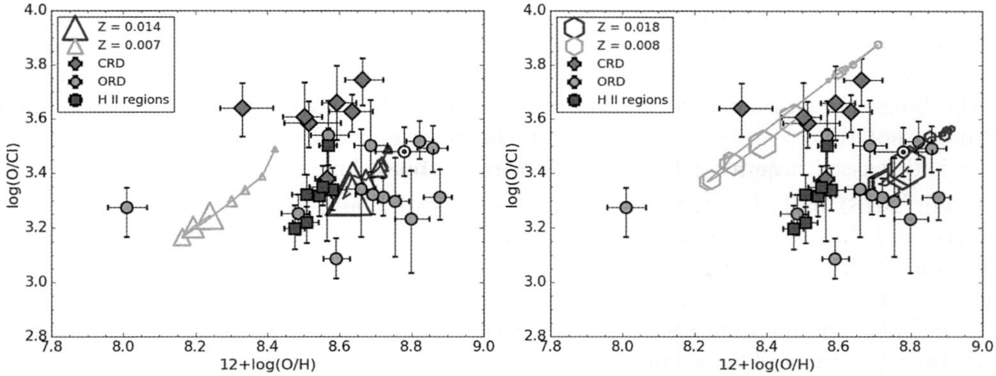

Figure 2. Values of O/Cl as a function of O/H for a sample of Galactic PNe with carbon-rich dust (red diamonds) and oxygen-rich dust (green circles) and Galactic H II regions (blue squares). Theoretical predictions by Karakas *et al.* (2016) (left) and Ventura *et al.* (2013) (right) are shown with connected symbols. The solar value from Lodders (2003) is presented for comparison. The figure has beed adapted from Delgado-Inglada *et al.* (2015).

stars seems to negligible in terms of the overall budget of oxygen in the Galaxy, some caution must be taken when using oxygen as a proxy for the initial composition of PN progenitors.

3.3. *Neutron-capture elements*

Around half of the elements beyond iron have been produced through slow neutron(n)-captures in AGB stars (Lattanzio & Karakas 2016) but there are still many uncertainties. In particular, it is important to know the efficiency of the mixing mechanism that carries the freshly synthesized elements to the surface and the exact mass at which the source of neutrons changes from the alpha captures of ^{12}C to the alpha captures of ^{22}Ne.

In the last few years, much progress has been done in the study of n-capture element abundances in PNe thanks to the combination of high quality spectra, new atomic data, and more reliable ICFs. The abundances of P, F, Ge, Se, Br, Kr, Rb, Ca, and Xe have been determined in several PNe (Otsuka *et al.* 2011; Otsuka & Tajitsu 2013; García-Rojas *et al.* 2015; Sterling *et al.* 2015; Sterling *et al.* 2016; Mashburn *et al.* 2016; Madonna *et al.* 2017, in prep.). Two of the main important results derived from these studies are: 1) A clear correlation between the C/O values in the PNe and the n-capture element enrichment, attributed to the fact that n-capture elements and carbon are brought together to the surface; 2) Different amounts of enrichment are found for different n-capture elements, and this is thought to be caused by the different progenitor mass. More discussion on the abundances of neutron-capture elements can be found in the articles by N. Sterling and M. Lugaro in this volume.

3.4. *Ages of stars in the Galactic bulge*

Nataf (2016) compiled a group of photometric and spectroscopic studies pointing to significantly different ages for the bulge stellar population. Photometric studies found an age of 8 Gyr whereas spectroscopic analysis found a significant population of stars with ages around 3 Gyr. This discrepancy results in the debate on whether the stars in the Galactic bulge are old or not, which has a crucial impact on the idea of how the bulge was formed. It is generally assumed that the bulge is an early product of the Galaxy formation but the presence of a young population would suggest that the formation of the bulge could have happened over a long period of time.

The analysis of AGB stars and PNe has played a role in this controversy. Gesicki *et al.* (2014) studied the star-formation history of the bulge through the analysis of the central star masses (and ages) of 31 bulge PNe. Their results point to an extended star formation in the bulge (i.e. a secular formation of the bulge). The authors concluded that bulge PNe are not representative of the most metal-poor and oldest stellar population. This issue is discussed in more detail by A. Zijlstra in this volume.

Buell (2013) find that the observed abundances of bulge PNe, the masses of white dwarfs, and the AGB luminosities, can be well reproduced using models with enhanced N/O and He/H values ($Y \sim 0.32$) and initial masses around 1.2–1.8 $M_{\odot}$ (implying that they were formed between 2 and 5 Gyr ago). The initial enrichment would come from intermediate- and high-mass stars formed earlier in the bulge that contaminated the ISM with their products. The assumption of a pre-enrichment of helium in the bulge would reduce the discrepancy between the photometric and spectroscopic age estimates of the bulge population (Nataf 2016).

3.5. *Extragalactic planetary nebulae*

PNe serve to study old stellar populations in external galaxies. Since PNe arise from stars formed at different epochs, their abundances trace the chemical composition of the ISM in the past. If these abundances are compared with those derived from H II regions which trace the present day abundances, one can study the chemical enrichment history of galaxies. Besides, the analysis of abundance gradients of PNe and H II regions contains information of various physical processes such as gas infall and outflow, star formation history, stellar migration, and the initial mass function.

A crucial issue before comparing the chemical abundances of extragalactic H II regions and PNe is to separate correctly these two types of objects. Stasińska *et al.* (2013) provided some useful diagnostics (involving, for example, the Hβ luminosities, N/O values, electron densities and temperatures, and ionized masses) for this purpose.

Magrini *et al.* (2016) study the oxygen abundances and gradients of four spiral galaxies, NGC 300, M33, M31, and M81 with PNe and H II regions. They find that the global oxygen enrichment increases with time in the four galaxies, being higher for the earlier type ones. They also find that the abundance gradients obtained from the older stellar population (PNe) are similar to or flatter than the ones derived for the younger stellar population (H II regions). Other studies of PNe in spiral galaxies are those from Kwitter *et al.* (2012); Sanders *et al.* (2012); Balick *et al.* (2013); Stasińska *et al.* (2013); Corradi *et al.* (2015b); Fang *et al.* (2015). Some of them are discussed in detail elsewhere in these proceedings.

Extragalactic PNe can also be used to study stellar nucleosynthesis at different metallicities or in different environments. For example, García-Rojas *et al.* (2016b) used deep and high resolution data obtained with GTC, VLT and CFH to study a sample of 22 PNe in the irregular and low-metallicity galaxy NGC 6822. The authors compare their derived chemical abundances with nucleosynthesis models from different authors to estimate the initial masses of the progenitor stars and find that the predictions differ from one model to another and they also depend on the element that is considered. Moreover, none of the nucleosynthesis models are able to reproduce the helium abundances. This kind of work illustrates that, despite the improvements achieved in observations and models, there is still much work to be done in the field of nucleosynthesis of low- and intermediate-mass stars.

3.6. *Dust*

The gaseous abundances of refractory elements in PNe provide useful information about the formation and evolution of dust grains in these ionized nebulae, and allow us to explore the different depletion patterns for dust forming elements.

More than 90% of the iron atoms present in Galactic PNe are found to be deposited into dust grains (e.g, Rodríguez & Rubin 2005; Delgado-Inglada *et al.* 2009). Combining the information of element depletions obtained from the gaseous abundances with the dust composition of the studied objects implied by the infrared dust features, Delgado-Inglada & Rodríguez (2014) find that PNe with carbon-rich dust tend to have higher iron depletions than those of PNe with oxygen-rich dust. This suggests that the formation of dust is probably more efficient in PNe with a carbon-rich environment. Recently, Delgado-Inglada *et al.* (2016) compare the depletion factors of iron and nickel in a group of Galactic PNe and H II regions and find that nickel atoms tend to be more attached to the grains than iron atoms at high depletions. This result is interesting because iron and nickel have similar chemical and physical properties and one would expect a similar behavior in their condensation into dust grains.

3.7. *Final remarks*

I would like to highlight that most of the results discussed here are based on high quality optical spectra, deep and with high resolution. Therefore, it is crucial to obtain more data of such quality, in our Galaxy and other galaxies, to get new and exciting results in the coming years.

Acknowledgements

I thank the organizers of the IAU Symp. 323 for the opportunity to present this work and for the warm hospitaly offered during the conference. I also want to thank the financial support received to attend this meeting from: Peking University, Conacyt grant no. CB-2014/241732, and PAPIIT (DGAPA-UNAM) grant no. 107215.

References

Aller, L. H. & Czyzak, S. J. 1968, *Planetary Nebulae*, 34, 209

Badnell, N. R., Ferland, G. J., Gorczyca, T. W., Nikolić, D., & Wagle, G. A. 2015, *ApJ*, 804, 100

Balick, B., Kwitter, K. B., Corradi, R. L. M., & Henry, R. B. C. 2013, *ApJ*, 774, 3

Buell, J. F. 2013, *MNRAS*, 428, 2577

Carigi, L., Peimbert, M., Esteban, C., & García-Rojas, J. 2005, *ApJ*, 623, 213

Corradi, R. L. M., García-Rojas, J., Jones, D., & Rodríguez-Gil, P. 2015a, *ApJ*, 803, 99

Corradi, R. L. M., Kwitter, K. B., Balick, B., Henry, R. B. C., & Hensley, K. 2015b, *ApJ*, 807, 181

Delgado-Inglada, G., Mesa-Delgado, A., García-Rojas, J., Rodríguez, M., & Esteban, C. 2016, *MNRAS*, 456, 3855

Delgado-Inglada, G., Morisset, C., & Stasińska, G. 2014, *MNRAS*, 440, 536

Delgado Inglada, G., Rodríguez, M., Mampaso, A., & Viironen, K. 2009, *ApJ*, 694, 1335

Delgado-Inglada, G., Rodríguez, M., Peimbert, M., Stasińska, G., & Morisset, C. 2015, *MNRAS*, 449, 1797

Fang, X., García-Benito, R., Guerrero, M. A., *et al.* 2015, *ApJ*, 815, 69

Ferland, G. J., Henney, W. J., O'Dell, C. R., & Peimbert, M. 2016, *RMxAA*, 52, 261

García-Rojas, J., Madonna, S., Luridiana, V., *et al.* 2015, *MNRAS*, 452, 2606

García-Rojas, J., Corradi, R. L. M., Monteiro, H., *et al.* 2016a, *ApJL*, 824, L27

García-Rojas, J., Peña, M., Flores-Durán, S., *et al.* 2016b, *A&A*, 586, A59

Gesicki, K., Zijlstra, A. A., Hajduk, M., & Szyszka, C. 2014, *A&A*, 566, A48

Henry, R. B. C., Kwitter, K. B., & Balick, B. 2004, *AJ*, 127, 2284
Henry, R. B. C., Speck, A., Karakas, A. I., Ferland, G. J., & Maguire, M. 2012, *ApJ*, 749, 61
Jacob, R., Sandin, C., Schönberner, D., & Steffen, M. 2012, *IAU Symposium*, 283, 398
Jones, D., Wesson, R., García-Rojas, J., Corradi, R. L. M., & Boffin, H. M. J. 2016, *MNRAS*, 455, 3263
Karakas, A. I. & Lattanzio, J. C. 2014, *PASA*, 31, 30
Karakas, A. I. & Lugaro, M. 2016, *ApJ*, 825, 26
Kingsburgh, R. L. & Barlow, M. J. 1994, *MNRAS*, 271, 257
Kwitter, K. B., Lehman, E. M. M., Balick, B., & Henry, R. B. C. 2012, *ApJ*, 753, 12
Lattanzio, J. & Karakas, A. 2016, *Journal of Physics Conference Series*, 728, 022002
Leisy, P. & Dennefeld, M. 2006, *A&A*, 456, 451
Liu, X.-W., Barlow, M. J., Zhang, Y., Bastin, R. J., & Storey, P. J. 2006, *MNRAS*, 368, 1959
Liu, X.-W., Storey P. J., Barlow M. J., Danziger I. J., Cohen M., Bryce M. 2000, *MNRAS*, 312, 585
Lodders, K. 2003, *ApJ*, 591, 1220
Luridiana, V., Morisset, C., & Shaw, R. A. 2015, *A&A*, 573, A42
Magrini, L., Coccato, L., Stanghellini, L., Casasola, V., & Galli, D. 2016, *A&A*, 588, A91
Marigo, P., Bressan, A., & Chiosi, C. 1996, *A&A*, 313, 545
Marigo, P., Bressan, A., & Chiosi, C. 1998, *A&A*, 331, 564
Mashburn, A. L., Sterling, N. C., Madonna, S., *et al.* 2016, *ApJL*, 831, L3
Mendoza, C. & Zeippen, C. J. 1982, *MNRAS*, 198, 127
Morisset, C., Delgado-Inglada, G., & Flores-Fajardo, N. 2015, *RMxAA*, 51, 103
Nataf, D. M. 2016, *PASA*, 33, e023
Nicholls D. C., Dopita M. A., Sutherland R. S., 2012, *ApJ*, 752, 148
Otsuka, M., Meixner, M., Riebel, D., *et al.* 2011, *ApJ*, 729, 39
Otsuka, M. & Tajitsu, A. 2013, *ApJ*, 778, 146
Peimbert, A., Peimbert, M., Delgado-Inglada, G., García-Rojas, J., & Peña, M. 2014, *RMxAA*, 50, 329
Peimbert, M., 1967, *ApJ*, 150, 825
Peimbert, M. & Costero, R. 1969, *Boletin de los Observatorios Tonantzintla y Tacubaya*, 5, 3
Peimbert, M. & Torres-Peimbert, S. 1971, *ApJ*, 168, 413
Péquignot, D., Zijlstra, A. A., Walsh, J., & Dudziak, G. 2000, *RMxAACS*, 9, 220
Portinari, L., Chiosi, C., & Bressan, A. 1998, *A&A*, 334, 505
Rodríguez, M. & Rubin, R. H. 2005, *ApJ*, 626, 900
Sanders N. E., Caldwell N., McDowell J., Harding P. 2012, *ApJ*, 758, 133
Shingles, L. J. & Karakas, A. I. 2013, *MNRAS*, 431, 2861
Smith, C. L., Zijlstra, A. A., & Dinerstein, H. L. 2014, *MNRAS*, 441, 3161
Stasińska, G., Peña, M., Bresolin, F., & Tsamis, Y. G. 2013, *A&A*, 552, A12
Sterling, N. C., Dinerstein, H. L., Kaplan, K. F., & Bautista, M. A. 2016, *ApJL*, 819, L9
Sterling, N. C., Porter, R. L., & Dinerstein, H. L. 2015, *ApJS*, 218, 25
Tayal, S. S. & Zatsarinny, O. 2010, *ApJS*, 188, 32
Torres-Peimbert, S., Peimbert, M., & Daltabuit, E. 1980, *ApJ*, 238, 133
Ventura, P., Di Criscienzo, M., Carini, R., & D'Antona, F. 2013, *MNRAS*, 431, 3642
Wyse, A. B. 1942, *ApJ*, 95, 356

Discussion

Stanghellini: Would you recommend a strong-line method? (applied usually to find abundances in H II regions)

Delgado Inglada: Strong-line methods should not be used to compute planetary nebula abundances because they have been calibrated for H II regions. The only possibility to obtain reliable abundances in PNe of the local group is to apply for telescope time in big telescopes with very efficient spectrographs.

Q: The sulfur abundance anomaly can indicate S depletion. If S is in the form of MgS, one expect some Mg anomaly, is this seen?

DELGADO INGLADA: I would expect other dust species such as GEMS.

BARLOW: Which ion stages of Fe and Ni did you have for your Ni/Fe abundance determinations?

DELGADO INGLADA: We used Ni^{++} and Fe^{++} ions to calculate the abundances.

SCHOENBERNER: PN ionization structure depends also on density structure. Hydrodynamic models develop negative density gradients which may help, e.g. to solve the sulfur problem. See the poster contribution by Jakob *et al.* in the Tenerife proceedings.

Planetary Nebulae: Multi-Wavelength Probes of Stellar and Galactic Evolution
Proceedings IAU Symposium No. 323, 2016
X. Liu, L. Stanghellini & A. Karakas, eds.

doi:10.1017/S1743921317000850

The kinematical behavior of ORLs and CELs in PNe with [WC] central star

Miriam Peña[1], Francisco Ruiz-Escobedo[1], Jackeline Rechy-García[1] and Jorge García-Rojas[2]

[1]Instituto de Astronomia, UNAM, México
email: miriam,fdruiz,jrechy@astro.unam.mx

[2]Instituto de Astrofisica de Canarias, La Laguna, Spain
email: jogarcia@iac.es

Abstract. In Planetary Nebulae (PNe) and HII regions ionic abundances can be derived by using collisionally excited lines (CELs) or recombination lines (ORLs). Such abundances do not coincide for the same ion and usually abundances from ORLs are larger than those from CELs by factors of 2 or larger. The origin of the discrepancy, known as the Abundance Discrepancy Factor is an open problem in astrophysics of gaseous nebulae. It has been attributed to temperature fluctuations in the plasma, tiny metal-rich inclusions embedded in the H-rich plasma, gas inhomogeneities or other processes. In this work we analyze the kinematical behavior of CELs and ORLs in two PNe ionized by [WC] stars, finding that kinematics of ORLS is incompatible with the kinematics of CELs. In particular the expansion velocities from CELs and ORLs for the same ion are different, indicating that ORLs seem to be produced in zones nearer the central star than CELs. This is in agreement with results found by other authors for individual PNe.

Keywords. planetary nebulae: general, kinematics and dynamics, planetary nebulae: individual: (Cn 1-5, PC 14)

1. Introduction

PNe are constituted by ionized gas surrounding a low-intermediate mass hot evolved star. The gas was ejected by the star, at low velocity (of about 10 km s^{-1}), in advanced stages of evolution (Giant or Asymptotic Giant Branch (AGB) stars). Thus the nebula is expanding away from the star with a velocity that depends on the ejection processes, the thermal processes and the interaction of the nebula with the interstellar medium.

The ionized gas emits collisionally excited lines (CELs) of ions of heavy element (ions of C, N, O, Ne, S, and others) and recombination lines (ORLs) of H, He and ions of heavy elements. Physical conditions (temperature and density) and ionic chemical abundances can be derived from both types of emission lines. Total abundances are obtained by considering all the ions detected in the gas and including the not visible ions by correcting the partial abundances with ionization correction factors (ICFs) or other procedures.

It is a well known fact that ionic abundances derived by using CELs or ORLs do not coincide for the same ion and usually the abundances from ORLs are larger than those from CELs by factors of 2 or larger. This occurs in HII regions and PNe as well (see e.g., García-Rojas & Esteban 2007). In some PNe such a factor, known as the Abundance Discrepancy Factor (ADF), can rise up to 100 or more but the mean value is around 3 (McNabb *et al.* 2013; García-Rojas *et al.* 2013). The origin of the ADF is, so far, an open problem in astrophysics of gaseous nebulae. It has been attributed to temperature fluctuations (e.g., Peimbert 1967, 1971), tiny metal-rich inclusions embedded in the H-rich plasma (Liu *et al.* 2006 and references therein), gas inhomogeneities or other processes. These proposals would imply that CELs and ORLs would be emitted in zones of different temperatures in the nebula, as CELs are highly dependent of the electron temperature, while ORLs are much less dependent of this parameter and they would be emitted mainly in low temperature zones.

Since some years ago some authors have indicated that the ORLs and CELs emitted by the same ion seem to show different kinematical behavior or different physical distribution in the nebula. In several objects recombination lines of O II appears to show lower expansion velocities than [O III] collisionally excited lines (e.g., Otsuka *et al.* 2010) or that the emission of ORLs from C^{++} and O^{++} strongly peak towards the nebular center, see e.g., Liu *et al.* (2000), Liu *et al.* (2006) and references therein.

More recently Richer *et al.* (2013) analyzed in detail the spatially and velocity-resolved spectroscopy of ORLs and CELs for the same ion in NGC 7009, finding that the lines show discrepant kinematics and location. ORLs kinematics is incompatible with the ionization structure given by CELs. These authors find that ORLs arise from a different volume than that giving rise to the forbidden emission from the same ion within this nebula, and suggest that there is an additional H-deficient plasma component.

Also García-Rojas *et al.* (2016) found that the O II λ4649+50 and the [OIII]λ5007 have different spacial distributions in NGC 6778. These authors claim that these differences are consistent with the presence of a H-poor gas where ORLs are being emitted.

We have analyzed the kinematics of ORLs and CELs for some PNe, for which we dispose of very high-resolution spectra, where the lines are well resolved. Most of spectra were obtained with the Magellan Inamori Kyocera Echelle spectrograph (MIKE; Berstein *et al.* 2003), attached to the 6.5-m Magellan telescope Clay, at Las Campanas Observatory, Chile. Most of these data have been already analyzed in order to derive physical conditions and chemical abundances from ORLs and CELs and to determine the ADFs for the objects (García-Rojas *et al.* 2012; García-Rojas *et al.* 2013). In these references the procedures for data reduction can be found. Data for other objects have been collected from the literature. The total sample includes PNe ionized by a [WC] central star, a *wels*, or a normal. The ADFs for the whole sample varies from 1.5 to 5.

In this article we present partial results for Cn 1-5 (PN G002.2-09.4) and PC 14 (PN G336.2-06.9), ionized by a [WC] central star, as examples of the whole sample, which will be presented elsewhere.

2. Data analysis

Due to the high spectral resolution of our data (MIKE has R $\sim$ 28,000 for a $1''$ slit width) the lines are well resolved and in several cases they are split showing blue and red components coming from the front and back zones of the expanding shell. In other cases the lines appear as a single component or show evidence of several overlapped components. Cn 1-5 and PC 14 present split lines and we measured both components with IRAF† routines by fitting a Gaussian profile to each line. Thus the observed central wavelength and Gaussian Full Width at Half Maximum ($gfwhm$) were determined for several ORLs and CELs in each object. The measured line widths are corrected for the instrumental width, considering that they add in quadrature. Thermal widths are not corrected for, therefore line widths include contributions from thermal broadening,

† IRAF is distributed by the National Optical Astronomy Observatories, which is operated by the Association of Universities for Research in Astronomy, Inc., under contract to the National Science Foundation.

Table 1. Central wavelength (Å), flux (erg cm^{-2} s^{-1}) and gfwhm (Å) for some CELs and ORLs in Cn 1-5 and PC 14. V$_{exp}$ in km s^{-1}.

	Cn 1-5			PC 14		
Ion	center	flux	gfwhm$_o$	center	flux	gfwhm$_o$
[O II]	3726.04	2.48E-13	0.25	3725.23	4.89E-14	0.29
[O II]	3726.69	1.64E-13	0.25	3725.90	4.74E-14	0.30
[O II]	3728.79	1.24E-13	0.24	3727.99	2.65E-14	0.29
[O II]	3729.44	8.04E-14	0.25	3728.66	2.49E-14	0.29
[O III]	4958.94	1.36E-12	0.47	4957.89	6.49E-13	0.50
[O III]	4959.80	1.09E-12	0.53	4958.63	6.69E-13	0.39
[O III]	5006.88	3.85E-12	0.46	5005.81	2.70E-12	0.46
[O III]	5007.76	3.21E-12	0.55	5006.56	2.96E-12	0.38
O II	4076.35	7.73E-15	0.35	4075.49	1.91E-15	0.48
O II	4077.04	6.02E-15	0.27	4076.21	8.53E-16	0.25
O II	4649.14	1.14E-15	0.37	4648.45	7.08E-16	0.30
O II	4649.91	1.11E-15	0.30	4649.94	3.19E-16	0.25
[N II]	6548.09	7.42E-13	0.54	6546.65	4.27E-14	0.61
[N II]	6549.21	5.59E-13	0.61	6547.84	3.87E-14	0.49
[N II]	6583.47	2.30E-12	0.54	6582.02	1.28E-13	0.61
[N II]	6584.60	1.73E-12	0.61	6583.21	1.15E-13	0.48
N I	8223.12	2.18E-16	0.75			
N I	8224.43	1.66E-16	0.80			
V$_{exp}$ [O II]		26.3±0.1			23.9±0.3	
V$_{exp}$ [O III]		26.4±1.2			22.3±0.4	
V$_{exp}$ (O II)		25.4±1.0			20.8±1.9	
V$_{exp}$ [N II]		25.7±0.2			27.2±1.2	

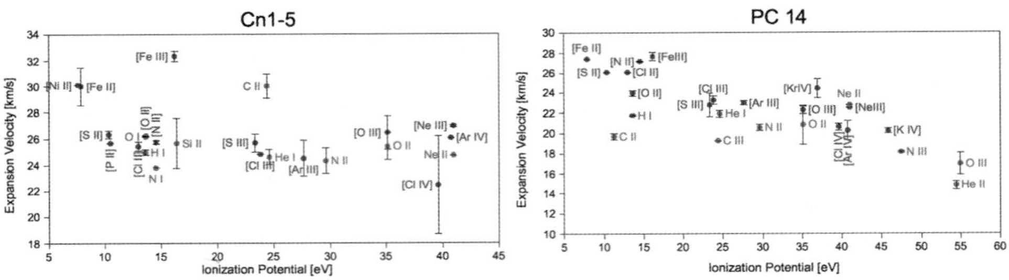

Figure 1. Expansion velocity vs. ionization potential for different ions as given by CELs (blue) and ORLs (red).

turbulence and structure. Partial results for the CELs and ORLs emitted by O^{+}, O^{++}, and N^{+} are presented in Table 1, where Column 1 gives the ion ID. In Columns 2, 3 and 4 the observed central wavelength, line flux, and $gfwhm$ for lines in Cn 1-5 are presented, and in Columns 5, 6 and 7, the same data are given for PC 14. As usual, CELs are represented with squared parenthesis, and ORLs without parenthesis. CELs and ORLs from several other ions (C^{+}, Fe^{+}, Fe^{++}, S^{+}, S^{++}, Ne^{++}, etc.) are also measured.

Expansion velocities (V$_{exp}$), as given by each CEL and each ORL, were calculated as half the difference between the red and blue components of each line. The final adopted V$_{exp}$ for each ion, is the average of velocities given by its different CELS and ORLs. They are given at the end of Table 1.

In Fig. 1 we present graphs showing the average V$_{exp}$ for each ion given by CELs (in blue) and ORls (in red) as a function of the ionization potential (IP) of the ion which, considering the stratification in the ionization structure usually found in a nebula, in some way represents the distance of the zone where the ion resides, relative to the central star.

3. Results

Fig. 1 shows evidence of the structure of the velocity field inside the expanding nebulae, in the sense that V_{exp} is lower near the central star and increases outwards. It is expected that low ionization species, which are farther away from the central star, have larger V_{exp} than highly ionized species which lie nearer the star. In Fig. 1 this is particularly true for velocities given by CELs, where a clear gradient is evident for both nebulae. In Cn 1-5 V_{exp} increases from about 24 km s^{-1} for the highly ionized species ([Cl IV], [Ar IV]) to about 30 km s^{-1} for [N II] and [Fe II] and the same happens with the CELs in PC 14. For the case of ORLs, there is either no gradient or a very flat one. Low- and high-ionization species show about the same V_{exp}. In addition, V_{exp} of ORLs are smaller than V_{exp} of CELs, at the same ionization potential. This is very clear if one considers for instance V_{exp}[O III] and V_{exp}(O II) in an object. In both cases presented here V_{exp}[O,III] > V_{exp}(O II), and the same occurs for V_{exp}[N II] and V_{exp}(N I).

This result recurs in most of the objects under study (Peña *et al.*, in preparation).

4. Discussion

From the results shown above, it seems evident that, as far as CELs and ORLs of the same ion do not share the same kinematics, these lines are not being emitted in the same zone of the nebula. Considering the expansion velocity field in the nebula and the fact that ORLs show lower V_{exp}, they seem to be emitted in an inner zone. In this sense we are finding the same results reported by Liu *et al.* (2000), Richer *et al.* (2013) and García-Rojas *et al.* (2016) for NGC 6153, NGC 7009 and NGC 6778 respectively, where ORLs are being mainly emitted in an inner zone than CELs. And that kinematics of ORLs is incompatible with that of CELs.

In Cn 1-5 and PC 14, the zones where ORLs are mainly emitted are not expanding in the same way as the zones where CELs are emitted, and show lower V_{exp}. It is tempting to say that ORLs come from H-deficient zones which were ejected by the star after the H-rich gas where CELs are emitted. However at this point this is mainly speculative. Much more analysis is needed, for a larger number of objects.

Acknowledgement

This work has received financial support from grant UNAM-PAPIIT IN109614.

References

Berstein, R. A., Shectman, S. A., Gunnels, S, *et al.* 2002, *Proc. SPIE* 4841
García-Rojas, J. & Esteban, C. 2007, *ApJ*, 670, 457
García-Rojas, J., Peña, M., Morisset, C. Mesa-Delgado, A., & Ruiz, M. T. 2012, *A&A*, 538, 54
García-Rojas, J., Peña, M., Morisset, C., *et al.* 2013, *A&A*, 558, 122
García-Rojas, J., Corradi, R. L. M., Monteiro, H., *et al.* 2016, *ApJ*, 824, 27
McNabb, I. A., *et al.* 2013, *MNRAS*, 428, 3443
Liu, X.-W., Sorey, P. J., Barlow, M. J., *et al.* 2000, *MNRAS*, 312, 585
Liu, X.-W., Barlow, M. J., Zhang, Y., *et al.* 2006, *MNRAS*, 368, 1959
Otsuka, M, Tajitsu, A., Hyung, S., & Izumiura, H. 2010, *ApJ*, 723, 658
Peimbert, M. 1967, *ApJ*, 150,825
Peimbert, M. 1971, *Bol. Observatorios Tonantzintla y Tacubaya*, 6, 29
Richer, M. G., Georgiev, L., Arrieta, A., & Torres-Peimbert, S. 2013, *ApJ*, 773, 133

Discussion

BARLOW: Are your spectra capable of being used to measure the ADFs as a function of angular radius in the nebulae?

PENA: It is possible to measure ADFs as a function of radius in extended nebulae. Anyway we are measuring an average over the line of sight.

LIU: Long-slit observations of NGC6153 (Liu *et al.* 2000) do show that ADF increases towards nebular centre.

PENA: Thank you. I will check it.

BARLOW: Two comments. 1) How valid is to plot H I velocity in your diagrams, considering that this ion is by definition present in all the region under consideration? 2) It is not surprising that the FWHM of ORL is lower than the one of CELs; this can be interpreted as corresponding to a lower temperature region.

PENA: 1) H I velocity represents the average over the nebula and it can be interpreted as the expansion velocity of the nebula, so it is important; 2) I agree. FWHM(CELs) are larger due to temperature, than FWHM(ORLs).

Planetary Nebulae: Multi-Wavelength Probes of Stellar and Galactic Evolution
Proceedings IAU Symposium No. 323, 2016
X. Liu, L. Stanghellini & A. Karakas, eds.

doi:10.1017/S1743921317000849

Imaging the elusive H-poor gas in planetary nebulae with large abundance discrepancy factors

Jorge García-Rojas[1,2], Romano L. M. Corradi[3,1], Henri M. J. Boffin[4], Hektor Monteiro[5], David Jones[1,2], Roger Wesson[6], Antonio Cabrera-Lavers[3,1] and Pablo Rodríguez-Gil[1,2]

[1]Inst. de Astrofísica de Canarias, Vía Láctea, s/n, La Laguna, E-38205, Spain,
email: jogarcia@iac.es

[2]Dept. de Astrofísica, Universidad de La Laguna, La Laguna, E-38205, Spain
[3]GRANTECAN, Cuesta de San José s/n, E-38712, Breña Baja, La Palma, Spain
[4]European Southern Observatory, Karl Schwarzschild Strasse 2, 85748 Garching, Germany
[5]Univ. Fed. Itajubá, Av. BPS 1303-Pinheirinho, 37500-903, Itajubá, Brazil
[6]Dept. of Physics & Astronomy, University College London, Gower Street, London WC1E 6BT, UK

Abstract. The discrepancy between abundances computed using optical recombination lines (ORLs) and collisionally excited lines (CELs) is a major, unresolved problem with significant implications for the determination of chemical abundances throughout the Universe. In planetary nebulae (PNe), the most common explanation for the discrepancy is that two different gas phases coexist: a hot component with standard metallicity, and a much colder plasma enhanced in heavy elements. This dual nature is not predicted by mass loss theories, and direct observational support for it is still weak. In this work, we present our recent findings that demonstrate that the largest abundance discrepancies are associated with close binary central stars. OSIRIS-GTC tunable filter imaging of the faint O II ORLs and MUSE-VLT deep 2D spectrophotometry confirm that O II ORL emission is more centrally concentrated than that of [OIII] CELs and, therefore, that the abundance discrepancy may be closely linked to binary evolution.

Keywords. binaries: close — ISM: abundances — (ISM:) planetary nebulae: general — stars: winds — stars: outflows

1. Introduction

The abundance discrepancy problem was first discovered in the physics of gaseous nebulae more than 70 years ago (Wyse 1942). It pertains to the fact that chemical abundances obtained from optical recombination lines (ORLs) are systematically larger than those obtained from collisionally excited lines (CELs) of the same ion. Generally, this discrepancy is parameterized by the abundance discrepancy factor (ADF), which for a given ion is defined as the ratio of ionic chemical abundances derived from ORLs and from CELs. Photoionized gaseous nebulae exhibit different ADF values. While H II regions have moderate and nearly constant values (ADF $\sim 2-3$; see García–Rojas & Esteban 2007), planetary nebulae (PNe) can reach values as large as 120 (Corradi *et al.* 2015). García–Rojas & Esteban (2007) concluded that the physical origin of the abundance discrepancy in PNe with extreme ADF values should be different to that of H II regions and the bulk of PNe, where moderate values are found.

Regarding PNe, in the last few years our group has focused on the possible link between the large ADFs found in some PNe and the binary nature of their central stars (Corradi *et al.* 2015; Jones *et al.* 2016). We have seen that it is relatively common to find extremely large ADFs in PNe with binary central stars that have experienced a common-envelope episode (post-CE PNe; e. g. Liu *et al.* 2006), reaching extreme values in the inner regions of several such objects (e. g. in the central 7″ of Abell 46, ADF∼300). Our recent observations of the large ADF PN NGC 6778 taken with the 10.4m Gran Telescopio Canarias (GTC) show that the spatial distribution of the O^{2+} ions producing the O II ORLs does not match that of the O^{2+} ions emitting the [O III] CELs; additionally, O II ORL emission is concentrated in the central parts of the nebula (García–Rojas *et al.* 2016). This is consistent with the presence of two distinct plasmas that are not well mixed, perhaps because they were produced in different ejection events. Our results are important because, for the first time, the evolution of close binary stars is clearly linked with the large ADFs in PNe. These results have encouraged us to acquire additional data in order to strengthen the link between the common-envelope process and the abundance discrepancy phenomenon. In particular, we have obtained high spatial resolution 2D spectroscopy with MUSE-VLT and direct imaging with a tunable filter using OSIRIS-GTC to study in detail the spatial distributions of both ORL and CEL emissions.

2. Observations

GTC observations. We obtained narrow-band images of Abell 46 using the blue tunable filter of the Optical System for Imaging and low Resolution Integrated Spectroscopy (OSIRIS) instrument at the 10.4m GTC (La Palma) on 20 September 2016. Since the central wavelength of the tunable filter is not constant across the size of Abell 46 (∼70″ diameter), a two-step scanning was required, centered at 4652 and 4656 Å, respectively. We took 7 exposures of 1900 s in each configuration to reach the signal-to-noise ratio needed to detect the emission from the faint O II ORLs. For details of the observing and reduction technique, see García–Rojas *et al.* (2016).

VLT observations. Observations were made with the Multi Unit Spectroscopic Explorer (MUSE) mounted on VLT UT4 on 6 July 2016. The extended wide field mode was used, with a field of 60 × 60″, a 0.2 spaxel size and a wavelength coverage of 4650–9300 Å at a mean spectral resolution of ∼2500, enough to isolate the emission of the brightest O II ORLs. We observed 5 large ADF PNe. For each object, we made 5 long exposures (between 150 s and 1800 s, depending on target) to have enough S/N ratio to detect the faint O II ORLs. We applied a dither pattern and a rotation of 90 degrees between the different exposures to avoid bad pixels and to get rid of any MUSE systematics. We also took two sky exposures to remove the sky-background emission from our final spectra. We followed the standard reduction procedures described in Walsh *et al.* (2016) using the instrument pipeline version 1.0 (Weilbacher *et al.* 2014). Each datacube was completely reduced, sky subtracted and wavelength and flux calibrated. Only a single datacube for each object was used in the present work.

3. Results

Abell 46 is the PN with the largest ADF ever measured in an integrated spectrum of a PN (∼120; see Corradi *et al.* 2015), and has been shown to harbor a post-common-envelope central binary star (Afsar & Ibanoglu 2008). In Fig. 1 we show the spatial distribution of the O II $\lambda\lambda$4649+50 ORLs emission (left), compared to that of the [O III] λ5007 CEL (right). Similar to what was observed in NGC 6778 by García–Rojas *et al.* (2016) using the same technique, the emission of ORLs is clearly produced closer to the binary central star than the CEL emission. This is consistent with the results of

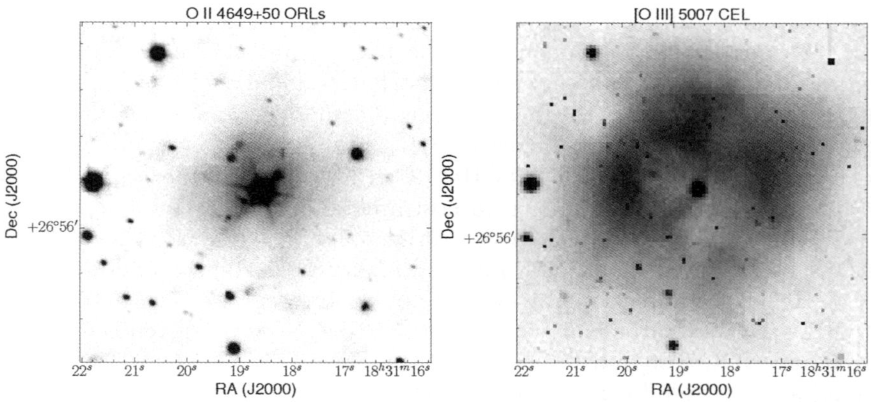

Figure 1. OSIRIS-GTC tunable filter image of Abell 46 in the O II 4649+50 ORLs (left), compared with an image taken with the 3.6m NTT in the [O III] 5007 CEL (right).

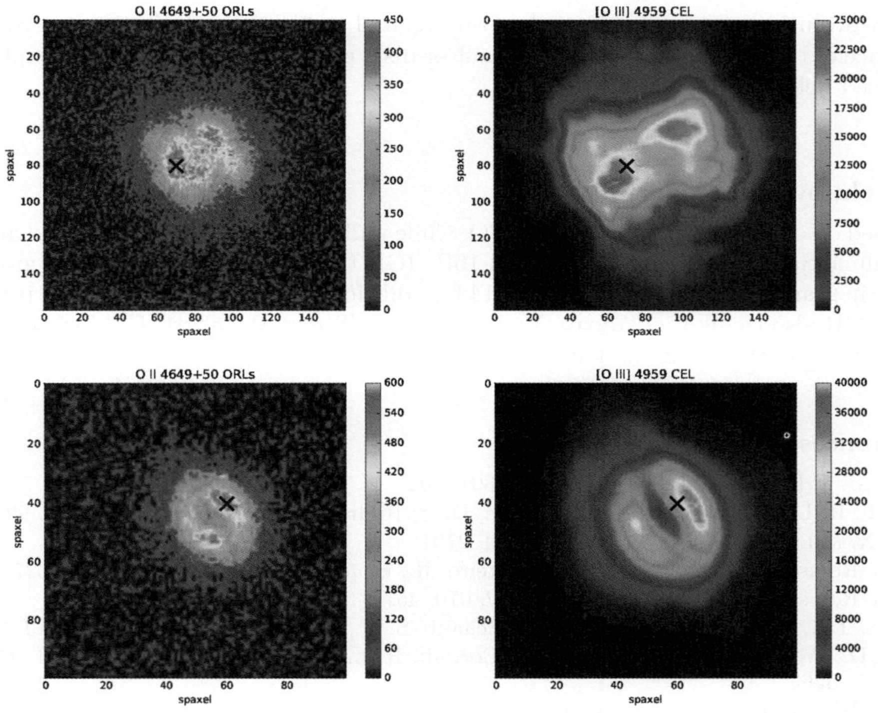

Figure 2. Preliminary maps of the O II $\lambda\lambda$4649+50 ORLs (left panels) and of the [O III] λ4959 CEL for two PNe in our sample: NGC 6778 (ADF$\sim$20; top panels), and M 1–42 (ADF$\sim$20; bottom panels). It is worth re-emphasising that the O II and the [O III] emission comes from the same ion: O^{2+}. The "x" marks a reference spaxel.

Corradi *et al.* (2015) for Abell 46, who analyze the emission spatial profiles with long-slit spectroscopy (see their Figure 4).

In Fig. 2 we show the continuum subtracted MUSE-VLT maps of the O II $\lambda\lambda$4649+50 ORLs (left panels) and of the [O III] λ4959 CEL (right panels) for NGC 6778 and M 1–42. It is clear that in both objects the emission of ORLs is more centrally concentrated than the emission of the strongest [O III] λ4959 CEL. Both PNe have very similar ADFs of

$\sim$20 (see Jones *et al.* 2016 and McNabb *et al.* 2016). NGC 6778 has a close binary central star that has undergone a common envelope phase (Miszalski *et al.* 2011). However, the central star of M 1–42 has not been reported as binary, although its asymmetric morphology, large ADF and the behaviour found in this work suggests that it is an excellent candidate to search for binarity.

We observed three additional objects. Hf 2–2, a PN with one of the largest known ADFs ($\sim$80; McNabb *et al.* 2016), is known to host a close binary central star (Hillwig *et al.* 2016). NGC 6153 and NGC 7009 are very bright and well-studied PNe with relatively large ADFs ($\sim$10) but without known central binary stars (Liu *et al.* 2000; Fang & Liu 2013). For Hf 2–2 there is also clear evidence of a more centrally concentrated emission of O II ORLs relative to the emission of [O III] CELs. However, given the brigthness of NGC 6153 and NGC 7009 and the relatively low S/N of O II ORLs with respect to the continuum, a very careful continuum subtraction must be performed to clearly see the different spatial distributions of the O II and [O III] emission.

Our observations have added a new, unexpected ingredient to understand the abundance discrepancy problem: large ADFs should be explained in the framework of close-binary evolution. A reasonable explanation related to binary evolution (nova ejecta, the presence of the common-envelope remnant or disk, etc.) should be invoked to explain the observed behavior.

4. Acknowledgements

Based on observations collected at ESO, Chile under proposal 097.D-0241, and at GTC, La Palma, Spain under proposal GTC3-16B. JGR thanks the IAU for a travel grant and Mexican grant UNAM-PAPIIT IN109614 for additional support of the research presented here. JGR also thank C. Esteban, M. Peimbert, S. Torres–Peimbert, C. Morisset and M. Peña for support and discussions.

References

Afsar, M. & Ibanoglu, C. 2008, *MNRAS*, 391, 802

Corradi, R. L. M., García–Rojas, J., Jones, D., & Rodríguez–Gil, P. 2015, *ApJ*, 803, 99

Fang, X. & Liu, X.-W. 2013, *MNRAS*, 429, 2791

García–Rojas, J., Corradi, R. L. M., Monteiro, H., *et al.* 2015, *ApJ* (Letters), 824, L27

García–Rojas, J. & Esteban, C. 2007, *ApJ*, 670, 457

Hillwig, T. C., Bond, H. E., Frew, D. J., Schaub, S. C., & Bodman, E. H. L. 2016 *AJ*, 152, 34

Jones, D., Wesson, R., García-Rojas, J., Corradi, R. L. M., & Boffin, H. M. J. 2016, *MNRAS*, 455, 3263

Liu, X.-W., Storey, P. J., Barlow, M. J., *et al.* 2000, *MNRAS*, 312, 585

Liu, X.-W., Barlow, M. J., Zhang, Y., Bastin, R. J., & Storey, P. J. 2006, *MNRAS*, 368, 1959

McNabb, I. A., Fang, X., & Liu, X.-W. 2016, *MNRAS*, 461, 2818

Miszalski, B., Jones, D., Rodríguez–Gil, P., *et al.* 2016, *MNRAS*, 531, 158

Walsh, J.-R., Monreal–Ibero, A., Barlow, M. J., *et al.* 2016, *A&A*, 588, A106

Weilbacher, P. M., Streicher, O., Urrutia, T., *et al.* 2014, in N. Manset, & P. Forshay (eds.), *Astronomical Data Analysis Software and Systems XXIII*, ASP Conf. Ser., 485, p. 451

Wyse, A. B. 1942, *ApJ*, 95, 356

Discussion

Liu: Liu *et al.* (2000; Figs. 17, 18) show that ORLs are more peaked towards nebular centre (i.e. higher ADFs). This is in contrast with your new MUSE imaging observation.

García-Rojas: You are right. In fact I was surprised about these results in NGC6153. However, these are preliminary results and a more detailed continuum subtraction should be done for the brightest objects, as the large exposure times make continuum subtraction difficult because continuum is very strong inthe O II ORL zone.

Balick: If ORLs arise closer to the nuclei than CELs, then those PNe with large central cavities should show relatively weak ORLs and lower ADFs. Have you corrected ADFs with nebular morphologies?

García-Rojas: We have not tried what you proposed, but we will explore your suggestion. Thanks.

Planetary Nebulae: Multi-Wavelength Probes of Stellar and Galactic Evolution
Proceedings IAU Symposium No. 323, 2016
X. Liu, L. Stanghellini & A. Karakas, eds.

doi:10.1017/S1743921317001958

Close binary central stars and the abundance discrepancy - new extreme objects

R. Wesson,[1,2]† D. Jones,[3,4] J. García-Rojas[3,4], R. L. M. Corradi,[3,4,5] and H. M. J. Boffin,[6]

[1]Department of Physics and Astronomy, University College London, Gower St, London WC1E 6BT, UK
[2]European Southern Observatory, Alonso de Córdova 3107, Casilla 19001, Santiago, Chile
[3]Instituto de Astrofísica de Canarias, E-38205 La Laguna, Tenerife, Spain
[4]Departamento de Astrofísica, Universidad de La Laguna, E-38206 La Laguna, Tenerife, Spain
[5]Gran Telescopio CANARIAS S. A., c/ Cuesta de San José s/n, Breña Baja, E-38712 Santa Cruz de Tenerife, Spain
[6]European Southern Observatory, Karl-Schwarzschild-Str. 2, 85738 Garching bei München, Germany

Abstract. Recent work (Corradi *et al.* 2015; Jones *et al.* 2016) has shown that the phenomenon of extreme abundance discrepancies, where recombination line abundances exceed collisionally excited line abundances by factors of 10 or more, seem to be strongly associated with planetary nebulae with close binary central stars. To further investigate, we have obtained spectra of a sample of nebulae with known close binary central stars, using FORS2 on the VLT, and we have discovered several new extreme abundance discrepancy objects. We did not find any non-extreme discrepancies, suggesting that a very high fraction of nebulae with close binary central stars also have an extreme abundance discrepancy.

Keywords. planetary nebulae: general – circumstellar matter – stars: mass-loss – stars: winds, outflows – binaries: close – ISM: abundances

1. Introduction

The abundances of heavy elements in photoionised nebulae may be determined from their strong, bright collisionally excited lines (CELs), whose emissivity depends exponentially on the temperature, or from the much weaker recombination lines (RLs), the emissivity of which has a weak power law dependence on the temperature. Abundances derived from RLs and CELs do not agree, with the values from RLs exceeding those from CELs by a factor most commonly around 2–3, but reaching extreme values of 10 or more in about 10 per cent of planetary nebulae (PNe). This abundance discrepancy problem has been known since the 1940s (Wyse 1942), but a full understanding of its causes remains elusive. The abundance discrepancy factor (*adf*) for an ion is defined as the ratio of its RL abundance to its CEL abundance. O^{2+} is much the easiest ion to measure an *adf* for, being usually the most abundant heavy element, and having RLs and CELs in the optical. In this article, unless otherwise specified, *adf* refers to *adf*(O^{2+}).

A number of plausible mechanisms have been postulated which could account for the abundance discrepancy. These include temperature fluctuations (Peimbert 1967), density fluctuations (Viegas & Clegg 1994), hydrogen-deficient clumps (Liu *et al.* 2000), X-ray illuminated quasi-neutral clumps (Ercolano 2009), and non-thermally distributed electrons (Nicholls *et al.* 2012). All except the hydrogen-deficient clump theory account for the discrepancy in the context of a chemically homogeneous nebula. Since 2006 there

† E-mail: rw@nebulousresearch.org

have been suggestions that whatever the mechanism, it may be related to central star binarity. Hf 2–2, one of the most extreme objects known with an abundance discrepancy of 80, has a binary central star with an orbital period so short that it must have undergone a common envelope (CE) phase (Liu *et al.* 2006). The difficulties in identifying such short period binaries meant that until recently, the numbers of planetary nebulae for which binarity was established and abundance discrepancy measured was small. However, recently the picture has become clearer, with Corradi *et al.* (2015) finding that three known post-CE binaries had nebulae with extreme abundance discrepancies, and then Jones *et al.* (2016) strengthening the link with a study of the post-CE nebula NGC 6778, finding an abundance discrepancy of a factor of nearly 20 on average, and peaking at around 40 in the centre of the nebula.

We have obtained new spectra of a sample of PNe known to have close binary central stars. The sample has revealed a number of objects with extreme abundance discrepancies, and no low-discrepancy objects. There thus appears to be an almost perfect correlation between close binarity and extreme abundance discrepancies.

2. Observations

We obtained spectra of about 25 planetary nebulae using FORS2 mounted on UT1 (Antu) at the Very Large Telescope in Chile, in ESO programmes 093.D-0038(A) and 096.D-0080(A). The spectra covered wavelengths from 3600-5000 and 5800-7200Å at a resolution of 1.5Å, sufficient to resolve recombination lines in the blended features at 4070Å and 4650Å. The target nebulae were southern objects with known close binary central stars, for which there were no recombination line abundances in the literature. The exception to that was Hf 2–2, known to have an extreme abundance discrepancy, which we included in our sample both as a control to verify our strategy and methodology, and with the hope that our spectra would be deep enough to permit a spatially resolved study.

The programme was designed as a filler to be carried out in almost any weather conditions, and so a number of objects in the sample were observed in less than ideal conditions, but several excellent spectra were obtained. The data on NGC 6778, obtained early in the programme, revealed a new extreme object with an abundance discrepancy of $\sim$20, presented in Jones *et al.* (2016). Upon the completion of the programme, we have detected recombination lines in seven further objects, including Hf 2–2.

3. Analysis

Spectra were reduced using standard STARLINK routines. Cosmic rays were removed from the 2D frames using a combination of STARLINK's figaro routines and a python implementation of the LAcosmic algorithm (van Dokkum 2001). Emission line fluxes were measured using ALFA (Wesson 2016), which we used in Jones *et al.* (2016) to perform a similar analysis for the high *adf* PN NGC 6778. ALFA derives fluxes by optimising the parameters of Gaussian fits to line profiles using a genetic algorithm, after subtracting a globally-fitted continuum.

NEAT (Wesson *et al.* 2012) was then used to derive final ionic and elemental abundances from these emission line fluxes. The code corrects for interstellar extinction using the ratios of Hγ and Hδ to Hβ (the Hα/Hβ ratio was not used to calculate the extinction but rather as a sanity check to ensure that line fluxes measured from the non-overlapping red and blue spectra were consistent) and the Galactic extinction law of Howarth (1983); temperatures and densities are then derived from the standard diagnostics (see Wesson

Object	$\mathbf{O}^{2+}_{RLs}$ / $\mathbf{O}^{2+}_{CELs}$	T_e([O III])	T_e(BJ)	T_e(He I)	T_e(O II)
Hf 2–2	80	8 800	800	2 000	2 000
MPA 1759	80	11 500	–	4 000	–
Pe 1-9	75	10 000	–	4 000	–
NGC 6326	50	14 500	8 000	4 000	–
NGC 6337	30	12 500	–	3 000	$<$1 000
NGC 6778	20	8 800	4 100	3 000	1 300
Fg 1	20	10 450	–	6 000	$<$1 000
Hen 2–283	13	8 800	–	–	3 000

Table 1. The abundance discrepancy for O^{2+} and temperature diagnostics in the objects where recombination lines were detected

et al. 2012, for full details). Ionic abundances are then calculated from flux-weighted averages of the emission lines of each species using the previously derived temperatures and densities, and total abundances estimated using the ionisation correction factors of Delgado-Inglada *et al.* (2014). The atomic data and ionisation correction factors used were as in Jones *et al.* (2016).

4. Abundance discrepancies and evidence for H-deficient material

We detected recombination lines in eight of our sample objects. In all eight cases, the abundance discrepancy was extreme. The values we obtained (including the Jones *et al.* 2016 value for NGC 6778) are listed in Table 1. Among the eight objects was Hf 2–2, for which we derive very similar results in the integrated spectrum to those of Liu *et al.* (2006).

We then estimated upper limits to the abundance discrepancy for the objects where recombination lines were not detected, using a spectral synthesis code to calculate the O^{2+} recombination line spectrum, and varying the abundance until we found the value for which the RLs would have been detectable. We found that the limits from our non-detections were not strong, and that an extreme abundance discrepancy could not be ruled out in any case.

Temperatures measured from various diagnostics support the picture of hydrogen-deficient material in these extreme objects. The standard [O III] line ratios give temperatures typical of a photoionised gas of 'normal' composition - 8–12kK. Three diagnostics give generally much lower temperatures: the Balmer jump (BJ), which lies close to the edge of our spectral coverage, but is detected with sufficient signal to noise in several nebulae; He I emission line ratios, and O II recombination lines, which should probe the coldest and most metal-rich regions of the gas (McNabb *et al.* 2013). The temperatures implied by the various diagnostics are listed in Table 1.

5. Discussion

The association between binarity and the abundance discrepancy suggests two hypotheses. Firstly, that all PNe with a close binary central star have an extreme abundance discrepancy. This work supports that hypothesis and almost doubles the number of objects known to have both a binary central star and an extreme abundance discrepancy. However, there is currently thought to be at least one object which contradicts that. NGC 5189 has a binary central star with a period of just over 4 days, but has an unusually low abundance discrepancy of 1.6 (García-Rojas *et al.* 2013), albeit measured from a very small region of the nebula. The Necklace is also a definite post-CE object, but no recombination lines were detected in deep spectra by Corradi *et al.* (2011). We have recently obtained observations of NGC 5189 covering most of the nebula, from which we will be able to see if the low value found by García-Rojas *et al.* (2013) is representative of the whole nebula or a chance observation of a low-*adf* region. If these two objects really have a low *adf*, then in total we now have 14 of 16 extreme-*adf* objects in the sample of close binaries with known chemistry.

The complemenary hypothesis, that all nebulae with extreme abundance discrepancies have a close binary central star, is not yet well tested, but a number of extreme *adf* objects whose central star status is not yet confirmed would be fruitful grounds for investigation: Abell 30, Abell 58, NGC 1501, M 1-42 and M 3-32 are among the most extreme *adf* objects whose central star status is still unknown. Spatially resolved studies of many high-*adf* objects have found that the RL abundances are strongly centrally peaked, further associating the phenomenon with the central star. See e.g. Liu *et al.* (2000), Jones *et al.* (2016) and García-Rojas *et al.* (2016, and this volume).

Whether the link between them is universal or not, it is nevertheless clear that central star binarity and nebular chemistry are strongly linked. Any explanation for the most extreme abundance discrepancy must account for the fact that they are preferentially found in objects with close binary central stars. The strong similarities long noted between high *adf* PNe and old nova shells such as CP Pup and DQ Her, which exhibit extremely low plasma temperatures and strong recombination lines, suggests that some kind of nova-like outburst from the close binary central star could be responsible for ejecting H-deficient material into the nebulae. The ~90% hit rate for close binaries having an extreme abundance discrepancy would suggest that this eruption must happen soon after the formation of the main nebula.

References

Corradi, R. L. M., *et al.*, 2011, in Asymmetric Planetary Nebulae 5 Conference.

Corradi, R. L. M., García-Rojas, J., Jones, D., & Rodríguez-Gil, P., 2015, *APJ*, 803, 99

Delgado-Inglada, G., Morisset, C., & Stasińska, G., 2014, *MNRAS*, 440, 536

Ercolano, B., 2009, *MNRAS*, 397, L69

García-Rojas, J., Peña, M., Morisset, C., Delgado-Inglada G., Mesa-Delgado, A., & Ruiz, M. T., 2013, *A&A*, 558, A122

García-Rojas, J., Corradi, R. L. M., Monteiro, H., Jones, D., Rodríguez-Gil, P., & Cabrera-Lavers, A., 2016, *ApJ*, 824, L27

Howarth, I. D., 1983, *MNRAS*, 203, 301

Jones, D., Wesson, R., García-Rojas, J., Corradi, R. L. M., & Boffin, H. M. J., 2016, *MNRAS*, 455, 3263

Liu, X.-W., Storey, P. J., Barlow, M. J., Danziger, I. J., Cohen, M., & Bryce, M., 2000, *MNRAS*, 312, 585

Liu, M. C., Leggett, S. K., Golimowski, D. A., Chiu, K., Fan, X., Geballe, T. R., Schneider, D. P., & Brinkmann, J., 2006, *ApJ*, 647, 1393

McNabb, I. A., Fang, X., Liu, X.-W., Bastin, R. J., & Storey, P. J., 2013, *MNRAS*, 428, 3443

Nicholls, D. C., Dopita, M. A., & Sutherland, R. S., 2012, *ApJ*, 752, 148

Peimbert, M., 1967, *ApJ*, 150, 825

Viegas, S. M. & Clegg, R. E. S., 1994, *MNRAS*, 271, 993

Wesson, R., 2016, *MNRAS*, 456, 3774

Wesson, R., Stock, D. J., & Scicluna, P., 2012, *MNRAS*, 422, 3516

Wyse, A. B., 1942, *ApJ*, 95, 356

van Dokkum, P. G., 2001, *PASP*, 113, 1420

Planetary Nebulae: Multi-Wavelength Probes of Stellar and Galactic Evolution
Proceedings IAU Symposium No. 323, 2016
X. Liu, L. Stanghellini & A. Karakas, eds.

doi:10.1017/S174392131700062X

Atomic Data and Neutron-Capture Element Abundances in Planetary Nebulae

N. C. Sterling

Department of Physics, University of West Georgia, 1601 Maple Street, Carrollton, GA 30118, USA
email: nsterlin@westga.edu

Abstract. Neutron(n)-capture elements are produced by s-process nucleosynthesis in low- and intermediate-mass AGB stars, and therefore can be enriched in planetary nebulae (PNe). In the last ten years, n-capture elements have been detected in more than 100 PNe in the Milky Way and nearby galaxies. In some objects, several different n-capture elements have been detected, providing valuable constraints to models of AGB nucleosynthesis and evolution. These detections have motivated theoretical and experimental investigations of the atomic data needed to derive accurate n-capture element abundances. In this review, I discuss the methods and results of these atomic data studies, and their application to abundance determinations in PNe.

Keywords. atomic data—planetary nebulae: general—nuclear reactions, nucleosynthesis, abundances—stars: AGB and post-AGB

1. Introduction

Neutron(n)-capture elements are heavier than zinc (atomic number $Z > 30$) and can be produced by slow n-capture nucleosynthesis, or the "s-process," during the asymptotic giant branch (AGB) evolutionary stage of planetary nebula (PN) progenitor stars. The chemical compositions of PNe bear the signatures of AGB nucleosynthesis, and thus can be enriched in n-capture elements. However, detecting the weak emission lines of these species in PNe poses a significant challenge, owing to the low cosmic abundances of n-capture elements (less than a few billionths of the H abundance in the Solar System; Asplund *et al.* 2009). The most easily detected trans-iron element ions are those with a small number of ground configuration levels (leading to a few relatively strong emission lines), such as those with $4p^k$ and $5p^k$ valence shells, where $k = 1$–5. Depletion into dust is another factor limiting the detectability of n-capture elements, many of which are among the most refractory elements in the Universe (Lodders 2003). Trans-iron elements that are observed in the spectra of AGB stars, such as Sr, Y, Zr, Ba, and La, are likely to be strongly depleted in PNe, making their detection unlikely.

Despite the observational challenges, nebular spectroscopy is a valuable tool for studying s-process nucleosynthesis, and reveals information independent of and complementary to stellar spectroscopy. For example, nebular spectroscopy provides access to elements that cannot be detected in cool giant stars (e.g., the lightest n-capture elements and the noble gases Kr and Xe), allowing the production and chemical evolution of these species to be studied. Moreover, nucleosynthesis is complete in PNe, whose chemical compositions reflect that of the progenitor's AGB envelope during the final few thermal pulses. Observations of PNe provide constraints to many poorly-understood processes in models of AGB nucleosynthesis, including the efficiency of dredge-up at low envelope masses, convective mixing and mass loss, and to stellar yields (e.g., Karakas & Lattanzio 2014).

Emission lines of n-capture elements had been detected in PNe as early as the mid-1970s (Treffers *et al.* 1976), but it was not until the pioneering work of Péquignot & Baluteau (1994, hereafter PB94) that features in PN spectra were associated with trans-iron elements. PB94 identified several lines of n-capture elements (including Se, Kr, Br, Rb, Xe, and possibly others) in the optical spectrum of the bright PN NGC 7027. While these identifications garnered some initial skepticism due to the relatively low spectral resolution and possible blending issues for some features, many were confirmed in subsequent studies (e.g., Sharpee *et al.* 2007).

Assessing n-capture element abundances required PB94 to make a number of approximations, as the only atomic data available to them were energy levels and transition probabilities (Biémont & Hansen 1986a,b, 1987; Biémont *et al.* 1988). For some ions (e.g., Se^{2+} and bromine ions) even the energy levels, compiled by Moore (1952) from spectroscopic measurements conducted in the early 20th century, were uncertain. (Recent measurements have greatly improved the accuracy of energy levels for Se^{2+} and Br ions, e.g., Tauheed *et al.* 2008; Tauheed & Hala 2012; Jabeen & Tauheed 2015.) PB94 used collision strengths extrapolated from those of $n = 2$ and $n = 3$ ions (where n is the principal quantum number of the valence shell) to estimate ionic abundances. To derive elemental abundances, they used ionization correction factors (ICFs) based on coincidences in the ionization potential ranges with light element ions. In spite of these approximations, PB94's conclusion that NGC 7027 exhibits large s-process enrichments has been verified in later studies (Sharpee *et al.* 2007; Sterling *et al.* 2015, hereafter SPD15; Sterling *et al.* 2016).

Nebular spectroscopy of n-capture elements is a field that has developed rapidly over the last 15 years, owing to the identification and detection of emission lines of these elements in several PNe, particularly in the near-infrared. Dinerstein (2001) identified two long-anonymous features at 2.199 and 2.287 μm as fine structure transitions of [Kr III] and [Se IV], respectively; these are the two most widely-detected n-capture element lines in PNe (Sterling & Dinerstein 2008). More recently, Sterling *et al.* (2016) discovered [Ge VI], [Rb IV], and [Cd IV] lines in the near-IR spectra of two PNe, while J band transitions of [Se III] and [Kr VI] have been identified by Sterling *et al.* (2017). To date, n-capture elements have been detected in more than 100 PNe in the Milky Way (e.g., Sharpee *et al.* 2007; Sterling & Dinerstein 2008; García-Rojas *et al.* 2015) and nearby galaxies (Otsuka *et al.* 2011; Mashburn *et al.* 2016).

These detections have motivated laboratory astrophysics efforts to determine the atomic data needed for accurate n-capture element abundance determinations in PNe. The atomic data fall into two categories: those needed for ionic abundances, and those for elemental abundances. To derive ionic abundances, transition probabilities and effective collision strengths for electron-impact excitation are required. Shortly after the publication of PB94, Biémont *et al.* (1995) calculated transition probabilities for $5p^k$ ions including Xe, while Schöning (1997) and Schöning & Butler (1998) computed effective collision strengths for various Kr, Xe, and Ba ions. For elemental abundance determinations, ionization correction factors (ICFs) must be used to account for the abundances of unobserved ions. The most reliable ICF prescriptions are determined from photoionization models (Delgado-Inglada *et al.* 2014; SPD15; see also the review of C. Morisset in this volume) that compute the ionization equilibrium of each element. Such models rely on the availability of accurate photoionization (PI) cross sections and rate coefficients for radiative recombination (RR), dielectronic recombination (DR), and charge transfer (CT) – none of which had been determined for n-capture element ions until recently.

In this review, I focus on atomic data developments for n-capture elements in the last 10 years, along with their application to observations.

2. Theoretical Methods and Results

Due to the atomic data needs for several ions over a large range of energies and temperatures, the majority of atomic data determinations for photoionization and spectral modeling are theoretical. The initial step in these calculations is modeling the structure of the ions of interest. This yields energy levels, transition probabilities, and radial orbitals. At present, the most widely-used codes for atomic structure calculations are AUTOSTRUCTURE (Badnell 2011) and GRASP (e.g., Jönsson *et al.* 2007). AUTOSTRUCTURE includes first-order relativistic corrections using Breit-Pauli formalism, while GRASP is based on the fully-relativistic Dirac-Fock method. Both approaches account for configuration interaction (mixing between electronic states) and allow for semi-empirical corrections to observed energy levels.

The orbital wave functions obtained in the structure calculations are essential ingredients to calculations of photoionization, electron-impact excitation, and recombination cross sections. Two general methods are used in such scattering calculations. The distorted wave (DW) method uses distorted wave functions to compute bound-free interactions. No correlation effects between bound and continuum states are taken into account, and hence the cross sections do not include resonances. However, radiative and autoionizing rates for levels above the ionization threshold can be calculated using the isolated resonance approximation, allowing DR rate coefficients to be computed with this method. DW calculations are computationally efficient and thus are a pragmatic approach for producing atomic data for large numbers of systems. Another benefit of the computational efficiency is that uncertainties in the resulting atomic data can be estimated (e.g., Sterling & Witthoeft 2011), for example by using different configuration expansions. The DW method is preferred for computing RR rate coefficients (Badnell 2006), and has been used to calculate DR rate coefficients for numerous ions (Badnell *et al.* 2003; Abdel-Naby *et al.* 2012, and references therein). For processes in which resonances play a crucial role and the isolated resonance approximation becomes inaccurate, such as photoionization and collisional excitation for low-charge ions, the R-matrix method is preferred. R-matrix calculations use the close-coupling approximation, which includes all correlation effects between bound and continuum states within the limits of the configuration expansion. The R-matrix method produces the most accurate data for resonant processes, but is significantly more computationally intensive than DW calculations.

PI, RR, and DR calculations for Se and Kr ions (up to five times ionized) have been computed in the DW approximation (Sterling & Witthoeft 2011; Sterling 2011). We have also completed calculations of these data for Br, Rb, and Xe ions, which are being prepared for publication. While the PI cross sections do not have accurate resonance energies, the direct cross sections (corresponding to direct ionization of a valence electron) agree well with experimental measurements (§3).

We estimated uncertainties for all processes, with typical error bars of 30–50% for PI, 10–30% for RR, and factors of 2–10 for DR. The large uncertainties for DR rate coefficients are important, since DR is the dominant recombination mechanism for low-charge n-capture element ions at the "low" temperatures of PNe (as illustrated for Br and Rb ions in Figure 1). Low-temperature DR rate coefficients of heavy elements cannot be precisely calculated even with the most sophisticated theoretical methods, due to the unknown energies of autoionizing states that lie just above the ionization threshold. This is an important uncertainty in ionization equilibrium solutions for all elements beyond the second row of the periodic table.

R-matrix PI cross section calculations have been performed for selected n-capture element ions: Se^+ (McLaughlin & Ballance 2012a), Se^{2+} (Macaluso *et al.* 2015), Kr^+

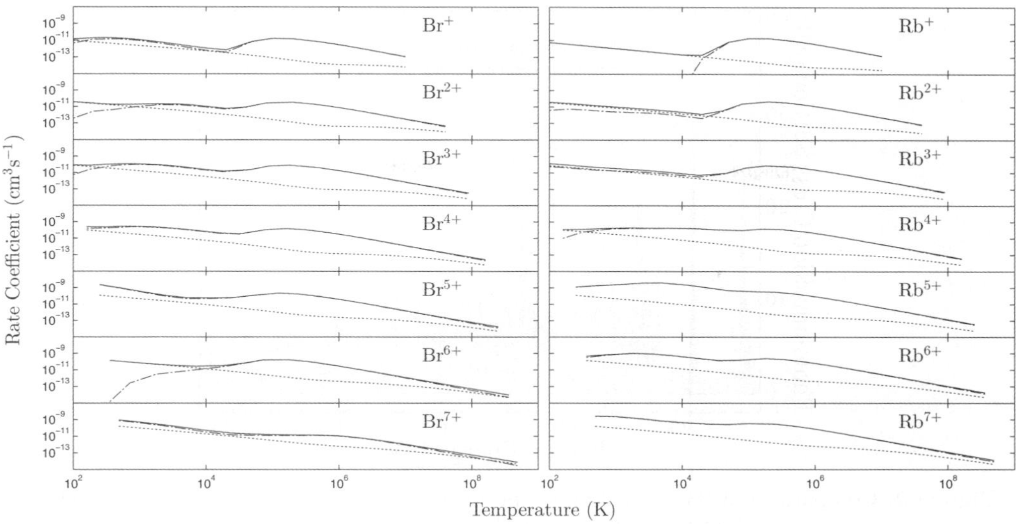

Figure 1. Comparison of RR (dotted lines), DR (dot-dashed lines), and total recombination (solid lines) rate coefficients for Br and Rb ions.

and Xe^{+} (McLaughlin & Ballance 2012b), and Rb^{2+} (Macaluso *et al.* 2016). These data agree well with experimental measurements, in terms of both the absolute cross sections and the resonance energies. R-matrix PI calculations are underway for Br and Rb ions.

The recent discovery of near-IR n-capture element transitions has spurred R-matrix calculations of effective collision strengths. In addition to the work of Schöning and Butler for Kr, Xe, and Ba (see §1), collision strengths have been computed for Se^{3+} (K. Butler 2007, private communication), Rb^{3+} (Sterling *et al.* 2016), and Se^{2+} and Kr^{5+} (Sterling *et al.* 2017). The relative intensities for different Rb^{3+} and Se^{2+} lines agree well with radiative-collisional models that incorporate the new collision strengths, supporting the accuracy of these calculations. No such comparison is possible for Se^{3+} and Kr^{5+}, whose $4s^2 4p$ ground configurations lead to a single transition. Additional collision strength calculations for Br and Rb ions detected in PNe are in progress.

Charge transfer recombination with neutral H atoms cannot be treated using the codes described above, because it is a quasi-molecular process. Using the Landau-Zener method (Butler & Dalgarno 1980; Janev *et al.* 1983) for multiply-charged atoms and the Demkov approximation (Demkov 1964; Swartz 1994) for singly-ionized species, Sterling & Stancil (2011) computed CT rate coefficients for Ge, Se, Br, Kr, Rb, and Xe ions up to five times ionized. These calculations are accurate to within a factor of three for systems with large rate coefficients (Butler & Dalgarno 1980), and are less accurate for those with smaller rate coefficients (although RR and DR typically dominate recombination for those ions). However, state-of-the-art CT calculations such as the quantum mechanical molecular-orbital close-coupling (QMOCC) method (e.g., Wang *et al.* 2004) are computationally intensive, and must be targeted for ions in which uncertainties in CT rate coefficients significantly affect abundance determinations.

3. Experimental Methods and Results

Although the majority of atomic data for n-capture elements are determined theoretically, it is critical to benchmark the results against experimental measurements when possible, due to the complexity of these many-electron systems. PI cross sections can be

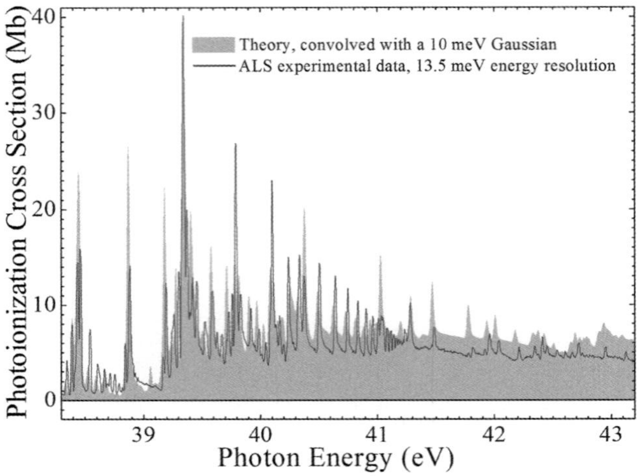

Figure 2. Comparison of theoretical and experimental Rb^{2+} photoionization cross section. Figure adapted from Macaluso *et al.* (2016).

measured via the merged beams technique (Lyon *et al.* 1986) at third-generation synchrotron radiation facilities. Experimental setups for measuring PI cross sections that are in current operation include the PIPE apparatus (Ricsóka *et al.* 2009) at PETRA III in Germany and the MAIA apparatus at SOLEIL in France (Bizau *et al.* 2016). Several of the results below were obtained with the now-defunct IPB apparatus (Covington *et al.* 2002) at the Advanced Light Source (ALS; USA) and a merged-beams setup on the Miyake undulator beamline at ASTRID (Denmark).

In the merged-beams method, atomic ion beams are produced with a plasma source, and the charge-to-mass ratio of the desired ion is selected with a dipole analyzing magnet. A series of electrostatic lenses and steering plates are used to focus the ion beam, which is merged with a counter-propagating beam of UV/soft X-ray photons. The resulting photoions are directed to a single-particle detector by a demerging magnet. Using monochromatic gratings to step the photon energy, photoionization yields can be measured as a function of energy. The relative cross sections are placed on an absolute scale at discrete energies by measuring the photoion yields produced in a well-defined volume, which entails characterizing the beam overlap with scanning slits. The disadvantage of the merged beams method is that the ion beams are not purely in the ground state (unless an ion trap is used; e.g., Bizau *et al.* 2011). The measured cross sections are linear combinations of those from the ground and metastable states. Therefore R-matrix calculations are needed to disentangle the contribution of each ground configuration state to the measured cross sections.

Experimental PI cross section measurements have been completed for ions X^{+q} ($q = 1$–5) of Se (Sterling *et al.* 2011; Esteves *et al.* 2012; Macaluso *et al.* 2015), Br, Kr (Lu *et al.* 2006a,b; Bizau *et al.* 2011), Rb (Macaluso *et al.* 2016), and Xe (Bizau *et al.* 2006, 2011). The experimental data for Br and Rb ions are being analyzed (and R-matrix calculations underway), with the exception of the recently-published results for Rb^{2+} (Figure 2).

Experimental measurements of other atomic processes have not been conducted for n-capture elements. However, the recently constructed Cryogenic Storage Ring (von Hahn *et al.* 2016) can be used to measure total recombination rate coefficients (for both RR and DR) of low charge-to-mass species such as the n-capture element ions detected in PNe. Given the large uncertainties in theoretical DR rate coefficients, such measurements would be particularly valuable for improving the accuracy of n-capture element abundance determinations.

4. Applying the New Atomic Data to Observations

Applications to observational data provide the ultimate test of the new atomic data. PNe have long served as "laboratories" for assessing the accuracy of atomic data. Atomic data for light elements have been improved through observations paired with increasingly sophisticated theoretical and experimental techniques, and one can expect a similar scenario to unfold for n-capture elements.

SPD15 added Se and Kr to the atomic database of Cloudy (Ferland *et al.* 2013) in order to determine ICFs for these elements. They modeled 15 PNe, with input parameters such as central star temperature and luminosity, nebular density, outer radius, and elemental abundances determined via optimization routines that provided the best fit to the observed emission line spectrum of each object. These 15 PNe were chosen because multiple Kr ions had been detected in their optical and near-IR spectra. It was found that Cloudy models were unable to reproduce the ionization balance of Kr in these PNe, with the Kr^{3+}/Kr^{2+} ratio systematically lower in the models than observed, regardless of the ionization level of the PN. To rectify this discrepancy, the PI cross sections of Kr^{2+}, Kr^{3+}, and Kr^{4+} were adjusted within their estimated uncertainties, and the DR rate coefficients of these ions by amounts slightly larger than the stated error bars. These empirical adjustments should be viewed as temporary measures that highlight the need for further laboratory astrophysics studies (e.g., experimental DR measurements) of the ions involved. More physically realistic 3D photoionization models, that take into account morphological and density structures, would also provide valuable tests of 1D model results for the Kr ionization balance.

SPD15 could not perform a similar analysis for the ionization equilibrium of Se, since only [Se IV] had been detected in PNe at the time (with the possible exception of [Se III] 8855.27 Å†, which is significantly blended with He I λ8854.20). However, observations of the uncontaminated [Se III] 1.0991 μm line identified by Sterling *et al.* (2017) can be used to test the atomic data and ICF schemes for Se.

SPD15 computed large grids of models that span the range of nebular densities, metallicities, and central star temperatures and luminosities of PNe in the Milky Way and nearby galaxies. Correlations between the fractional abundances of Kr/Se ions and those of routinely detected ions of O, S, Cl, and Ar were fitted with analytical functions, the inverses of which serve as ICFs. These ICFs were applied to a sample of 120 PNe (Sterling & Dinerstein 2008) and produced Se and Kr abundances 0.1–0.3 dex lower than previous estimates. However, the new abundances did not affect the conclusions of Sterling & Dinerstein (2008) or the correlations they found between Se and Kr abundances and other nebular and stellar properties. Recent results from deep, high-resolution optical spectra of PNe (García-Rojas *et al.* 2015; Madonna *et al.* 2017) show good agreement between the Kr abundances derived with the various ICF prescriptions of SPD15. New Kr ICFs that include Kr^{5+} have been computed by Sterling *et al.* (2017).

Efforts are underway to add Br, Rb, and Xe to the atomic database of Cloudy. The atomic data for Xe will be tested as for Kr, via comparisons of modeled and observed line intensities for different Xe ions (unfortunately, multiple Br and Rb ions have been detected in too few PNe thus far to rigorously test their atomic data). ICFs for these elements will be derived from grids of models.

† Air wavelengths for [Se III] computed from the energy levels of Tauheed & Hala (2012)

5. Future Work

Atomic physics forms the foundation on which spectroscopy is built. The range of energies and temperatures for which atomic data must be determined necessitates a predominantly theoretical approach. However, all theoretical treatments of several-electron ions are approximations, and therefore observations are needed to assess the accuracy of the atomic data and ICF formulae. Detections of multiple Br and Rb ions in the optical and near-IR spectra of PNe are sorely needed to test atomic data for these elements. Observations of the newly-identified [Se III] 1.0991 μm line in PNe with a range of ionization levels are required to test the Se ICFs computed by SPD15.

The accuracy of ionization equilibrium solutions (and hence ICFs) can also be improved via further experimental and theoretical atomic data investigations. Specifically, the large uncertainties of theoretical DR rate coefficients for n-capture elements can be addressed with experimental measurements at the new Cryogenic Storage Ring facility in Germany, and more accurate CT rate coefficient calculations (e.g. with the QMOCC method) may be needed for specific ions. Monte Carlo routines are being developed to test the sensitivity of emission line strengths (and hence abundances) to uncertainties in PI, RR, DR, and CT atomic data for individual ions. The results will help identify the ions and atomic processes that most urgently require further investigation.

The number of n-capture element detections and new identifications will continue to grow, which will lead to new atomic data needs. The synthesis of observations, atomic physics, and photoionization modeling is essential to continue honing nebular spectroscopy into an effective tool for studying s-process enrichments in PNe and hence AGB nucleosynthesis and evolution.

6. Acknowledgments

I am grateful to M. A. Bautista for his helpful comments on §2, and J.-M. Bizau and R. C. Bilodeau for sending information about currently operating instrumentation for experimental PI measurements. I also thank my undergraduate students who have contributed to the research described herein: John Harrison, Austin Kerlin, Amanda Mashburn, Cameroun Sherrard, and Courteney Spencer. I acknowledge support from NSF grant AST-1412928.

References

Abdel-Naby, S. A., Nikolić, D., Gorczyca, T. W., Korista, K. T., & Badnell, N. R. 2012, *A&A*, 537, A40

Asplund, M., Grevesse, N., Sauval, A. J., & Scott, P. 2009, *ARAA*, 47, 481

Badnell, N. R., *et al.* 2003, *A&A*, 167, 334

Badnell, N. R. 2006, *ApJS*, 406, 1151

Badnell, N. R. 2011, *Comp. Phys. Comm.*, 182, 1528

Biémont, E., Cowan, R. D., & Hansen, J. E. 1988, *Phys. Scr.*, 37, 850

Biémont, E. & Hansen, J. E. 1986a, *Phys. Scr.*, 34, 116

—. 1986b, *Phys. Scr.*, 33, 117

—. 1987, *Nuc. Inst. Meth. Phys. Res.*, 23, 274

Biémont, E., Hansen, J. E., Quinet, P., & Zeippen, C. J. 1995, *A&AS*, 111, 333

Bizau, J. M., *et al.* 2006, *Phys. Rev. A*, 73, 022718

Bizau, J. M., *et al.* 2011, *J. Phys. B: At. Mol. Opt. Phys.*, 44, 055205

Bizau, J. M., *et al.* 2016, *J. Elect. Spect. Rel. Phen.*, 210, 5

Butler, S. E. & Dalgarno, A. 1980, *ApJ*, 241, 838

Covington, A. M., *et al.* 2002, *Phys. Rev. A*, 66, 062710

Delgado-Inglada, G., Morisset, C., & Stasińska, G. 2014, *MNRAS*, 440, 536
Demkov, Y. N. 1964, *Sov. Phys.-JETP*, 18, 138
Dinerstein, H. L. 2001, *ApJ*, 550, L223
Esteves, D. A., *et al.* 2012, *J. Phys. B: At. Mol. Opt. Phys.*, 45, 115201
Ferland, G. J., *et al.* 2013, *Rev. Mexicana AyA*, 49, 137
García-Rojas, J., *et al.* 2015, *MNRAS*, 452, 2606
Jabeen, S. & Tauheed, A. 2015, *J. Quant. Spect. Rad. Tran.*, 154, 9
Janev, R. K., Belic, D. S., & Bransden, B. J. 1983, *Phys. Rev. A*, 28, 1293
Jönsson, P., He, X., Froese Fischer, C., & Grant, I. P. 2007, *Comp. Phys. Comm.*, 177, 597
Karakas, A. I. & Lattanzio, J. C. 2014, *PASA*, 31, 30
Lodders, K. 2003, *ApJ*, 591, 1220
Lu, M., *et al.* 2006a, *Phys. Rev. A*, 74, 062701
Lu, M., *et al.* 2006b, *Phys. Rev. A*, 74, 012703
Lyon, I. C., Peart, B., West, J. B., & Dolder, K. 1986, *J. Phys. B: At. Mol. Opt. Phys.*, 19, 4137
Macaluso, D. A., *et al.* 2015, *Phys. Rev. A*, 92, 063424
Macaluso, D. A., *et al.* 2016, *J. Phys. B: At. Mol. Opt. Phys.*, 49, 235002
Madonna, S., *et al.* 2017, *MNRAS*, submitted
Mashburn, A. L., *et al.* 2016, *ApJ*, 831, L3
McLaughlin, B. M. & Ballance, C. P. 2012a, *J. Phys. B: At. Mol. Opt. Phys.*, 45, 095202
McLaughlin, B. M. & Ballance, C. P. 2012b, *J. Phys. B: At. Mol. Opt. Phys.*, 45, 085701
Moore, C. E. 1952 *Atomic Energy Levels, Vol. II*, NBS Circular 467, Washington, DC: NBS
Otsuka, M., *et al.* 2011, *ApJ*, 729, 39
Péquignot, D. & Baluteau, J. P. 1994, *A&A*, 283, 593 (PB94)
Ricsóka, T., *et al.* 2009, *J. Phys. Conf. Ser.*, 194, 142012
Schöning, T. 1997, *A&AS*, 122, 277
Schöning, T. & Butler, K. 1998, *A&AS*, 128, 581
Sharpee, B., *et al.* 2007, *ApJ*, 659, 1265
Sterling, N. C. 2011, *A&A*, 533, A62
Sterling, N. C. & Dinerstein, H. L. 2008, *ApJS*, 174, 158
Sterling, N. C., Dinerstein, H. L., Kaplan, K. F., & Bautista, M. A. 2016, *ApJ*, 819, L9
Sterling, N. C., *et al.* 2011, *J. Phys. B: At. Mol. Opt. Phys.*, 44, 025701
Sterling, N. C., *et al.* 2017, *ApJ*, submitted
Sterling, N. C., Porter, R. L., & Dinerstein, H. L. 2015, *ApJS*, 218, 25 (SPD15)
Sterling, N. C. & Stancil, P. C. 2011, *A&A*, 535, A117
Sterling, N. C. & Witthoeft, M. C. 2011, *A&A*, 529, A147
Swartz, D. A. 1994, *ApJ*, 428, 267
Tauheed, A. & Hala. 2012, *Phys. Scr.*, 85, 025304
Tauheed, A., Joshi, Y. N., & Jabeen, S. 2008, *Phys. Scr.*, 78, 065305
Treffers, R. R., Fink, U., Larson, H. P., & Gautier, III, T. N. 1976, *ApJ*, 209, 793
von Hahn, R., *et al.* 2016, *Rev. Sci. Inst.*, 87, 063115
Wang, J. G., Stancil, P. C., Turner, A. R., & Cooper, D. L. 2004, *Phys. Rev. A*, 69, 062702

Discussion

Q: In your last slide you show the comparison with nucleosynthesis models. Rb is different than the noble gasses as it will deplete in the gas phase. Do you need to correct for that?

STERLING: Yes, Rb can be depleted, however it has a lower condensation temperature than Na or K (depleted by factors of 2 - 4 in PNe), so it should be less depleted. However, this does need to be taken into consideration when interpreting nebular Rb abundances.

Planetary Nebulae: Multi-Wavelength Probes of Stellar and Galactic Evolution
Proceedings IAU Symposium No. 323, 2016
X. Liu, L. Stanghellini & A. Karakas, eds.

doi:10.1017/S1743921317000631

Abundances of Iron-Group Elements in Planetary Nebulae and Consequences for Chemical Enrichment

Harriet L. Dinerstein[1], T. R. Geballe[2] and N. C. Sterling[3]

[1]Department of Astronomy, University of Texas at Austin,
2515 Speedway, Stop C1400, Austin, TX 78712-1205, U.S.A.
email: harriet@astro.as.utexas.edu

[2]Gemini Observatory, 670 N Aohoku Place, Hilo, HI 96720-2700, U.S.A.
email: tgeballe@gemini.edu

[3]Deptartment of Physics, University of West Georgia, 105 Boyd Building
1601 Maple Street, Carrollton, GA 30118-0001, U.S.A.
email: nsterlin@westga.edu

Abstract. We have developed a method for determining elemental Fe-group abundances in planetary nebulae using an infrared emission line of Zn, the least refractory Fe-group species. Many planetary nebulae, particularly those of the Milky Way's thick disk and bulge, display subsolar [Fe/H] (as inferred from Zn) although their abundances of α elements such as O, S, and Ar are nearly solar. We discuss the implications for determining enhancements of species synthesized by the progenitor star during the AGB (e.g., *s*-process products), and for galactic chemical evolution in view of the metallicity dependence of AGB nucleosynthetic yields.

Keywords. planetary nebulae: general, stars: AGB and post-AGB, nuclear reactions, nucleosynthesis, abundances

1. Introduction

The primary index of metallicity in stars has historically been the Fe abundance expressed as [Fe/H], the base 10 logarithmic difference between Fe/H in the star and the Sun. Recent spectroscopic surveys include more elements but continue to use Fe as a reference for normalizing the abundances of other elements X; the values of [X/Fe] specify abundance patterns in 'chemical tagging' (e.g., Freeman & Bland-Hawthorn 2002).

Meanwhile, the element that is most accurately and easily measured in ionized nebulae is O. If O and Fe always varied together during galactic chemical evolution things would be simple, but this is not the case since they have different primary sites of origin. It has long been recognized that enrichment of α species such as O, Ne, S, and Ar proceeds rapidly after a stellar generation forms, since the massive-star supernovae that disperse them occur promptly, while the enrichment of Fe-group elements by Type Ia supernovae is necessarily time-delayed since it requires the pre-existence of white dwarfs (Tinsley 1979). Indeed, [α/Fe] ratios are widely used to set constraints on star formation histories of stellar ensembles, with high values indicating 'bursty' pasts and low ones reflecting a more continuous, quiescent history of star formation.

Multiwavelength spectroscopic observations of planetary nebulae (PNe), especially in the UV and infrared, have recently expanded the palette of elements for which abundances can be determined. They include species affected by nuclear processing in the progenitor star as well as elements that retain their initial abundances and thus provide information on the star's parent population. For example, advances in measuring self-enrichments of trans-iron nuclei synthesized by slow neutron-capture reactions (the '*s*-process') have been largely propelled by infrared spectroscopy, which has enabled determinations of abundances for several elements with atomic numbers in the 30s and above: Ge, Se, Kr, Rb, & Cd (Dinerstein 2001, Sterling & Dinerstein 2008, Sterling *et al.* 2015, Sterling *et al.* 2016; also see the review by Sterling, this volume).

A crucial missing ingredient in our picture of the composition of PNe has been the abundance of Fe. This is not due to a lack of spectral features from multiple Fe ions. In fact, Fe^{++} alone is observable from the far-UV (in absorption, e.g., Sterling *et al.* 2005) to the infrared (e.g., Likkel *et al.* 2006), as well as in optical emission lines. However, these trace only the gas phase component, whereas most Fe atoms are in the dust (Delgado-Inglada & Rodríguez 2014). To be an effective proxy for Fe, a non-depleted species is required. Unfortunately, most of the other members of the Fe-group of elements are also highly refractory, Zn being the only exception (e.g., Welty *et al.* 1999). Consequently, the discovery of an infrared emission line of its dominant ion, Zn^{+3}, was immediately recognized by Dinerstein & Geballe (2001) as having unique potential for determining the first truly elemental abundances for the Fe group in PNe.

2. Results: Fe Abundances from Zn

At 3.625 μm, the [Zn IV] line lies in the thermal infrared, where ground-based measurements are challenging. Our first targeted observations of [Zn IV], made at Gemini South, included one PN each in the Large Magellanic Cloud (SMP LMC 62) and Sgr dSph (Hen 2-436) as well as a few Milky Way PNe to serve as control objects (Wood *et al.* 2007, Dinerstein *et al.* 2007). As additional Galactic disk PNe were observed with UKIRT and IRTF over the next few years, it quickly became apparent that a substantial fraction of them showed Zn deficiencies, with a typical value of [Zn/H] ≈ -0.6. Most of these are Type III objects in the classification system of Peimbert (1978), which are distinguished by having high radial velocities and belong to the population now called the thick disk. Type II PNe, which have radial velocities consistent with the thin disk population, show a range of [Zn/H] values but are more concentrated towards the solar abundance. Our sorting of PNe into these two categories is imperfect since there is limited information on their space motions. At present only the radial velocities are known, hence some thick disk PNe may be misclassified as thin disk objects due to unfavorable directions of motion (this situation should be partly remedied by GAIA). More recently, Zn measurements were acquired for several PNe in the Galactic bulge (Smith 2014, Smith *et al.* 2014), which display a wide range of [Zn/H] values.

A second major finding is that most PNe with low [Zn/H], hence by inference [Fe/H], have elevated [α/Zn]. This is particularly true for O, although other α species, for example Ar, are also enhanced albeit to a lesser degree. As described above, elevated [α/Fe] values are characteristic of some stellar populations in the Milky Way and are a natural consequence of rapid modes of star formation. This creates ambiguity for studies of PNe in which the only observed elements with abundances unchanged from their original values are α species. To be specific, the α abundances of an Fe-deficient PN with enhanced [α/Fe] may be the same as those of an object of solar composition. Unless [α/Fe] is known, abundances of the α elements alone may be unreliable or even misleading indicators of metallicity (Dinerstein *et al.* 2011, Dinerstein *et al.* 2014, Dinerstein *et al.* 2015).

Figure 1 presents a comparison of our results for PNe with abundances for stars of the same Galactic populations, in the [O/Fe] vs [Fe/H] diagnostic plane (right and top axis labels). We use Zn as a proxy for Fe in the PNe (left and bottom axes). The stellar data are from Bensby *et al.* (2014, and earlier works referenced therein). Their O abundances were

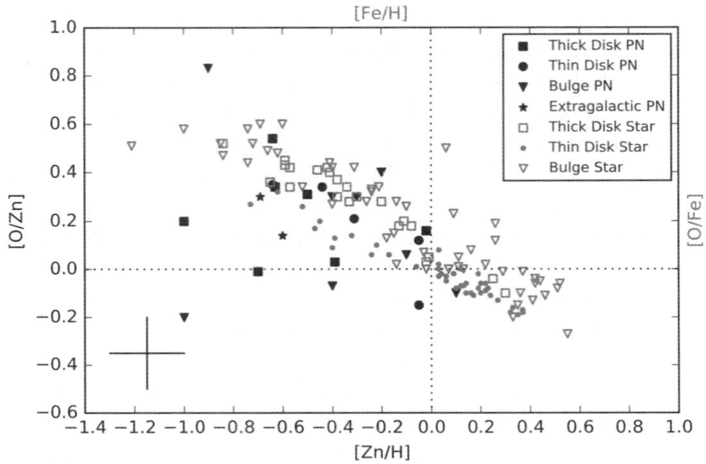

Figure 1. Abundance ratios [O/Zn] vs. [Zn/H] for the observed PNe (filled symbols, larger dots) overplotted on [O/Fe] vs. [Fe/H] for stars of several Milky Way populations (open symbols, small dots). Most PNe follow the same trend as the stellar data, which are from Bensby *et al.* (2014, and references therein). The asterisks are SMP LMC 62 and Hen 2-436.

derived using the most reliable method, based on forbidden [O I] lines, and the bulge values are for gravitationally-lensed dwarf stars. It can be seen that the PNe occupy similar regions of the diagram as the corresponding stars, although there are fairly large uncertainties for individual nebulae. While this diagram seems to indicate that a larger fraction of PNe than stars have solar [O/Fe] at low [Fe/H], low [α/Fe] disk stars do exist although they are not represented in this particular plot. The general agreement between the PNe and the (unevolved) stars argues that the enhancements in [O/Zn] are primarily a population effect rather than due to self-enrichment in O, which can occur at low metallicities (Karakas & Lattanzio 2014).

3. Conclusions and Consequences

We have demonstrated the utility of an infrared line of Zn as a tracer of the abundances of Fe-group elements in gaseous media. Zn is lightly depleted out of the gas phase, in contrast to all other Fe-group species, especially in warm and ionized ISM phases (by ≈ -0.1 to -0.2 dex, Welty *et al.* 1999). Any small residual depletions of Zn are likely to be offset by other minor systematic effects (to be discussed more fully in Dinerstein *et al.*, in preparation). At present, and likely permanently, Zn offers the best available opportunity for determining initial [Fe/H] values for PN progenitor stars.

Many PNe are found to be moderately subsolar in [Zn/H], and since Zn is a surrogate for Fe, this means they are [Fe/H]-deficient. As is the case for many metal-poor stars, this Fe deficiency is often accompanied by elevated values of [O/Fe] and other [α/Fe] ratios. As a result, the fact that a PN is descended from an Fe-poor star may be masked by an α enhancement, leading to incorrect inferences about the properties of the progenitor star. The larger scatter exhibited by PNe (compared to stars) in the [O/(Fe,Zn)] vs. [(Fe,Zn)/H] diagram is probably mainly due to measurement uncertainties, but the general consistency between stars and PNe of the same stellar population is clear.

Knowing the initial [Fe/H] of a PN is critical for obtaining an accurate assessment of the nucleosynthesis that occurred within the star during its overall lifetime, especially during the AGB phase, when some of the most dramatic alterations in composition occur

(Karakas & Lattanzio 2014). In order to derive quantitative enrichments for an element, for example an *s*-process product, we must adopt an initial abundance. This is usually assumed to be the solar value, but that is unlikely to hold for metal-poor stars. Sterling *et al.* (2015) attempted to take this into consideration in interpreting Se enrichments in thick vs. thin disk stars, but were limited to a statistically-based assessment. To address this for individual PNe would require measurements of many additional Zn abundances.

The above consideration affects the empirically-based yields for elements made within the star. Fe abundances are particularly relevant when making comparisons of observations and theory for neutron-capture species because Fe is the main 'seed' that initiates the neutron-capture reactions. Consequently, the amounts of trans-iron elements and ratios among different nuclides synthesized in AGB stars are sensitive to the star's initial [Fe/H] (Karakas & Lattanzio 2014). Fixing the value of [Fe/H] enables other properties, such as initial stellar mass and the efficiency of *s*-processing, convective mixing, and dredge-up, to be constrained by observational results. Furthermore, accurate yields of nucleosynthetic products per star must be known in order to calculate the cumulative contributions of stars of the mass range that gives rise to PNe to the ISM of galaxies, which are essential inputs for models of galactic chemical evolution.

This research received support from NSF grants AST-0708429 and 0901432.

References

Bensby, T., Feltzing, S., & Oey, M. S. 2014, *A&A*, 562, 71

Delgado-Inglada, G. & Rodrìguez, M. 2014, *ApJ*, 784, 173

Dinerstein, H. L. 2001, *ApJ* (Letters), 550, L223

Dinerstein, H. L. & Geballe, T. R. 2001, *ApJ*, 562, 515

Dinerstein, H. L., Geballe, T. R., & Sterling, N. C. 2011, *AAS*, id. 217.256.04

Dinerstein, H. L., Geballe, T. R., & Sterling, N. C. 2014, *AAS*, id. 223.353.29

Dinerstein, H. L., Geballe, T. R., & Sterling, N. C. 2015, *IAUGA*, Meeting 29, id. 2255718

Freeman, K. & Bland-Hawthorn, J. 2002, *ARAA*, 40, 487

Karakas, A. I. & Lattanzio, J. C. 2014, *Pub. Astr. Soc. Australia*, 31, id. 30

Likkel, L., Dinerstein, H. L., Lester, D. F., Kindt, D. F., & Bartig, K. A. 2006, *AJ*, 131, 1515

Peimbert, M. 1978, in: Y. Terzian (ed.), *Planetary Nebulae: Observations and Theory*, IAU Symposium No. 76 (Dordrecht: Reidel), p. 216

Smith, C. L. 2014, Ph.D., University of Manchester

Smith, C. L., Zijlstra, A. A., & Dinerstein, H. L. 2014, *MNRAS*, 441, 3161

Sterling, N. C., Dinerstein, H. L., Bowers, C. W., & Redfield, S. 2005, *ApJ*, 625, 368

Sterling, N. C. & Dinerstein, H. L. 2008, *ApJS*, 174, 158

Sterling, N. C., Dinerstein, H. L., Kaplan, K. F., & Bautista, M. A. 2016, *ApJ* (Letters), 819, L9

Sterling, N. C., Porter, R. L., & Dinerstein, H. L. 2015, *ApJS*, 218, id. 25

Tinsley, B. M. 1979, *ApJ*, 229, 1046

Welty, D. E., Hobbs, L. M., Lauroesch, J. T., Morton, D. C., Spitzer, L., & York, D. G. 1999, *ApJS*, 124, 465

Wood, J. L., Dinerstein, H. L., & Geballe, T. R. 2007, *BAAS*, 39, 113, id. 210.16.06

Discussion

STANGHELLINI: Have you tried to look at the behavior of your iron-probe against distance from the Galactic centre, to see if there is a gradient?

Planetary Nebulae: Multi-Wavelength Probes of Stellar and Galactic Evolution
Proceedings IAU Symposium No. 323, 2016
X. Liu, L. Stanghellini & A. Karakas, eds.

doi:10.1017/S1743921317001247

Stellar yields and abundances: new directions from planetary nebulae

Maria Lugaro[1,2], Amanda I. Karakas[2,3], Marco Pignatari[4,5] and Carolyn L. Doherty[1]

[1]Konkoly Observatory, Research Centre for Astronomy and Earth Sciences, Hungarian Academy of Sciences, H-1121 Budapest, Hungary
email: maria.lugaro@csfk.mta.hu, carolyn.doherty@csfk.mta.hu

[2]Monash Centre for Astrophysics, School of Physics and Astronomy, Monash University, VIC 3800, Australia
email: amanda.karakas@monash.edu

[3]Research School of Astronomy and Astrophysics, Australian National University, Canberra, ACT 2611, Australia

[4]E. A. Milne Centre for Astrophysics, University of Hull, Cottingham Road Hull HU6 7RX United Kingdom
email: m.pignatari@hull.ac.uk

[5]The NuGrid collaboration (www.nugridstars.org)

Abstract. Planetary nebulae retain the signature of the nucleosynthesis and mixing events that occurred during the previous AGB phase. Observational signatures complement observations of AGB and post-AGB stars and their binary companions. The abundances of the elements heavier than iron such as Kr and Xe in planetary nebulae can be used to complement abundances of Sr/Y/Zr and Ba/La/Ce in AGB stars, respectively, to determine the operation of the *slow* neutron-capture process (the *s* process) in AGB stars. Additionally, observations of the Rb abundance in Type I planetary nebulae may allow us to infer the initial mass of the central star. Several noble gas components present in meteoritic stardust silicon carbide (SiC) grains are associated with implantation into the dust grains in the high-energy environment connected to the fast winds from the central stars during the planetary nebulae phase.

Keywords. nuclear reactions, nucleosynthesis, abundances, stars: AGB and post-AGB, stars: abundances, stars: chemically peculiar

1. Introduction

The evolutionary track of the life of a star of 2 $M_\odot$ after core H exhaustion proceeds to the red giant phase, until core He burning stars. After core He exhaustion, the star becomes a red giant again and climbs the asymptotic giant branch (AGB). During the AGB phase, the stars loses most of its envelope via strong stellar winds, until mainly the C-O degerate core is left, with a thin layer of H ($\sim 10^{-3}$ $M_\odot$) still burning. At this point the star travels back in the H-R diagram towards hotter temperatures. If it becomes hot enough before all the material expelled in the previous AGB phase is dispersed, the UV photons from the central star illuminate the circumstellar material producing a planetary nebula. It follows that in planetary nebulae we observe the chemical abundances produced during the AGB phase.

During the AGB the structure of the star can be described by two components: the entended convective H-rich envelope and the compact core. The core is hot and dense, and nuclear reactions can occur there that produce and modify the chemical abundances. Two burning shells are present: the H and the He burning shells. These shells act alternately, with recurrent episodes of He burning driving convective motions (a thermal pulse, TP) in the whole region in-between the two shells (the intershell). As the star expands and cools down, H burning is extinguished, the TP ceases, and He continues to burn radiatively at the bottom of the intershell. At this point the base of the convective envelope may sink deep inside the intershell (third dredge-up, TDU) and carry to the stellar surface the products of H and He burning. After the TDU, He burning extinguishes and H burning starts up again. This cycle of H burning, TP, and TDU, repeats until the envelope mass is lost. The material brought to stellar surface is carried into the surroundings by the AGB winds and contributes to making up the chemical enrichment of the Galaxy. For reviews on AGB stars, see Herwig (2005) and Karakas & Lattanzio (2014).

Four main chemical signatures result from the AGB evolution:

(*a*) During the TPs, partial He burning makes carbon, and not much oxygen, in the intershell and stars in the mass range roughly 1.5 to 4 $M_{\odot}$(depending on the initial composition) can eventually become the observed C-rich stars (see, e.g., the middle panels of Figs. 2-4 of Karakas & Lugaro 2016).

(*b*) The base of the convective envelope can be hot enough to drive proton captures on the isotopes of the elements from C to Al (hot bottom burning, HBB). They transform, for example, C into N in stars of initial mass larger than roughly 4 $M_{\odot}$(see, e.g., the bottom panels of Figs. 2-4 of Karakas & Lugaro 2016).

(*c*) The ingestion of H-burning ashes in the TPs results in the production of specific rare isotopes, such as ^{19}F and ^{22}Ne (e.g., Lugaro *et al.* 2004; Abia *et al.* 2015).

(*d*) Finally, AGB stars are the site of origin of half of the cosmic abundances of the elements heavier than iron, via the operation of the *slow* neutron capture process (the *s* process) in the intershell.

The last of these four points will be discussed in more detail in the present paper, together with implications from the observations of planetary nebulae.

1.1. *The s process*

Because of their high number of protons (>26) the elements heavier than Fe have a large Coulomb barrier and can be produced only by capturing neutrons. The *s* process occurs typically in hydrostatic conditions at relatively low neutron densities (typical 10^8 n/cm^3, but up to 10^{13} n/cm^3 in the more massive AGB stars). The path of neutron captures proceeds mostly through stable nuclei, since unstable nuclei decay to their corresponding isobar before they can capture another neutron.

Along the path of stable nuclei, the flux of neutron captures encounters three *bottlenecks* corresponding to the magic number of neutrons N = 50, 82, 126 (Fig. 1). Nuclei with such numbers of neutrons are relatively more stable against neutron captures than other nuclei, and tend to accumulate until the neutron flux reaches a large enough value to make the probability of neutron captures on these nuclei significant. The total, time-integrated neutron flux is the crucial quantity that determines if the bottlenecks are bypassed and if the production flux can proceed to the next bottleneck. The bottleneck nuclei determine the final *s*-process distribution and correspond to the so-called first *s*-process peak – or *light* s-process elements (*ls*) with N=50, Sr, Y, Zr; second *s*-process peak – or *heavy* s-process elements (*hs*) with N=82, Ba, La, Ce; and third *s*-process peak at N=126 (Pb), which can only accumulate since the *s*-process path ends there. The traditional *s*-process *indexes* are defined as [ls/Fe]=([Sr/Fe]+[Y/Fe]+[Zr/Fe])/3 and [hs/Fe]=([Ba/Fe]+[La/Fe]+[Ce/Fe])/3; although other choices have also been used in the literature (see discussion in Lugaro *et al.* 2012).

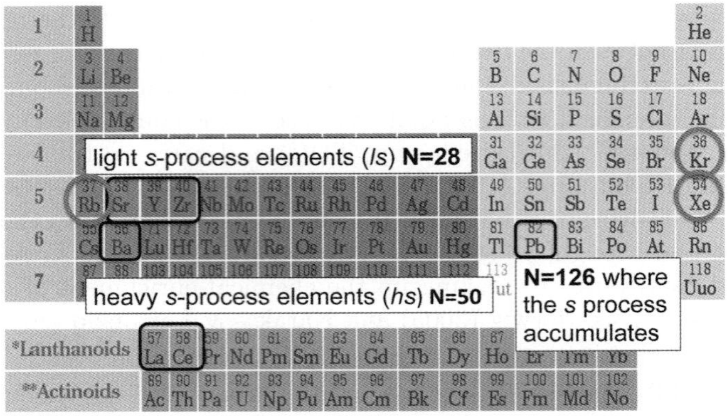

Figure 1. Periodic table of the elements with the locations of the first, second and third s process peaks marked along with the corresponding neutron numbers. Also highlighted are the noble gases Kr and Xe and the element Rb, which can be observed in planetary nebulae.

1.2. *The s-process in AGB stars*

In the intershell of AGB stars free neutron are produced by two reactions. The ^{13}C(α,n)^{16}O reaction is the main neutron surce in AGB stars of mass lower than roughly 4 $M_{\odot}$. A region rich in ^{13}C, the ^{13}C *pocket*, is assumed to form via mixing of protons from the envelope into the intershell at the deepest extent of the TDU (Gallino *et al.* 1998; Goriely & Mowlavi 2000; Lugaro *et al.* 2003b; Cristallo *et al.* 2009). The protons react with the abundant ^{12}C to produce ^{13}C via ^{12}C(p,γ)^{13}N(β^+)^{13}C. The ^{13}C pocket controls the total number of neutrons and thus the observed s-process distribution, i.e., the [hs/ls] ratio. The absolute s-process abundances [ls/Fe] and [hs/Fe] are controlled also by the TDU and other dilution effects. For example, binary transfer plays a crucial role in interpreting the data derived from an AGB star companion (a so-called *extrinsic* s-process enhanced star), rather than an AGB star itself (an *intrinsic* s-process enhanced star). Because both the formation of the ^{13}C pockets and the effects of stellar rotation and magnetic fields on the neutron flux are very uncertain, observations of [hs/ls] represent one of a few viable ways to understand the s process in low-mass AGB stars.

For AGB stars more massive than roughly 4 $M_{\odot}$, the main source of neutrons is instead the ^{22}Ne(α,n)^{25}Mg reaction, which operates at higher temperatures (above 300 MK) inside the convective TPs. This neutron source produces a lower total number of neutrons but a higher neutron density than the ^{13}C source. As such, it controls the possible operation of branching points, introducing deviations from the main s-process path. For instance, the abundance ratio [Rb/Zr] is an indicator of the neutron density and of the activation of the ^{22}Ne source. The production of Rb is much increased when the neutron magic isotope ^{87}Rb (N=50) is reached by the s-process path via the activation of the branching points at the unstable ^{85}Kr and ^{86}Rb (Abia *et al.* 2001; van Raai *et al.* 2012). Large uncertainties are present in the operation of the ^{22}Ne neutron source including the rate of the neutron source reaction, the efficiency of the TDU, and the mass loss rate in massive AGB stars. The difficulty in constraining the operation of this neutron source using observations is that the [Rb/Zr] ratio in massive AGB stars is highly dependent on the model atmosphere used to analyse the spectra (Zamora *et al.* 2014). Independent constraints are crucial to settle the question of how much Rb is produced by massive AGB stars.

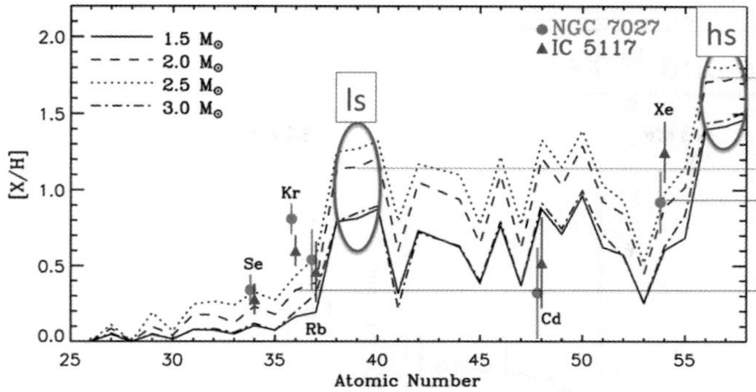

Figure 2. Figure adapted from Sterling *et al.* (2016) (©AAS. Reproduced with permission) showing the comparison between abundances observed in two planetary nebulae (red and blue circle symbols, with error bars) and results from models of the *s* process in AGB stars from the FRUITY database (Cristallo *et al.* 2015) for different masses and metallicity [Fe/H]=−0.37. Also indicated are the location of the first and second *s*-process peaks, labelled as ls and hs, respectively. The green and black arrows on the right side of the figure measure the [hs/ls] and the [Xe/Kr] ratios, respectively, predicted by the 2 $M_{\odot}$ model. It can be seen that [hs/ls]$\simeq$[Xe/Kr].

2. New directions

Observations of the elements heavier than Fe in planetary nebulae (see Sterling *et al.*, these proceedings) provide us complementary indicators and independent constraints for the *s*-process in AGB stars. As it can be seen from Fig. 1 the noble gases Kr and Xe, which are observable in planetary nebulae, are located very close to the first and second *s*-process peaks, respectively. Interestingly, only a small fraction of the cosmic abundances of these noble gases are produced by the *s*-process in AGB stars (Arlandini *et al.* 1999; Bisterzo *et al.* 2014): Xe is a typical element made by the *rapid* neutron captures in explosive environments (Thielemann *et al.* 2011). Krypton, on the other hand, is mostly produced in massive stars by a multitude of processes, including the weak *s* process and different explosive mechanisms (e.g., Roberts *et al.* 2010; Pignatari *et al.* 2016, and references therein).

Nevertheless, the *s* process in AGB stars contributes roughly 30% and 15% of the solar Kr and Xe, respectively (Bisterzo *et al.* 2014). This production, although much lower in absolute terms, follows the production of the *s*-process first and second peaks. This means that effectively the *s*-process models predict [Xe/Kr] values comparable to [hs/ls] values. The observed [Xe/Kr] ratios in planetary nebulae can be used to further constrain the total number of free neutrons and consequently the ^{13}C pocket (Sharpee *et al.* 2007; Pignatari *et al.* 2008). Furthemore, Rb can be directly seen in planetary nebulae, providing us with a constrain on the initial mass of the parent star and the opportunity to disentangle the effects of hot bottom burning in massive AGB stars (potentially Rb-rich) and of *extra* mixing in low-mass AGB stars (certainly Rb-poor) on the He and N excesses observed in Type I planetary nebulae.

From a preliminary analysis by Pignatari *et al.* (2008), the observations of the s-process [Xe/Kr] in PNe seem to be consistent with the theoretical predictions of AGB models. However, Sterling *et al.* (2016) (see Fig. 2) show a mismatch between models and observations for Kr, with the observations being higher than the model predictions. This problem needs to be analyzed in more details, also in the light of the other independent s-process observations of noble gases available from meteoritic presolar silicon carbide (SiC) grains, which formed in the winds of C-rich AGB stars.

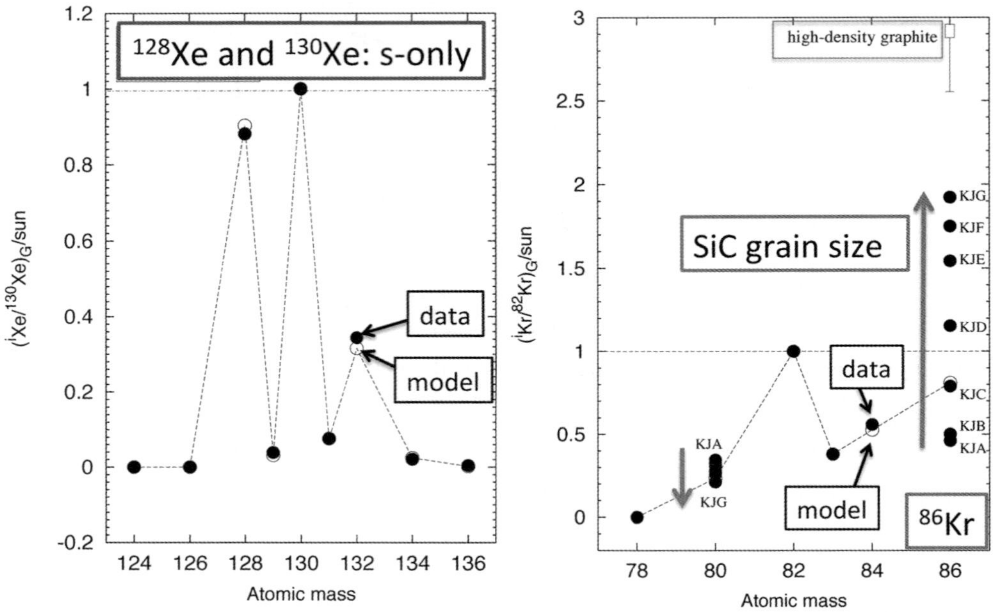

Figure 3. Figures adapted from Lugaro (2005) showing a comparision between the isotopic composition of Xe and Kr (normalised to ^{130}Xe and ^{82}Kr, respectively, and to the solar composition) measured in collections of millions of SiC grains (in *bulk* by Lewis *et al.* 1994) and predicted by AGB models for the typical parent stars of the grains (1.5–3 $M_\odot$ and solar metallicity).

2.1. *The record of planetary nebula processes in stardust SiC*

Meteoritic stardust SiC grains contain noble gases that carry the signature of processes occurring in AGB stars and planetary nebulae. One of the first evidences that most of these grains originated from the envelopes of C-rich AGB star comes from the composition of the noble gas Xe (left panel of Fig. 3), which shows the indisputable signature of the *s* process in AGB stars. The Xe isotopic ratios produced by the *s* process are predominantly controlled by the neutron-capture cross sections of the different Xe isotopes, which strongly favour the production of ^{128}Xe and ^{130}Xe. These are the two Xe *s*-only isotopes that are shielded from *r*-process production by their isobars in Te and are predominantly produced by the *s* process.

Very intriguing is the abundance trend of the different noble gases, He, Ne, Ar, Kr, and Xe with the size of the grains. Since noble gases are chemically inert, they are believed to have been incorporated in the SiC grains via an implantation process. Verchovsky *et al.* (2004) analysed the data from Lewis *et al.* (1994) in the light of an implantation model and discovered that the two different patterns (highlighted in green and blue in Fig. 4) can be reproduced by the models by varying the energy per nucleon, an input parameter of the model. A relatively low energy per nucleon, of the order of 0.5 keV, can reproduce the pattern observed in Xe. This energy can be associated with turbulence in the winds during the AGB phase, which have typical speeds of 10-30 km/s (the *cold* component, Table 1). As a consequence, Xe atoms must have been implanted carrying the signature of the matter in the winds of the AGB star. On the other hand, a much higher energy per nucleon ($\simeq$40 keV) is required to reproduce the pattern observed in He, Ne, Ar. This is because if the atom has too much energy, it cannot be stopped inside small grains. Post-AGB/planetary nebula fast winds ($\simeq$700 km/s) can imprint such an energy in the noble gases atoms (the *hot* component, Table 1). As a

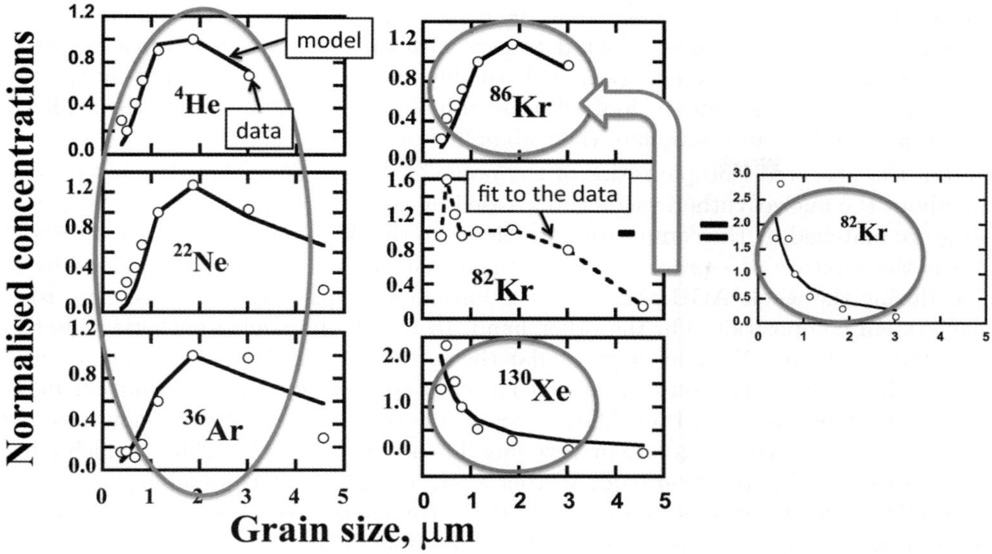

Figure 4. Figure adapted from Verchovsky *et al.* (2004) (©AAS. Reproduced with permission) and showing a comparison between the abundances of the different noble gases in meteoritic SiC as a function of the grain size (from Lewis *et al.* 1994) and the implantation model. Two clearly different patterns can be identified: ^{4}He, ^{22}Ne, ^{36}Ar, and ^{86}Kr are most abundant in large grains (highlighted in green), ^{130}Xe is most abundant in the small grains (highlighted in blue), while the ^{82}Kr abundance is made by a mixture of the two patterns, i.e., the blue pattern can be obtained by subtracting the ^{86}Kr green pattern from the total ^{82}Kr.

Table 1. Qualitive description of the two components of noble gases implanted in meteoritic SiC grains.

Component	cold	hot
Grain size (μm)	< 1	> 1
^{4}He	no[1]	yes
^{22}Ne	no[1]	yes
^{36}Ar	no[1]	yes
^{82}Kr	yes	yes
^{86}Kr	no[2]	yes
^{130}Xe	yes	no[3]
Energy/nucleon (keV)	0.5	40
Implantation winds	AGB	post-AGB
Wind speed (km/s)	10-30	700
AGB nucleosynthetic signature	envelope	intershell (final)

Notes:
[1] Not ionised.
[2] Nucleosynthetic signature not present.
[3] Much less abundant than the cold component.

consequence, He, Ne, and Ar must have been implanted carrying the signature of matter in the post-AGB winds, the same winds invoked as one of the possible factors involved in the production of planetary nebulea. Finally, Kr is a mixute of the two components, some of it being implanted in the AGB phase, and some during the planetary nebula phase.

The question arises of why He, Ne and Ar show only the fast wind/planetary nebula component, Xe shows only the AGB wind component, and Kr shows them both.

Verchovsky *et al.* (2004) proposed that these signatures are due to the different ionisation energy of the noble gases, which decreases with the number of electrons: Xe has the lowest ($\simeq$ 12 eV) ionization energy, and He the highest ($\simeq$ 25 eV) ionisation energy. If we assume that the atoms must be ionized to be implanted in the grains, then their different ionization energies can fractionate their abundances.

From the observed isotopic ratios of the noble gases we can gather detailed information about the nucleosynthetic signature carried by the two components. The Xe isotopic ratios are matched by the composition predicted in the AGB winds (left panel of Fig. 3). This is the mixture of *s*-process-rich intershell material carried to the envelope by the TDU during the whole AGB phase and the initial composition of the stars, close to the solar system composition. On the other hand, to match the nucleosynthetic signatures presented by He and Ne, and in particular the so-called Ne-E(H) component extremely enriched in ^{22}Ne, almost pure intershell material is required, with no dilution by the original envelope material. Post-AGB winds can carry such almost pure signature only in the case some flavour of a late or very late TP occurred (Blöcker 2001), which affects ~20% of the AGB population (e.g., Werner & Herwig 2006). It is possible that a nucleosynthetic signature such as that of the H-rich envelope is also present for He and Ne in the hot component, but, it is not seen because its abundance is much lower than that of the intershell in the late or very late TP scenario.

The ^{82}Kr abundance shows both the cold and the hot components (as Kr may be ionised in both cases), while ^{86}Kr shows only the fast wind/planetary nebula hot component. The most likely explanation is that the nucleosynthetic signature of the cold component carries a ^{82}Kr but not a ^{86}Kr excess, while the nucleosynthetic signature of the hot component carries both. The explanation for this difference lies in the fact that ^{82}Kr is located on the main *s*-process path, while ^{86}Kr is produced by the activation of the branching point at ^{85}Kr, for neutron densities exceeding few 10^8 n/cm^3. The hot component recorded in large grains points to high ^{86}Kr/^{82}Kr ratios in the intershell (right panel of Fig. 3). The stellar origin of such signature remains debated. Potential proposed explanations are models of metallicity lower than solar and either a higher efficiency of the activation of ^{13}C pocket (Pignatari *et al.* 2006) or the ingestion of the ^{13}C pocket inside the TPs (Raut *et al.* 2013), which can occur if the initial mass is lower than 1.5 $M_\odot$. However, such models predict Sr and Ba isotopic ratios inconsistent with measurements of the same bulk and single SiC grains (Gallino *et al.* 1997; Lugaro *et al.* 2003a; Barzyk *et al.* 2007; Ávila *et al.* 2013; Liu *et al.* 2015). The impact of convective-boundary mixing processes at the bottom of the convective thermal pulse (e.g., Lugaro *et al.* 2003b; Battino *et al.* 2016) needs to be also evaluated for Kr and Xe. Another possibility would be that the production of extra ^{86}Kr occurs during the post-AGB phase in association with a late or a very late TP that is already required to carry to the stellar surface the almost pure intershell material seen in the hot component (e.g., as in the case of the *intermediate* neutron-capture process seen in Sakurai's object; Herwig *et al.* 2011).

Finally, it would be important to fully define the connection between the excess ^{86}Kr in the stardust data and the Kr excess seen in the planetary nebulae observations (Fig. 2). From a back-of-the-envelope calculation, if we increase by a factor of 4 the ^{86}Kr abundance in the 2.5 $M_\odot$ model shown in Fig. 2, to simulate an extreme composition such as that shown by the high-density graphite grains (Fig. 3), the [Kr/Fe] ratio would increase from its current value of 0.43 to 0.66, which is a much better match with the data. As we mentioned ealier, the capability of theoretical AGB models to reproduce PN *s*-process observations of Kr and Xe is still matter of debate, and a consistent scenario should be derived that works for both PNe and measurements from presolar grains.

3. Summary and conclusions

Nucleosynthesis in AGB stars shapes the abundances observed in planetary nebulae via a complex interplay of nuclear reactions and mixing processes such as the third dredge-up and hot bottom burning. As such, planetary nebula abundances provide an independent, complementary testbed for all these processes, including the production of the elements heavier than rion via the *s* process. Conversely, the observed chemical abundance can be used to constraint the evolutionary path leading to the formation of planetary nebulae and the initial mass range. Specifically, we confirm that observations of [Xe/Kr] are a potential proxy for the traditional [hs/ls] *s*-process index and that the abundance of Rb can provide us crucial further constraints on the initial masses of planetary nebulae and the nucleosynthesis and mixing processes in massive AGB stars. Furthermore, noble gas abundances observed in meteoritic stardust grains are tracers of processes in planetary nebulae, from ionisation and implantation, to nucleosynthesis in the central star. The combination of all these different independent constraints will allow us to open new ways to understand planetary nebulae and the link to their parent AGB stars.

References

Abia, C., Busso, M., Gallino, R., Domínguez, I., Straniero, O., & Isern, J. 2001, Astrophys. J., 559, 1117

Abia, C., Cunha, K., Cristallo, S., & de Laverny, P. 2015, Astron. Astrophys., 581, A88

Arlandini, C., Käppeler, F., Wisshak, K., Gallino, R., Lugaro, M., Busso, M., & Straniero, O. 1999, Astrophys. J., 525, 886

Ávila, J. N., Ireland, T. R., Gyngard, F., Zinner, E., Mallmann, G., Lugaro, M., Holden, P., & Amari, S. 2013, Geochim. Cosmochim. Acta, 120, 628

Barzyk, J. G., Savina, M. R., Davis, A. M., Gallino, R., Gyngard, F., Amari, S., Zinner, E., Pellin, M. J., Lewis, R. S., & Clayton, R. N. 2007, Meteoritics and Planetary Science, 42, 1103

Battino, U., Pignatari, M., Ritter, C., Herwig, F., Denisenkov, P., Den Hartogh, J. W., Trappitsch, R., Hirschi, R., Freytag, B., Thielemann, F., & Paxton, B. 2016, Astrophys. J., 827, 30

Bisterzo, S., Travaglio, C., Gallino, R., Wiescher, M., & Käppeler, F. 2014, Astrophys. J., 787, 10

Blöcker, T. 2001, Astrophys. Space Sci., 275, 1

Cristallo, S., Straniero, O., Gallino, R., Piersanti, L., Domínguez, I., & Lederer, M. T. 2009, Astrophys. J., 696, 797

Cristallo, S., Straniero, O., Piersanti, L., & Gobrecht, D. 2015, Astrophys. J. Suppl., 219, 40

Gallino, R., Arlandini, C., Busso, M., Lugaro, M., Travaglio, C., Straniero, O., Chieffi, A., & Limongi, M. 1998, Astrophys. J., 497, 388

Gallino, R., Busso, M., & Lugaro, M. 1997, in American Institute of Physics Conference Series, Vol. 402, American Institute of Physics Conference Series, ed. T. J. Bernatowicz & E. Zinner, 115–153

Goriely, S. & Mowlavi, N. 2000, Astron. Astrophys., 362, 599

Herwig, F. 2005, Ann. Rev. Astron. Astrophys., 43, 435

Herwig, F., Pignatari, M., Woodward, P. R., Porter, D. H., Rockefeller, G., Fryer, C. L., Bennett, M., & Hirschi, R. 2011, Astrophys. J., 727, 89

Karakas, A. I. & Lattanzio, J. C. 2014, Publ. Astron. Soc. Aust., 31, 30

Karakas, A. I. & Lugaro, M. 2016, Astrophys. J., 825, 26

Lewis, R. S., Amari, S., & Anders, E. 1994, Geochim. Cosmochim. Acta, 58, 471

Liu, N., Savina, M. R., Gallino, R., Davis, A. M., Bisterzo, S., Gyngard, F., Käppeler, F., Cristallo, S., Dauphas, N., Pellin, M. J., & Dillmann, I. 2015, Astrophys. J., 803, 12

Lugaro, M. 2005, Stardust from meteorites. An introduction to presolar grains Copyright@2005 World Scientific

Lugaro, M., Davis, A. M., Gallino, R., Pellin, M. J., Straniero, O., & Käppeler, F. 2003a, Astrophys. J., 593, 486

Lugaro, M., Herwig, F., Lattanzio, J. C., Gallino, R., & Straniero, O. 2003b, Astrophys. J., 586, 1305

Lugaro, M., Karakas, A. I., Stancliffe, R. J., & Rijs, C. 2012, Astrophys. J., 747, 2

Lugaro, M., Ugalde, C., Karakas, A. I., Görres, J., Wiescher, M., Lattanzio, J. C., & Cannon, R. C. 2004, Astrophys. J., 615, 934

Pignatari, M., Gallino, R., Amari, S., & Davis, A. M. 2006, Mem. Soc. Astron. It., 77, 897

Pignatari, M., Gallino, R., Cristallo, S., & Straniero, O. 2008, in American Institute of Physics Conference Series, Vol. 1001, Evolution and Nucleosynthesis in AGB Stars, ed. R. Guandalini, S. Palmerini, & M. Busso, 154–159

Pignatari, M., Herwig, F., Hirschi, R., Bennett, M., Rockefeller, G., Fryer, C., Timmes, F. X., Ritter, C., Heger, A., Jones, S., Battino, U., Dotter, A., Trappitsch, R., Diehl, S., Frischknecht, U., Hungerford, A., Magkotsios, G., Travaglio, C., & Young, P. 2016, Astrophys. J. Suppl., 225, 24

Raut, R., Tonchev, A. P., Rusev, G., Tornow, W., Iliadis, C., Lugaro, M., Buntain, J., Goriely, S., Kelley, J. H., Schwengner, R., Banu, A., & Tsoneva, N. 2013, Physical Review Letters, 111, 112501

Roberts, L. F., Woosley, S. E., & Hoffman, R. D. 2010, Astrophys. J., 722, 954

Sharpee, B., Zhang, Y., Williams, R., Pellegrini, E., Cavagnolo, K., Baldwin, J. A., Phillips, M., & Liu, X.-W. 2007, Astrophys. J., 659, 1265

Sterling, N. C., Dinerstein, H. L., Kaplan, K. F., & Bautista, M. A. 2016, Astrophys. J. Lett., 819, L9 *Title: Discovery of Rubidium, Cadmium, and Germanium Emission Lines in the Near-infrared Spectra of Planetary Nebulae*, http://iopscience.iop.org/article/10.3847/2041-8205/819/1/L9/meta

Thielemann, F.-K., Arcones, A., Käppeli, R., Liebendörfer, M., Rauscher, T., Winteler, C., Fröhlich, C., Dillmann, I., Fischer, T., Martinez-Pinedo, G., Langanke, K., Farouqi, K., Kratz, K.-L., Panov, I., & Korneev, I. K. 2011, Progress in Particle and Nuclear Physics, 66, 346

van Raai, M. A., Lugaro, M., Karakas, A. I., García-Hernández, D. A., & Yong, D. 2012, Astron. Astrophys., 540, A44

Verchovsky, A. B., Wright, I. P., & Pillinger, C. T. 2004, Astrophys. J., 607, 611 *Title: Astrophysical Significance of Asymptotic Giant Branch Stellar Wind Energies Recorded in Meteoritic SiC Grains* http://iopscience.iop.org/article/10.1086/383230/meta

Werner, K. & Herwig, F. 2006, Publ. Astron. Soc. Pac., 118, 183

Zamora, O., García-Hernández, D. A., Plez, B., & Manchado, A. 2014, Astron. Astrophys., 564, L4

Discussion

MUTHUMARIAPPAN: Ne abundances in the H-poor knots of A30, A78 are several percentage more than expected from AGB nucleosynthesis. How do we account the same?

DE MARCO: There are too many Type I PN to be explained by hot bottom burning, i.e. stars with initial mass > 4 M$\odot$.

LUGARO: We think there could be processes to increase N/O even in lower mass stars. The Rb abundance is a better indicator of mass. Finding low Rb (i.e. lower mass) and high N/O would be an indication that such additional processes operate (even if we do not yet know what might be).

Planetary Nebulae: Multi-Wavelength Probes of Stellar and Galactic Evolution
Proceedings IAU Symposium No. 323, 2016
X. Liu, L. Stanghellini & A. Karakas, eds.

doi:10.1017/S174392131700059X

Understanding Galactic planetary nebulae with precise/reliable nebular abundances

D. A. García-Hernández[1,2], P. Ventura[3], G. Delgado-Inglada[4], F. Dell'Agli[1,2,3], M. Di Criscienzo[3] and A.Yagüe[1,2,3]

[1]Instituto de Astrofísica de Canarias, C/ Vía Láctea s/n, E−38205 La Laguna, Spain
[2]Departamento de Astrofísica, Universidad de La Laguna, E−38206 La Laguna, Spain
[3]INAF - OAR, Via Frascati 33, 00040, Monte Porzio Catone, Italy
[4]Instituto de Astronomía, UNAM, Apdo. Postal 70264, 04510, México D. F., México

Abstract. We compare recent precise/reliable nebular abundances - as derived from high-quality optical spectra and the most recent ICFs - in a sample of Galactic planetary nebulae (PNe) with nucleosynthesis predictions (HeCNOCl) from asymptotic giant branch (AGB) ATON models in the metallicity range $Z_\odot/4 < Z < 2Z_\odot$. According to the infrared dust features, the sample is divided among carbon-, oxygen-, and double-dust chemistry (CC, OC, and DC, respectively), providing an independent proxy for the nature of the PNe progenitors. Our AGB models, with diffusive overshooting from all the convective borders, nicely reproduce the O overabundances observed in CC PNe, indicating that they evolve from low-Z low-mass (∼1−3 $M_\odot$) AGB stars. This indicates that O is not always a good indicator of the original ISM metallicity and that the O production by low-mass stars should be considered in galactic-evolution models. The lowest metallicity OC PNe evolve from low-mass (∼1 $M_\odot$) O-rich AGBs, while the higher metallicity ones (all with uncertain dust classifications) display a chemical pattern similar to the DC PNe. In agreement with the recent literature, the DC PNe mostly descend from high-mass (M > 3.5 $M_\odot$) solar/supersolar metallicity AGBs that experience hot bottom burning (HBB), but other formation channels in low-mass AGBs like extra mixing, stellar rotation, binary interaction, or He pre-enrichment cannot be disregarded until more accurate C/O ratios can be obtained. Two DC PNe show the imprint of advanced CNO processing and deep second dredge-up, suggesting progenitors masses close to the limit to evolve as core collapse supernovae (above 6 $M_\odot$). Their actual C/O ratios, if confirmed, indicate contamination from the third dredge-up, rejecting the hypothesis that the chemical composition of such high-metallicity massive AGBs is modified exclusively by HBB.

Keywords. Nuclear reactions, nucleosynthesis, abundances, ISM: abundances, H II regions, planetary nebulae: general, Galaxy: abundances.

---

1. Introduction

The chemical composition of planetary nebulae (PNe) proves an extremely useful and valuable tool to constrain asymptotic giant branch (AGB) models (see e.g., Ventura *et al.* 2015). The chemistry of PNe is the outcome of the combination of the different processes that contribute to alter the surface chemical composition during the whole AGB life. In addition, PNe - because of their emission-line nature - are easily observed at very large distances, and the gas chemical composition can be derived. Some chemical elements (e.g., Ar and Cl) may remain practically unchanged in PNe, reflecting the primordial composition of the interstellar medium (ISM) where their progenitor stars were born. Other elements like C and N (to a lesser extent O), however, may be strongly modified during the previous AGB phase. The O abundance in PNe has been widely used as a metallicity indicator of the ISM where they originated; e.g., to derive metallicity gradients in our own Galaxy and other nearby galaxies (see e.g., Richer 2015). This is because "standard" (with no extra-mixing processes) nucleosynthesis theoretical models of low- and intermediate-mass stars available in the literature (e.g., Karakas 2010) do not predict significant O enrichment (or destruction) at near-solar metallicities during the AGB phase.

However, by using high-quality optical spectra in conjunction with the best available ionization correction factors (ICFs), Delgado-Inglada *et al.* (2015) have very recently reported the first observational evidence of O self-enrichment (by ~0.3 dex) in Galactic PNe with C-rich dust; the expected outcome of low-mass (~1.5−3 $M_{\odot}$) stars. Models that include diffusive convective overshooting (e.g., Herwig *et al.* 1997)† predict a significant production of oxygen, even at solar metallicities. This mixing process in combination with an efficient third dredge-up (TDU), may produce an increase of O abundance for low-mass (~1.5−3 $M_{\odot}$) stars. Rotation, magnetic fields, and thermohaline mixing are alternative mechanisms, even less understood, that may cause extra-mixing (see e.g., Karakas & Lattanzio 2014 for a review). Here we present the main results from the comparison of self-consistent AGB nucleosynthesis predictions (based on the ATON code) for the CNO elements and He in the metallicity range $Z_{\odot}/4 < Z < 2Z_{\odot}$ with the Delgado-Inglada *et al.* (2015) sample of Galactic PNe (see García-Hernández *et al.* 2016a,b).

2. The AGB ATON models

Briefly, the AGB ATON models are calculated using the following physical ingredients: (i) The convection is modelled according to the full spectrum of turbulence (FST) model. In regions unstable to convection, mixing of chemical and nuclear burning are coupled by a diffusion–like equation. Overshoot of convective eddies into the radiatively stable regions follows an exponential decay of velocities from the border of the convective zones (which is fixed via the Schwarzschild criterion) with the e–folding distance of the decay given by ζH_p=0.002; (ii) Different mass-loss prescriptions are used for O- and C-rich AGB stars; (iii) The molecular opacities at low temperatures (<10^4K) are calculated with the AESOPUS tool (Marigo & Aringer 2009), which is especially important for the description of the C-rich phase. We refer the reader to García-Hernández *et al.* (2016a,b) for more details about the AGB ATON models as well as for an overview of the AGB ATON abundances predictions.

3. Understanding the Galactic PNe sample with precise/reliable nebular abundances

Fig. 1 compares the Delgado-Inglada *et al.* (2015) sub-sample (7 objects) of lower metallicity Galactic PNe with C-dust features in their mid-IR spectra (García-Hernández *et al.* (2016a). By comparing the observed chemistry (HeCNOCl) with the AGB ATON models, García-Hernández *et al.* (2016a) concluded that they mostly descend from low-metallicity ($Z = 8 \times 10^{-3}$) low-mass (~1−3 $M_{\odot}$) AGB stars that produce O, confirming that O is not always a good metallicity indicator (especially for low-metallicity PNe with C-rich dust) and that O production by low-mass stars should be thus considered in the Galactic chemical evolution models.

On the other hand, Fig. 2 compares the Delgado-Inglada *et al.* (2015) sub-samples (13 objects) of Galactic PNe with O-rich and/or C-rich dust features in their mid-IR spectra

† The Pignatari *et al.* (2016) nucleosynthesis models (only very recently accepted for publication) also predict O production in low-mass stars.

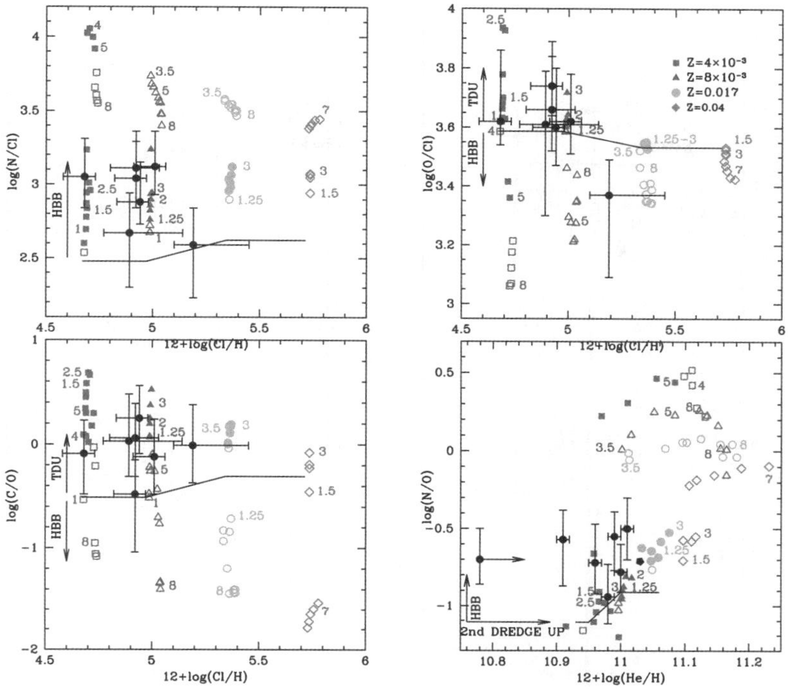

Figure 1. Chemical abundances of Galactic carbon-dust PNe (solid circles, Delgado-Inglada *et al.* 2015) vs. the AGB ATON model predictions for different masses (a few relevant masses are marked) and metallicities: Cl vs. N/Cl (top, left), Cl vs. O/Cl (top, right), Cl vs. C/O (bottom, left), He vs. N/O (bottom, right). The thin, solid lines indicate the assumed initial abundances, while the arrows indicate the qualitative effects of the HBB, SDU and TDU. Filled and open symbols correspond to C- and O-rich stars, respectively. Updated from García-Hernández *et al.* (2016a).

(García-Hernández *et al.* 2016b). Unlike the Delgado-Inglada *et al.* (2015) subsample of carbon-dust chemistry (CC) PNe (García-Hernández *et al.* 2016a), here the individual abundance errors (in particular that of the C/O ratio), the wide metallicity range, and the uncertain dust types/subtypes in some objects do not permit a clear determination of the AGB initial mass for both PNe samples (see García-Hernández *et al.* 2016b for a full discussion). The PNe observations versus AGB models comparison is thus more focussed on a star-by-star basis. By comparing the models vs. observations (Fig. 2), García-Hernández *et al.* (2016b) concluded that: i) The two lowest metallicity OC PNe are interpreted as the descendants of low-mass ($\sim$1 $M_\odot$) AGB stars that did not reach the C-rich phase, while the three PNe of higher metallicities in this group have uncertain dust classifications and are chemically indistinguishable from the rest of DC PNe in our sample, being interpreted as truly DC PNe; ii) The DC PNe in our sample are best separated (in terms of progenitor AGB masses) in the He vs. N/O plane, which otherwise suggests that they mostly descend from high mass (M $\geqslant$3.5 $M_\odot$) HBB AGB stars of solar/supersolar metallicities. This is consistent with the recent work in the literature, and also we cannot discard alternative formation channels in low-mass non-HBB stars such as extra mixing, stellar rotation, binary interaction, or He pre-enrichment (more precise C/O ratios turn out to be fundamental to learn about the stellar origin of DC PNe); iii) Two DC PNe are likely descendants of the more massive AGB stars of close-to-solar chemistry, with masses $\sim 6-7\ M_\odot$. The C/O ratios of these objects ($C/O \sim 0.5$),

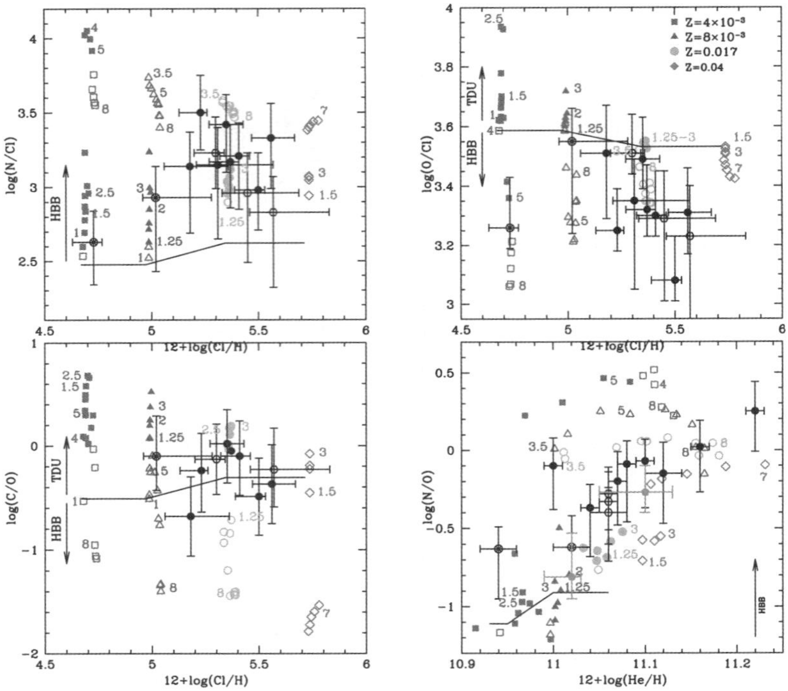

Figure 2. Same as Fig. 1 but for the Galactic OC (open-crossed circles) and DC (solid circles) PNe of Delgado-Inglada *et al.* (2015). The three OC PNe with more uncertain dust classifications (see García-Hernández *et al.* 2016b for more details) are shown with open hexagons. Updated from García-Hernández *et al.* (2016b).

if confirmed, suggests a role of the TDU in the contamination of the surface chemistry during the previous AGB phase, being at odds with our AGB model predictions.

Acknowledgments

DAGH acknowledges support provided by the RYC−2013−14182 fellowship and the MINECO AYA−2014−58082-P grant.

References

Delgado-Inglada, G., Rodríguez, M., Peimbert, M. *et al.* 2015, *MNRAS*, 449, 1797
García-Hernández, D. A., Ventura, P., Delgado-Inglada, G. *et al.* 2016a, *MNRAS*, 458, L118
García-Hernández, D. A., Ventura, P., Delgado-Inglada, G. *et al.* 2016b, *MNRAS*, 461, 542
Herwig, F., Blöcker, T., Schönberner, D., El Eid, & M. 1997, *A&A*, 324, L81
Karakas A. I. 2010, *MNRAS*, 403, 1413
Karakas A. I., Lattanzio J. C. 2014, *PASA*, 31, 30
Marigo P. & Aringer B. 2009, *A&A*, 508, 1538
Pignatari, M., Herwig, F., Hirschi, R., *et al.* 2016, *ApJS*, 225, 24
Ventura P., Stanghellini, L., Dell'Agli, F. *et al.* 2015, *MNRAS*, 452, 3679
Richer, M. G. 2015, *in IAU General Assembly*, Meeting #29, #2256079

Planetary Nebulae: Multi-Wavelength Probes of Stellar and Galactic Evolution
Proceedings IAU Symposium No. 323, 2016
X. Liu, L. Stanghellini & A. Karakas, eds.

doi:10.1017/S1743921317000655

^{3}He Abundances in Planetary Nebulae

Lizette Guzman-Ramirez[1,2]†

[1]European Southern Observatory, Alonso de Córdova 3107, Casilla 19001, Santiago, Chile
[2]Leiden Observatory, Leiden University, Niels Bohrweg 2, 2333 CA Leiden, The Netherlands

Abstract. Determination of the ^{3}He isotope is important to many fields of astrophysics, including stellar evolution, chemical evolution, and cosmology. The isotope is produced in stars which evolve through the planetary nebula phase. Planetary nebulae are the final evolutionary phase of low- and intermediate-mass stars, where the extensive mass lost by the star on the asymptotic giant branch is ionised by the emerging white dwarf. This ejecta quickly disperses and merges with the surrounding ISM.

^{3}He abundances in planetary nebulae have been derived from the hyperfine transition of the ionised ^{3}He, ^{3}He$^+$, at the radio rest frequency 8.665 GHz. ^{3}He abundances in PNe can help test models of the chemical evolution of the Galaxy.

Many hours have been put into trying to detect this line, using telescopes like the Effelsberg 100m dish of the Max Planck Institute for Radio Astronomy, the National Radio Astronomy Observatory (NRAO) 140-foot telescope, the NRAO Very Large Array, the Arecibo antenna, the Green Bank Telescope, and only just recently, the Deep Space Station 63 antenna from the Madrid Deep Space Communications Complex.

Keywords. circumstellar matter – radio: abundances, planetary nebulae.

1. Introduction

Our Universe has been evolving for 13.8 Gyr. Over these years many stars formed and ended their lives enriching the interstellar medium (ISM), and in consequence enriching the Universe (Planck Collaboration *et al.* 2014). Very few elements have been around since the beginning, formed by the Big Bang nucleosynthesis (BBN). BBN is responsible for the formation of most of the helium isotope (^{4}He) in the Universe, along with small amounts of deuterium (D), the helium isotope (^{3}He), and a very small amount of the lithium isotope (^{7}Li) (Alpher *et al.* 1948).

The predicted abundance (by number) of ^{3}He (relative to H) formed by the BBN (just after a few minutes after the Big Bang) is 1.0×10^{-5} (Karakas & Lattanzio 2014). This abundance depends mainly on the parameter of the current density of baryonic matter. The present interstellar ^{3}He abundance, as per all the light elements, comes from a combination of BBN and stellar nucleosynthesis (Wilson & Rood 1994). H II regions are young objects compared with the age of the Universe, and represent zero-age objects. Their ^{3}He abundance is the result of 13.8 Gyr of Galactic chemical evolution. In between is the Solar System, which traces abundances at the time of its formation, 4.6 Gyr ago.

Observed values in pre-solar material (Geiss 1993) and the ISM (Gloeckler & Geiss 1996) imply that ^{3}He/H=$(2.4\pm0.7)\times10^{-5}$. These values from the ISM and pre-solar material are approximately twice that produced by the BBN, implying that the ^{3}He abundance has increased a little in the last 13.8 Gyr. On the other hand any hydrogen-burning zone of a star which is not too hot ($>7\times10^6$ K) will produce ^{3}He via the *p-p* chain, implying that stars of masses $<2.5\,M_\odot$ are net producers of ^{3}He. For those stars,

† E-mail: guzmanl@strw.leidenuniv.nl

p-p burning is rapid enough to produce D *in situ*, and enables the production of ^{3}He (D+ p $\rightarrow$ ^{3}He + γ). Stellar evolution models indeed predict the formation of ^{3}He in significant amounts by stars of 1–2.5 $M_\odot$, yielding an abundance of ^{3}He/H $> 10^{-4}$ (Bania *et al.* 2010), which would have raised the current ^{3}He abundance to ^{3}He/H$\sim 5 \times 10^{-5}$, substantially higher than observed (Karakas & Lattanzio 2014).

Galli *et al.* (1995) presented "The ^{3}He Problem". According to standard models of stellar nucleosynthesis there should be a ^{3}He/H abundance gradient in the Galactic Disk and the pre-solar ^{3}He/H value should be less than what is found in the present ISM. Observations of the ^{3}He abundance in H II regions show almost no enrichment above the BBN value (Rood *et al.* 1979; Bania *et al.* 2007). For the ^{3}He problem to be solved, the vast majority of low-mass stars should fail to enrich the ISM. One suggestion to solve this problem is by adding extra mixing in the red giant branch (RGB) stage. This extra mixing during the standard first dredge-up phase may modify the surface abundances. Eggleton. *et al.* (2006) estimated that while 90% of the ^{3}He is destroyed in 1 $M_\odot$ stars, only 40-60% is destroyed in a 2 $M_\odot$ star model, depending on the speed of mixing.

2. Detecting the ^{3}He$^+$ emission line

The abundance of ^{3}He has been derived from the hyperfine transition at the rest frequency of 8.665 GHz. Detecting ^{3}He$^+$ in PNe challenges the sensitivity limits of all the existing radio telescopes. Composite ^{3}He$^+$ average spectra for six planetary nebulae (NGC 3242, NGC 6543, NGC 6720, NGC 7009, NGC 7662, and IC 289), were collected using the Effelsberg, Arecibo and Green Bank telescopes. They all consistently show possible ^{3}He$^+$ emission at the $\sim$1 mK level. Using the same telescopes later on, a detection was obtained in NGC 3242 and J320. Those detections and the composite spectrum measurements translate into an abundance of ^{3}He/H of a few 10^{-4} (Balser *et al.* 1997, 1999, 2006).

Later on more effort was put into this problem, using the Very Large Array, for several tens of hours 3 planetary nebulae were observed (IC 418, NGC 6572, and NGC 7009). Unfortunately no line was detected and only upper limits in the abundance were obtained (Guzman-Ramirez *et al.* 2013). Recently (beginning of this year) Guzman-Ramirez *et al.* (2016), using the Deep Space Station 63 antenna from the Madrid Deep Space Communications Complex, another ^{3}He$^+$ detection was obtained in the planetary nebula IC 418. With more than 20 years of looking for such detections, so far we have only 3 objects with more than a 3σ detection of ^{3}He: J320, NGC 3242, and IC 418.

Two aspects of these detections needs to be considered. Firstly, the derived ^{3}He$^+$ abundance is well above the model expectations for all the three objects. This tells us that the stars do produce and release ^{3}He into the interstellar medium. Secondly, the ^{3}He$^+$ line profile for all the three objects shows a double peaked shape. Although the full width of the ^{3}He$^+$ line is consistent with the optical expansion velocity, the profile differs from that of the recombination lines, peaking at larger velocities.

Two other planetary nebulae have reported ^{3}He$^+$ detections: J320 (Balser *et al.* 2006) and NGC 3242 (Balser *et al.* 1997, 1999). Both have double peaked profiles, similar to IC 418 (Fig. 1). Both objects have haloes, and in the case of NGC 3242 possibly as large as 18 by 24 arcmin in size. Whether helium in such a halo could be photo-ionized by the star is not clear. A double-peaked profile could arise from an expanding detached shell, which is larger than the beam of the telescope. In this case, the outer components of the profile, which differs that of the recombination lines, come from this large region, whilst the central part of the profile arises from the inner, ionized nebula. Balser *et al.* (1999) also proposed a contribution from a large, low density halo. The emission in the ^{3}He$^+$

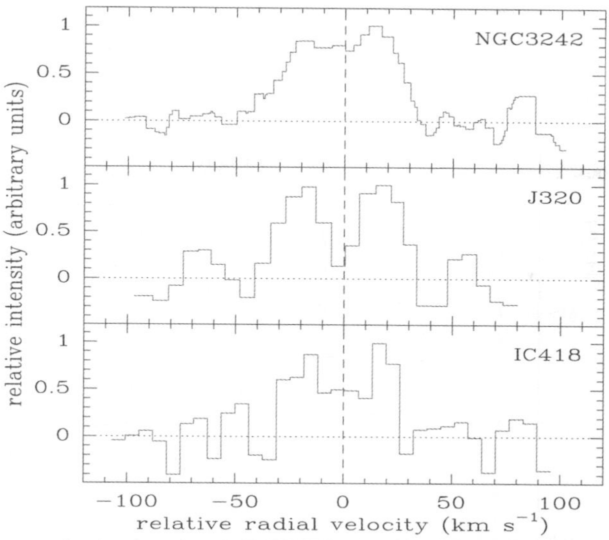

Figure 1. All the ^{3}He$^+$ line detected in planetary nebulae. The sources are NGC 3242 (Balser *et al.* 1999), J320 (Balser *et al.* 2006), and IC 418 (this work). In order to facilitate the comparison, the abscissa is presented as radial velocities relative the LSR systemic velocity of each source (dashed-line), and the intensity scale as fractions of the peak intensity of each source.

hyperfine line scales with $\int n\, dr$ (column density), while that of the recombination lines with $\int n^2\, dr$. Therefore, a low density but high mass halo could explain the difference in profiles.

3. Stellar evolution

The contribution of PNe to the ^{3}He abundance is crucial for understanding the Galactic chemical evolution. The number density of ^{3}He$^+$ atoms, $n(^3\mathrm{He}^+)$, can be obtained by dividing the column density $N(^3\mathrm{He}^+)$ by the averaged optical path, $< \Delta s >$, through the source. Representing a PN as a homogeneous sphere of radius $R = \theta_s D$, where D is the distance, the optical path at an angle θ from the centre is $\Delta s(\theta) = 2\sqrt{R^2 - (\theta D)^2}$, and the optical path averaged over the source becomes $< \Delta s >= \pi R/2 = \pi\theta_s D/2$. To calculate the fractional ^{3}He abundance we divide $n(^3\mathrm{He}^+)$ thus obtained by the H$^+$ density.

Figure 2 shows the ^{3}He abundances of PN IC 418 (the two purple crosses, taken from Guzman-Ramirez *et al.* (2016)). For comparison the upper limits calculated from Guzman-Ramirez *et al.* (2013) observations are presented using the red arrows. Balser *et al.* (1997) observations are shown in green: the cross represents J320 and the arrows are the upper limits, where the mass estimates are from Galli *et al.* (1997). The stellar evolution models are also presented in the same figure.

4. Future

The values observed in these objects proves that ^{3}He is produced at the centre of low-mass stars, and is ejected into the interstellar medium at the end of their lives. However, the large amounts of ^{3}He produced in these models is at odds with the abundance of ^{3}He observed in the interstellar medium and the Solar System. Further research on other planetary nebulae will be needed to solve the ^{3}He problem. As the emission is extremely

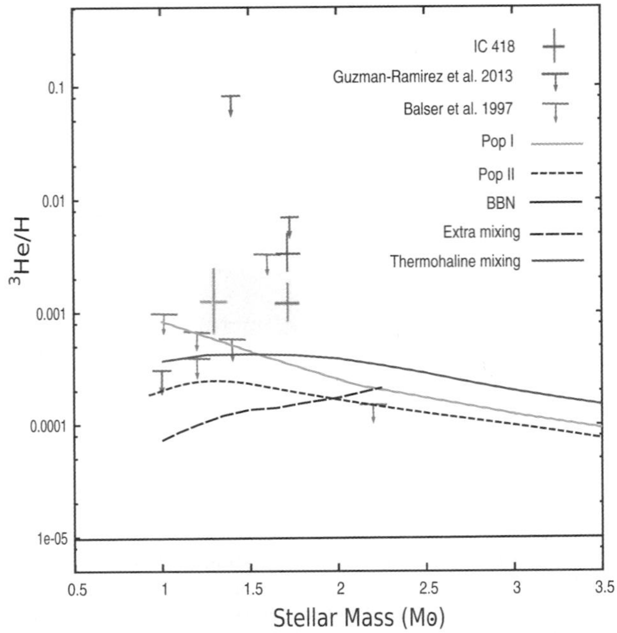

Figure 2. Abundances of ^{3}He (relative to H) for the PN IC 418 (purple crosses, for upper and lower estimates), the size of the bars corresponds to the uncertainty. Red arrows are the upper limits from Guzman-Ramirez *et al.* (2013). The green cross represents J320 and the green arrows represent the sample of 6 PNe from Balser *et al.* (1997). The curves for Pop I (light-blue line) and II (red dashed-line) show the standard abundance of ^{3}He taken from Weiss *et al.* (1996). The purple dashed-line labelled Extra mixing represents the results of models including deep mixing by Boothroyd & Sackmann, (1999), and the blue line shows the stellar models using thermohaline mixing (Charbonnel & Zahn 2007). The black line is the primordial value of ^{3}He/H from the BBN (Karakas & Lattanzio 2014).

weak and hard to detect, many observing hours are needed. Telescopes of high sensitivities like the Square Kilometre Array will be very useful for this research.

References

Alpher, R. A., Bethe, H., & Gamow, G 1948, Physical Review, 73, 803
Balser, D. S., Bania, T. M., Rood, R. T., & Wilson, T. L. 1997, ApJ, 483, 320
Balser, D. S., Rood, R. T., & Bania, T. M. 1999, ApJL, 522, L73
Balser, D. S., Goss, W. M., Bania, T. M., & Rood, R. T. 2006, ApJ, 640, 360
Bania, T. M., Balser, D. S., Rood, R. T., Wilson, T. L., & LaRocque, J. M. 2007, ApJ, 664, 915
Bania, T. M., Rood, R. T., & Balser, D. S. 2010, IAU Symposium, 268, 81
Boothroyd, A. I. & Sackmann, I.-J. 1999, ApJ, 510, 232
Charbonnel, C. & Zahn, J.-P. 2007, A&A, 467, L15
Eggleton, P. P., Dearborn, D. S. P., & Lattanzio, J. C. 2006, Science, 314, 1580
Galli, D., Palla, F., Ferrini, F., & Penco, U. 1995, ApJ, 443, 536
Galli, D., Stanghellini, L., Tosi, M., & Palla, F. 1997, ApJ, 477, 218
Geiss, J. 1993, Origin and Evolution of the Elements, 89
Gloeckler, G. & Geiss, J. 1996, Nature, 381, 210
Guzman-Ramirez, L., Pineda, J. E., Zijlstra, A. A., Stancliffe, R., & Karakas, A. 2013, MNRAS, 432, 793
Guzman-Ramirez L., Rizzo J. R., Zijlstra A. A., *et al.* 2016, MNRAS, 460, L35
Hjellming, R. M., Andrews, M. H., & Sejnowski, T. J. 1969, Astrophys. Lett, 3, 111

Karakas, A. I. & Lattanzio, J. C. 2014, Publications of the Astronomical Society of Australia, 31, 30
Morisset, C. 2006, IAU Symposium, 234, 467
Planck Collaboration, *et al.* 2014, A&A, 571, A1
Rood, R. T., Wilson, T. L., & Steigman, G. 1979, ApJL, 227, L91
Weiss, A., Wagenhuber, J., & Denissenkov, P. A. 1996, A&A, 313, 581
Wilson, T. L. & Rood, R. 1994, Annual Review of Astron and Astrophys, 32, 191

Discussion

MENDEZ: What is the source of your distance to IC418? If the distance is a bit larger, your abundance would go down a little.

GUZMAN-RAMIREZ: The distance was measured using the expansion parallax technique that gives you a very accurate distance. But yes if the distance is higher the abundance goes lower.

GUERRERO: There is not a known halo around IC418. What's the beam size of Robledo antennae? The $^3He^+$ line probably probing a different component than other ionized species do.

GUZMAN-RAMIREZ: The beam size is 115″ which is very big. So yeah the ^{3}He emission is probably coming from a different component.

Q: Are there destruction mechanisms (e.g. spallation) that could reduce the interstellar ^{3}He abundances from the PN values to what is observed in H IIregions?

GUZMAN-RAMIREZ: Yes, spallation is a possible mechanism. The problem is that it hasn't been tested in the lab yet, so we don't know how much ^{3}He can bed estroyed in the ISM.

Q: 1) Have you considered observing the ^{3}He line at Arecibo? 2) The line widths of the ^{3}He recombination line are >25 km/s – too large to be associated with a halo. Is there another explanation for the location of ^{3}He emission?

GUZMAN-RAMIREZ: 1) Arecibo is a good option that will be considered; 2) Yes, we are not sure about the halo so we are exploring other possibilities.

Planetary Nebulae: Multi-Wavelength Probes of Stellar and Galactic Evolution
Proceedings IAU Symposium No. 323, 2016
X. Liu, L. Stanghellini & A. Karakas, eds.

doi:10.1017/S1743921317000965

X-ray Studies of Planetary Nebulae

Rodolfo Montez Jr.

Smithsonian Astrophysical Observatory
Cambridge, MA 02138 USA
email: rmontez@cfa.harvard.edu

Abstract. X-ray emission from planetary nebulae (PNe) provides unique insight on the formation and evolution of PNe. Past observations and the ongoing *Chandra* Planetary Nebulae Survey (ChanPlaNS) provide a consensus on the two types of X-ray emission detected from PNe: extended and compact point-like sources. Extended X-ray emission arises from a shocked "hot bubble" plasma that resides within the nebular shell. Cooler than expected hot bubble plasma temperatures spurred a number of potential solutions with one emerging as the likely dominate process. The origin of X-ray emission from compact sources at the location of the central star is less clear. These sources might arise from one or combinations of the following processes: self-shocking stellar winds, spun-up binary companions, and/or accretion, perhaps from mass transfer, PN fallback, or debris disks. In the discovery phase, X-ray studies of PNe have mainly focused on the origin of the various emission processes. New directions incorporate multi-wavelength observations to study the influence of X-ray emission on the rest of the electromagnetic spectrum.

Keywords. planetary nebulae: general, binaries: close, X-rays: general

1. Introduction

The modern era of X-ray studies of Planetary Nebulae (PNe) began with the launches of the NASA-led *Chandra* X-ray Observatory and the ESA-led X-ray Multiple Mirror Newton Observatory (XMM). Both satellites launched in 1999 and continue observing today. These observatories are equipped with imaging spectrometers that provide spatial and spectral (energy) information for each detected X-ray photon. The information-rich X-ray data are essential for distinguishing between point-like and extended sources of X-ray emission; the former typically associated with central stars and/or their companions and the latter typically associated with hot bubble emission. The modest energy resolution of X-ray spectra are used to estimate the characteristics like plasma temperatures, densities, and fluxes of X-ray emission. As the impact of X-ray studies of PNe continue to build, this review presents the current state of these studies and the emerging directions for future research.

2. Insights from X-ray Emission from Planetary Nebulae

Big Trouble in Hot Bubbles: One of the first modern X-ray observations of a PNe was by *Chandra* of the young PN BD+30°3639 (Kastner *et al.* 2000). This observation used the high-spatial resolution ($\sim 0.3''$) of *Chandra* to provide the first unambiguous detection of extended "hot bubble" emission. In the interacting stellar wind theory of PN formation, as the fast stellar wind collides with the slower AGB wind, AGB material is swept up to form the nebula while a reverse shock travels back towards the central star creating a post-shock region within the nebula. The post-shock region should be filled with hot (10^7 K), low-density ($n_e < 1$ cm^{-3}) plasma, the so-called "hot bubble". The spatial extent of the

X-ray emission detected from the early *Chandra* observation of BD+30°3639 fits within the nebula (as imaged in Hα by HST), suggesting the nebula is filled with the X-ray emitting plasma. But the energy spectrum of the X-ray photons suggests that the hot bubble is only $\sim 3 \times 10^6$ K.

Lower than expected hot bubble temperatures, first observed in BD+30°3639, emerge as a trend in all subsequent observations of hot bubbles (Chu *et al.* 2001, Guerrero, Gruendl, & Chu 2002, Montez *et al.* 2005, Ruiz *et al.* 2013), including those from the *Chandra* Planetary Nebulae Survey (ChaPlaNS; Kastner *et al.* 2012, Freeman *et al.* 2014). Several solutions have been proposed to resolve the hot bubble temperature discrepancy:

• *Early-phase stellar winds* during the formation of the nebula are expected to have lower velocities than present-day winds. Soker & Kastner (2003) study the formation of hot bubbles and argue that the lower wind speeds dictate the hot bubble temperatures not the present-day winds.

• *Collimated winds or outflows* driven by companions and studied with hydrodynamic simulations (Akashi, Meiron, & Soker 2008) are found to produce plasmas with characteristics (temperature and spatial distribution) consistent with some sources of extended X-ray emission.

• *Heat conduction* across the nebula-hot bubble interface are considered in Steffen, Schönberner, & Warmuth (2008) and the process is shown to efficiently regulate the hot bubble temperature from 10^7 K to the observed range of $1-3\times10^6$ K. Heat conduction acts by "evaporating" nebular material into the hot bubble, which has the effect of increasing the hot bubble density and lowering its temperature.

Heat conduction appears to be the dominate process that explains the lower than expected hot bubble temperatures. Supporting evidence for heat conduction has been found from multi-wavelength studies of the 10^5 K transition region (see Ruiz *et al.* 2013 and the contribution by Guerrero in this volume). Studies of heat conduction in nebula with distinct chemical compositions are ongoing (see Sandin *et al.* 2016 and the contribution by Schoenberner in this volume). Other predictions of heat conduction models, such as the radial distribution of hot bubble emission (Freeman & Kastner 2016) and mixing of nebular and fast wind elemental abundances (Maness & Vrtilek 2003) are limited by the capabilities of the two modern X-ray observatories.

Colliding Winds, Hidden Companions: The perplexing characteristics of point-like X-ray emission from central stars of PNe (CSPNe) began with early *Chandra* observations of the Cat's Eye and Helix Nebulae (Guerrero *et al.* 2001). The spatial resolution of *Chandra* allows us to pin-point the location of the point-like sources to the central stars. Spectral characteristics of the X-ray emission indicate higher temperatures ($> 10^6$ K) than expected from the CSPNe. Such X-ray emission could arise from wind shocks, accretion, or binary companions. Montez *et al.* (2010) argue that short-period (< 1 day) post-common envelope binary companions are spun-up during the common envelope phase resulting in rejuvenated coronal activity. It is the companions and not the CSPNe that are the source of hard, point-like X-ray emission from a sample of short-period (< 1 day) post-common envelope binary CSPNe.

With the *Chandra* Planetary Nebulae Survey (ChanPlaNS), it becomes clear that X-ray emission of these known binary systems have distinct spectral characteristics than some of the other detected point-like sources (Montez *et al.* 2015). Specifically, known close binary CSPNe tend to harbor hard X-ray emitting sources with $T_X \geqslant 10^7$ K, while other CSPNe with X-ray emission and no known close companions tend to harbor softer X-ray sources with $T_X \sim 10^6$ K. These softer CSPNe tend to be high-luminosity sources driving fast stellar winds, suggesting that their X-ray emission might be analogous to the self-shocking winds observed from O-stars. Such a hypothesis is supported by the

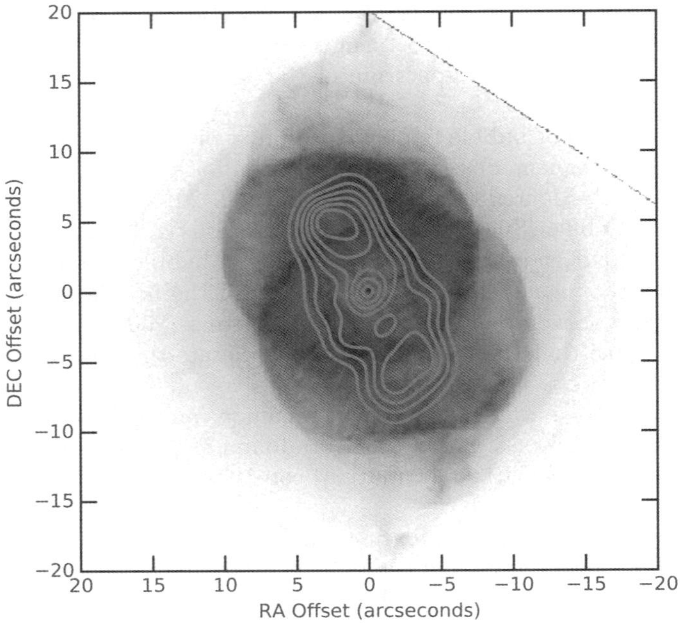

Figure 1. Cat's Eye Nebula showing the two types of X-ray emission detected from PNe: extended and point-like emission. Contours of the smoothed X-ray emission from *Chandra* are overlaid upon an Hα image acquired by HST.

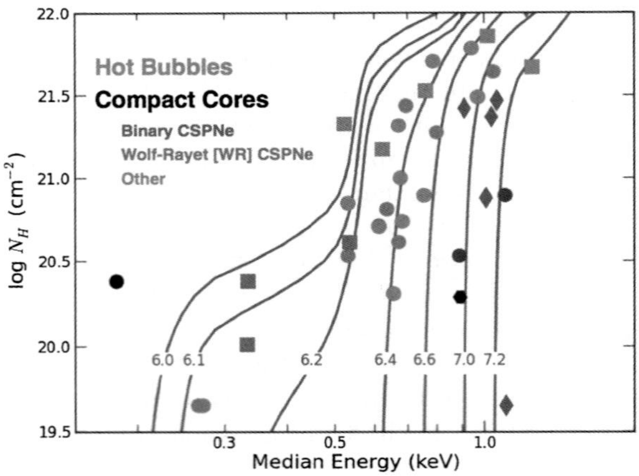

Figure 2. Median energy of detected X-ray photons from PNe in ChanPlaNS and column densities, N_H, derived from nebular observations of the Balmer decrement. Red symbols indicate hot bubbles, other colors indicate point sources. PNe with Wolf-Rayet central stars, H-rich central stars, or close binary companions are represented by square, circles, and diamonds, respectively. The Helix central star is represented by the octagon shape. The blue lines indicate the loci for plasma models with labels in $\log T_X$.

fact that L_X/L_{bol} ratios of CSPNe (10^{-7}) are similar to ratios of O-stars (Montez *et al.* 2015).

A few point-like sources without known companions feature the hard X-ray emission associated with close binary systems. This includes the Cat's Eye and Eskimo Nebulae.

If the hard X-ray emission is indicative of spun-up companions, the X-ray emission might betray the hidden binary companions.

3. Future of X-ray Astronomy

The studies made possible with modern X-ray observations of PNe have provided insight into nebular formation, wind shocks, and binary interactions. These observations maximize the capabilities of the present observatories. Future X-ray missions, such as eRosita (2017), Athena+ (2028), and Lynx/X-ray Surveyor (beyond 2030), will offer improved sensitivities and capabilities to address the open questions regarding the origin of X-ray emission from PNe. The all-sky imaging spectroscopy survey offered by eRosita has the potential to expand the number X-ray detections of PNe and improve signal to noise for presently known sources with modest spatial resolution. Presently, only BD+30°3639 has been a suitable target for high-resolution spectroscopic X-ray observations Yu *et al.* (2009) but the high-throughput, high-resolution spectroscopy offered by Athena+ will allow for comprehensive survey of PNe X-ray emitting plasma, including elemental abundances that may provide evidence for the mixing of nebular and fast wind material.

References

Akashi, M., Meiron, Y., & Soker, N. 2008, NewA, 13, 563
Chu, Y.-H., Guerrero, M. A., Gruendl, R. A., Williams, R. M., & Kaler, J. B. 2001, ApJ, 553, L69
Freeman, M., Montez, R., Jr., Kastner, J. H., *et al.* 2014, ApJ, 794, 99
Freeman, M. J. & Kastner, J. H. 2016, ApJS, 226, 15
Guerrero, M. A., Gruendl, R. A., & Chu, Y.-H. 2002, A&A, 387, L1
Guerrero, M. A., Chu, Y.-H., Gruendl, R. A., Williams, R. M., & Kaler, J. B. 2001, ApJ, 553, L55
Kastner, J. H., Montez, R., Jr., Balick, B., *et al.* 2012, AJ, 144, 58
Kastner, J. H., Soker, N., Vrtilek, S. D., & Dgani, R. 2000, ApJ, 545, L57
Maness, H. & Vrtilek, S. D. 2003, PASP, 115, 1002
Montez, R., Jr., Kastner, J. H., Balick, B., *et al.* 2015, ApJ, 800, 8
Montez, R., Jr., De Marco, O., Kastner, J. H., & Chu, Y.-H. 2010, ApJ, 721, 1820-1828
Montez, R., Jr. & Kastner, J. H. 2013, ApJ, 766, 26
Montez, R., Jr., Kastner, J. H., De Marco, O., & Soker, N. 2005, ApJ, 635, 381
Nordon, R., Behar, E., Soker, N., Kastner, J. H., & Yu, Y. S. 2009, ApJ, 695, 834
Ruiz, N., Chu, Y.-H., Gruendl, R. A., *et al.* 2013, ApJ, 767, 35
Sandin, C., Steffen, M., Schönberner, D., & Rühling, U. 2016, A&A, 586, A57
Soker, N. & Kastner, J. H. 2003, ApJ, 583, 368
Steffen, M., Schönberner, D., & Warmuth, A. 2008, A&A, 489, 173
Yu, Y. S., Nordon, R., Kastner, J. H., *et al.* 2009, ApJ, 690, 440

Discussion

Muthumariappan: How do the hot bubble in [WR]PNe differ with normal PNe: Are they hotter and possess He, C abundances of their H-poor central star photosphere?

Barlow: To explain the excess X-ray emission from the Helix Nebular central star, are you predicting that it is a binary?

Montez: Yes, I think the X-ray emission is consistent with all the close binary X-ray characteristics so it suggests there is a companion.

GUZMAN-RAMIREZ: For the point source X-ray emission, how does the binary produce the hard X-ray emission? And do you need a binary to produce this?

MONTEZ: For late type spun up companions, the idea is that the convection and differential rotation is enhanced leading to increased magnetic activity and so flares and coronal emission in X-rays. A binary companion can do this but a CSPN cannot because they are convective.

MENDEZ: When you say "binary" do you mean specifically a close binary, with P < 1 day?

MONTEZ: Yes for a spun-up companion. But late type dwarfs can remain active formost of their lives so you could get away with a wide binary in that case but we should have found a companion to the Helix central star.

Planetary Nebulae: Multi-Wavelength Probes of Stellar and Galactic Evolution
Proceedings IAU Symposium No. 323, 2016
X. Liu, L. Stanghellini & A. Karakas, eds.

doi:10.1017/S1743921317001223

Analysis of the X-ray spectrum of the hot bubble of BD+30°3639

Detlef Schönberner, Ralf Jacob, René Heller† and Matthias Steffen

Leibniz-Institut für Astrophysik (AIP), An der Sternwarte 16, 14482 Potsdam, Germany
email: deschoenberner@aip.de

Abstract. We developed a model for wind-blown bubbles with temperature and density profiles based on self-similar solutions including thermal conduction. We constructed also heat-conduction bubbles with chemical discontinuities. The X-ray emission is computed using the well-documented CHIANTI code (v6.0.1). These bubble models are used to (re)analyse the high-resolution X-ray spectrum of the hot bubble of BD+30°3639, and they appeared to be much superior to constant temperature approaches.

We found for the X-ray emission of BD+30°3639 that temperature-sensitive and abundance-sensitive line ratios computed on the basis of heat-conducting wind-blown bubbles and with abundances as found in the stellar photosphere/wind can only be reconciled with the observations if the hot bubble of BD+30°3639 is chemically stratified, i.e. if it contains also a small mass fraction ($\simeq 3\,\%$) of hydrogen-rich matter immediately behind the conduction front. Neon appears to be strongly enriched, with a mass fraction of at least about 0.06.

Keywords. conduction, planetary nebulae: individual: BD+30°3639, stars: abundances, X-rays: stars

1. Introduction

The cavities of elliptical/roundish planetary nebulae (PNe) are not empty but instead filled-up with hot gas originating from the shock-heated fast stellar wind. Space-born X-ray telescopes (ROSAT, XMM-Newton, Chandra) revealed that these wind-blown "bubbles" are the source of rather soft X-ray emission, with typical plasma temperatures between 1 and 3 MK (see Kastner *et al.* 2012). The low plasma temperature observed is substantially *below* the post-shock temperature of about 10 MK expected from the high observed velocities of the stellar wind. Possible "cooling" processes are either heat conduction in the absence of magnetic fields (Soker 1994; Steffen *et al.* 2008) and/or dynamical mixing of matter across the bubble/nebula interface (Toalá & Arthur 2016). The latest 2D simulations by Toalá & Arthur (2016) suggest that heat conduction is indispensable for achieving the observed high X-ray emission measures.

The case of BD+30°3639 is special in three aspects: i) A hydrogen-poor wind interacts with a nebular shell of normal, hydrogen-rich composition; ii) it is the brightest PN X-ray source, allowing to take high-resolution spectra (Yu *et al.* 2009); iii) it is a young, still unevolved nebula, obviously shortly after the conversion to a hydrogen-poor central star has taken place. A detailed analysis of the X-ray spectrum with respect to elemental abundance ratios and the existence of a possible abundance discontinuity within the bubble produced by heat conduction is thus very rewarding.

All the existing spectral analyses of this nebula are based on simple plasma models with constant densities and temperatures. In their analysis of the bubble of BD+30°3639, Yu *et al.* (2009) and Nordon *et al.* (2009) needed a two-temperature plasma (1.9 and 3.0 MK) to account for the emission from species with different degree of ionisation. The derived chemical abundances appeared to be rather unusual: very high ratios of C/O and Ne/O, exceeding by far the corresponding solar ratios.

† Present address: Max-Planck-Institut für Sonnensystemforschung, Justus-von-Liebig-Weg 3, 37077 Göttingen, Germany

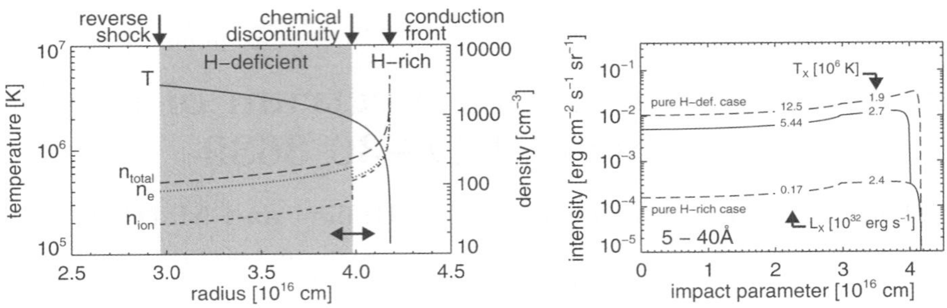

Figure 1. *Left*: physical structure of a ZP96 bubble of age = 500 yr with a stratified composition such that $\omega \equiv M_{\rm PN}/(M_{\rm WR} + M_{\rm PN}) = 0.03$. Shown are radial runs of electron, ion, and total particle densities, and the electron temperature. The reverse wind shock, the chemical discontinuity, and the conduction front are marked by vertical arrows. The condition of constant pressure across the bubble leads to the density jumps at the chemical boundary at $r \simeq 4{\times}10^{16}$ cm, or 2 MK. *Right*: the model's X-ray surface brightness integrated over 5–40 Å. The dashed curves serve for comparisons with the homogeneous WR (H-poor) and PN (H-rich) cases only. Remarkably is the high emission of the WR matter which is exclusively residing in the hotter bubble regions. Thus, $T_{\rm X}$ is increased from 1.9 to 2.7 MK for the $\omega = 0.03$ case shown here, while the X-ray luminosity is more than halved because of the low PN-matter emissivity.

From the physical point of view, such a low-temperature plasma is fully inconsistent with its outer and inner boundary conditions, viz. the nebular shell and the fast stellar wind. In the following we report briefly on our new approach to analyse the X-ray emission from hot bubbles by means of a physically more sound model. Details will be published in a forthcoming journal paper.

2. Bubble models

Our newly developed analysis tool is based on self-similar solutions for wind-blown bubbles formed by the interaction of two spherical winds with different densities and velocities, as developed by Zhekov & Perinotto (1996, 1998). These so-called ZP96-bubbles contain heat conduction across the bubble from the reverse wind shock to the contact discontinuity/conduction front. The general properties of these bubbles can be characterised as follows:

Inner boundary condition. Power-law representation of the fast stellar wind from an evolving 0.595 $\rm M_\odot$ central star, time-dependent mass-loss rate and velocity, adapted for BD+30°3639 according to Sandin *et al.* (2016).

Outer boundary condition. Constant slow wind with explored parameter ranges $10^{-7} \ldots 10^{-4}$ $\rm M_\odot\,yr^{-1}$ and $10 \ldots 40$ $\rm km\,s^{-1}$.

Chemical composition. Each bubble has either a hydrogen-deficient ("WR"; Marcolino *et al.* 2007) or hydrogen-rich ("PN") chemical composition, with appropriate conduction coefficients. Also, chemically inhomogeneous bubbles are constructed to allow for "evaporated" PN matter behind the conduction front. The WR composition is extremely helium-, carbon-, and oxygen-rich: He:C:O = 0.43 : 0.51 : 0.06 by mass (Marcolino *et al.* 2007).

X-ray emission. The X-ray spectrum is computed by means of the well-documented CHIANTI code, v6.0.1 (Dere *et al.* 2009).

Altogether, about 1000 bubbles have been generated, spanning ages from 200 to 1000 years, corresponding characteristic X-ray temperatures, $T_{\rm X}$ (Eq. 17 in Steffen *et al.* 2008), between 1.1 and 3.9 MK (WR) or between 1.1 and 6.8 MK (PN). WR-bubbles with ages of 400–500 yr have sizes and X-ray luminosities which correspond well to the respective values observed for BD+30°3639. The structure of a bubble with a chemical discontinuity is displayed in Fig. 1.

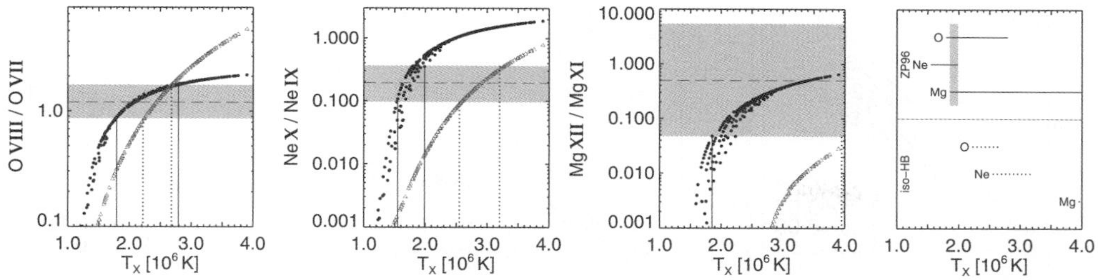

Figure 2. Line ratios for two ionisation stages of oxygen (*left*), neon (*next to left*), and magnesium (*next to right*) vs. T_X as predicted by our ZP96 bubble models with hydrogen-deficient WR composition (dots) and constant-temperature bubbles (triangles) with the same chemical composition. The horizontal long-dashed lines give the measured values for BD+30°3639, the shaded regions indicate their uncertainties. The *right* panel summarises the T_X determinations from the various line ratios and for the ZP96-bubbles and the constant-temperature approach.

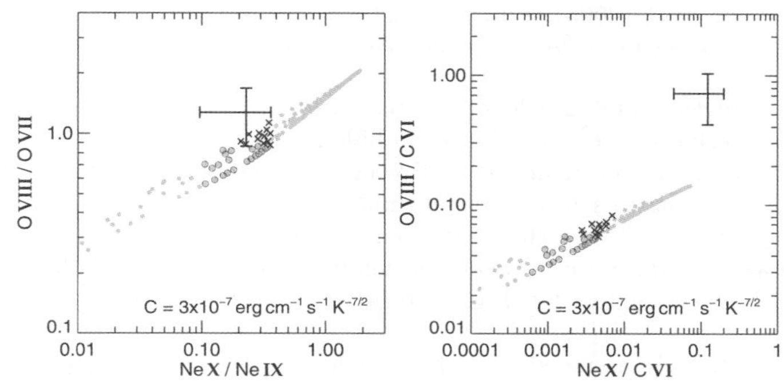

Figure 3. Temperature sensitive line ratios of oxygen and neon (*left*) and abundance sensitive line ratios (*right*) as predicted by bubbles with homogeneous WR composition and observed values for BD+30°3639 (big error crosses). Open symbols represent bubbles which fulfil the neon criterion, while crosses belong to those bubbles that fulfil additionally the oxygen temperature criterion.

3. Application to the case of BD+30°3639

Our analysis of the X-ray emission from BD+30°3639's bubble rests exclusively on the high-resolution observations reported in Yu *et al.* (2009). We first used our ZP96-bubbles with WR composition to fix the characteristic plasma temperature T_X. The result is seen in Fig. 2.

One sees that all bubbles nearly degenerate so that a unique plasma temperature can easily be derived from the oxygen and neon line ratios: $T_X = 1.9^{+0.3}_{-0.2}$ MK. This temperature value, obtained from the ZP96 bubbles, is also marginally consistent with magnesium. We emphasise that constant-temperature plasma models *cannot* provide a unique value for T_X (Fig. 2, right)!

For the further analysis, we have to check whether those bubbles from our grid that have the right temperature T_X can also reproduce the observed line ratios of carbon and neon with respect to oxygen. However, it turns out that *none* of our bubbles with homogeneous WR composition is able to reproduce the observed line ratios Ne/C and O/C (Fig. 3). The discrepancies amount to one order of magnitude, at least!

The large discrepancy between models and observation seen in the right line ratio diagram in Fig. 3 cannot be remedied by changing the input abundances of our WR composition because they are fixed by the analyses of the stellar wind of BD+30°3639 (e.g. Marcolino *et al.* 2007), with the exception of neon whose abundance has been chosen by us. Instead, one has to conclude that the bubble of BD+30°3639 contains, as the consequence of "evaporation", some hydrogen-rich PN matter immediately behind the conduction front. The PN-matter contains much less carbon (and oxygen as well), so that even small amounts of PN matter will drastically change

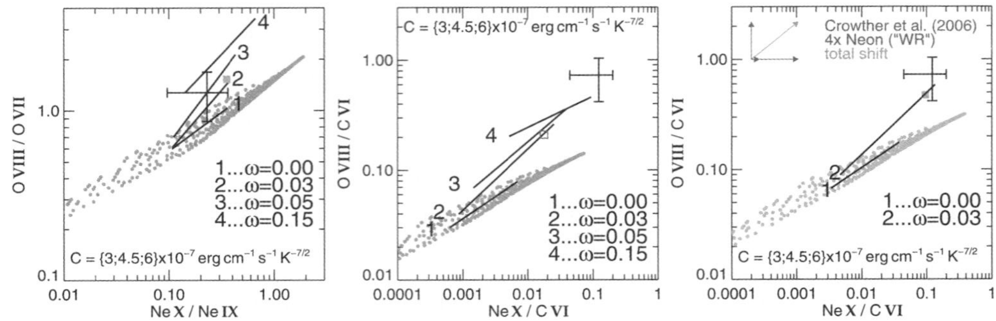

Figure 4. *Left and middle*: same as in Fig. 3 but with chemically inhomogeneous bubbles (black bars) added, labelled according to their ω-values. The bars represent the mean positions of all bubbles with the same ω. The square seen in all panels represent our "best-fit" model from the $\omega = 0.03$ cloud (cf. Fig. 1). *Right*: positions of all models assuming a higher oxygen/carbon ratio as found by Crowther *et al.* (2006) and the neon/oxygen ratio increased by a factor of 2.5. The arrows indicate the shifts caused by the respective abundance changes.

the carbon line strengths, and to a lesser degree those of oxygen and neon lines because the corresponding ions reside in the inner, hotter bubble regions.

Figure 4 illustrates how the inclusion of PN matter changes the relevant line ratios. The discrepancy in the abundance sensitive panel (middle) is substantially reduced, although there is still some disagreement between the models and the observations. The square represents our "best-fit" model as a compromise between the temperature criteria and the abundance ratios: $\omega = 0.03^{+0.02}_{-0.01}$ with age = 500 yr (cf. Fig. 1), together with appropriate bubble size and X-ray luminosity.

So far we have implicitly assumed that the bubble's Ne abundance, which is not really constrained by photospheric/wind analyses, comes from the complete processing of CNO matter into N, followed by conversion to Ne during a thermal pulse on the AGB. This leads to a mass fraction of 0.022. By comparing the observed Ne/O line ratio with the model predictions, we conclude that the Ne abundance in the bubble of BD+30°3639 must be higher by a factor of about 2.5 (compared to oxygen), i.e. we have about equal amounts (by mass) of O and Ne (at least $\simeq 0.06$)! Combined with a higher O/C ratio as proposed by Crowther *et al.* (2006), full consistency between both the temperature sensitive and abundance sensitive diagrams in Fig. 4 is achieved. In this case, the Ne content would be even higher (up to 0.09 by mass).

4. Summary and conclusion

We have developed a new tool for analysing the X-ray emission from wind-blown bubbles with thermal conduction. Application to existing high-resolution data for BD+30°3639 shows that its X-ray line emission can only be explained if the bubble contains i) helium-, carbon-, and oxygen-enriched matter as found in the stellar photosphere/wind, especially enriched by neon, and ii) a shell of hydrogen-rich PN matter of about 3 % by mass behind the conduction front.

It appears that heat conduction is a viable option for explaining the diffuse X-ray emission from wind-blown bubbles. More high-resolution X-ray observations are urgently needed for further testing our analytical bubble models. Of course, new analyses of existing low-resolution X-ray spectra would also benefit from our heat-conduction chemically stratified bubble models.

References

Crowther, P. A., Morris, P. W., & Smith, J. D. 2006, *ApJ*, 636, 1033

Dere, K. P., Landi, E., Young, P. R., *et al.* 2009, *A&A*, 498, 915

Kastner, J. H., Montez, R.,Jr., Balick, B., *et al.* 2012, *AJ*, 144, 48

Marcolino, W. L. F., Hillier, D. J., de Araujo, F. X., & Pereira, C. B. 2007, *ApJ*, 654, 1068

Nordon, R., Behar, E., Soker, N., Kastner, J. H., & Yu, Y. S. 2009, *ApJ*, 695, 834
Sandin, C., Steffen, M., Schönberner, D., & Rühling, U. 2016, *A&A*, 586, A57
Soker, N. 1994, *AJ*, 107, 276
Steffen, M., Schönberner, D., & Warmuth, A. 2008, *A&A*, 489, 173
Toalá, J. A. & Arthur, S. J. 2016, *MNRAS*, 463, 4438
Yu, Y. S., Nordon, R., Kastner, J. H., *et al.* 2009, *ApJ*, 690, 440
Zhekov, S. A. & Perinotto, M. 1996, *A&A*, 309, 648
Zhekov, S. A. & Perinotto, M. 1998, *A&A*, 334, 239

Discussion

De Marco: Some born agains/[WR] have Ne mass fraction well above the value expected (few percent by mass). Is the Ne mass fraction were high enough to suspect a high mass progenitor? (higher than $\sim$2 $M_\odot$)?

Schoenberner: No. The mass fraction of Ne in BD +30 3639 is $\sim$6% by mass and can be explained by evolution with overshoot in a $\sim$3 $M_\odot$ star.

Torres-Peimbert: How important is the mixing of the envelope and the stellar wind at the conductive interface? Will it affect the overall nebula?

Schoenberner: Heat conduction across the bubble/nebula interface has an egligible influence on the nebular evolution. The evaporated mass is very small for the nebula, but important for the bubble to provide the observed X-ray luminosity, see Steffen *et al.* (2008, A&A, 489, 173).

Planetary Nebulae: Multi-Wavelength Probes of Stellar and Galactic Evolution
Proceedings IAU Symposium No. 323, 2016
X. Liu, L. Stanghellini & A. Karakas, eds.

doi:10.1017/S1743921317002320

Revealing the Location of the Mixing Layer in a Hot Bubble

Guerrero, M. A.[1], **Fang, X.**[1,2], **Chu, Y.-H.**[3] **Toalá, J. A.**[1,3] **and Gruendl, R. A.**[4]

[1]Instituto de Astrofísica de Andalucía (IAA-CSIC), Glorieta de la Astronomía s/n, E-18008 Granada, Spain
email: mar@iaa.es

[2]Department of Physics, The University of Hong Kong, Pokfulam Road, Hong Kong
email: fang@hku.hk

[3]Institute of Astronomy and Astrophysics, Academia Sinica (ASIAA), Taipei 10617, Taiwan
email: yhchu@asiaa.sinica.edu.tw, toala@asiaa.sinica.edu.tw

[4]Department of Astronomy, University of Illinois, 1002 West Green Street, Urbana, IL 61801, USA
email: gruendl@astro.illinois.edu

Abstract. The fast stellar winds can blow bubbles in the circumstellar material ejected from previous phases of stellar evolution. These are found at different scales, from planetary nebulae (PNe) around stars evolving to the white dwarf stage, to Wolf-Rayet (WR) bubbles and up to large-scale bubbles around massive star clusters. In all cases, the fast stellar wind is shock-heated and a hot bubble is produced. Processes of mass evaporation and mixing of nebular material and heat conduction occurring at the mixing layer between the hot bubble and the optical nebula are key to determine the thermal structure of these bubbles and their evolution. In this contribution we review our current understanding of the X-ray observations of hot bubbles in PNe and present the first spatially-resolved study of a mixing layer in a PN.

Keywords. stars: winds, outflows – (ISM:) planetary nebulae: individual (NGC 6543) – ultraviolet: ISM

---

1. Introduction

Planetary nebulae (PNe) are the progeny of low- and intermediate-mass stars with initial masses $\leqslant$8–10 $M_{\odot}$. As such stars ascend the asymptotic giant branch (AGB), they experience episodes of heavy mass loss through a slow (v_{∞} ~10 km s^{-1}) wind. Once the stellar envelope is lost and the stellar core is exposed, the stellar wind increases its terminal velocity up to 1000–4000 km s^{-1} (Cerruti-Sola & Perinotto 1985; Guerrero & de Marco 2013). This fast wind sweeps up the slow AGB wind, which is further photoionized by the central star (CSPN) to form a PN (Kwok 1983; Frank *et al.* 1990).

The interaction between these two stellar winds produces hot gas that fills a hot bubble in the inner region of the PN. Extended X-ray emission has now been detected inside the inner cavities of nearly 30 PNe with plasma temperatures of (1–3)$\times$10^{6} K and electron densities of 1–10 cm^{-3} (e.g., Guerrero *et al.* 2000, 2002, 2005; Kastner *et al.* 2000, 2001, 2008, 2012; Chu *et al.* 2001; Freeman *et al.* 2014). The *"low"* temperatures and *"high"* densities of the X-ray-emitting plasma in PNe, as compared to those predicted by simple models of adiabatically shocked stellar winds, suggests that some mechanism is reducing the temperature of the hot bubble and raising its density. Thermal conduction (Soker 1994; Steffen *et al.* 2008, and references therein) and/or hydrodynamical instabilities (e.g., Toalá & Arthur 2014, 2016) in the windwind interaction zone can inject material

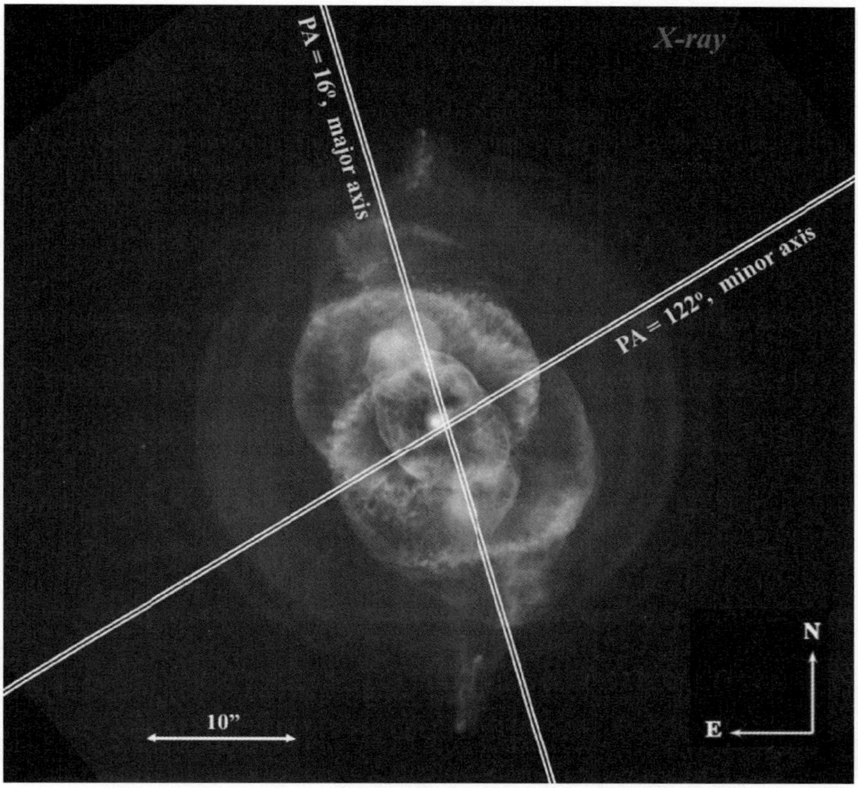

Figure 1. *HST* optical (red and purple) and *Chandra* X-ray (blue) color–composite image of NGC 6543. Image adopted from the *Chandra* X-ray Center (http://chandra.harvard.edu/photo/2008/catseye/). The positions of the *HST* STIS $52'' \times 0.2''$ long slit are marked with white lines.

into the hot bubble, creating a mixing layer of gas with intermediate temperatures ($\sim 10^5$ K) between the hot bubble and the optical nebular shell.

X-ray observations of NGC 6543 (a.k.a. the Cat's Eye Nebula) reveal a physical structure qualitatively consistent with the ISW models (Chu *et al.* 2001). The *Chandra* image of NGC 6543 (Figure 1) shows simple limb-brightened diffuse X-ray emission confined within the bright inner shell and two blisters at the tips of its major axis, in sharp contrast to its complex optical morphology (Balick 2004). The observed X-ray temperature (1.7×10^6 K; Chu *et al.* 2001) is much lower than that expected for a stellar wind of $v_\infty \sim 1400$ km s^{-1} (Prinja *et al.* 2007). The morphology and physical conditions of the hot plasma are suggestive of the presence of a mixing layer in NGC 6543.

2. Observations and Data Analysis

HST STIS UV and optical spectroscopic observations of NGC 6543 (PI: M. A. Guerrero, GO prop. ID 12509, Cycle 19) were carried out on 2012 July 3 and October 21. The observations aimed to detect and trace the spatial extent of the interface layer and compare it with the layers of the nebular shell and hot bubble. The $52'' \times 0.2''$ long slit was placed at position angles (PAs) of 16° and 122° along the major and minor axes of the inner nebular shell (Figure 1), respectively. The G140M grating and STIS/FUV-MAMA

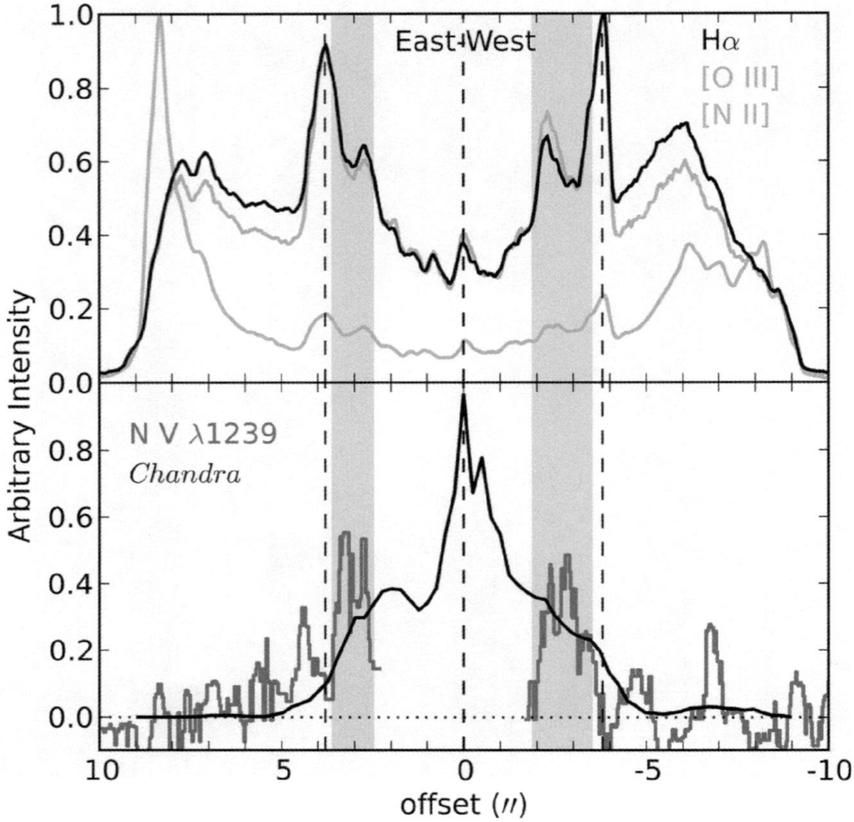

Figure 2. Spatial emission profiles of the Hα, [O III] λ5007, and [N II] λ6583 emission lines from the optical nebular shell (top panel), and the N V λ1239 UV resonance line from the interface layer and the *Chandra* X-ray emission from the hot bubble (bottom panel) along the minor axis of NGC 6543 at PA 122° (Figure 1). The offset is relative to the CSPN. Gray shaded areas mark the positions of the interface layer, as indicated by the N V line.

detector were used to acquire spectra of the N V λλ1239,1243 and C IV λλ1548,1551 emission lines. The G430L and G750M gratings and STIS/CCD detector were used to obtain information of the [O III], Hα, and [N II] lines from the optical nebular shell.

The two-dimensional (2D) STIS spectra were used to extract spatial profiles of emission along the minor axis of the photoionized innermost nebular shell for the Hα, [O III] λ5007, and [N II] λ6583 lines (Figure 2, top panel) and the collisionally excited N V λ1239 UV line (Figure 2, bottom panel). The spatial profile of the X-ray emission, as derived from the *Chandra* observations, is also plotted in the bottom panel of Figure 2.

3. Results and Discussion

The spatial profiles along the minor axis of NGC 6543 (Figure 2) reveal the location of the mixing layer gas originating from very different processes. The brightest emission peaks in Hα and [O III] at $\sim 4''$ from the CSPN mark the location of the $\sim 10^4$ K swept-up inner nebular shell. On the other hand, the profile of the X-ray emission from the $\geqslant 10^6$ K hot gas shows an eastern peak at $\sim 2''$ and a shoulder of declining emission toward the west. This irregular profile suggests that the X-ray-emitting gas is confined within a region of radius $\leqslant 3''$. Therefore, the X-ray-emitting gas is confined within the cool nebular shell.

Interestingly, the spatial profile of the N V emission peaks at intermediate positions (gray shades in Figure 2), between the optical lines from the optical nebular shell and the X-ray emission from the hot bubble. We remind that the N V ion cannot be produced by photoionization because the effective temperature of the CSPN of NGC 6543 is 50,000 K; thus it must be produced by thermal collisions at temperatures of $\sim 10^5$ K, as expected for the mixing layer.

The spatial profile of the N V λ1239 emission line has been used to estimate an outer radius and thickness of the mixing layer of $\sim 3.7''$ and $1.0''$, respectively. At a distance of 1.0$\pm$0.3 kpc (Reed *et al.* 1999), this implies a thickness of 1.5×10^{16} cm.

The density of the mixing layer can also be estimated to be ~ 180 cm^{-3}, assuming a simple geometry for the N V λ1239-emitting region and a N^{4+}/H^+ ionic abundance ratio close to the nebular nitrogen abundance (N/H) of 2.3×10^{-4} (Bernard-Salas *et al.* 2003). This density and the adopted temperature of 2×10^5 K imply a thermal pressure of 2×10^{-8} dyn cm^{-2} in the mixing layer, which agrees with the pressure of the hot bubble and the ionized swept-up shell (Gruendl *et al.* 2004).

To sum up, we have presented high-spatial resolution *HST* STIS UV and optical spectroscopy of the Cat's Eye Nebula (NGC 6543). These STIS observations have enabled the first view of the spatial distribution of the mixing layer gas, as probed by the collisionally ionized N V UV emission line. The mixing layer is located exactly between the optical nebular rim and the X-ray-emitting hot bubble as previously detected by *Chandra*. The hot gas in the mixing layer is in hydrodynamical equilibrium with the hot bubble and the ionized nebular rim. More extended details of this work are reported in Fang *et al.* (2016).

References

Balick B. 2004, AJ, 127, 2262

Bernard-Salas J., Pottasch S. R., Wesselius P. R., Feibelman W. A. 2003, A&A, 406, 165

Cerruti-Sola M. & Perinotto M. 1985, ApJ, 291, 237

Chu Y.-H., Guerrero M. A., Gruendl R. A. *et al.* 2001, ApJL, 553, L69

Fang, X., Guerrero, M. A., Toalá, J. A., Chu, Y.-H., & Gruendl, R. A. 2016, ApJL, 822, L19

Frank A., Balick B., Riley J. 1990, AJ, 100, 1903

Freeman M., Montez R. J.r., Kastner J. H. *et al.* 2014, ApJ, 794, 99

Gruendl R. A., Chu Y.-H., Guerrero M. A., 2004, ApJL, 617, L127

Guerrero M. A., Chu Y.-H., Gruendl R. A. 2000, ApJS, 129, 295

Guerrero M. A., Chu Y.-H., Gruendl R. A., Meixner M. 2005, A&A, 430, L69

Guerrero M. A., de Marco O. 2013, A&A, 553, A126

Guerrero M. A., Gruendl R. A., Chu Y.-H. 2002, A&A, 387, L1

Kastner J. H., Montez R. Jr, Balick B., & De Marco O. 2008 ApJ 672 957

Kastner J. H., Montez R. J.r, Balick B. *et al.* 2012, AJ, 144, 58

Kastner J. H., Soker N., Vrtilek S. D., Dgani R. 2000, ApJL, 545, L57

Kastner J. H., Vrtilek S. D., Soker N. 2001, ApJ, 550, 189

Kwok S. 1983 Proc. IAU Symp. 103, Planetary Nebulae ed D. R. Flower, 293 (Dordrecht: D. Reidel)

Prinja R. K., Hodges S. E., Massa D. L., Fullerton A. W., Burnley A. W. 2007, MNRAS, 382, 299

Reed D. S., Balick B., Hajian A. R. *et al.* 1999, AJ, 118, 2430

Soker N. 1994, AJ, 107, 276

Steffen M., Schönberner D., Warmuth A. 2008, A&A, 489, 173

Toalá J. A., Arthur S. J. 2014, MNRAS, 443, 3486
Toalá J. A., Arthur S. J. 2016, MNRAS, 463, 4438

Discussion

Q: Refer to the "mixing layer" that you showed in a slide. What is mixing with what? As time proceeds, does some of the colder gas in the shell find itself inside the hot bubble?

SCHOENBERNER: How do you know that mixing is at work and not heat conduction? The hydro-models give comparable temperatures and densities immediately behind the conductive front.

Session 3: PNe as a versatile laboratory II
Chair: Alberto Lopez

Planetary Nebulae: Multi-Wavelength Probes of Stellar and Galactic Evolution
Proceedings IAU Symposium No. 323, 2016
X. Liu, L. Stanghellini & A. Karakas, eds.

doi:10.1017/S1743921317002034

Dust in planetary nebulae

G. C. Sloan[1,2,3]

[1]Space Telescope Science Institute, 3700 San Martin Dr., Baltimore, MD 21218, USA
[2]Department of Physics and Astronomy, University of North Carolina, Chapel Hill, NC 27599-3255, USA
[3]Cornell Center for Astrophysics and Planetary Science, Cornell University, Ithaca, NY 14850-6801, USA

Abstract. Infrared spectra from the *Spitzer Space Telescope* trace the evolution of carbon-rich dust from the asymptotic giant branch (AGB) to young planetary nebulae (PNe). On the AGB, amorphous carbon dominates the dust, but SiC and MgS also appear. In more evolved systems with warmer central stars, the spectra reveal the unidentified 21 μm feature, features from aliphatic hydrocarbons, and spectra from polycyclic aromatic hydrocarbons (PAHs), often with shifted feature positions indicative of the presence of aliphatics. More evolved systems with hot central stars show more typical PAH spectra, along with fullerenes and/or an emission feature known as the big-11 feature at $\sim$11 μm. This features arises from a combination of SiC and PAHs, and it is usually accompanied by a shoulder at 18 μm, which while unidentified might be from cool silicate grains. The strong emission from MgS and SiC in young PNe probably arises from coatings on carbonaceous grains.

Keywords. stars: AGB and post-AGB, planetary nebulae: general, dust

1. Introduction

In an effort to better understand the role of metallicity in the death of intermediate-mass stars, several projects used the Infrared Spectrograph (IRS; Houck *et al.* 2004) on the *Spitzer Space Telescope* (Werner *et al.* 2004) to obtain infrared spectra of post-main-sequence objects in the Large and Small Magellanic Clouds (LMC and SMC). The observed sample included dozens of post-AGB objects and planetary nebulae (PNe). Carbon-rich dust dominated the sample, and the nature of that dust will be the focus of this contribution.

The story of carbon-rich dust in PNe begins with the discovery of the unidentified infrared (UIR) emission features in the spectra of two young Galactic PNe by Gillett *et al.* (1973). The carrier of these spectral features is best described as polycyclic aromatic hydrocarbons (PAHs; Leger & Puget 1984; Allamandola, Tielens, & Barker 1985), because it must consist of hydrocarbons, those hydrocarbons need to be aromatic to withstand the harsh radiation fields where they are found, and they must contain enough carbon atoms to be at the boundary between large molecules and very small dust grains. Aromaticity requires that the hydrocarbons are built of many benzene rings, and the size constraint means that they must be polycyclic. Thus, they must be polycyclic, predominantly aromatic, and hydrocarbon, or, to put it simply, PAHs. (See Allamandola, Tielens, & Barker 1989 for a more thorough argument.)

PNe excite PAH emission in a thin photo-dissociation region at the boundary between an ionized bubble and the surrounding dust shell produced when the central star was still on the asymptotic giant branch (AGB). As the ionized region expands into the circumstellar dust shell, the PNe slowly but steadily destroys the dust grains. The PAHs are just the final stage for that dust as the ionization front reaches it.

The IRS sample of evolved stellar objects in the Magellanic Clouds gives us insights into the nature of carbon-rich dust and how it changes as the central star from a carbon star on the AGB to a hot white dwarf at the center of a mature PN. Sloan *et al.* (2014; hereafter S14) examined the full IRS sample of 27 carbon-rich post-AGB objects in the LMC and 15 in the SMC.

2. The asymptotic giant branch

The AGB is the starting point for the dust. To investigate the large samples of carbon-rich AGB stars observed by the IRS, Sloan *et al.* (2006) and Zijlstra et al. (2006) introduced the Manchester Method. Amorphous carbon dominates the dust they produce, but it does not generate any identifiable emission features in the infrared. The Manchester Method uses the [6.4]−[9.3] color, which samples the "continuum" from star plus dust in spectral regions unaffected by absorption bands or emission features, as a proxy for the dust-production rate. The [6.4]−[9.3] color reddens as the spectral contribution from amorphous carbon dust grows and buries the contribution from the central star. At the same time, the emission feature from SiC dust at ~11.3 μm increases in strength, but the strongest SiC emitters have largely vanished at redder [6.4]−[9.3] colors where MgS dust emission at ~30 μm grows strong.

This behavior suggests that MgS, which has a relatively low condensation temperature, condenses onto pre-existing grains in more optically thick dust shells (Lagadec *et al.* 2007; Leisenring *et al.* 2008), generating a MgS feature as it masks the emission from SiC. Zhukovska & Gail (2008) reached a similar conclusion independently from a theoretical perspective.

MgS has an interesting story. Forrest *et al.* (1981) first discovered the 30 μm emission feature in spectra of carbon-rich PNe and dust-enshrouded objects like IRC +10216 from the *Kuiper Airborne Observatory.* Goebel & Moseley (1985) identified the carrier as MgS. In some PNe, the 30 μm feature is so strong, MgS cannot be the carrier without violating abundance constraints, at least if the grains are solid MgS (Zhang *et al.* 2009). However, Lombaert et al. (2012) showed that a layer of MgS can reproduce the observed 30 μm feature and avoid the abundance problem, and as pointed out the IRS observations suggest exactly this geometry. S14 considers other objections to MgS and concludes that it is still a strong candidate for the 30 μm carrier.

3. Aliphatic hydrocarbons

If PAHs are the form that carbonaceous grains take when they are heavily photo-processed in a UV field, what do the grains look like in milder conditions? Peeters *et al.* (2002) provided some important clues by classifying all of the PAH spectra observed with the Short-Wavelength Spectrometer (SWS) on the *Infrared Space Observatory* into three groups. Figure 1 illustrates these, plus some newer classifications. The Class A spectrum is typical for H II regions around young stars, and Class B is typical for PNe. The spectra differ most notably in the 7–9 μm region, with Class A PAHs showing a feature at 7.65 μm, while in Class B PAH spectra, the feature at 7.85 μm is stronger. In two spectra, both post-AGB objects, an emission feature at ~8.2 μm dominates the emission at 7.65 and 7.85 μm.

The IRS observed several more Class C PAH spectra, all of which are excited by central objects with temperatures $\lesssim$ 6500 K (Sloan *et al.* 2007). The PAH features in the 3 μm regime provide more clues. Usually, the 3.29 μm C–H stretching mode from PAHs dominates the 3.40 μm feature attributed to an aliphatic C–H stretching mode, but some post-AGB objects show much stronger 3.40 μm features (e.g. Geballe & van

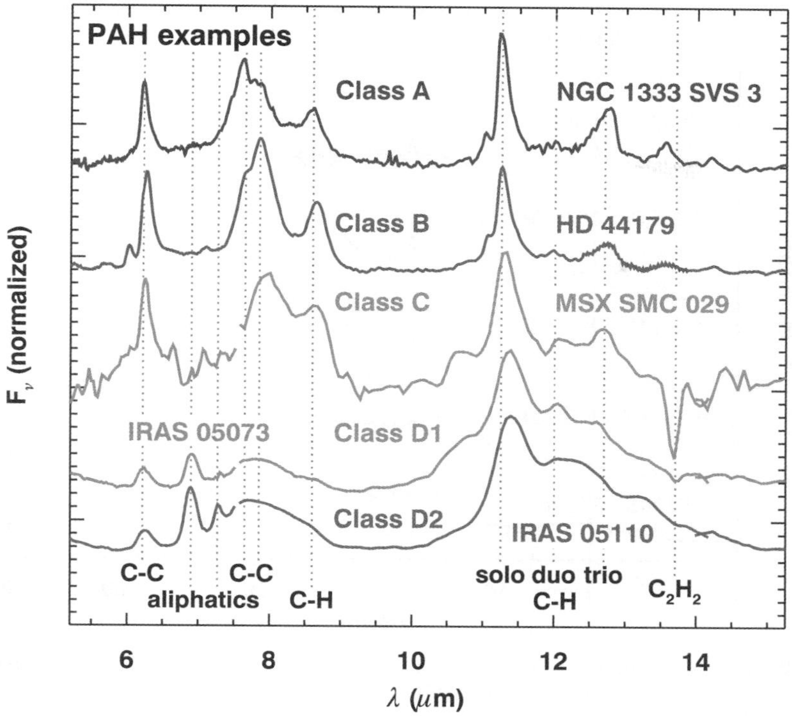

Figure 1. Sample spectra for each PAH class. An in-plane C–H bending mode produces the 8.6 μm feature, while out-of-plane C–H bending modes produce the 11.2, 11.9, and 12.7 μm features, with the number of adjacent H atoms in an aromatic ring determining the wavelength. Aliphatic hydrocarbons produce the features at 6.85 and 7.25 μm, and acetylene absorption appears at 13.7 μm in many post-AGB objects.

der Veen 1990; Geballe *et al.* 1992), in one case even in absorption (Chiar *et al.* 1998). Together, the evidence suggests that aliphatic hydrocarbons can survive alongside PAHs if the radiation field is cool enough. In the hotter radiation fields associated with Class A and Class B PAHs, the relatively fragile aliphatics are destroyed. Laboratory analysis by Pino *et al.* (2008) confirmed that the presence of aliphatics shifts the PAH features to longer wavelengths. When aliphatics can survive, we observe the PAHs as a Class C spectrum.

4. The 21 μm sources

Carbon-rich objects which have evolved past the AGB show a rich variety of spectra, as Figure 2 illustrates. Post-AGB objects can show the 21 μm feature, which was discovered by Kwok *et al.* (1989) in the Galaxy. Volk *et al.* (2011) found several more sources with the still-unidentified 21 μm feature in the Magellanic Clouds. Figure 3 plots two typical spectra. The 21 μm feature always appears with several other features, although any given feature may be absent from a given spectrum. Generally, most of the 21 μm spectra show emission features at 6.85 and 7.25 μm attributed to aliphatic hydrocarbons. They also show features at 15.4 and 17.1 μm which S14 attributes to aliphatic hydrocarbons with triple C–C bonds. Chains which end with this bond produce the 15.4 μm feature. Displacing the triple bond one atom inward shifts the feature to 17.1 μm. A triple bond suggests some dehydrogenation, which could indicate increasing photoprocessing as the central source reveals itself.

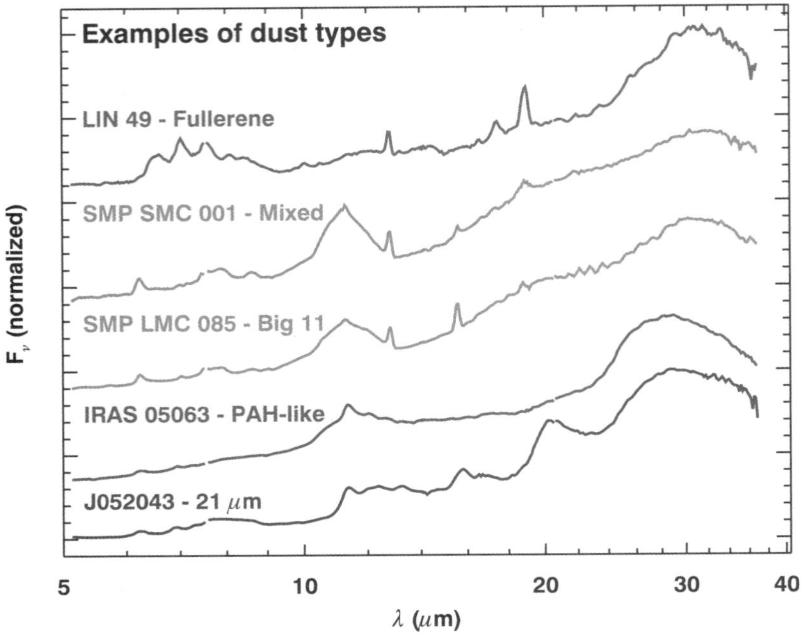

Figure 2. Examples of the well-populated classes of dust spectra identified by S14 in the sample of carbon-rich post-AGB spectra in the Magellanic Clouds.

Cerrigone *et al.* (2011) noted an association between the 21 μm feature and PAHs, but the additional evidence points more to aliphatics. An association does not necessarily imply an identification; it could be that the carrier of the 21 μm feature appears in conditions which are also favorable for aliphatic hydrocarbons.

Matsuura *et al.* (2014) introduced a new class of PAH spectra, Class D (Figure 1). In Class D PAHs, emission fills what is usually a valley between the C–C modes at 6–8 μm and the 8.6 μm in-plane C–H bending mode. In addition, the shape of the 11–13 μm PAH complex changes, with a much stronger contribution between 11.2 and 12.7 μm. S14 broke the Class D PAHs into two groups, with D2 distinguished from D1 by a shift of the 11.9 and 12.7 μm features to ~12.4 and 13.2 μm. Most of the 21 μm sources are associated with some form of Class D PAH emission. As Figure 3 shows, the classification at 6–9 μm does not always agree with that at 11–13 μm. What produces Class D PAH spectra is not known, but aliphatic hydrocarbons seem to be a likely suspect.

5. Young carbon-rich PNe

Young PNe show spectra that S14 described as “Fullerene,” “Mixed,” and “Big-11”. Cami (2017) will address the fullerene sources in a separate contribution to these proceedings. The fullerene class blends smoothly into the mixed class, which show features from both fullerenes and PAHs (typically Class B). These spectra blend smoothly into the Big-11 class, which shows a strong feature at ~11 μm (hence the name) and a shoulder at ~18 μm.

Figure 4 illustrates two Big-11 spectra. S14 showed that the big-11 feature is mostly SiC emission, with some PAH emission superimposed. Bernard-Salas *et al.* (2009) noted that if the big-11 feature were SiC, it would be puzzling why it is so much stronger in Magellanic PNe than in Galactic PNe, given the lower metallicities in the LMC and SMC.

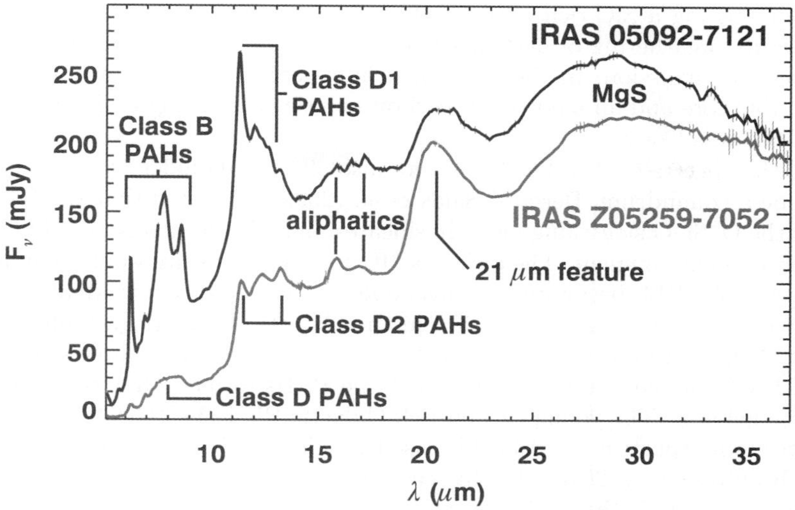

Figure 3. The IRS spectra of two spectra with 21 μm feautures. Both spectra also show strong MgS emission at $\sim$30 μm, and they show features at 15.4 and 17.1 μm which S14 have identified with aliphatics. They also show unusual PAHs, with IRAS Z05259 showing Class D2 PAHs and IRAS 05092 showing a mixture of Class D1 and Class B PAHs. These combinations of different PAHs at different wavelengths are not uncommon in the 21 μm sources.

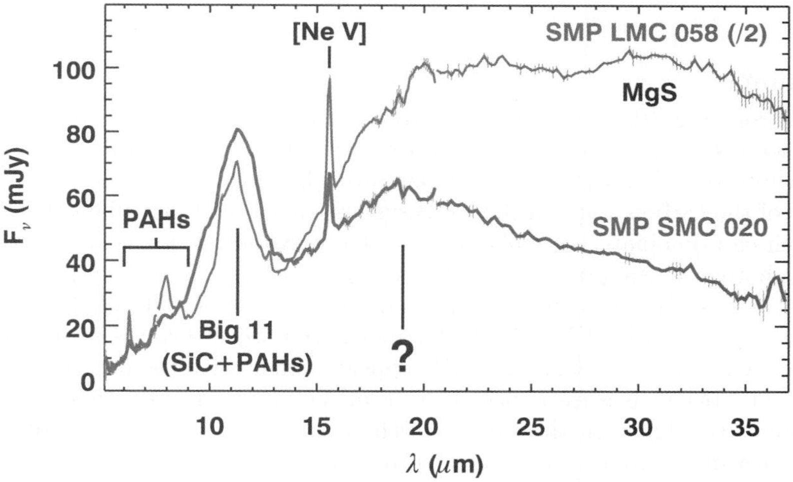

Figure 4. The IRS spectra of two Big-11 sources. The spectrum of SMP LMC 058 (divided by two here) is typical, with clear PAH emission in the 6–9 μm region, a big-11 feature from a combination of SiC and PAHs, a shoulder in the spectrum $\sim$18 μm, and an MgS feaure at $\sim$30 μm. The spectrum of SMP SMC 020 is a more extreme case, with a much more pronounced big-11 feature due to a lack of cool dust. The carrier of the 18 μm shoulder is not known, but if it were really cool silicates, as originally suggested by Bernard-Salas *et al.* (2009), that raises some interesting possibilities.

S14 found that the big-11 feature is dominated by SiC, with a smaller contribution from PAHs, and to solve the abundance problem, they suggested that the feature could arise from a layer of SiC on the grains, much like MgS. If grains composed of both amorphous carbon and SiC are exposed to a hot radiative environment, as happens in young PNe, one might expect the aliphatic hydrocarbons to be removed preferentially on the surface of the grain, leaving a SiC-rich coating which would enhance the 11.3 μm SiC feature. In other words, weathering of carbonaceous grains in harsh radiation fields would lead to enhanced SiC emission. In the Galaxy, it is expected that the carbonaceous grains contain an SiC core and an amorphous carbon mantle (Leisenring *et al.* 2008), making a coating of SiC less likely.

Most of the spectra with a big-11 feature also show a shoulder at 18 μm, which has proven to be a conundrum. Bernard-Salas *et al.* (2009) suggested that the feature could arise from the O–Si–O stretching mode in silicates, but silicate grains would be surprising in a carbon-rich environment. The lack of a 10 μm silicate feature would require that the silicates were cold. S14 discounted the likelihood, as it would require dual chemistry.

PNe with dual chemistry are plentiful in the Galactic Bulge (e.g., Gutenkunst *et al.* 2008), but Guzman-Ramirez *et al.* (2014) found that these sources show silicate emission inside the PAH emission and suggested that the PAHs formed after the dissociation of CO in these nebulae. The young carbon-rich Galactic PN BD+30 3639 also shows PAHs and silicates in its spectrum, but in this case the silicates appear to be *outside* the PAHs (Guzman-Ramirez *et al.* 2015). In this case, the silicates could be dust which formed before dredge ups made the AGB star carbon-rich.

It is interesting to speculate that the carbon-rich Magellanic PNe with the 18 μm shoulder in their spectra might be similar to BD+30. Cold silicates would be far from the central star, which would be consistent with their having formed before dredge-ups made the stars carbon-rich. However, this is only one possibility. For the moment, the 18 μm shoulder remains unidentified.

6. Summary

Spectra from the IRS on *Spitzer* of evolved carbon-rich objects in the Magellanic Clouds trace the chemical evolution of dust as the central star evolves from the AGB to a hot white dwarf and the circumstellar dust shell evolves into a PN. The spectral properties of the carbon stars on the AGB suggest that the grains might be layered with MgS, which can dominate the emission from more evolved objects. As the central star heats up, photoprocessing of the carbonaceous grains increases, leading to spectra with 21 μm features, aliphatic hydrocarbons, and Class D PAHs. The photoprocessing can also produce objects with Class C PAHs, and as the object evolves into a young PNe, more typical emission from Class B PAHs appears, along with the big-11 feature and/or fullerenes. The big-11 feature is primarily from SiC, and its strength suggests the SiC fraction might be higher on the surface of the grains, perhaps due to photoprocessing. The big-11 feature is usually accompanied by a shoulder at 18 μm, for which the carrier is not known. If the 18 μm feature arose from cool silicate grains, it could point to dual chemistry in Magellanic PNe.

References

Allamandola, L. J., Tielens, A. G. G. M., & Barker, J. R. 1985, *ApJ*, 290, L25
Allamandola, L. J., Tielens, A. G. G. M., & Barker, J. R. 1989, *ApJS*, 71, 733
Bernard-Salas, J., Peeters, E., Sloan, G., *et al.* 2009, *ApJ*, 699, 1541
Cami, J. 2017, these proceedings
Cerrigone, L., Hora, J. L., Umana, G., *et al.* 2011, *ApJ*, 738, 121
Chiar, J. E., Pendleton, Y. J., Geballe, T. R., & Tielens, A. G. G. M. 1998, *ApJ*, 507, 281
Forrest, W. J., Houck, J. R., & McCarthy, J. F. 1981, *ApJ*, 248, 195
Geballe, T. R., Tielens, A. G. G. M., Kwok, S., & Hrivnak, B. J. 1992, *ApJ*, 387, L89

Geballe, T. R. & van der Veen, W. E. C. J. 1990, *A&A*, 235, L9
Gillett, F. C., Forrest, W. J., & Merrill, K. M. 1973, *ApJ*, 183, 87
Goebel, J. H. & Moseley, S. H. 1985, *ApJ*, 290, L35
Gutenkunst S., Bernard-Salas J., Pottasch S. R., Sloan G. C., Houck J. R. 2008, *ApJ*, 680, 1206
Guzman-Ramirez, L., Lagadec, E., Jones, D., Zijlstra, A. A., & Gesicki, K. 2014, *MNRAS*, 441, 364
Guzman-Ramirez, L., Lagadec, E., Wesson, R., *et al.* 2015, *MNRAS*, 451, L1
Houck, J. R., Roellig, T. L., van Cleve, J., *et al.* 2004, *ApJS*, 154, 18
Kwok, S., Hrivnak, B. J., & Volk, K. M. 1989, *ApJ*, 345, L51
Lagadec, E., Zijlstra, A. A., Sloan, G. C., *et al.* 2007, *MNRAS*, 376, 1270
Leger, A. & Puget, J. L. 1984, *A&A*, 137, L5
Leisenring, J. M., Kemper, F., & Sloan, G. C. 2008, *ApJ*, 681, 1557
Lombaert, R., de Vries, B. L., de Koter, A., *et al.* 2012, *A&A*, 544, L18
Matsuura, M., Bernard-Salas, J., Lloyd Evans, T., *et al.* 2014, *MNRAS*, 439, 1472
Peeters, E., Hony, S., Van Kerckhoven, C., *et al.* 2002, *A&A*, 390, 1089
Pino, T., Dartois, E., Cao, A.-T., *et al.* 2008, *A&A*, 490, 665
Sloan, G. C., Jura, M., & Duley, W. W., *et al.* 2007, *ApJ*, 664, 1144
Sloan, G. C., Kraemer, K. E., Matsuura, M., *et al.* 2006, *ApJ*, 645, 1118
Sloan, G. C., Kraemer, K. E., McDonald, I., *et al.* 2016, *ApJ*, 826, 44
Sloan, G. C., Lagadec, E., Zijlstra, A. A., *et al.* 2014, *ApJ*, 791, 28
Volk, K., Hrivnak, B. J., Matsuura, M., *et al.* *ApJ*, 735, 127
Werner, M. W., Roellig, T. L., Low, F. J., *et al.* 2004, *ApJS*, 154, 1
Zhang, K., Jiang, B. W., & Li, A. 2009, *ApJ*, 702, 680
Zhukovska, S. & Gail, H.-P. 2008, *A&A*, 486, 229
Zijlstra, A. A., Matsuura, M., Wood, P. R., *et al.* 2006, *MNRAS*, 370, 1961

Discussion

PAGOMENOS: Could you comment on the 21μm feature only being observed in the presence of the 30μm feature, which you are convinced can be attributed to MgS?

SLOAN: I should have mentioned that in these samples, there always seems to be one source that violates whatever trend we see. The geometries are complex, and we can expect inclination, evolutionary, and chemical effects. Nevertheless, we usually see MgS with the 21μm feature, and usually with a specific shape and wavelength position.

VAN WINCKEL: If you explain the changing chemistry of BD +30.3639 due to the thermal pulses (cold O-rich dust in the outside, C-rich in the inner part which was expelled when the star turned C-star), you need heavy mass loss both when the star was an M star and when it was a C star. Does this matches with the ages of the different circumstellar components? A star needs probably more than 1 thermal pulse to go from M to C. Could the O-rich dust not be stored in a stable disc? (e.g. the Red Rectangle mix?) This could also explain the processing and high crystallinity.

SLOAN: It would only take 1 thermal pulse to cross the C/O = 1 boundary and cross from O-rich to C-rich dust. But I agree that we should not rule out disks, especially when the crystalline silicates are present.

Q: 1) What is the carrier of the 18-μm feature in C-rich objects? 2) Can we include/consider silicate grains in SED modeling for the 18-μm feature showing C-rich objects?

Planetary Nebulae: Multi-Wavelength Probes of Stellar and Galactic Evolution
Proceedings IAU Symposium No. 323, 2016
X. Liu, L. Stanghellini & A. Karakas, eds.

doi:10.1017/S1743921317001971

Herschel observations of planetary nebulae†

Griet C. Van de Steene

Royal Observatory of Belgium, Department Astronomy and Astrophysics
Ringlaan 3, BE-1180, Brussels, Belgium
email: g.vandesteene@oma.be

Abstract. This article presents an overview of the published results for planetary nebulae based on images and spectroscopy from the PACS, SPIRE, and HIFI instruments on board the Herschel satellite.

Keywords. ISM: abundances, atoms, dust, globules, molecules, planetary nebulae: general, infrared: ISM, stars: circumstellar matter

1. Introduction

Grains play an important role in many environments, including planetary nebulae (PNe), because of extinction, photoelectric heating, their influence on the charge and ionization balance of the gas, as catalysts for grain-surface chemical reactions and as seeds for freeze-out of molecules. Previous satellite missions such as IRAS, ISO, Spitzer, and AKARI have allowed us to study the dust in PNe, but unfortunately the angular resolution of these instruments was too low to get detailed information on the spatial distribution of the dust. This has changed with the Herschel satellite, which has allowed us to study the spatial structures in unprecedented detail.

2. Herschel and its instruments

The Herschel Space Observatory (Pilbratt *et al.* 2010) was launched on May 14 2009 and operated for nearly four years. It carried the largest, most powerful infrared telescope ever flown in space and three sensitive scientific instruments. Herschel's observations finished on April 29 2014 when the tank of liquid helium used to cool the instruments finally ran dry.

The three instruments on board were: PACS (Photoconductor Array Camera and Spectrometer), SPIRE (Spectral and Photometric Imaging REceiver), and HIFI (Heterodyne Instrument for the Far Infrared), a high-resolution spectrometer. These instruments were designed for deep, wideband photometry with high spatial resolution and full spectral coverage, making Herschel the first space facility to completely cover the far infrared and submillimeter range from 55 to 672 μm.

PACS (Photodetecting Array Camera and Spectrometer) (Poglitsch *et al.* 2010) was an imaging camera and low-resolution spectrometer covering wavelengths from 55 to 210 μm. The spectrometer had a spectral resolution between R = 1000 and R = 5000. It operated as an integral field spectrograph, combining spatial and spectral resolution. The imaging camera was able to image simultaneously in two bands (either 60 − 85 / 85 − 130 μm and 130 − 210 μm) with a detection limit of a few mJy.

† *Herschel* is an ESA space observatory with science instruments provided by European-led Principal Investigator consortia and with important participation from NASA.

SPIRE (Spectral and Photometric Imaging Receiver) (Griffin *et al.* 2010) was an imaging camera and low-resolution spectrometer covering 194 to 672 μm wavelength. The spectrometer had a resolution between R = 40 and R = 1000 at a wavelength of 250 μm and was able to image point sources with brightnesses around 100 mJy and extended sources with brightnesses of around 500 mJy. The imaging camera observed simultaneously in three bands, centred at 250, 350 and 500 μm, each with 139, 88 and 43 pixels respectively. It was able to detect point sources with brightness above 2 mJy.

HIFI (Heterodyne Instrument for the Far Infrared) (de Grauw *et al.* 2010) is a heterodyne detector able to electronically separate radiation of different wavelengths, giving a spectral resolution up to R = 10^7 . The spectrometer was operated within two wavelength bands, from 157 to 212 μm and from 240 to 625 μm.

3. PACS and SPIRE imaging results

The extended circumstellar envelopes of evolved low-mass AGB stars display a large variety of morphologies. Understanding the various mechanisms that give rise to these extended structures is important to trace their mass-loss history. The data presented by Cox *et al.* (2012) showed for the first time the variety of interaction between the circumstellar shell and the interstellar medium, which can be divided in roughly four categories: "fermata", "eyes", "irregular", and "rings". In particular the star's peculiar space velocity and the density of the ISM appear decisive in detecting emission from bow shocks or detached rings. Tentatively, the "eyes" class objects are associated with (visual) binaries, while the "rings" generally do not appear to occur for M-type stars, only for C or S-type objects that have experienced a thermal pulse. The occurrence of the observed eye-shape of AGB detached shells is most strongly influenced by the interstellar magnetic field, the stellar space motion, and density of the interstellar medium (van Marle *et al.* 2014). Observability of this transient phase is favoured for lines-of-sight perpendicular to the interstellar magnetic field direction. The simulations of van Marle *et al.* (2014) indicate that "eye" shapes of such pre-PN circumstellar shells can strongly affect the shape and size of PNe.

A total of 18 well known PNe have been imaged with PACS and SPIRE instruments in the framework of MESS and HerPlans programs. Seven PNe (NGC 6720, NGC 650, NGC 7293, NGC 6853, NGC 3587, NGC 7027) were imaged by the MESS team. Mass loss of Evolved StarS (MESS) was a Guarenteed Time Key Programme to study the circumstellar environment of evolved post main sequence stars. A detailed description of the program can be found in Groenewegen *et al.* (2011). An overview of the Herschel observations of PNe in the MESS program was presented in van Hoof *et al.* (2012). Eleven PNe (NGC 40, NGC 2392, NGC 3242, NGC 6445, NGC 6543, NGC 6720, NGC 6781, NGC 6826, NGC 7009, NGC 7026, Mz 3) have been observed in the Herschel Planetary Nebulae Survey (HerPlanS). A data overview and first analysis was presented in Ueta *et al.* (2014). HerPlaNS obtained far-infrared broadband images and spectra of eleven well-known PNe with the PACS and SPIRE instruments. The target PNe all have distances less than ∼1.5 kpc and are dominated by relatively high-excitation nebulae as they were selected from the Chandra Planetary Nebula Survey (ChanPlaNS; Kastner *et al.* 2012)

Herschel PACS and SPIRE imaging showed that the dust emission in PNe has a very clumpy structure for all nebula. There is excellent agreement between the H_2 images and the PACS 70 μm maps. For the Ring nebula (NGC 6720, Figure 1) it appears to be the first observational evidence that H_2 forms on oxygen rich dust grains. van Hoof *et al.* (2010) developed a photoionisation model of the Ring nebula with `Cloudy` to investigate

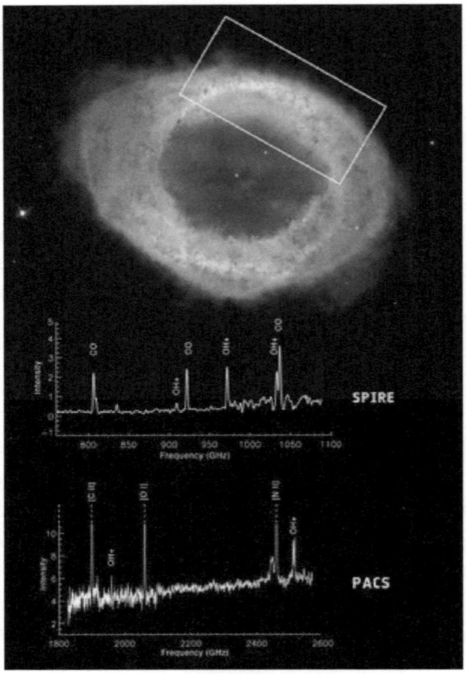

(a) SPIRE and PACS spectra of molecules in the Ring nebula NGC 6720.

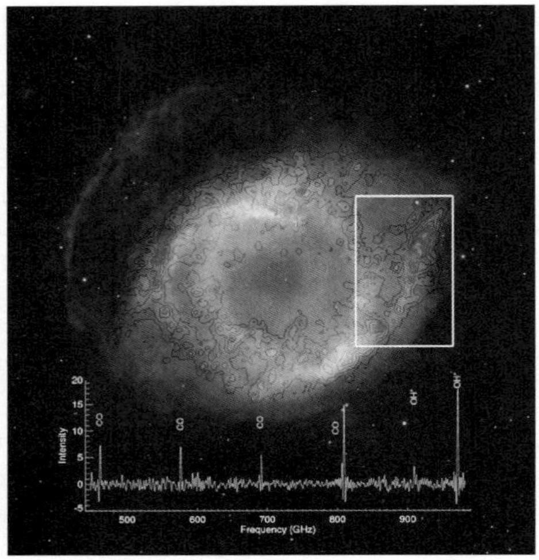

(b) SPIRE 250 μm image and spectrum of molecules in the Helix nebula NGC 7293.

Figure 1. Herschel spectra of molecules in the Ring and Helix nebula

possible formation scenarios for H_2. They concluded that the most plausible scenario is that the H_2 resides in high density knots which were formed after the recombination of the gas started, when the central star luminosity dropped steeply after the central star entered the cooling track. H_2 formation may still be ongoing at this moment, depending on the density of the knots and the properties of the grains in the knots (van Hoof *et al.* 2010). This is also a possible scenario for the formation of high density clumps in other evolved nebula with a central star on the cooling track such as the Helix (NGC 7293, Figure 1) and the Dumbell (NGC 6853).

Comparison between the 70 μm Herschel and corresponding optical maps showed that they are very similar indicating that there is a very steep temperature gradient from the ionized region to the dusty photodissociation region. For NGC 6781 the PACS 70 μm map, showing the distribution of thermal dust continuum is very similar to what is seen in the [NII]λ658.4 nm image (Ueta *et al.* 2014). For the Helix it was also observed and shown that the radiation field decreases rapidly outwards through the barrel wall (Fig. 9, Van de Steene *et al.* 2015).

Previous knowledge of the 3D structure of the nebula is extremely useful to correctly interpret the far infrared images. For instance, the PACS and SPIRE images of the Helix nebula could be understood based on the kinematic model of Zeigler *et al.* (2013) and the Herschel images of NGC 6781 with the 3D model of Schwarz & Monteiro (2006). Both show bipolar, barrel-like structures inclined to the line of sight, a frequent morphology in PNe.

Herschel PACS imaging photometry was obtained for these 17 different PNe in the MESS and HerPlanS projects (van Hoof *et al.* 2010, van Hoof *et al.* 2013, Van de Steene *et al.* 2015,Van de Steene *et al.* in preparation, Ueta *et al.* 2014, Ueta *et al.* in preparation) with the PACS and SPIRE instruments at 70, 160, 250, 350, and 500 μm. This photometry was complemented with photometry obtained from the literature at many other wavelengths from the UV to radio wavelengths to construct full SEDs. The modified black body fit to these SEDs revealed that the emission factor β is always close to 1.0, indicating that the dust grains are mainly amorphous carbon (Menella *et al.* 1995, Boudet *et al.* 2005). The fit to the SED also showed that the flux emitted in the far infrared is significant: without far-IR data fitting constraints the dust mass is underestimated by 40%.

The dust temperature obtained from the SED fits and the temperature maps made, showed that the cool dust temperature of the PNe is around 30 to 100 K. For the Helix nebula the gas kinetic temperature T_k was determined to be about 20 to 40 K (Zack & Ziurys 2013, Etxaluze *et al.* 2014), which is similar the Helix' dust temperature (Van de Steene *et al.* 2015). The gas density of the H_2 cometary globules is on the order of $n(H_2) \sim$ (1-5) 10^5 cm^{-3}. Goldsmith (2001) found that for gas densities higher than $10^{4.5}$ cm^{-3} the dust and gas temperatures will be closely coupled, also for the dust temperatures determined for the Helix nebula.

The dust masses found so far for NGC 6781, NGC 7293, and NGC 650 are all a few thousandths of solar masses. By integrating over the entire nebula, the dust column mass density map the total mass of far emitting dust mass was determined to be 4×10^{-3} M$_\odot$ for NGC 6781 at a distance of 950 pc and 3.5×10^{-3} M$_\odot$ for the Helix nebula at distance of 216 pc, while for NGC 650 the dust mass is about 1.4×10^{-3} M$_\odot$ at a distance of 1200 pc based on Cloudy modeling (Ueta *et al.* 2014, Van de Steene *et al.* 2015, van Hoof *et al.* 2013)

One of the goals of HerPlaNS is to empirically obtain spatially resolved gas-to-dust mass ratio distribution maps by deriving both the dust and gas column mass distribution maps directly from observational data. For NGC 6781 direct comparison of the dust and gas column mass maps constrained data allowed to construct an empirical gas-to-dust mass ratio map, which showed a large range of ratios with the median of 195$\pm$110 and hence, is generally consistent with the typical spatially-unresolved ratio between 100 and 400 widely used in the literature for the case of PNe and AGB stars (Ueta *et al.* 2014).

The MESS and HerPlaNS teams have collected not only photometry, but also other spectroscopic data from the literature over the whole spectral range from X-rays to radio to make the most comprehensive Cloudy models ever made of NGC 650 (van Hoof *et al.* 2013) and NGC 6781 (M. Otsuki, this volume). For NGC 650 the `Cloudy` model showed that the grains in the ionized nebula are large (assuming single-sized grains, they would have a radius of 0.15 μm). Most likely these large grains were inherited from the asymptotic giant branch phase. However the PACS 70/160 μm temperature map showed evidence of two radiation components heating the grains. The first component is direct emission from the central star, while the second component is diffuse emission from the ionized gas (mainly Lyα). Unlike what was thought before, the neutral material resides in dense clumps inside the ionized region. These may also harbour stochastically heated very small grains in addition to the large grains. This is unusual for such a highly evolved PN.

In the past, far-IR SED fitting with broadband fluxes were performed under the assumption of negligible line contamination. With the Herschel data and `Cloudy` modeling we verified that the degree of line contamination is approximately 8-20% (Ueta *et al.* 2014, van Hoof *et al.* 2013) and does not significantly affect the fitting results.

4. PACS and SPIRE spectroscopic results

4.1. *NGC 6781*

The Herschel spectra obtained at various locations within NGC 6781 revealed both the physical and chemical nature of the nebula. The spectra showed a number of ionic and atomic lines such as [O III] 52, 88 μm, [N III] 57 μm, [N II] 122, 205 μm, [C II] 158 μm, and [O I] 63, 146 μm, as well as various molecular lines, in particular, high-J CO rotational transitions, OH, and OH^+ emission lines. Thermal dust continuum emission was also detected in most bands in these deep exposure spectra. On average, the relative distributions of emission lines of various nature suggested that the barrel cavity in NGC 6781 is uniformly highly ionized, with a region of lower ionization delineating the inner surface of the barrel wall. The least ionic and atomic gas, molecular, and dust species are concentrated in the cylindrical barrel structure.

Based on the PACS IFU spectral cube data, Ueta *et al.* (2014) derived line maps in the detected ionic and atomic fine-structure lines. Next diagnostics of the electron temperature and density using line ratios such as [O III] 52/88 μm and [N II] 122/205 μm, resulted in (T_e, n_e) and ionic/elemental/relative abundance profiles for the first time in the far-IR for any PN. The derived T_e profile substantiated the typical assumption of uniform $T_e = 10^4$K in the main ionized region, while showing an interesting increase in the barrel (up to 10% higher), followed by a sudden tapering off toward the halo region. The n_e profile of high-excitation species is nearly flat across the inner cavity of the nebula, whereas the n_e profile of low excitation species exhibits a radially increasing tendency with a somewhat complex variation around the barrel wall. In fact, this n_e[N II] profile is reflected in the physical stratification of the nebula revealed by the ionic/elemental abundance analysis. The detected stratification is consistent with the previous inferences made from the past optical imaging observations in various emission lines of varying levels of excitation. The derived relative elemental abundance profiles showed uniformly low N and C abundances, confirming the low initial mass (< 2 $M_\odot$) and marginally carbon-rich nature of the central star. However, the profiles did not appear to reveal variations reflecting the evolutionary change of the central star, such as a radially increasing carbon abundance.

4.2. *SPIRE spectroscopy: OH^+ and CO*

Etxaluze *et al.* (2014) and Aleman *et al.* (2014) reported the first detection of extended OH^+ lines in emission in 5 PNe observed as part of the HerPlans in NGC 6445, NGC 6720 (Fig. 1a), and NGC 6781 and MESS in NGC 7293 (Fig. 1b). Also NGC 6853 shows OH^+ in emission (Van de Steene *et al.*, in preparation). All five PNe are molecule rich, with dense clumpy structures and hot central stars (Teff $>$ 100000 K). The OH^+ emission is most likely due to excitation in a photodissociation region. Although other factors such as high density and low C/O ratio may also play a role in the enhancement of the OH^+ emission. The fact that OH^+ is not detected in objects with $T_{eff} < 100000$ K suggests that the hardness of the ionising central star spectra could be an important factor in the production of OH^+ emission in PNe.

The Herschel spectra towards the Helix nebula also show, besides OH^+, CO emission lines (from J = 4 to 8), [N II] at 1461 GHz from ionized gas, and [C I] ($^3P_2-^3P_1$). The SPIRE spectral maps suggest that CO arises from dense and shielded clumps in the western rims of the Helix nebula, whereas OH^+ and [C I] lines trace the diffuse gas and the UV and X-ray illuminated clump surfaces where molecules reform after CO photodissociation. The [N II] line traces a more diffuse ionized gas component in the interclump medium (Etxaluze *et al.* 2014).

For NGC 6781 the CO observations and analysis with higher-J transitions sampled a much warmer CO gas component in the cylindrical barrel structure, probably located closer to the equatorial region along the line of sight, compared with the previous CO measurements and diagnostics by Bachiller *et al.* (1993). However, the amount of this warm component was determined to be an order of magnitude smaller than the cold component (Ueta *et al.* 2014).

4.3. *Crystalline olivine*

Blommaert *at al.* (2014) (GT1 "Forsterite dust in the circumstellar environment of evolved stars") presented 48 PACS spectra of evolved stars in the wavelength range of $67 - 72\,\mu$m, covering the $69\,\mu$m band of crystalline olivine ($Mg_{22x}Fe(_{2x})SiO_4$). For 27 objects in the sample, they detected the $69\,\mu$m band of crystalline olivine ($Mg_{(22x)}Fe_{(2x)}SiO_4$). The $69\,\mu$m band showed that all the sources produce pure forsterite grains containing no iron in their lattice structure. They fit the $69\,\mu$m band and used its width and wavelength position to probe the composition and temperature of the crystalline olivine. The fits showed that on average the temperature of the crystalline olivine is highest in the group of OH/IR stars and the post-AGB stars with confirmed Keplerian disks. The temperature is lower for the other post-AGB stars and lowest for the PNe. A couple of the detected $69\,\mu$m bands are broader than those of pure magnesium-rich crystalline olivine, which can be due to a temperature gradient in the circumstellar environment of these stars.

5. HIFI

5.1. *HIFISTARS*

The Herschel guaranteed time key programme HIFISTARS (Bujarrabal *et al.* 2012) aimed to study the physical conditions, particularly the excitation state, of the intermediate-temperature gas in proto-PNe and young PNe. The information that the observations of the different components deliver is of particular importance for the wind-shock interaction and hence understanding the evolution and shaping of PNe. They performed Herschel/HIFI observations of intermediate-excitation molecular lines in the far-infrared range of a sample of ten nebulae. The high spectral resolution provided by HIFI allows the accurate measurement of the line profiles. The dynamics and evolution of these nebulae are known to result from the presence of several gas components, notably fast bipolar outflows and slow shells (that often are the fossil AGB shells), and the interaction between them. Because of the diverse kinematic properties of the different components, their emission can be identified in the line profiles. The observation of these high-energy transitions allows an accurate study of the excitation conditions, particularly in the warm gas, which cannot be properly studied from the low-energy lines. They detected far infrared lines of several molecules, in particular of ^{12}CO, ^{13}CO, and H_2O. Emission from other species, like NH_3, OH, $H_2{}^{18}O$, HCN, SiO, etc., has been also detected. Wide profiles showing sometimes spectacular line wings have been found. In the case of CRL 618 the ^{12}CO and ^{13}CO high excitation line profiles present a composite structure showing spectacular wings in some cases, which become dominant as the energy level increases (Soria-Ruiz *et al.* 2013). Bujarrabal *et al.* (2012) mainly studied the excitation properties of the high-velocity emission, which is known to come from fast bipolar outflows. From comparison with general theoretical predictions, they find that CRL 618 showed a particularly warm fast wind ∼300 K, hotter than previously estimated (Soria-Ruiz *et al.* 2013). In contrast, the fast winds in OH 231.8+4.2 and NGC 6302 are cold, $T_k \sim 30$ K.

Other nebulae, like CRL 2688, show intermediate temperatures, with characteristic values around 100 K. They argue that the differences in temperature in the different nebulae can be caused by cooling after the gas acceleration (that is probably caused by shocks). For instance, CRL 618 is a case of very recent acceleration of the gas by shocks, less than ~100 yr ago, while the fast gas in OH 231.8+4.2 was accelerated ~1000 yr ago. They also find indications that the densest gas tends to be cooler, which may be explained by the expected increase of the radiative cooling efficiency with density. The dense central core of CRL 618 is characterised by a very low expansion velocity, ~5 km s^{-1}, and a strong velocity gradient. This component is very likely to be the unaltered circumstellar layers that are lost in the last AGB phase, where the ejection velocity is particularly low. The physical properties of the diffuse halo and the double empty shell, contribute to its line profiles mainly in the low-J CO transitions (Soria-Ruiz *et al.* 2013).

5.2. *Shapemol*

Herschel/HIFI has opened a new window for probing molecular warm gas. On the other hand, the software SHAPE (Steffen & Lopez 2006, Steffen *et al.* 2011) has emerged in the past few years as a standard tool for determining the morphology and velocity field of different kinds of gaseous emission nebulae via spatio-kinematical modelling. SHAPE implements radiative transfer solving, but it is only available for atomic species and not for molecules. SHAPEMOL (Santander-Garcia *et al.* 2015) is a complement to SHAPE which enables user-friendly, spatio-kinematic modelling with accurate non-LTE calculations of excitation and radiative transfer in CO lines. Shapemol is a plug-in completely integrated within SHAPE v5 . It allows radiative transfer solving in the ^{12}CO and ^{13}CO J = 10 to J = 1716 lines, but its implementation easily permits extending the code to different transitions and other molecular species, either by the code developers or by the user. Used along Shape, Shapemol allows easily generating synthetic maps and synthetic line profiles to match against observations.

As an example of the power and versatility of Shapemol, a model of the molecular envelope of the planetary nebula NGC 6302 was made and compared with ^{12}CO and ^{13}CO J = 21 interferometric maps from SMA and high-J transitions from HIFI. Santander-Garcia *et al.* (2015) found that its molecular envelope has a complex, broken ring-like structure with an inner, hotter region and several fingers and high-velocity blobs, emerging outwards from the plane of the ring. The Herschel spectra are extremely rich, especially in terms of molecular line transitions.

HIFI data have also allowed a very detailed description of the young PN NGC 7027. Santander-Garcia *et al.* (2012) also used Shapemol for radiative transfer, spatio-kinematic modeling of the molecular envelope of the young planetary nebula NGC 7027 in several high- and low-J ^{12}CO and ^{13}CO transitions observed by HIFI and the IRAM 30 m radio telescope, and discussed the structure and dynamics of the molecular envelope. They used this code to build a Russian doll model to account for the physical and excitation conditions of the molecular envelope of NGC 7027. The model nebula consisted of four nested, mildly bipolar shells plus a pair of high-velocity blobs. The innermost shell is the thinnest and showed a significant increase in physical conditions (temperature, density, abundance, and velocity) compared to the adjacent shell. This is a clear indication of a shock front in the system, which may have played a role in shaping the nebula. Each of the high-velocity blobs is divided into two sections with considerably different physical conditions. The striking presence of H_2O in NGC 7027, a C-rich nebula, is likely due to photo-induced chemistry from the hot central star, although formation of water by shocks cannot be ruled out.

6. Outlook

Soon, the THROES atlas (Garcia-Lario *et al.* 2016, this volume) will be publicly available through the Herschel science archive. THROES is a catalogue of fully reprocessed, homogenously reduced PACS spectra of all evolved stars from the AGB to the PN stage, including some massive red supergiants and LBVs, complemented with ancillary data taken by other facilities. SPIRE spectra will be added later. The catalog will contain more than 200 sources, originally part of more than 40 different research programs. This will hopefully trigger additional research by the community.

A lot of Herschel observations of PNe are available to be exploited and more interesting, scientific results await discovery.

Acknowledgements

I thank the SOC for inviting me to do this review talk. I am indebted to my colleagues of the MESS and HerPlanS consortia who have closely collaborated with her on the Herschel data. G. Van de Steene wishes to acknowledge support from FWO through travel grant K1C8716N. G. Van de Steene and the MESS consortium wish to acknowledge support from the Belgian Science Policy office through the ESA PRODEX programme.

References

Aleman, I., Ueta, T., & Ladjal, D., 2014, *A&A*, 566, A79
Bachiller, R., Huggins, P. J., Cox, P., Forveille, T., 1993, *A&A*, 267, 177
Bujarrabal, V., Alcolea, J., Soria-Ruiz, R., *et al.*, 2012, *A&A*, 537, A8
Blommaert, J. A. D. L., de Vries, B. L.,Waters, L. B. F. M., *et al.*, 2014, *A&A*, 565, A109
Boudet, N., Mutschke, H., Nayral, C., *et al.*, 2005, *ApJ.*, 633, 272
Cox N. L. J., Kerschbaum F., van Marle A.-J. *et al.*, 2012, *A&A*, 537, 35
de Graauw, T., Helmich, F. P., Phillips, T. G. *et al.*, 2010, *A&A*, 518, L4
Etxaluze, M., Cernicharo, J., Goicoechea, J. R., *et al.*, 2014, *A&A*, 566, A78
Goldsmith, P. F., 2001, *ApJ*, 557, 736
Griffin, M. J., Abergel, A., Abreu, A., *et al.*, 2010, *A&A*, 518, L3
Groenewegen, M. A. T., Waelkens, C., Barlow, M. J., *et al.*, 2011, *A&A*, 526, A162
Kastner, J. H., Montez, R., Jr., Balick, B., *et al.*, 2012, *AJ*, 144, 58
Mennella, V., Colangeli, L., & Bussoletti, E., 1995, *A&A*, 295, 165
Santander-Garcia M., Bujarrabal V., Alcolea J., 2012, *A&A*, 545, 114
Santander-Garcia M., Bujarrabal V., Koning N., Steffen W., 2015, *A&A*, 573, 56
Schwarz, H. E. & Monteiro, H., 2006, *ApJ*, 648, 430
Soria-Ruiz R., Bujarrabal V., Alcolea J., 2013, *A&A*, 559, 45
Steffen W. & Lopez J. A., 2006 *Rev. Mexicana AyA* 26 30
Steffen W., Koning N., Wenger S., Morisset C. and Magnor M., 2011, *IEEE Transactions on visualisation and computer graphics*, 17, 454
Pilbratt, G. L., Riedinger, J. R., Passvogel, T., *et al.*, 2010, *A&A*, 518, L1
Poglitsch, A., Waelkens, C., Geis, N., *et al.*, 2010, *A&A*, 518, L2
Ueta, T., Ladjal, D., Exter, K. M., *et al.*, 2014, *A&A*, 565, A36
Van de Steene G. C., van Hoof P. A. M., Exter K. M., *et al.*, 2015, *A&A*, 574, 134
van Hoof, P. A. M., Van de Steene, G. C., Barlow, M. J., *et al.*, 2010, *A&A*, 518, L137
van Hoof, P. A. M., Barlow, M. J., Van de Steene, G. C., *et al.*, 2012, *IAU Symp.*, 283, 41
van Hoof, P. A. M., Van de Steene, G. C., Exter, K. M., *et al.*, 2013, *A&A*, 560, A7
van Marle A. J., Cox N. L. J., Decin L., 2014, *A&A*, 570, 131
Zack, L. N. & Ziurys, L. M., 2013, *ApJ*, 765, 112
Zeigler, N. R., Zack, L. N., Woolf, N. J., & Ziurys, L. M., 2013, *ApJ*, 778, 16

Planetary Nebulae: Multi-Wavelength Probes of Stellar and Galactic Evolution
Proceedings IAU Symposium No. 323, 2016
X. Liu, L. Stanghellini & A. Karakas, eds.

doi:10.1017/S1743921317000874

Magnetic fields in PNe and other evolved low-mass and intermediate-mass stars

Laurence Sabin[1], Qizhou Zhang[2], Gregg A. Wade[3], Agnès Lèbre[4] and Roberto Vázquez[1]

[1]Instituto de Astronomía, Universidad Nacional Autónoma de México, Apdo. Postal 877, 22800 Ensenada, B. C, Mexico.
email: lsabin@astro.unam.mx,vazquez@astro.unam.mx

[2]Harvard-Smithsonian Center for Astrophysics, 60 Garden Street, Cambridge, MA 02138, USA.
email: qzhang@cfa.harvard.edu

[3]Department of Physics, Royal Military College of Canada, PO Box 17000, Station Forces, Kingston, Ontario K7K 7B4, Canada.
email: Gregg.Wade@rmc.ca

[4]Laboratoire Univers et Particules de Montpellier - CNRS and Université Montpellier - Place E. Bataillon, 34 090 Montpellier, France
email: agnes.lebre@umontpellier.fr

Abstract. Magnetic fields are likely to be an efficient mechanism which can affect evolved intermediate mass stars (i.e. post-AGB stars and planetary nebulae) in different ways such as via the shaping of their envelope. However, observational probes for the presence of those fields are still scarce. I will present a summary of the works, including those from our group, on the detection and measurement of magnetic fields in various evolved objects.

Keywords. magnetic fields, polarization, stars: AGB and post-AGB, (ISM:) planetary nebulae: general

---

1. Introduction

One of the reasons why we investigate magnetism in evolved intermediate mass stars (i.e. post-AGBs stars (PAGBs), proto-planetary nebulae (PPNe) and planetary nebulae (PNe)) is to understand its role in the structure and dynamics of these objects. Indeed, we are looking for magnetic field-related effects such as the collimation of outflows or the generation and launching of jets which are expected to occur at this phase of stellar evolution, as they are found happening in other objects like the Sun, T Tauri stars or Herbig-Haro objects for example. In a similar vein we also want to establish a direct correlation between the magnetic fields (MFs) and the different morphological structures found in PPNe and PNe. While it has been emphasized in recent years that these fields are not likely the main shaping agent (binary systems are more likely the principal actor), this absolutely does not discard the action of MFs at other scales.

MFs have been searched for in circumstellar envelopes (CSE) via the polarized emission from aligned dust grains, molecular lines and MASERs; and at the surface of central stars via the direct measurement of Zeeman splitting of atomic lines.

2. Review of searches for magnetic fields in circumstellar envelopes

Using the theory of dust alignment by MFs, Greaves (2002) and Sabin *et al.* (2007, 2014, 2015) have shown the presence of MFs and traced their structure in a small set of evolved stars. Large scale toroidal and polar MF structures were found in CRL 2688, NGC 7027, NGC 6302 and NGC 6537 (resolution ∼7"– 14") and a toroidal or helicoidal MF geometry was seen in OH231.8+4.2 at medium scales (resolution ∼2") using interferometric arrays. For this bipolar object a magnetic launching mechanism has also been invoked.

Assuming the Goldreich-Kylafis effect and that the linear polarization of the thermal emission of molecules (CO, SiO, CS ..) traces the MF, Vlemmings *et al.* (2012) and Girart *et al.* (2012) investigated two AGB stars (IK Tau and IRC+10216 respectively) and presented first time detection of molecular line polarization (not MASERS). The polarized emission distribution might indicate a complex MF geometry associated with the CSE geometry. However a full radiative transfer analysis is required to define the actual direction of the MF (see Kylafis 1983 & Morris *et al.* 1985).

In an analysis to be published, Sabin *et al.*, investigated the PPN Frosty Leo using both dust and molecular polarization and found an overall marginal MF detection.

Aside from those results, Pérez Sánchez *et al.* (2013) detected synchrotron emission towards the post-AGB star IRAS 154455449 tracing the bipolar shape: this tends to indicate that the jet is likely collimated by MFs.

Finally, several works have been carried out over the years to investigate MASERs in evolved stars and the correlation with MFs via circular and linear polarization (e.g. Herpin *et al.* 2006, Vlemmings *et al.* 2007, Etoka *et al.* 2009, Leal Ferreira *et al.* 2013, Gomez *et al.* 2016). Ordered field structures have been found (e.g. the red supergiant (RSG) VX Sgr). And, in PNe field strengths between ∼0.8 and 24 mG have been measured via the analysis of OH masers in K3-35 (Gómez *et al.* 2009), IRAS 16333-4807 (Qiao *et al.* 2016), JaSt 23 and IRAS 17393-2727 (Gómez *et al.* 2016). In addition to the fact that masers, being compact and bright sources, can give both the strength and direction of the MF in CSE; surface MFs can be extrapolated based on different distribution laws ($1/r$, $1/r^2$,$1/r^3$) representative of the different MF geometries (toroidal, poloidal/solar or dipole structure respectively).

3. Review of searches for magnetic fields in central stars

The method applied is based on the detection of circularly polarised Zeeman splitting patterns in atomic lines using medium to high resolution spectropolarimeters. It is then clear that objects showing many emission lines in their spectrum (due to the nebular contribution) as well as the faint ones i.e. $\geqslant$ 9–10 mag in B- or V-band are not the best candidates. Hence there is no clear and definite detection of MFs at the stellar surface of PNe so far (see Leone *et al.* 2011,2014; Todt *et al.* 2014, Asensio Ramos *et al.* 2014, Steffen *et al.* 2014). Our group's (LS, GAW, AL) attempt to look for MFs at the stellar surface of the PN NGC 6826 with the high resolution spectropolarimeter ESPaDOnS used the least squares deconvolution technique (LSD, Donati *et al.* 1997) to investigate the presence of a field. A subsequent analysis using the first-order moment method (Rees & Semel 1979) yielded a value for the longitudinal field B_l = 620±500 G. No magnetic field was unveiled for this PN (see Fig.1).

PAGBs appear to be the most suitable objects as they have fewer emission lines and the brightest objects have sharp absorption lines. Sabin *et al.* (2014) studied a group of seven Post-AGB stars with the spectropolarimeters ESPaDOnS and Narval (R=65000), and found a definite MF detection for the RV Tauri stars U Mon (10.2±1.7 G) and R Sct (0.9±0.5 G). The latter result was confirmed and extended by Tessore *et al.* (2015).

Table 1 summarizes the field strengths measured at the surface (or photosphere) of different stars along the HR diagram, from the literature. One of the least studied and therefore rather unknown groups consists of the AGB/Post-AGB/PPNe/PNe objects.

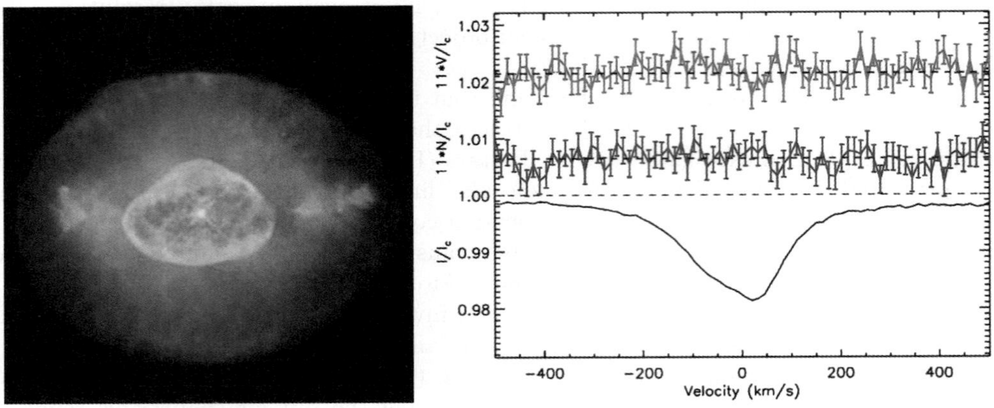

Figure 1. Stellar surface magnetic field analysis of the PN NGC 6826 with ESPaDOnS. The red line (top) indicates the Stokes V spectrum, the blues line (middle) is for the null spectrum an the black line (bottom) indicates the Stokes I spectrum. No MF detection is seen in this case.

Table 1. MFs strength along the HR diagram

Evolutionary phase	MFs range
T Tauri and low mass stars on the Main Sequence (MS)	Up to ∼ 2 kG
Cool MS solar type stars	∼10 G and ∼ kG in the spots
Intermediate and massive stars on the PreMS and MS	∼100 G up to 20–30 kG
Post-MS stars (RSGs, AGBs, PostAGBs and PNe)	**∼ 1–10 G ?**
White Dwarfs	$\sim 10^3$-10^9 G

The difficulty in measuring surface MFs for these objects is likely related to the fact that their fields are both weak and complete, and consequently are below the detection limit.

4. Future and prospects

The detection of MFs in the envelopes of PPNe and PNe is now possible with ALMA via dust (since cycle 2) and line polarization (since cycle 4). This polarimetric array will allow us to reach higher resolutions and then trace the field closer to the central star to detect any magnetic launching mechanism. In cycle 4, three programs targeting evolved stars involve the use of polarization. The measurement of MFs at the stellar surface is an ongoing investigation with the currently available facilities.

In particular, several projects aiming at the search for MFs are now underway in Mexico and will help in our studies of evolved stars: I) a fiber-fed polarimeter which can be coupled with medium and high resolution optical spectrographs on the 2.1m telescope of the Observatorio Astronómico Nacional is set to be operational for 2017 and plans are already made to also equip the future 6.5m San Pedro Martír Telescope (with an expected first light in 2023); II) TolTEC, a polarimeter on the Large Millimeter Telescope is also expected to be available around 2019.

5. Conclusions

• Magnetic fields are found in PNe, although they are more easily accessible in the envelope than in the central star. This is likely due to sensitivity issues for the detection. Spectropolarimetry of these evolved stars is a delicate and time consuming work. Hence, an upgrade in the instrumentation (resolution and permitted waveband) and telescope (size) would certainly help to resolve this issue.

• A magnetic launching mechanism seems to be a viable explanation for the morphology found in some objects but this would need confirmation with deeper data from ALMA for example and modelling with MHD tools.

• Although the detection rate is rather slow and the detection numbers still pretty low, we are entering in an exciting time for polarization studies with much needed and more sensitive instrumentation either already available or about to become available!

6. Acknowledgements

LS acknowledges support from the Mexican grant UNAM-PAPIIT IA101316.

References

Asensio Ramos, A., Martínez González, M. J., Manso Sainz, R., Corradi, R. L. M., & Leone, F. 2014, *ApJ*, 787, 111

Donati, J.-F., Semel, M., Carter, B. D., Rees, D. E., & Collier Cameron, A. 1997, *MNRAS*, 291, 658

Etoka, S., Zijlstra, A., Richards, A. M., Matsuura, M., & Lagadec, E. 2009, *The Eighth Pacific Rim Conference on Stellar Astrophysics: A Tribute to Kam-Ching Leung*, 404, 311

Girart, J. M., Patel, N., Vlemmings, W. H. T., & Rao, R. 2012, *ApJ*, 751, L20

Gómez, J. F., Uscanga, L., Green, J. A., *et al.* 2016, *MNRAS*, 461, 3259

Gómez, Y., Tafoya, D., Anglada, G., *et al.* 2009, *ApJ*, 695, 930

Greaves, J. S. 2002, *A&A*, 392, L1

Herpin, F., Baudry, A., Thum, C., Morris, D., & Wiesemeyer, H. 2006, *A&A*, 450, 667

Kylafis, N. D. & Shapiro, P. R. 1983, *ApJ*, 272, L35

Leal-Ferreira, M. L., Vlemmings, W. H. T., Kemball, A., & Amiri, N. 2013, *A&A*, 554, A134

Leone, F., Corradi, R. L. M., Martínez González, M. J., Asensio Ramos, A., & Manso Sainz, R.2014, *A&A*, 563, A43

Leone, F., Martínez González, M. J., Corradi, R. L. M., Privitera, G., & Manso Sainz, R. 2011, *Ap. Lett.*, 731, L33

Morris, M., Lucas, R., & Omont, A. 1985, *A&A*, 142, 107

Pérez-Sánchez, A. F., Vlemmings, W. H. T., Tafoya, D., & Chapman, J. M. 2013, *MNRAS*, 436, L79

Qiao, H.-H., Walsh, A. J., Gómez, J. F., *et al.* 2016, *ApJ*, 817, 37

Rees, D. E. & Semel, M. D. 1979, *A&A*, 74, 1

Sabin, L., Hull, C. L. H., Plambeck, R. L., *et al.* 2015, *MNRAS*, 449, 2368

Sabin, L., Wade, G. A., & Lèbre, A. 2015, *MNRAS*, 446, 1988

Sabin, L., Zhang, Q., Zijlstra, A. A., *et al.* 2014, *MNRAS*, 438, 1794

Sabin, L., Zijlstra, A. A., & Greaves, J. S. 2007, *MNRAS*, 376, 378

Steffen, M., Hubrig, S., Todt, H., *et al.* 2014, *A&A*, 570, A88

Tessore, B., Lèbre, A., & Morin, J. 2015, *SF2A-2015: Proceedings of the Annual meeting of the French Society of Astronomy and Astrophysics*, 429

Todt, H., Steffen, M., & Hubrig, S., *et al.* 2014, *Asymmetrical Planetary Nebulae VI Conference*, 103

Vlemmings, W. H. T., Ramstedt, S., Rao, R., & Maercker, M. 2012, *A&A*, 540, L3

Vlemmings, W. H. T. 2007, *Astrophysical Masers and their Environments*, 242, 37

Discussion

De Marco: Many AGB have now detected magnetic fields (MF). Only a minority of AGB stars are part of an interacting binary. So are we saying that single AGB stars can sustain long term MF, contrary to what theory predicts (Nordhaus *et al.* 2007)?

Zijlstra: The AGB star B fields measure B fields in the outflows and, with masers, in clumps where the MF could be intensified. Hence it is hard to determine what these measurements mean for the global MF in the AGB star.

Q: In systems like U Mon is the magnetic field detectable in the central star? Is it strong enough to affect the stellar system, e.g. the orbital separation?

Zijlstra: Magnetic fields should be detectable in principle: in stars like UMon and R Sct we suspect that the shocks in the stellar atmospheres play a role in detectability as they could amplify the fields by compression. In the weakfield regime, you need to combine high resolution spectropolarimetry, bright stars and good timing (when stars are at maximum). We have not yet conducted any study on the effects of a strong magnetic field. The current work (e.g. Ramirez Velez *et at al.* 2016) are aimed at recovering aspects such such as the geometry of the magnetic field.

Q: The key issue is whether the fields shape the gas or vice versa. When will we get an answer, and how will we be able to tell?

Planetary Nebulae: Multi-Wavelength Probes of Stellar and Galactic Evolution
Proceedings IAU Symposium No. 323, 2016
X. Liu, L. Stanghellini & A. Karakas, eds.

doi:10.1017/S1743921317000497

Molecular studies of Planetary Nebulae

Yong Zhang[1,2]

[1]Department of Physics, The University of Hong Kong, Pokfulam Road, Hong Kong, China
[2]Laboratory for Space Research, Faculty of Science, The University of Hong Kong, Pokfulam Road, Hong Kong, China
email: zhangy96@hku.hk

Abstract. Circumstellar envelopes (CEs) around evolved stars are an active site for the production of molecules. After evolving through the Asymptotic Giant Branch (AGB), proto-planetary nebula (PPN), to planetary nebula (PN) phases, CEs ultimately merge with the interstellar medium (ISM). The study of molecules in PNe, therefore, is essential to understanding the transition from stellar to interstellar materials. So far, over 20 molecular species have been discovered in PNe. The molecular composition of PNe is rather different from those of AGB and PPNe, suggesting that the molecules synthesized in PN progenitors have been heavily processed by strong ultraviolet radiation from the central star. Intriguingly, fullerenes and complex organic compounds having aromatic and aliphatic structures can be rapidly formed and largely survive during the PPN/PN evolution. The similar molecular compositions in PNe and diffuse clouds as well as the detection of C_{60}^{+} in the ISM reinforce the view that the mass-loss from PNe can significantly enrich the ISM with molecular species, some of which may be responsible for the diffuse interstellar bands. In this contribution, I briefly summarize some recent observations of molecules in PNe, with emphasis on their implications on circumstellar chemistry.

Keywords. planetary nebulae, circumstellar matter, astrochemistry, molecular processes, ISM: molecules, infrared: ISM

---

1. Introduction

The mass loss of evolved stars with intermediate-mass progenitors creates an expanding circumstellar envelope (CE), in which a variety of molecules are synthesized. During the CE evolution from the Asymptotic Giant Branch (AGB), proto-planetary nebula (PPN), to planetary nebula (PN) phases, the physical conditions dramatically alter, resulting in considerable changes in molecular composition. The yielded molecules eventually disperse into the interstellar medium (ISM), and may act as the seed of complex organic molecules discovered in molecular clouds (Ziurys 2008). Therefore, the investigation of molecules in PNe provides essential information on the material cycle in galaxies. The very existence of molecular gas in PNe can be inferred from the large discrepancy between the mass of the PN progenitor (1–8 $M_\odot$) and the masses of PN central stars and ionized nebulae ($\sim 0.9\,M_\odot$), known as the missing mass problem (Kwok 1994).

Up to now more than 80 different molecular species have been detected in C-rich AGB envelopes and PPNe. Numerous models have been developed to simulate the chemical processes in AGB envelopes (see Millar 2016, for a recent review). The standard picture is that the parent molecules are formed through equilibrium chemistry near the stellar photosphere, and then accelerated by radiation pressure on dust into outer regions where more complex molecules are synthesized by the chemistry induced by the external interstellar ultraviolet radiation field. The chemical evolution during the AGB-PPN-PN transitions has been investigated through systematic molecular line surveys (see Bujarrabal 2006, Cernicharo *et al.* 2011, for detailed reviews), which clearly imply dramatic changes of molecular composition at the PPN phase (e.g. Pardo *et al.* 2007, Park *et al.*

Table 1. The molecules detected in PNe.

2 atoms	3 atoms	4 atoms	5 atoms	> 5 atoms
H_2	C_2H	NH_3	c-C_3H_2	C_{60}
CH	HCN	H_2CO	HC_3N	C_{60}^+
CH^+	HCO^+			C_{70}
CN	H_2O			
CO	HNC			
CO^+	N_2H^+			
OH	SO_2			
OH^+	HCS^+?			
SiO	SiC_2?			
CS				
SO				
NS				

Note: Not including the molecular-rich pre-PN CRL 618 and solid-phase species.

2008, Zhang *et al.* 2013). The dynamical timescales of the three evolutionary stages can impose constraints on the efficiencies of chemical reactions. Thus CEs of evolved stars can serve as a laboratory for studying the chemical processes under extreme conditions.

With the evolution to the PN phase, increasing central-star temperature results in a photoionized region surrounded by a photon-dominated region (PDR), where the chemistry is dominated by the photons of energy from 6 to 13.6 eV, which contribute significantly to gas heating by ejecting electrons from dust grains and induce ion-neutral reactions. Shocks may play an important role in circumstellar chemistry by inducing endothermic reactions and destructing dust into molecules, though strong shocks might destroy molecules. Although high photodissociation and dissociative recombination rates render the environments of PNe hostile for the survival of molecules, the ultraviolet radiation can be largely attenuated by clumps, density-enhanced tori, as well as self- and mutual-shielding of molecules. These optically thick regions and the extensive gas envelopes presumably have different chemical evolution routes from one another. The first chemical model of PNe was developed by Black (1978) who predicted the existence of simple molecules such as H_2, H_2^+, HeH^+, OH, and CH^+ in PDRs. Although most of them were confirmed in subsequent observations, HeH^+, a molecule of cosmological interest, remains undetected. To date, various models, including equilibrium and time-dependent models, have been constructed to study the chemistry in envelopes and clumps, but none of them is fully satisfying (Kimura 2012, and reference therein).

Early observational studies of molecules in PNe mainly focused on H_2 and CO (e.g. Zuckerman & Gatley 1988, Huggins *et al.* 1996). Later on, millimetre observations were undertaken to trace the chemical variations of PNe (Bachiller *et al.* 1997, Josselin & Bachiller 2003), which indicated an appreciable enrichment of HCO^+ and CN radicals in PNe with respect to AGB envelopes. Infrared observations have proven to be useful for searching for new small molecules (Cernicharo *et al.* 1997, Liu *et al.* 1997) and fullerenes (Cami *et al.* 2010, García-Hernández *et al.* 2010) in PNe. The list of molecules in PNe continues to lengthen with more sensitive radio observations (e.g. Zhang *et al.* 2008, Tenenbaum *et al.* 2009, Edwards & Ziurys 2013,2014, Schmidt & Ziurys 2016) and the launch of the *Herschel* Space Observatory (e.g. Bujarrabal *et al.* 2012, van de Steene, this volume). Table 1 shows the molecules discovered so far in PNe. Although most of them are simple molecules, a big gap exists between the molecules with 5 and 60 atoms. It is reasonable to hypothesize that complex molecules ($\geqslant 6$ atoms) are generally present in PNe, and it is worthwhile to pursue future observations to search for the missing molecules. The aim of this paper is to selectively review the studies of molecules observed in PNe, mostly focusing on the results published in the last few years.

2. Small gas-phase molecules

H_2 *and CO*. The number of PNe harboring H_2 has recently been considerably increased by the UWISH2 survey (Froebrich *et al.* 2015, Gledhill, this volume), in which 284 extended H_2 sources are identified as PNe or PN candidates. H_2 emission appears stronger in bipolar PNe with equatorial rings (Marquez-Lugo *et al.* 2013, and reference therein). Manchado *et al.* (2015 and this volume) presented high-resolution H_2 images showing that the H_2 emission in the central torus appears clumpy rather than uniform, and suggested that those clumps might eventually populate the ISM and contribute to the baryonic dark matter. Based on the *Herschel* observations of an evolved PN NGC 6720, van Hoof *et al.* (2010) concluded that H_2 could be reformed on dust grains after the central star enters the cooling track. So far more than 80 PNe have been detected with CO (Huggins *et al.* 1996, 2005). Early observations revealed large variations in the CO properties from object to object, with young PNe exhibiting larger CO mass than evolved ones (Bujarrabal 2006). However, recent observations show that the CO abundance does not significantly vary during PN evolution (Edwards *et al.* 2014), suggesting that substantial amount of molecules can be expelled from PNe into the ISM.

H_2O *and OH*. When CEs enter the PN phase, H_2O and OH masers will rapidly disappear ($< 10^3$ yrs), and thus can trace young PNe. To date, five PNe with H_2O and seven with OH maser emission have been confirmed, among which only two exhibit both maser emission (Gomez *et al.* 2015a, Uscanga *et al.* 2012, Qiao *et al.* 2016). Gomez *et al.* (2015b) discovered the first "water fountain" PN that was speculated to be the youngest PN known so far. H_2O and OH infrared emission has been detected in the O-rich PN NGC 6302 (Bujarrabal *et al.* 2012) and the C-rich PN NGC 7027 (Liu *et al.* 1996, Wesson *et al.* 2010). A possible explanation for the presence of oxygen compounds in C-rich environments is that the PN might have experienced a transition from O- to C-rich chemistry following a helium shell flash. However, a more plausible explanation is that they origin from an ongoing photo-induced process in PDRs (Santander-García *et al.* 2012).

OH^+*,* CO^+*, and* HCO^+. The three cationic molecules can be formed in PDRs through the path $CO \xrightarrow{h\nu} O \xrightarrow{H^+} O^+ \xrightarrow{H_2} OH^+ \xrightarrow{H_2} H_2O^+ \xrightarrow{H_2} H_3O^+ \xrightarrow{e} OH \xrightarrow{C^+} CO^+ \xrightarrow{H_2} HCO^+$. For the first time, OH^+ was discovered in both young and evolved PNe by the *Herschel* Space Observatory (Aleman *et al.* 2014, Etxaluze *et al.* 2014). In evolved PNe, direct photoionization of CO provides an alternative route to produce CO^+. Bell *et al.* (2007) detected CO^+ in a sample of PNe with HCO^+ emission. The correlation between the CO^+ and HCO^+ line intensities provides support for the formation route of HCO^+ from CO^+. HCO^+ is a commonly detected molecule in PNe. Bachiller *et al.* (1997) found that HCO^+ abundance increases by two orders of magnitude during the PPN-PN transition. Recent observations suggest that the HCO^+ abundance is not closely related to the ages of PNe (Fig. 1). In highly HCO^+ or OH^+ enriched PNe, soft X-ray from the center may play an important role in chemistry, which results in abundant H_3^+ and then enhances the reaction $CO \xrightarrow{H_3^+} HCO^+$ or $O \xrightarrow{H_3^+} OH^+$.

CH and CH^+. Methylidyne radical and its cation, the important parent species leading to the formation of CN and HCN, have been detected in the infrared spectrum of NGC 7027 (Cernicharo *et al.* 1997, Liu *et al.* 1997, Wesson *et al.* 2010). They can be formed through the endothermic reaction $C^+ + 0.4\text{eV} \xrightarrow{H_2} CH^+$ followed by $CH^+ \xrightarrow{H_2}$

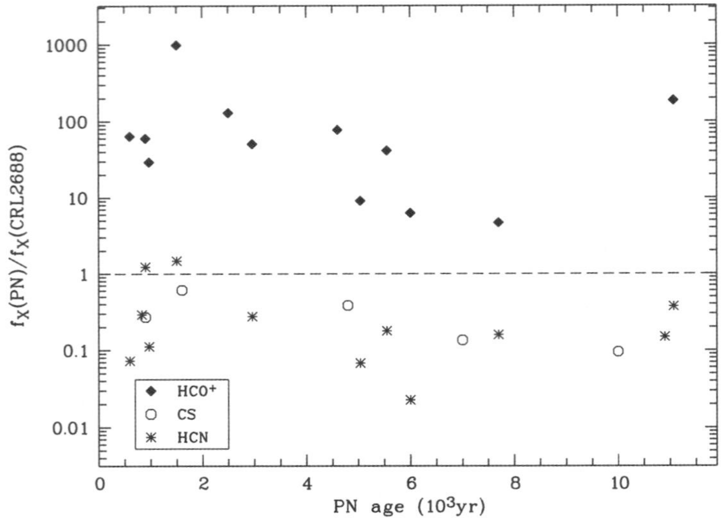

Figure 1. Molecular abundance (wrt H_2) ratios between PNe and the PPN CRL 2688 versus PN ages. The data are taken from Bachiller *et al.* (1997), Zhang *et al.* (2008,2013), Edwards *et al.* (2014), and Schmidt & Ziurys (2016). The dash line represents equivalent abundances.

$CH_2{}^+ \xrightarrow{e} CH$. The activation barrier for the formation of CH^+ may be overcome by the energy carried by shocks or vibrationally excited H_2 in PDRs.

N-bearing molecules. CN has a relatively high abundance in PNe (Bachiller *et al.* 1997), which can be attributed to the photodissociation of HCN. Chemical models predict that, even considering the shielding of clumps, significant destruction of HCN in PNe occurs at a timescale of $\sim 10^4$ yrs (Redman *et al.* 2003). However, this was not supported by recent observations that suggested an approximately constant HCN abundance during the PN evolution (Fig. 1, Schmidt & Ziurys 2016). HNC, a less stable isomer of HCN, has generally increasing abundance in PNe with respect to PPNe (Bachiller *et al.* 1997). However, the HNC/HCN isomer ratio in NGC 7027 appears abnormally low, which has not been well understood. HC_3N and NH_3, unlike those in the AGB and PPNe, are rarely detected in PNe. HC_3N was detected in NGC 7027 with low abundance (Zhang *et al.* 2008). NGC 6302 shows NH_3 emission (Bujarrabal *et al.* 2012). The two molecules can either be the remnant of circumstellar chemistry, or be reformed through neutral reactions in the dense, warm post-shock layers of PNe. The abundance of N_2H^+ in NGC 7027 is extremely high (Josselin & Bachiller 2003, Zhang *et al.* 2008). This might be due to the strong X-ray radiation which enhances the reaction $N_2 \xrightarrow{H_3^+} N_2H^+$. This molecule was detected in NGC 6537 and NGC 6302 as well (Edwards & Ziurys 2013, Hebden 2014).

S- and Si-bearing molecules. PNe are characterized with a lack of S- and Si-bearing molecules, presumably because of depletion onto dust grains. CS rotational lines are typically strong in C-rich AGB envelopes and PPNe, but never detected in C-rich PNe. Only a few O-rich PNe exhibit weak CS emission (Fig. 1, Edwards *et al.* 2014, Smith *et al.* 2016), probably suggesting that oxidation environments are favorable for sulfur in grains to be released into the gas phase. HCS^+ and SiC_2 were only tentatively detected in NGC 7027 and NGC 6302, respectively (Zhang *et al.* 2008, Hebden 2014). The other uncommon molecules SO, SO_2, and SiO were detected in M 2-28 (Edwards & Ziurys

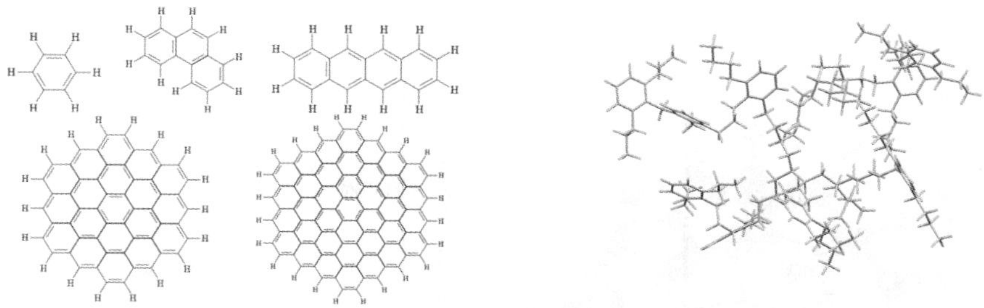

Figure 2. Schematic chemical structures of PAHs (left) and MAON (right).

2014). SiO, SO, and NS were revealed in the dense molecular regions of NGC 6302 by the SMA and ALMA observations (Hebden 2014, Santander-García *et al.* 2016).

C_2H, c-C_3H_2, and H_2CO. The three compounds are present in young and evolved PNe (e.g. Woods & Nyman 2005, Tenenbaum *et al.* 2009), suggesting that they can survive intense radiation fields. Initiating with acetylene, C_2H and the cyclic molecule *c*-C_3H_2 can be formed through photochemistry. However, Fuente *et al.* (2003) found that the large *c*-C_3H_2 abundance in NGC 7027 is beyond the predictions of gas-phase chemistry models, and might arise from the photodestruction of large aromatic/aliphatic compounds. Although the formation of H_2CO in PDRs might result from the radical reaction between O and CH_3, it is unclear whether dust surface reactions ($CO \xrightarrow{H} HCO \xrightarrow{H} H_2CO$) followed by photodesorption also play a role.

3. Complex organic species

The presence of complex organics can be inferred from the unidentified infrared emission (UIE) bands that mainly manifest themselves as aromatic features at 3.3, 6.2, 7.7, 8.6, and 11.3 μm, aliphatic features at 3.4, and 6.9 μm, and plateau emission around 8, 12, and 17 μm (Tielens 2008, Kwok 2011, and reference therein). UIE bands have been observed in a variety of astronomical environments, and their fluxes show that the carrier might be the most abundant molecules in the ISM. Both low- and high-excitation PNe exhibit UIE bands, indicating that their carrier is a significant nebular component. A wide variety of materials has been proposed as potential carriers of the UIE bands, among which polycyclic aromatic hydrocarbon (PAH) is the most commonly accepted one (see Peeters 2013, for a recent review). However, the PAH hypothesis is not without problems (Kwok & Zhang 2013, Zhang & Kwok 2015). Kwok & Zhang (2011) invoked an alternative model, mixed aromatic/aliphatic organic nanoparticles (MAONs), to interpret the UIE phenomenon. Fig. 2 shows the chemical structures of the PAHs and MAON. The spectral analysis shows that the aromatic/aliphatic component ratio increases during the transition from PPN to PN, suggesting that the aliphatic component may be processed to more stable aromatic rings by ultraviolet photons. The destruction of MAONs can provide a top-down route for the formation of smaller molecules. Nevertheless, no consensus has been reached concerning the UIE carrier. The MAON model was recently questioned (e.g. Li & Draine 2012, Yang *et al.* 2016).

Recent observations of NGC 7027 provide support for the formation of the UIE carrier through grain-grain collisions in post-shock regions (Lau *et al.* 2016). UIE bands were also discovered in O-rich PNe. Infrared images of O-rich PNe show that UIE primarily originates from PDRs associated with the outer edge of central tori or dense knots,

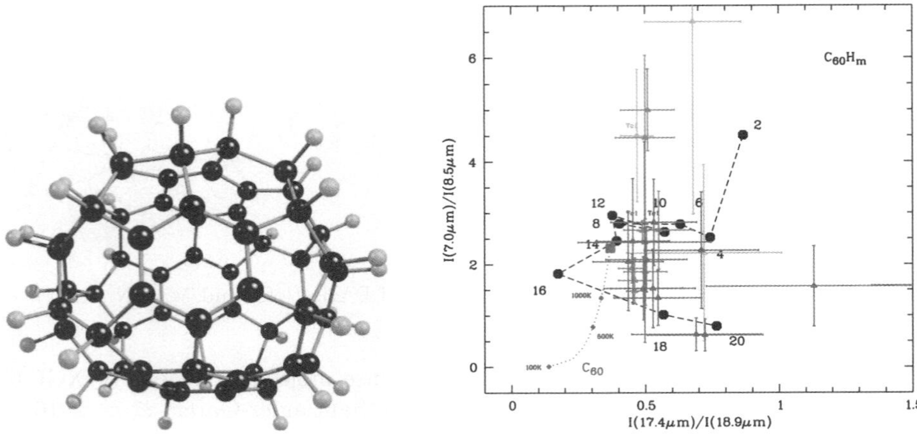

Figure 3. *Left*: Optimized structures of a $C_{60}H_{36}$ isomer. *Right*: Theoretical intrinsic strength ratios of the four $C_{60}H_m$ backbone vibrations (filled circles connected by dashed line, with the m values marked). The triangles with error bars are the observed values taken from Bernard-Salas *et al.* (2012), García-Hernández *et al.* (2012), and Otsuka *et al.*(2014). The dotted curve represents the predicted ratios based upon the thermal excitation model of C_{60}.

suggesting that the UIE carrier can be formed through a bottom-up process following the photodissociatoin of CO (Guzman-Ramirez *et al.* 2014, Cox *et al.* 2016).

4. Fullerene-related compounds

Given its high stability, C_{60}, the best-known fullerene member, has been long believed to exist throughout circumstellar and interstellar environments (Kroto *et al.* 1985). Because of high symmetry, the vibrational modes of C_{60} are degenerated into four individual infrared features, making the astronomical search possible. Soon after the first detection in a young PN (Cami *et al.* 2010), C_{60} was discovered in various astronomical environments, suggesting that it could survive in very harsh conditions, though the formation route is under a controversial debate (see Cami *et al.* 2011 and this volume, Zhang & Kwok 2013, Zhang *et al.* 2016, for the details). So far, 11 Galactic PNe and 11 Magellanic Cloud PNe have been detected with C_{60} (García-Hernández *et al.* 2012, Sloan *et al.* 2014, Otsuka *et al.* 2016). The detection rate is higher in more metal-poor environments. The presence of C_{60} in a PPN suggests that it can be formed within a timescale of $< 10^3$ yrs (Zhang & Kwok 2011). Given its rather low ionization potential, C_{60} presumably exists largely in the cation form in the ISM. The notion of C_{60}^+ as one of the DIB carriers (Foing & Ehrenfreund 1994) was recently verified by gas-phase experimental data, that is, five near-infrared DIBs were convincingly assigned to C_{60}^+ (Campbell *et al.* 2015,2016, Walker *et al.* 2015). A few DIBs were found to be enhanced in C_{60}-containing nebulae (Iglesias-Groth & Esposito 2013, Díaz-Luis *et al.* 2015), suggesting that they might have a circumstellar origin and probably be related to fullerenes or derivatives. Given their chemical activity and physical stability, fullerene-related compounds are good candidates for the DIB carrier (see Omont 2016, for a comprehensive discussion). Future studies should examine the spectra of fullerene compounds including hydrogenated fullerenes ($C_{60}H_m$), C_{60} adducts, endofullerenes, hererofullerenes, buckyonions, and so on.

If mixed with H atoms, C_{60} can be readily hydrogenated into $C_{60}H_{36}$ (Fig. 3) in laboratory environments. Thus $C_{60}H_m$ is highly likely to be present in PNe, and may contribute to the formation of H_2 by dehydrogenation (Cataldo & Iglesias-Groth 2009). A tentative detection of $C_{60}H_m$ in a C_{60}-containing PPN has been presented by Zhang & Kwok (2013). However, two C_{60}-rich PNe do not exhibit any detectable C-H feature (Díaz-Luis *et al.* 2016). Theoretical investigation indicated that with slight hydrogenation, the four C_{60}-backbone vibration features at 7.0, 8.5, 17.4, and 18.9 μm are still visible but diminish their intensities in different degrees (Zhang *et al.* 2016). Therefore, C_{60}-poor PNe might be more ideal targets for the search of $C_{60}H_m$. The excitation mechanism of C_{60} bands is still unclear, hindering their use as a probe of PN environments (Brieva *et al.* 2016). As shown in Fig. 3, the observed intensity ratios are too dispersed to accord with the excitation model characterized with one single free parameter (i.e. fluorescence or thermal excitation models governed by the photon field energy and temperature, respectively), but can be well interpreted in terms of hydrogenation (Zhang *et al.*, in preparation). This might provide circumstantial evidence for the presence of $C_{60}H_m$ in PNe.

5. Conclusion

A wealth of observations have demonstrated that molecules are an essential component in PNe. Even large molecules can be formed and/or survive during the PN evolution. The complex molecules blown from PNe into the ISM might contribute to the DIBs and seed the chemistry of molecular clouds. However, the complete picture of the chemistry in PNe is still far from clear. Systematic molecular line surveys of PNe are relatively scarce. Many molecules are discovered in only one or a few PNe. There is an absence of thorough understanding of the connection between the molecular composition and the properties of PNe. It is expected that new and important insights will be offered by the next generation instruments. Searching for fullerene-related compounds in PNe might provide crucial clues toward solving the DIB enigma. Further theoretical and experimental studies are badly needed.

Acknowledgements

I thank the SOC for inviting me to do this review talk. I am indebted to my colleagues who have closely collaborated with me in the study of circumstellar molecules, including Sun Kwok, Jun-ichi Nakashima, and Seyedabdolreza Sadjadi. Financial support for this work was provided by the Research Grants Council of the Hong Kong under grants HKU7073/11P and HKU7062/13P.

References

Aleman, I., Ueta, T., Ladjal, D., *et al.* 2014, *A&A*, 566, A79
Bachiller, R., Forveille, T., Huggins, P. J., & Cox, P. 1997, *A&A*, 324, 1123
Bell, T. A., Whyatt, W., Viti, S., & Redman, M. P. 2007, *MNRAS*, 382, 1139
Bernard-Salas, J., Cami, J., Peeters, E., *et al.* 2012, *ApJ*, 757, 41
Black, J. H. 1978, *ApJ*, 222, 125
Brieva, A. C., Gredel, R., Jäger, C., Huisken, F., & Henning, T. 2016, *ApJ*, 826, 122
Bujarrabal, V. 2006, in IAU Symp., 234, 193
Bujarrabal, V., Alcolea, J., Soria-Ruiz, R., *et al.* 2012, *A&A*, 537, A8
Cami, J., Bernard-Salas, J., Peeters, E., & Malek, S. E. 2010, *Science*, 329, 1180
Cami, J., Bernard-Salas, J., Peeters, E., & Malek, S. E. 2011, in IAU Symp., 280, 216
Campbell, E. K., Holz, M., Gerlich, D., & Maier, J. P. 2015, *Nature*, 523, 322
Campbell, E. K., Holz, M., Maier, J. P., *et al.* 2016, *ApJ*, 822, 17
Cataldo, F. & Iglesias-Groth, S. 2009, *MNRAS*, 400, 291
Cernicharo, J., Liu, X.-W., González-Alfonso, E., *et al.* 1997, *ApJL*, 483, L65
Cernicharo, J., Agúndez, M., & Guélin, M. 2011, in IAU Symp., 280, 237

Cox, N. L. J., Pilleri, P., Berné, O., Cernicharo, J., & Joblin, C. 2016, *MNRAS*, 456, L89
Díaz-Luis, J. J., García-Hernández, D. A., Kameswara Rao, N., *et al.* 2015, *A&A*, 573, A97
Díaz-Luis, J. J., García-Hernández, D. A., Manchado, A., & Cataldo, F. 2016, *A&A*, 589, A5
Edwards, J. L. & Ziurys, L. M. 2013, *ApJL*, 770, L5
Edwards, J. L., Cox, E. G., & Ziurys, L. M. 2014, *ApJ*, 791, 79
Edwards, J. L. & Ziurys, L. M. 2014, *ApJL*, 794, L27
Etxaluze, M., Cernicharo, J., Goicoechea, J. R., *et al.* 2014, *A&A*, 566, A78
Foing, B. H. & Ehrenfreund, P. 1994, *Nature*, 369, 296
Froebrich, D., Makin, S. V., Davis, C. J., *et al.* 2015, *MNRAS*, 454, 2586
Fuente, A., Rodrıguez-Franco, A., Garcıa-Burillo, S., *et al.* 2003, *A&A*, 406, 899
García-Hernández, D. A., Manchado, A., García-Lario, P., *et al.* 2010, *ApJL*, 724, L39
García-Hernández, D. A., Villaver, E., García-Lario, P., *et al.* 2012, *ApJ*, 760, 107
Gómez, J. F., Rizzo, J. R., Suárez, O., *et al.* 2015a, *A&A*, 578, A119
Gómez, J. F., Suárez, O., Bendjoya, P., *et al.* 2015b, *ApJ*, 799, 186
Guzman-Ramirez, L., Lagadec, E., Jones, D., *et al.* 2014, *MNRAS*, 441, 364
Hebden, K., 2014, *Ph.D. Thesis*
Huggins, P. J., Bachiller, R., Cox, P., & Forveille, T. 1996, *A&A*, 315, 284
Huggins, P. J., Bachiller, R., Planesas, P., Forveille, T., & Cox, P. 2005, *ApJS*, 160, 272
Iglesias-Groth, S. & Esposito, M. 2013, *ApJL*, 776, L2
Josselin, E. & Bachiller, R. 2003, *A&A*, 397, 659
Kimura, R. K., Gruenwald, R., & Aleman, I. 2012, *A&A*, 541, A112
Kroto, H. W., Heath, J. R., Obrien, S. C., Curl, R. F., & Smalley, R. E. 1985, *Nature*, 318, 162
Kwok, S. 1994, *PASP*, 106, 344
Kwok, S. 2011, in IAU Symp., 280, 203
Kwok, S. & Zhang, Y. 2011, *Nature*, 479, 80
Kwok, S. & Zhang, Y. 2013, *ApJ*, 771, 5
Lau, R. M., Werner, M., Sahai, R., & Ressler, M. E. 2016, *ApJ*, in press
Li, A. & Draine, B. T. 2012, *ApJL*, 760, L35
Liu, X.-W., Barlow, M. J., Nguyen-Q-Rieu, *et al.* 1996, *A&A*, 315, L257
Liu, X.-W., Barlow, M. J., Dalgarno, A., *et al.* 1997, *MNRAS*, 290, L71
Manchado, A., Stanghellini, L., Villaver, E., *et al.* 2015, *ApJ*, 808, 115
Marquez-Lugo, R. A., Ramos-Larios, G., Guerrero, M. A., *et al.* 2013, *MNRAS*, 429, 973
Millar, T. J. 2016, *JPhCS*, 728, 052001
Omont, A. 2016, *A&A*, 590, A52
Otsuka, M., Kemper, F., Cami, J., Peeters, E., & Bernard-Salas, J. 2014, *MNRAS*, 437, 2577
Otsuka, M., Kemper, F., Leal-Ferreira, M. L., *et al.* 2016, *MNRAS*, 462, 12
Pardo, J. R., Cernicharo, J., Goicoechea, J. R., Guélin, M., & Ramos, A. A. 2007, *ApJ*, 661, 250
Park, J. A., Cho, S.-H., Lee, C. W., & Yang, J. 2008, *AJ*, 136, 2350
Peeters, E. 2013, in IAU Symp., 297, 187
Qiao, H.-H., Walsh, A. J., Gómez, J. F., *et al.* 2016, *ApJ*, 817, 37
Redman, M. P., Viti, S., Cau, P., & Williams, D. A. 2003, *MNRAS*, 345, 1291
Santander-García, M., Bujarrabal, V., & Alcolea, J. 2012, *A&A*, 545, A114
Santander-García, M., Bujarrabal, V., Alcolea, J., *et al.* 2016, *A&A*, in press
Schmidt, D. R. & Ziurys, L. M. 2016, *ApJ*, 817, 175
Sloan, G. C., Lagadec, E., Zijlstra, A. A., *et al.* 2014, *ApJ*, 791, 28
Smith, C. L., Zijlstra, A. A., & Fuller, G. A. 2015, *MNRAS*, 454, 177
Tenenbaum, E. D., Milam, S. N., Woolf, N. J., & Ziurys, L. M. 2009, *ApJL*, 704, L108
Tielens, A. G. G. M. 2008, *ARA&A*, 46, 289
Uscanga, L., Gómez, J. F., Suárez, O., & Miranda, L. F. 2012, *A&A*, 547, A40
van Hoof, P. A. M., van de Steene, G. C., Barlow, M. J., *et al.* 2010, *A&A*, 518, L137
Walker, G. A. H., Bohlender, D. A., Maier, J. P., & Campbell, E. K. 2015, *ApJL*, 812, L8
Wesson, R., Cernicharo, J., Barlow, M. J., *et al.* 2010, *A&A*, 518, L144
Woods, P. M. & Nyman, L.-Å. 2005, IAU Symp., 235, 326
Yang, X. J., Glaser, R., Li, A., & Zhong, J. X. 2016, *MNRAS*, 462, 1551

Zhang, Y. & Kwok, S., Dinh-V-Trung 2008, *ApJ*, 678, 328-346
Zhang, Y. & Kwok, S. 2011, *ApJ*, 730, 126
Zhang, Y. & Kwok, S. 2013, *EP&S*, 65, 1069
Zhang, Y., Kwok, S., Nakashima, J.-i., & Chau, W., Dinh-V-Trung 2013, *ApJ*, 773, 71
Zhang, Y. & Kwok, S. 2015, *ApJ*, 798, 37
Zhang, Y., Kwok, S., & Sadjadi, S. 2016, *JPhCS*, 728, 052004
Ziurys, L. M. 2008, IAU Symp., 251, 147
Zuckerman, B. & Gatley, I. 1988, *ApJ*, 324, 501

Discussion

CHEN: How do you prepare hydragenated fullerenes?

ZHANG: We did not prepare hydragenated fullerenes in experiments. These spectra were from theoretical calculations. We calculated the spectra of many isomers for each hydragenated C60, selected those with low energy, and combined them. The combined spectra can be used to compare with the observations.

Planetary Nebulae: Multi-Wavelength Probes of Stellar and Galactic Evolution
Proceedings IAU Symposium No. 323, 2016
X. Liu, L. Stanghellini & A. Karakas, eds.

doi:10.1017/S1743921317001478

GTC/CanariCam mid-IR imaging of the fullerene-rich planetary nebula IC 418

J. J. Díaz-Luis[1,2], D. A. García-Hernández[1,2], A. Manchado[1,2,3], P. García-Lario[4], E. Villaver[5] and G. García-Segura[6]

[1]Instituto de Astrofísica de Canarias, C/ Vía Láctea s/n, E−38205 La Laguna, Spain
email: jdiaz@iac.es

[2]Departamento de Astrofísica, Universidad de La Laguna (ULL), E−38206 La Laguna, Spain

[3]Consejo Superior de Investigaciones Científicas, Madrid, Spain

[4]Herschel Science Centre, ESAC-ESA, E−28691 Villanueva de la Cañada, Madrid, Spain

[5]UAM, Departamento de Física Teórica C-XI, E−28049 Madrid, Spain

[6]Instituto de Astronomía, UNAM, 22860 Ensenada, México

Abstract. The formation process(es) of fullerenes in space is still uncertain and several mechanisms have been proposed in the literature. In particular, the most accepted idea to explain the simultaneous presence of fullerenes and PAH-like emission in the H-rich circumstellar envelopes of PNe is that these molecular species may be formed from the photochemical processing of a carbonaceous compound with a mixture of aromatic and aliphatic structures, which should be a major constituent of their circumstellar envelopes. Here we present seeing-limited narrow-band mid-IR GTC/CanariCam images of the fullerene-containing PN IC 418. The narrow-band images cover the 9−13, 11.3, and 17.4 μm emission features (and their adjacent continua) in this extended PN. We study the relative sub-arcsecond spatial distribution of the nebula in these filters with the intention of getting some clues about the formation process of fullerenes in H-rich circumstellar environments.

Keywords. Astrochemistry, molecular processes, stars: evolution, planetary nebulae: individual (IC 418), AGB and post-AGB, dust.

1. Introduction

The detection of fullerenes around old stars together with the recent identification of $C_{60}{}^{+}$ as a diffuse interstellar band (DIB) carrier (the only DIB carrier known to date; Campbell *et al.* 2015) indicate that fullerenic molecular nanostructures should be ubiquitous in the interstellar medium (ISM), playing an important role in many aspects of circumstellar/interstellar Chemistry and Physics.

The formation process of fullerenes in space is still uncertain and several mechanisms have been proposed in the literature. The most suitable ones being: i) the formation in H-poor environments (Goeres & Sedlmayr 1992); ii) high-temperature formation in C-rich environments (Jäger *et al.* 2009); iii) photochemical processing of hydrogenated amorphous carbon grains (HACs; García-Hernández *et al.* 2010); and iv) photochemical processing of large PAHs (Berné & Tielens 2012).

At present, the most likely explanation for the simultaneous presence of fullerenes and PAH-like emission in the H-containing circumstellar environments of planetary nebulae (PNe) is that they may be formed by the decomposition of a carbonaceous material with a mixture of aromatic and aliphatic structures; e.g., similar to hydrogenated amorphous carbon grains (HAC-like hereafter; García-Hernández *et al.* 2010; Micelotta *et al.* 2012). Such carbon compounds should be a major constituent of the circumstellar envelope of

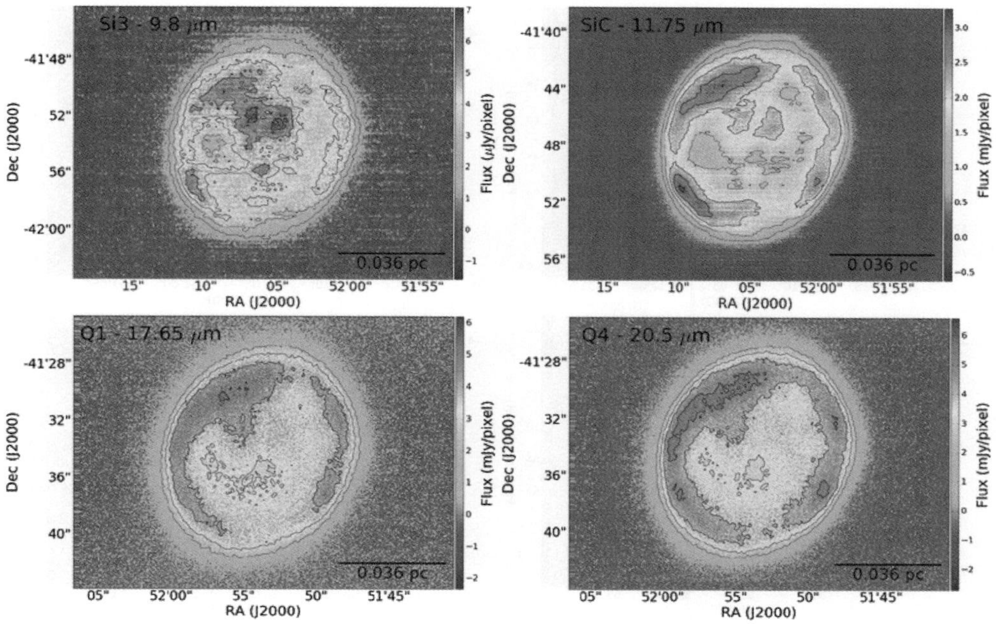

Figure 1. Flux-calibrated GTC/CanariCam mid-IR images of the C_{60}-PN IC 418 in the Si3, SiC, Q1, and Q4 filters.

C-rich evolved stars. This fullerene formation scenario (the HAC scenario) is suggested by the coexistence of a complex mix of aliphatic and aromatic species such as HACs, PAH clusters, fullerenes, and small dehydrogenated carbon clusters (possibly planar C_{24} or a small fragment of a graphene sheet) in PNe of the Magellanic Clouds and our own Galaxy (García-Hernández *et al.* 2011, 2012).

Mid-IR imaging observations of the different molecular/dust species in PNe with fullerenes may help to understand the formation process of fullerenes in these objects.

2. GTC/Canaricam mid-IR observations

Seeing-limited (FWHM ~0.5") mid-IR images of the extended fullerene-rich PN IC 418 were obtained with GTC/Canaricam in January and December 2013 by using narrow-band filters centered at 11.30 μm (PAH2, $\Delta\lambda$ = 0.6 μm), 11.75 μm (SiC, $\Delta\lambda$ = 2.5 μm), and 17.65 μm (Q1, $\Delta\lambda$ = 0.9 μm), as well as in their adjacent continua at 9.8 μm (Si3, $\Delta\lambda$ = 1.0 μm) and 20.50 μm (Q4, $\Delta\lambda$ = 1.0 μm). The signal-to-noise ratio in the final combined images is ~5-16. Figure 1 shows the final flux-calibrated mid-IR GTC/Canaricam images of IC 418.

The PAH2 and SiC filters cover the PAH-like feature at 11.3 μm and the broad 9−13 μm band (possibly related with HAC-like dust grains; e.g., García-Hernández *et al.* 2010, 2011), while the Q1 filter covers the C_{60} emission feature at 17.4 μm. The Si3 and Q4 filters cover the dust continuum emission at 9.8 and 20.5 μm, respectively.

3. Results

The Si3 image displays the maximum emission near the IC 418 central star, while the SiC, Q1, and Q4 images show a ring-like structure (Fig. 1). The similar spatial

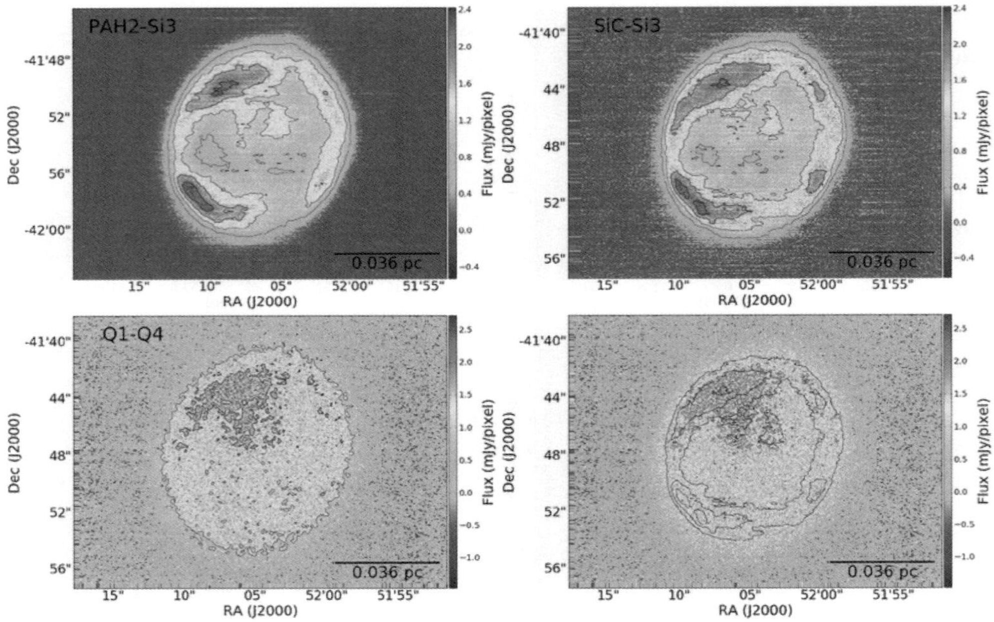

Figure 2. Continuum-subtracted GTC/Canaricam mid-IR images of IC 418: PAH2−Si3, SiC−Si3, and Q1−Q4. The bottom-right panel displays the Q1−Q4 continuum-subtracted image (C_{60} emission) with overlaid contours of the 9−13 μm + 11.3 μm PAH-like (SiC−Si3) emission.

distribution seen in the Q1 and Q4 filters suggests that fullerenes may be co-spatial with the dust continuum emission (e.g., they could be attached to the dust grains) or that both images are dominated by the dust continuum emission (e.g., because of the low feature-to-continuum ratio). Also, the similar ring-like structure seen in the SiC image (which is dominated by the broad 9−13 μm feature) indicates that the 9−13 μm band carrier is co-spatial with the dust grains emitting at 20.5 μm (Q4). On the other hand, by comparing the Si3 image with public Hubble Space Telescope optical nebular emission images of IC 418, we find that the dust continuum emission at 9.8 μm is very similar to the [O III] nebular emission, indicating that dust grains coexist with ionized material.

In order to study the relative spatial distribution of the C_{60} molecules, the PAH-like molecules, and the 9−13 μm band carrier, we have used the Si3 and Q4 images to subtract the dust continuum contribution. Figure 2 shows the resulting PAH2−Si3, SiC−Si3, and Q1−Q4 continuum-subtracted images as well as a comparison of the spatial distribution of the C_{60} (Q1−Q4) and 9−13 μm + 11.3 μm PAH-like (SiC−Si3) emissions.

The PAH2 and SiC images cover the weak 11.3 μm and the broad 9-13 μm features (the PAH+HAC-like emission) with different filter widths; moreover, the PAH2−Si3 and SiC−Si3 continuum-subtracted images display, as expected because of the weak 11.3 μm PAH-like feature, an almost identical spatial distribution. The 9−13 μm emission is mainly located in the outer regions of the nebula, defining a ring-like structure with the stronger emission peaks at the northeast and southeast regions. The continuum-subtracted fullerene emission (the Q1−Q4 image in Fig. 2), however, is very weak and mainly located at the northeast, extending from the inner (even near the central star) to the outer regions of the nebula. The weak fullerene emission is quite similar to that of the dust continuum emission at 9.8 μm, which also coincides with the spatial distribution of the [O III] nebular emission (see above). At the northeast outer regions of the nebula,

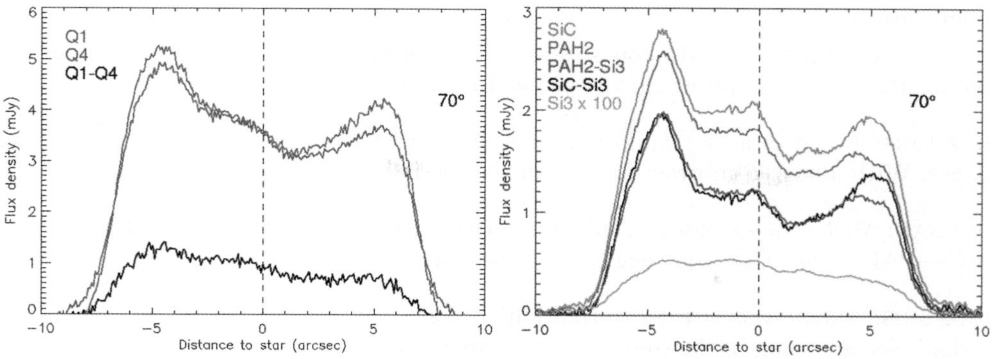

Figure 3. Intensity profiles along the IC 418 nebula at 70° with respect to the equatorial plane as a function of the distance to the central star. Left panel: intensity profiles in the Q1 (C_{60} + continuum at 17.65 μm), Q4 (continuum at 20.5 μm), and continuum-subtracted Q1−Q4 (C_{60}) images. Right panel: intensity profiles in the Si3 (continuum at 9.8 μm), SiC and PAH2 (both containing 9-13 μm carrier + 11.3 μm PAH-like emission + continuum), and continuum-subtracted SiC−Si3 and PAH2−Si3 (9-13 μm carrier + 11.3 μm PAH-like emission) images.

fullerenes seem to coexist with the 9−13 μm carrier, while they are co-spatial with the dust grains and ionized material in the inner regions closer to the central star.

Figure 3 shows intensity profiles along the IC 418 nebula at 70° with respect to the equatorial plane as a function of the distance to the central star. The 9−13 μm emission peaks at ~4−5" from the central star (at both northeast and southwest sides of the nebula), while the weaker fullerene emission mainly extends from the central star to ~5" at the northeast of the nebula (~6300 au).

The interpretation of the present images regarding the dominant formation process of fullerenes in PNe is somewhat contradictory. The presence of fullerene emission near the central star, where the 9−13 μm emission is much weaker, would suggest that C_{60} may be a photo-product of the 9-13 μm band carrier. However, fullerenes seem to coexist with the 9-13 μm band carrier at the northeast outer regions of the nebula; something that could indicate that fullerenes may be attached to dust grains or well mixed with the 9-13 μm band carrier. Another possible interpretation may be that fullerenes are not directly linked to the 9-13 μm band carrier. We conclude that higher sensitivity mid-IR images as well as spatially resolved mid-IR spectra along the nebula would be desirable in order to reach a definite answer.

References

Berné, O. & Tielens, A. G. G.. M. 2012, *Proceedings of the National Academy of Science*, 109, 401

Campbell, E. K., Holz, M., Gerlich, D., & Maier, J. P. 2015, *Nature*, 523, 322

García-Hernández, D. A., Manchado, A., García-Lario, P. *et al.* 2010, *ApJ*, 724, L39

García-Hernández, D. A., Iglesias-Groth, S., Acosta-Pulido, J. A. *et al.* 2011, *ApJ*, 737, L30

García-Hernández, D. A., Villaver, E., García-Lario, P. *et al.* 2012, *ApJ*, 760, 107

Goeres, A. & Sedlmayr, E. 1992, *A&A*, 265, 216

Jäger, C., Huisken, F., Mutschke, H., Jansa, I. L., & Henning, Th. 2009, *ApJ*, 696, 706

Micelotta, E. R., Jones, A. P., Cami, J., Peeters, E. *et al.* 2012, *ApJ*, 761, 35

Discussion

GUZMAN-RAMIREZ: How do you subtract the continuum images? Did you take into account the size of the filter and the slope of the continuum?

DIAZ-LUIS: No, it is a first approximation and now we will have to flux-calibrate our images. With a simple subtraction we can see the different distributions.

HRIVNAK: You see association of fullerenes with HACs+PAH in NE region but not with HACs+PAH in SE. Can you suggest why the difference?

DIAZ-LUIS: Maybe the process (formation) takes place in that part of the nebula and we don't see anything in the south. Photochemical processing of HACs has to take place under certain conditions.

SLOAN: What is the basis for identifying the "9 - 13 μm emission" as HAC? Sloan *et al.* (2014) showed that the broad feature $\sim$11.5 μm was a combination of SiC+PAHs, perhaps Class D PAHs in some cases, but still SiC+PAHs.

DIAZ-LUIS: The 12 μm plateau is certainly not SiC. SiC feature has been detected in over 700 carbon stars and the SiC feature does not look like the 12 μm plateau. This is due to superposition of C-H out-of plane bending mode of aliphatic side groups.

Session 4: From the AGB to the WD stellar phases
Chair: Miriam Pena

Planetary Nebulae: Multi-Wavelength Probes of Stellar and Galactic Evolution
Proceedings IAU Symposium No. 323, 2016
X. Liu, L. Stanghellini & A. Karakas, eds.

doi:10.1017/S1743921317000977

Progress and Application of The Initial-Final Mass Relation

Jeffrey D. Cummings

Center for Astrophysical Sciences, Johns Hopkins University
3400 N Charles Street
Baltimore, MD 21218, USA
email: jcummi19@jhu.edu

Abstract. The stellar life cycle is dominated by phases such as the hydrogen-burning stage and the remnant white dwarf cooling phase. However, between these two stages, stars dramatically transform themselves by losing the bulk of their mass. Planetary nebulae (PNe) provide a powerful clue to the processes involved in this transformation, but they are very complex. Over the past 15 years, a new wave of imaging and spectroscopy programs have uncovered the remnants of PNe, white dwarfs, in a wide range of well-measured environments. With this we can map the masses and temperatures of the stellar remnants to the properties of their progenitors. This work has now led to the first uniform mapping of the initial-final mass relation from 1.5 to 7 $M_{\odot}$. The resulting relation is a fundamental input to our understanding of stellar evolution for low and intermediate-mass stars that produce PNe and has a wide range of applications.

Keywords. White Dwarfs, AGB and post-AGB, Planetary Nebula, Mass Loss

1. Introduction

White dwarfs provide tools to help understand the late stages of stellar evolution. They help most critically with the challenging asymptotic giant branch (AGB) stars and planetary nebulae (PNe) phase. AGB stars and PNe are very bright but extremely brief stages in stellar evolution ($\sim$a few million years or less and tens of thousand years, respectively), and their complexity makes both modeling them and observationally disentangling their characteristics challenging. White dwarfs, while significantly fainter, are extremely long lived and relatively numerous. Additionally, we can directly observe them without having to consider significant effects of the surrounding nebulosity from mass loss.

Deriving the intrinsic properties of white dwarfs can be done with low resolution spectra of their Balmer lines or He features. A majority of white dwarfs have pure Hydrogen (DA) but an important fraction have pure Helium (DB) atmospheres. This results from the rapid gravitational settling of all heavier elements. This gives white dwarfs simple spectra with multiple strong and broad Balmer features (e.g., Bergeron *et al.* 1992) or Helium features (e.g., Beauchamp *et al.* 1996) in the blue (3700 to 5200 Å). These features are pressure sensitive and fitting them all simultaneously can precisely derive both the $T_{\rm eff}$ and surface gravities of these very compact and typically very hot objects. This technique has become far more common than other methods to estimate parameters, those being estimates from multi-color photometry (e.g., Wegner 1983), gravitational redshift (e.g., Wegner 1989), and white dwarfs in binaries. These methods, however, still provide important consistency checks and in certain contexts can be very useful.

Direct application of the $T_{\rm eff}$ and surface gravities to white dwarf cooling models (e.g., Fontaine *et al.* 2001) derives the mass, cooling age, luminosity, and color of the white dwarf. The cooling models are relatively simple, but they have important sensitivity to the white dwarf compositions. For example, the most massive white dwarfs are believed to have oxygen-neon (ONe) cores (Althaus *et al.* 2007) instead of the more common carbon-oxygen (CO) cores formed from lower-mass stars that cannot ignite off-center carbon burning in their cores (Garcia-Berro *et al.* 1997). ONe core white dwarfs cool more rapidly and have higher surface gravities than CO white dwarfs of the same mass would. The white dwarf mass at which this transition occurs remains uncertain, but it is approximately 1.1 $M_{\odot}$.

Being able to measure spectroscopically both the intrinsic white dwarf masses and cooling ages provides many tools to studying the historical characteristics and distributions of stellar populations (e.g., Tremblay *et al.* 2014; 2016), and to infer information about each white dwarf's progenitor. One of the most critical relations for understanding stellar evolution is the initial-final mass relation (IFMR), which is a direct comparison of the mass that a star forms with on the main sequence to its final mass that it will have as a white dwarf. The IFMR provides a direct measurement of the total mass loss during a star's life and how it varies with initial mass, and with more detailed analysis, how it varies with metallicity. This gives critical constraints to AGB stars (e.g., Kalirai *et al.* 2014), PNe (e.g., Ciardullo 2010), and to our understanding of supernovae and their rates (e.g., Greggio *et al.* 2010). The IFMR, however, remained poorly constrained for many years. After significant work on the relation in the past 15 years its applications are now more reliable, but it still has several important limitations.

2. The Initial-Final Mass Relation

2.1. *A Brief History*

One of the first detailed looks at the IFMR was the analysis of Weidemann (1977). Weidemann merged the results of a small number of studied white dwarfs at the time. These being LB 1497 from the Pleiades (Schulz 1976), seven white dwarfs from the Hyades (Eggen & Greenstein 1965; van Altena 1969; Sweeney 1976), and three well studied white dwarfs in binaries. Weidemann found a generally consistent trend with increasing $M_{\rm initial}$ creating more massive white dwarfs. Weidemann's primary goal in this analysis was to compare this semi-empirical relation to the evolutionary and mass-loss models of the time (e.g., Paczyński 1970; Fusi-Pecci & Renzini 1976; Wood & Cahn 1977). It was consistently found that to recreate the IFMR of Weidemann (1977) required higher mass loss than nearly all models of the time predicted. However, the relation was still limited by very small numbers and limitations in the techniques used to derive both initial and final masses.

In the 80s and 90s progress with the IFMR was steady but slow due to the challenges of finding new cluster white dwarfs and getting deep spectroscopy for precision analysis. Focus was put on the Hyades (e.g., Reid 1996), NGC 3532 (e.g., Reimers & Koester 1989; Koester & Reimers 1993), and NGC 2516 (Koester & Reimers 1996). With significantly improved models and techniques to determine white dwarf parameters (Bergeron *et al.* 1992), the masses and cooling ages were meaningfully adjusted. For example, spectroscopic redeterminations of the mass of the Pleiades's LB 1497 increased its previously measured 0.9 $M_{\odot}$ to 1.08 $M_{\odot}$ (Bergeron *et al.* 1995). With these data Weidemann *et al.* (2000) found a more precise but still sparse and limited IFMR.

With the advent of wide-field imagers and multi-object spectrographs on larger telescopes over the past 15 years, there has been an explosion of new white dwarf data. Namely in NGC 2099 (Kalirai *et al.* 2005), Praesepe (Dobbie *et al.* 2006a; Casewell *et al.* 2009), NGC 6791 (Kalirai *et al.* 2007), NGC 6819 and NGC 7789 (Kalirai *et al.* 2008), NGC 2287 and NGC 3532 (Dobbie *et al.* 2009), NGC 2618 (Williams *et al.* 2004; 2009),

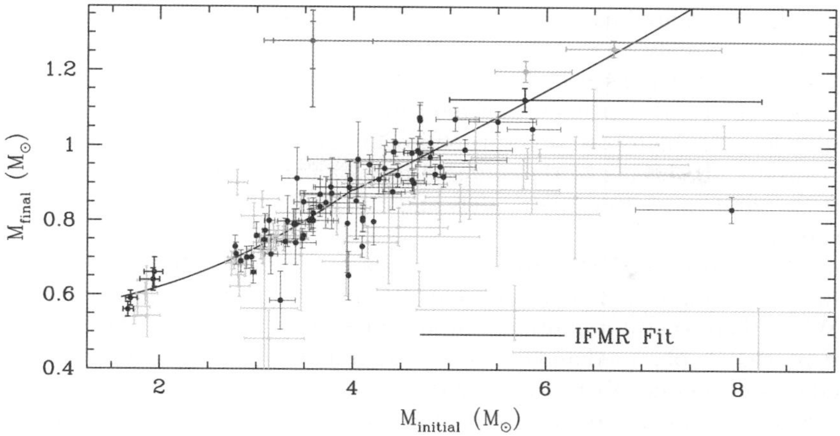

Figure 1. The IFMR combining the work of Kalirai *et al.* (2014) and Cummings *et al.* (2015; 2016a; 2016b), which analyzed newly discovered white dwarfs and uniformly analyzed previous data publicly available. Green points show the Pleiades white dwarfs connected only kinematically to the cluster (Dobbie *et al.* 2006b). The blue point shows the ultramassive NGC 2099 white dwarf that deviates significantly, likely due to its old age introducing significant errors (blue 1σ, red 2σ). Comparison to the IFMR of Salaris *et al.* (2009; light blue) shows that newly discovered white dwarfs and precision analysis of the photometry and spectroscopy increases the sample by ~50% and reduces the scatter by ~50%. (See online version for Figure 1 in color.)

and the globular cluster NGC 6121 (Kalirai *et al.* 2009). This more than doubled the number of analyzed cluster white dwarfs in the IFMR from 2000 to 2009.

2.2. *Recent Progress*

Salaris *et al.* (2009), Catalán *et al.* (2008), and Kalirai *et al.* (2009) each looked at the now large samples of known white dwarfs in open clusters, combining the work of all previous studies. There remained two critical limitations in these semi-empirical relations. First, there were two large gaps in the relation at $M_{initial}$ of 2 to 2.75 $M_\odot$ and for masses greater than 6 to 7 $M_\odot$ (e.g., $M_{final} > 1.05$ $M_\odot$). This intermediate mass range requires observations of white dwarfs in star clusters of ~1 Gyr, where stars of these masses have had enough time to fully evolve and the resulting white dwarfs have not significantly cooled. For the highest-mass gap, the question remained, where are the high-mass white dwarfs? The second limitation was that the scatter in the relations were large (±0.10 $M_\odot$). Because these white dwarfs come from multiple clusters and studies from multiple independent researchers, is this scatter simply the result of systematic differences between the analyses? Salaris *et al.* (2009) considered in more depth the adopted ages of the host clusters, but the scatter remained. Or is this persistent scatter indicative of true variations in mass-loss rates at constant $M_{initial}$? Do differences in composition or environment between these clusters affect mass-loss rates by up to 10%? Or is mass loss stochastic in nature?

To begin to address both of these limitations, Cummings *et al.* (2015; 2016a; 2016b) have searched for new massive white dwarfs and have done detailed self-consistent analysis of both the white dwarf spectroscopy and open cluster photometry. For the well-studied brightest white dwarfs, however, the parameters from the identical spectroscopic analysis in Gianninas *et al.* (2011) were used. Cummings *et al.* (2016a) focused in particular on self-consistently analyzing all publicly available raw spectroscopy of intermediate (0.7 to 1.0 $M_\odot$) and high-mass (>1.0 $M_\odot$) white dwarfs using the white dwarf atmospheric models of Tremblay *et al.* (2011) with the Stark profiles of Tremblay & Bergeron

(2009) and the automated fitting techniques of Bergeron *et al.* (1992). The inferred progenitor masses are highly sensitive to systematic errors in the host cluster ages. Therefore, Cummings *et al.* (2016a) used multi-color UBV analysis, with detailed consideration of peculiar stars, the color dependence of reddening (see Fernie 1963), and the metallicity when available, to do high-precision turnoff isochronal ages of a wide number of star clusters. To consider the effects of the adopted evolutionary models, isochrones from both the Yi *et al.* (2001, hereafter Y^2) models and the PARSEC models from Bressan *et al.* (2012) were used.

Figure 1 compares the IFMR from Cummings *et al.* (2015; 2016a; 2016b) and newer data from M7 (Cummings *et al.* in prep.) for Y^2 cluster ages to the IFMR of Salaris *et al.* (2009). While there is still potentially important scatter in the relation, it is ~50% less than that previously observed and can now predominantly be explained by the remaining errors and cluster non-member contamination. This suggests that the source of scatter in previous IFMR was dominated by systematics and that stellar mass loss is not stochastic.

Figure 1 also shows that progress has been made with finding white dwarfs in star clusters at masses greater than that of the long known LB 1497 in the Pleiades (>1.046 $M_\odot$). These include several from 1.05 to 1.1 $M_\odot$ in NGC 2323 (Cummings *et al.* 2016a), NGC 2168 (Williams *et al.* 2009), and NGC 2287 (Dobbie *et al.* 2012). Then most recently in NGC 3532 at 1.123 $M_\odot$ (Raddi *et al.* 2016) and in NGC 2099 at 1.28 $M_\odot$ (Cummings et al. 2016b; in blue). However, the very large sensitivity of cooling rate on mass at these highest masses makes the cooling age errors too large in this ultramassive white dwarf from NGC 2099. To reliably apply ultramassive white dwarfs to the IFMR, they must be hot and young before cooling age uncertainties have time to grow with age (Althaus *et al.* 2007; private communication). The most massive white dwarfs that can be applied to the IFMR are PG 0136+251 (1.20 $M_\odot$) and GD50 (1.26 $M_\odot$) and are both shown in green. They are young but do not remain in the Pleiades; Dobbie *et al.* (2006b) have argued kinematically that they are coeval and connected to the Pleiades.

2.3. *Remaining Challenges*

While important progress has been made in our understanding of the IFMR, several key limitations still remain. First, the evolutionary timescales of lower-mass stars (<4 $M_\odot$) are characterized well enough that using comparable evolutionary models to both derive isochronal cluster ages and interpret the resulting evolutionary timescales gives consistent progenitor masses. Any variations in the models are mitigated by the progenitor mass's lower sensitivity to evolutionary lifetime. However, when it comes to higher-mass progenitors the inferred progenitor masses are increasingly sensitive to the evolutionary models adopted. To illustrate this, in Figure 2 we compare the IFMR from Figure 1, which derived cluster ages from the Y^2 isochrones, to the IFMR derived from PARSEC isochronal ages. For the typically young clusters these higher-mass white dwarfs are observed in, the PARSEC isochrones derive younger ages and infer more massive progenitors for the most massive cluster white dwarfs.

Second, many of the clusters analyzed for the IFMR are lacking precision spectroscopic metallicities. For clusters without metallicity information, the isochronal studies typically assume solar [Fe/H]. Therefore, this overestimates the isochronal ages in metal-rich clusters and underestimates them in metal-poor clusters. This will artificially increase the observed scatter in the IFMR. Furthermore, metallicity itself plays an important role in both mass loss and core mass growth during stellar evolution. This leads to more metal-rich stars losing more mass throughout their evolution giving lower mass white dwarfs, introducing an intrinsic scatter in the IFMR (see Marigo *et al.* 2007; Meng *et al.* 2008; Doherty *et al.* 2015; Romero *et al.* 2015). Detection of this scatter's correlation with

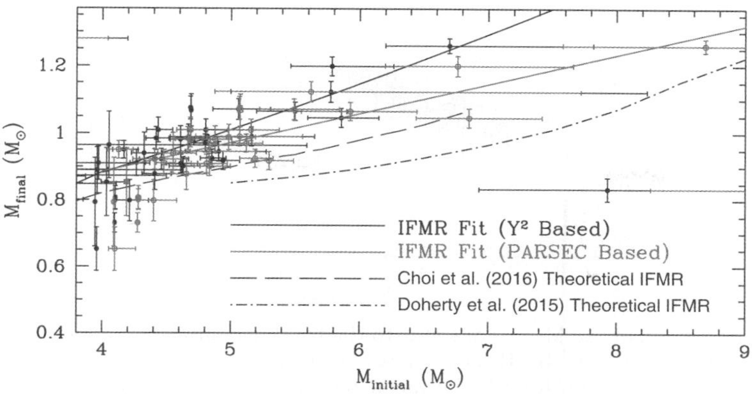

Figure 2. A look at the effects of adopting the Y^2 and PARSEC stellar evolutionary models to fit the isochronal ages of the star clusters and infer progenitor masses. For clarity the focus is placed on the higher-mass stars where the effects are important, with the PARSEC models finding increasingly more massive progenitors. Comparisons to theoretical IFMR at solar metallicity are shown from Choi *et al.* (2016) and Doherty *et al.* (2015). Important systematic differences between the semi-empirical and theoretical IFMRs remain. (See online version for Figure 2 in color.)

cluster metallicity would provide critical constraints on our understanding of metallicity's effects on stellar evolution.

Lastly, the highest mass white dwarfs are still elusive. This is for a number of reasons. First, they are far rarer because there are very few stars with masses greater than 6 $M_\odot$. Second, for compact objects like white dwarfs, the most massive ones also have the smallest radius, making them far less luminous than the most common intermediate-mass white dwarfs. Third, observations of white dwarfs in the field have found that in addition to the high-mass ejected white dwarfs from the Pleiades (Dobbie *et al.* 2006b), there are six intermediate-mass white dwarfs that are kinematically connected to the Hyades (see Tremblay et al. 2012). Are white dwarfs prone to being ejected out of open clusters after formation? For example, asymmetric mass loss during their formation could provide strong enough velocity kicks to eject many white dwarfs soon after their formation (Fellhauer *et al.* 2003). To best overcome these issues and the very large cooling age errors in older ultramassive white dwarfs, the search for ultramassive white dwarfs should focus on young 50 to 100 Myr old clusters. These white dwarfs will be hotter and brighter, have minimal cooling age errors, and will not have had significant time to be ejected from their host cluster.

3. Applications of the IFMR

The most direct application of the IFMR is to measure total integrated mass loss across a star's lifetime and how it is sensitive to initial mass (and metallicity). Figure 3 looks at what percentage of a star's mass is lost throughout its lifetime versus $M_{\rm initial}$. This shows that from $M_{\rm initial}$ of 1.75 to 7 $M_\odot$ the total mass loss rapidly increases from 65% to over 80%. Furthermore, to take into consideration the sensitivity to the adopted stellar model, there are two sets of data at the highest masses. The PARSEC models infer lower white dwarf masses for a given $M_{\rm initial}$, and hence a marginally higher mass loss. Because stellar mass loss occurs predominantly during the AGB phase, this provides critical information for understanding this complex stellar phase. Additionally, with PNe it gives a detailed

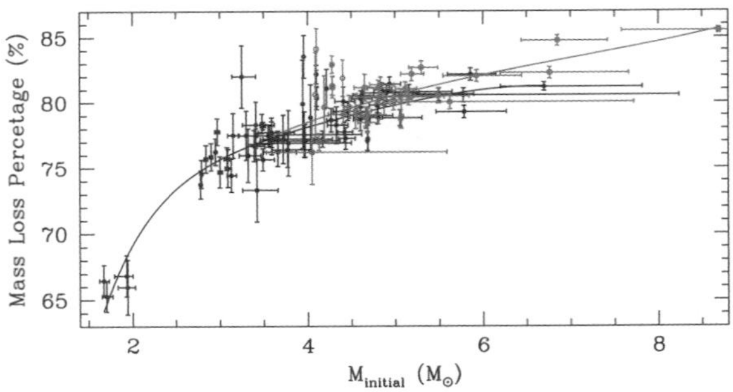

Figure 3. Comparison of the $M_{initial}$ of a star to the total percentage of mass lost throughout its lifetime. As in Figure 2, the Y^2 IFMR is in black and the PARSEC IFMR (shown for higher-mass stars only) is in open red, both with the mean relation shown in solid lines of corresponding color. (See online version for Figure 3 in color.)

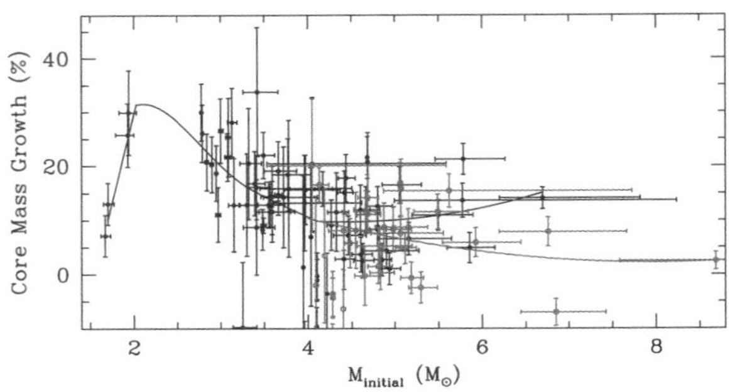

Figure 4. Comparison of $M_{initial}$ of a star to how much the core mass grows during the TP-AGB phase. As in Figures 2 and 3, the Y^2 IFMR is in black and the PARSEC IFMR (shown for higher-mass stars only) is in open red, both with the mean relation shown in solid lines of corresponding color. (See online version for Figure 4 in color.)

measurement of the amount of mass that will surround (but not completely be ionized by) the hot central stars.

In the AGB a poorly constrained parameter is the amount of core mass growth that occurs during the thermally pulsing AGB (TP-AGB) phase. Modern stellar evolutionary models have reassuring agreement on the core masses at the first thermal pulse of the AGB (M_{1stTP}). However, note that variations that are seen in published models are more the result of differences in how different groups define this point in evolution and what layers are included in the core, rather than resulting differences in the predictions of stellar evolution (e.g., Marigo *et al.* 2013 versus Choi *et al.* 2016).

Figure 4 looks at this resulting core mass growth, in terms of (M_{final} - M_{1stTP})/M_{1stTP}. Again, the IFMR from both evolutionary models are considered, and while this resulted in important differences in the previous diagrams, here the differences are more telling. The Y^2 models predict that the core masses grow significantly in the higher-mass stars, while the PARSEC models predict that the core masses do not grow and at the highest masses may decrease in mass during the TP-AGB. What is telling about this is that for the highest-mass AGB stars the thermally pulsing phase is very rapid. The phase is rapid enough that, unlike in lower mass stars, there is little to no time for the core masses to grow (e.g., Doherty *et al.* 2015). Any theoretical uncertainties remaining are likely not large enough to produce $\gtrsim$10% core mass growth at $M_{initial}$>6 $M_\odot$. This strongly suggests that the Y^2 evolutionary models are underestimating the progenitor masses for the highest mass white dwarfs, while the PARSEC models are more appropriate.

4. Conclusions

The IFMR is valuable for our understanding of stellar evolution. It was a longstanding challenge to find and spectroscopically observe a large population of white dwarfs within star clusters. With the advent of larger telescopes, wide-field imaging and multi-object spectrographs, the white dwarf sample greatly increased but significant scatter still remained in the relation. Within the past 5 years the sample size in the IFMR has continued to increase, and with the more advanced models and analysis techniques uniformly applied to both the spectroscopy and photometry, there has been a great reduction in the scatter.

While the highest-mass white dwarfs still remain elusive, addressing the remaining challenges of the IFMR will be more successful with the greatly diminished effects of systematics. These challenges include the remaining systematics between models used to infer progenitor masses, and the effects of metallicity for precise age dating and to test for metallicity's effect on the IFMR.

The diminished scatter of the current IFMR also makes the application of the relation more instructive. The applications of the IFMR to mass-loss rates across a broad range of $M_{\rm initial}$ shows that from $M_{\rm initial}$ of 1.75 $M_\odot$ to 7 $M_\odot$ the total mass loss percentage increased from 65% to greater than 80%. Comparison of the IFMR to the $M_{\rm 1stTP}$ also helps to illustrate the growth of core mass during the complex TP-AGB phase.

References

Althaus, L. G., García-Berro, E., Isern, J., Córsico, A., & Rohrmann, R. 2007, A&A, 465, 249

Beauchamp, A., Wesemael, F., Bergeron, P., Liebert, J., & Saffer, R. A. 1996, Hydrogen Deficient Stars, 96, 295

Bergeron, P., Saffer, R. A., & Liebert, J. 1992, ApJ, 394, 228

Bergeron, P., Liebert, J., & Fulbright, M. S. 1995, ApJ, 444, 810

Bressan, A., Marigo, P., Girardi, L., *et al.* 2012, MNRAS, 427, 127

Casewell, S. L., Dobbie, P. D., Napiwotzki, R., *et al.* 2009, MNRAS, 395, 1795

Catalán, S., Isern, J., García-Berro, E., & Ribas, I. 2008, MNRAS, 387, 1693

Choi, J., Dotter, A., Conroy, C., *et al.* 2016, ApJ, 823, 102

Ciardullo, R. 2010, PASA, 27, 149

Cummings, J. D., Kalirai, J. S., Tremblay, P.-E., & Ramirez-Ruiz, E. 2015, ApJ, 807, 90

Cummings, J. D., Kalirai, J. S., Tremblay, P.-E., & Ramirez-Ruiz, E. 2016a, ApJ, 818, 84

Cummings, J. D., *et al.* 2016b, ApJ, 820, L18

Dobbie, P. D., Napiwotzki, R., Burleigh, M. R., *et al.* 2006a, MNRAS, 369, 383

Dobbie, P. D., Napiwotzki, R., Lodieu, N., *et al.* 2006b, MNRAS, 373, L45

Dobbie, P. D., Day-Jones, A., Williams, K. A., *et al.* 2012, MNRAS, 423, 2815

Doherty, C., Gil-Pons, P., Siess, L., Lattanzio, J., & Lau, H. 2015, MNRAS, 446, 2599

Eggen, O. J. & Greenstein, J. L. 1965, ApJ, 141, 83

Fellhauer, M., Lin, D. N. C., Bolte, M., Aarseth, S. J., & Williams, K. A. 2003, ApJ, 595, L53

Fernie, J. D. 1963, AJ, 68, 780

Fontaine, G., Brassard, P., & Bergeron, P. 2001, PASP, 113, 409

Fusi-Pecci, F. & Renzini, A. 1976, A&A, 46, 447

Garcia-Berro, E., Isern, J., & Hernanz, M. 1997, MNRAS, 289, 973

Gianninas, A., Bergeron, P., & Ruiz, M. T. 2011, ApJ, 743, 138

Greggio, L. 2010, MNRAS, 406, 22

Kalirai, J. S., Richer, H. B., Reitzel, D., *et al.* 2005, ApJ, 618, L123

Kalirai, J. S., Bergeron, P., Hansen, B. M. S., *et al.* 2007, ApJ, 671, 748

Kalirai, J. S., Hansen, B. M. S., Kelson, D. D., *et al.* 2008, ApJ, 676, 594
Kalirai, J. S., Saul Davis, D., Richer, H. B., *et al.* 2009, ApJ, 705, 408
Kalirai, J. S., Marigo, P., & Tremblay, P.-E. 2014, ApJ, 782, 17
Koester, D. & Reimers, D. 1993, A&A, 275, 479
Koester, D. & Reimers, D. 1996, A&A, 313, 810
Marigo, P. & Girardi, L. 2007, A&A, 469, 239
Marigo, P., Bressan, A., Nanni, A., Girardi, L., & Pumo, M. L. 2013, MNRAS, 434, 488
Meng, X., Chen, X., & Han, Z. 2008, A&A, 487, 625
Paczyński, B. 1970, AcA, 20, 47
Raddi, R., Catalán, S., Gänsicke, B. T., *et al.* 2016, MNRAS, 457, 1988
Reid, I. N. 1996, AJ, 111, 2000
Reimers, D. & Koester, D. 1989, A&A, 218, 118
Romero, A. D., Campos, F., & Kepler, S. O. 2015, MNRAS, 450, 3708
Salaris, M., Serenelli, A., Weiss, A., & Miller Bertolami, M. 2009, ApJ, 692, 1013
Schulz, H. 1976, Paper presented at Second Europ. Workshop on White Dwarfs, Rome
Sweeney, M. A. 1976, A&A, 49, 375
Tremblay, P.-E. & Bergeron, P. 2009, ApJ, 696, 1755
Tremblay, P.-E., Bergeron, P., & Gianninas, A. 2011, ApJ, 730, 128
Tremblay, P.-E., Schilbach, E., Röser, S., *et al.* 2012, A&A, 547, A99
Tremblay, P.-E., Kalirai, J., Soderblom, D., Cignoni, M., & Cummings, J. 2014, ApJ, 791, 92
Tremblay, P.-E., Cummings, J., Kalirai, J. S., *et al.* 2016, MNRAS, 461, 2100
van Altena, W. F. 1969, AJ, 74, 2
Wegner, G. 1983, AJ, 88, 109
Wegner, G., Reid, I. N., & McMahan, R. K. 1989, LNP, 328, 378
Weidemann, V. 1977, A&A, 59, 411
Weidemann, V. 2000, A&A, 363, 647
Williams, K. A., Bolte, M., & Koester, D. 2004, ApJ, 615, L49
Williams, K. A., Bolte, M., & Koester, D. 2009, ApJ, 693, 355
Wood, P. R. & Cahn, J. H. 1977, ApJ, 211, 499
Yi, S., Demarque, P., Kim, Y.-C., *et al.* 2001, ApJS, 136, 417

Discussion

Muthumariappan: How do the binary interaction and common envelope evolution influence the initial and final mass relation?

Q: I am surprised to see that your plots start at 1.5 - 1.75 $M_\odot$ in initial mass. If you consider the peak mass of 0.6 $M_\odot$ observed in WDs, it seems that most PNe should come from 2 $M_\odot$ progenitor stars. Is this the case?

Cummings: The mean mass is roughly 0.6 $M_\odot$, but the peak (most common) massis roughly 0.56 to 0.58 $M_\odot$. This gives progenitors of $\sim$1.5 $M_\odot$ as likely the most common PN progenitors.

Mendez: What is the final mass for an initial mass of 1 $M_\odot$ and what would be the error bar?

Cummings: White dwarfs with progenitors of this mass have been observed in M4. While its very low metallicity may affect our comparison of this to a solar metallicity star, but it suggests that they will for 0.55 to 0.50 $M_\odot$ white dwarfs. See Kalirai *et al.* (2009).

Q: What is the metallicity range you are sampling with these clusters and can the mass-loss rate dependency on metallicity explain the scatter or some of the scatter you are seeing in the IFMR?

Planetary Nebulae: Multi-Wavelength Probes of Stellar and Galactic Evolution
Proceedings IAU Symposium No. 323, 2016
X. Liu, L. Stanghellini & A. Karakas, eds.

doi:10.1017/S1743921317000953

Modelling dust production in AGB stars

Flavia Dell'Agli[1,2,3]

[1]Instituto de Astrofísica de Canarias, Vía Láctea s/n, E-38200 La Laguna, Tenerife, Spain
[2]Departamento de Astrofísica, Universidad de La Laguna (ULL), E-38206 La Laguna, Spain
[3]Dipartimento di Fisica, Università di Roma "La Sapienza", P.le Aldo Moro 5, 00143, Roma, Italy

Abstract. Asymptotic giant branch (AGB) stars are among the most important gas and dust polluters of the Universe. The latest AGB evolutionary models take into account dust production in the circumstellar envelope of these stars, starting from a detailed computation of the main physical processes and chemical surface variations occurring in this evolutionary phase. Following the formation and growth of dust particles, they provide the unique possibility of interpreting the AGB population observed in resolved galaxies. The first application was for the Spitzer observations of dusty AGBs in the Magellanic Clouds, characterising carbon-rich and oxygen-rich stars in terms of initial mass, epoch of star formation, evolutionary time on the AGB and dust contribution. The same set of models are able to interpret the CNO surface abundances observed for the PNe of the same galaxies.

Keywords. Stars: abundances, AGB and post-AGB. ISM: dust, planetary nebulae: individual

1. Dust production during the AGB evolution

In this contribuiton we present the dust yields from low and intermediate mass stars, obtained considering the dust formation in AGB winds, starting from a self-consistent description of AGB evolution (see e.g. Ventura *et al.* 2014). This approach allows us to follow the evolution of the dust grain size distribution, optical properties of the wind and the rate of dust formation and to discuss the effects of the main physical processes involved in the AGB evolution: third drudge up (TDU) and hot bottom burning (HBB). The present analysis is used to interpret the infrared excess observed in AGB stars, and to infer their role in the dust pollution of the host environment; on the other hand, the comparison with the observations is important to draw information on the still poorly known physical mechanisms affecting the description of the AGB phase.

The evolutionary models are calculated with the ATON code (Canuto & Mazzitelli 1999) from the pre-main sequence up to the end of the AGB phase; the range of masses explored is $1M_{\odot} \leqslant M \leqslant 8M_{\odot}$ with metallicities $3 \times 10^{-4} \leqslant Z \leqslant 18 \times 10^{-3}$. The main physical (luminosity, effective temperature, mass–loss rate) and chemical (surface abundances) output of the evolutionary computations are used as input to calculate dust formation and wind dynamics, following the model proposed by Ferrarotti & Gail (2006).

In Fig. 1 we show the total amount of dust produced by AGB stars of different metallicities. Stars with masses below $3-4M_{\odot}$, after several TDU episodes, increase their carbon surface abundance and eventually become carbon stars (C/O $>$ 1). In a carbon–rich environment the main dust species that can be formed are silicon carbide (SiC), closer to the surface of the star, and solid carbon, in a more external region. Carbon grain sizes span the range $0.1\mu m \leqslant a_C \leqslant 0.25\mu m$, while SiC particles are smaller ($\sim 0.07\mu m$). Dust production depends on the availability of carbon molecules, therefore on the efficiency of TDU. More massive stars reach large temperatures ($>$ 30 million K (MK)) at the base of the convective envelope, thus being exposed to HBB, with the consequent depletion

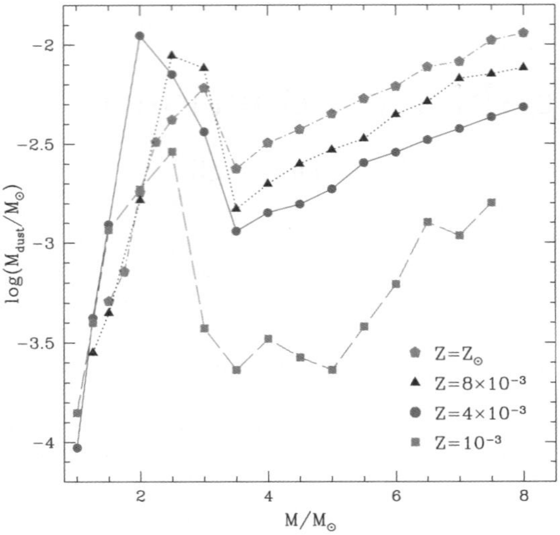

Figure 1. Total dust mass produced as a function of the initial mass and metallicity models.

of the surface carbon (C/O < 1) and increase in the nitrogen abundances. In an oxygen–rich environment, the main dust compounds formed are silicates (olivine, pyroxene and quartz) and alumina dust (Al_2O_3). Alumina dust grains form in an internal layer of the circumstellar envelope with respect to silicates; however, silicates reach larger dimensions (0.07-0.13 μm) compared to alumina dust (0.05-0.07 μm). Higher metallicity stars produce more dust, owing to the larger amount of silicon available for condensation. More massive stars undergo a more efficient HBB: this favours higher mass-loss rates, which increase the density of the wind and the dust formation rate.

2. Modelling vs. observations

We use the wide Spitzer-SAGE surveys of AGB stars in the Magellanic Clouds (Riebel *et al.* 2012, Boyer *et al.* 2011) to test the theoretical predictions and to assess the role played by AGB stars in polluting the host environment where they form. By using the DUSTY code (Nenkova *et al.* 1999), we follow the variation of the synthetic emerging spectra during the AGB phase for various masses and metallicities, starting from the outputs of the AGB+dust models. It is possible to construct evolutionary tracks, to reproduce the evolution of these objects in the colour-colour and colour-magnitude diagrams (CCD and CMD, respectively). In order to obtain a quantitive description of the AGB contribution to the total dust budget, we based the analysis on a population synthesis approach, using the star formation history (SFH) and age–metallicity relations by Harris & Zaritsky (2004, 2009). The comparison with the observations allows a full characterisation of the AGB stars in these galaxies, in terms of mass, formation epoch and chemical composition of the progenitors, and of the dust nowadays present in the circumstellar envelope (Dell'Agli *et al.* 2015a,b).

2.1. *AGB stars in the MCs*

The two panels of Fig. 2 show the CCDs of the AGB population of the LMC and SMC, in the Spitzer bands (Riebel *et al.* 2012, Boyer *et al.* 2011). We present some of the evolutionary tracks, which better represent the main protagonists of the SFH and dust pollution in each galaxy.

In the LMC we can identify three main groups of stars, which occupy well defined regions of the CCD. The obscured carbon stars (OCS) and the hot bottom burning stars (HBBS), which constitute only the $\sim 20\%$ and $\sim 1\%$ of the total sample, respectively, are responsible for almost the totality of the dust pollution. The third group is a miscellaneous group of carbon and oxygen-rich stars (CMS) and represents the majority of the AGB population of the LMC; it is mainly composed of low-mass stars ($< 3M_{\odot}$), with a low metallicity component ($Z < 4 \times 10^{3}$) and it is characterised by a low infrared excess.

The OCS group is entirely composed of carbon stars, populating a diagonal band in the CCD, with $0.2 < [3.6] - [4.5]$ and $0.2 < [5.6] - [8.0] < 1.6$. They are the progeny of stars with mass $1M_{\odot} \leqslant M \leqslant 3M_{\odot}$, and formed $3 \times 10^{8} - 3 \times 10^{9}$ years ago. As more carbon is accumulated on the surface, the stars produce more dust, moving towards redder colours, with a corresponding increase in the optical depth ($0.01 < \tau_{10} < 3.5$). The stars formed during the peak in the SFH which occurred $\sim 5 \times 10^{8}$ year ago, and descend from stars with initial mass in the range $2M_{\odot} \leqslant M \leqslant 3M_{\odot}$ of higher metallicity ($Z \geqslant 4 \times 10^{3}$) and are the only ones which evolve redder than $[3.6] - [4.5] > 1$.

Stars clustering around $[3.6] - [4.5] \sim 0.2$, $[5.8] - [8.0] \sim 0.8$ belong to the HBBS group. These stars have initial masses in the range $3.5M_{\odot} \leqslant M \leqslant 7.5M_{\odot}$ and are experiencing HBB, forming mainly silicates and alumina dust. The lower extinction coefficients of silicates compared to carbon dust determine a smaller optical depth of the circumstellar envelope in these stars (τ_{10} is below 2 in all cases) compared to their C–rich counterparts. They belong to the more metal–rich population ($Z \geqslant 4 \times 10^{3}$) and formed during the burst in the SFH which occurred $\sim 10^{8}$ years ago (Harris & Zaritsky 2009).

In the SMC we distinguish a population of stars with negligible infrared emission, similar to the CMS group in the LMC. It is composed mainly of oxygen–rich, low mass stars of various metallicities, in the AGB phases previous to or at the beginning of the C–star phase. In analogy to the LMC case, in the SMC we identify a sequence of dusty AGB stars with infrared emission that extends diagonally towards the reddest region of the CCD. We interpret this group of stars, with $[3.6] - [4.5] > 0.2$, as C–rich objects producing mainly carbon dust. The majority of the stars in this group have initial mass $1.5 - 1.75M_{\odot}$ and metallicity $Z = 4 \times 10^{-3}$, formed between $1 - 1.5$ Gyr ago. They correspond to the objects with the largest optical depth, formed during the secondary peak in the SFH of the SMC. The comparison between the evolutionary tracks of the 1.5 $M_{\odot}$ model of $Z = 4 \times 10^{3}$ and the $3M_{\odot}$ model of $Z = 8 \times 10^{3}$ shows that the latter evolves to much redder colours, which is the reason for the difference in the extension of the C–star sequences observed in the LMC and SMC. Turning to more massive stars, unlike the LMC, the SFR of the SMC presents a minimum at ~ 100 Myr, when stars with initial mass $3.5M_{\odot} \leqslant M \leqslant 7.5M_{\odot}$ formed. This explain the paucity of AGB stars in the region around $[3.6] - [4.5] \sim 0.2$, $[5.8] - [8.0] \sim 0.8$ in the SMC CCD diagram.

2.2. *Interpreting CNO abundances in MCs PNe*

The final surface composition of the AGB models previously described have been recently used to characterize the MCs planetary nebulare (PNe) in terms of age, chemical composition and mass of the progenitors, on the base of the CNO abundances observed (Ventura *et al.* 2015, 2016). A dichotomy in the distribution of the nitrogen abundances is observed in both galaxies, a clear signature of the HBB activation in stars of initial mass above $\sim 3M_{\odot}$. The objects enriched in nitrogen, with $\log(\mathrm{N/H}) + 12 \geqslant 8$, are the progeny of stars with mass higher than the aforementioned threshold, and formed 50–200 Myr ago. PNe with $log(N/H) + 12 < 7.5$ are interpreted as the final stages of stars that did not experience any HBB. These objects, with initial mass below $\sim 3M_{\odot}$, formed between 500 Myr and 10 Gyr ago, with a different spread in carbon determined by the

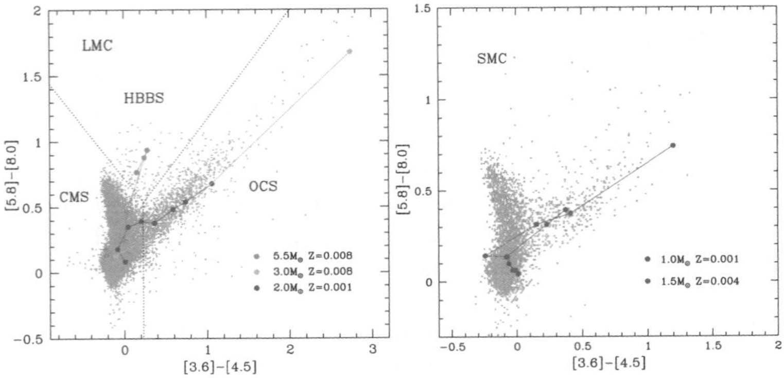

Figure 2. Evolutionary tracks of AGB stars of different initial mass and metallicity and the AGB stars observed in the LMC (left panel) from Riebel *et al.* (2012) and in the SMC (right panel) from Boyer *et al.* (2011).

different number of TDU episodes experienced. The highest carbon abundances found in both the samples is $\log(C/H) + 12 < 9$. If confirmed by future observations, it indicates a limited number of TDU episodes was experienced, probably due to a rapid loss of the external envelope.

3. Conclusions

We present models of AGB stars which consider the formation of dust in the circumstellar envelope and discuss the trend of the amount of dust produced with mass and metallicity. The results are compared with the infrared observations of the AGB stars in the MCs and with the chemical composition of PNe in the MCs. In both cases the agreement is satisfactory, adding more robustness to the present findings.

Acknowledgments

FD acknowledges support from LOC/SOC of the IAU323 and from "La Sapienza", University of Rome under the grant "Avvio alla ricerca 2015".

References

Boyer M. L., *et al.* 2011, *AJ*, 142, 103 *et al.* 2014a, *MNRAS*, 442, L38
Dell'Agli F., Ventura P., Schneider R., Di Criscienzo M., *et al.* 2015a, *MNRAS*, 447, 2992
Dell'Agli F., Ventura P., García-Hernández D. A., Schneider R. *et al.* 2015b, *MNRAS*, 454, 4235
Ferrarotti A. D. & Gail H. P., 2006, *A&A*, 553, 576
Harris J., Zaritsky D. 2004, *AJ*, 127, 1531
Harris J., Zaritsky D. 2009, *ApJ*, 138, 1243
Mazzitelli I., D'Antona F., & Ventura P., 1999, A&A, 348, 846
Nenkova M., Ivezic Z., & Elitzur M., 1999, in LPIContributions 969, 20
Riebel D., Srinivasan S., Sargent B., Meixner M. 2012, *AJ*, 753, 71
Ventura P., Dell'Agli F., Di Criscienzo M., Schneider R., *et al.* 2014a, *MNRAS*, 439, 977
Ventura P., Stanghellini L., Dell'Agli F., García-Hernández D. A., Di Criscienzo M. 2015, *MNRAS*, 452, 3679
Ventura P., Stanghellini L., Di Criscienzo M., García-Hernández D. A., Dell'Agli F. 2015, *MNRAS*, 460, 3940

Discussion

SCHOENBERNER: Can you say something about the mass-loss rate?

Planetary Nebulae: Multi-Wavelength Probes of Stellar and Galactic Evolution
Proceedings IAU Symposium No. 323, 2016
X. Liu, L. Stanghellini & A. Karakas, eds.

doi:10.1017/S1743921317000862

Central Stars of Planetary Nebulae

David Jones[1,2]

[1]Instituto de Astrofísica de Canarias,
E-38205 La Laguna, Tenerife, Spain
email: djones@iac.es

[2]Departamento de Astrofísica, Universidad de La Laguna,
E-38206 La Laguna, Tenerife, Spain

Abstract. In this brief invited review, I will attempt to summarise some of the key areas of interest in the study of central stars of planetary nebulae which (probably) will not be covered by other speakers' proceedings. The main focus will, inevitably, be on the subject of multiplicity, with special emphasis on recent results regarding triple central star systems as well as wide binaries which avoid a common-envelope phase. Furthermore, in light of the upcoming release of Kepler's Campaign 11 data, I will discuss a few of the prospects from that data including the unique possibility to detect merger products.

Keywords. planetary nebulae: general, white dwarfs, binaries: general, binaries: eclipsing, binaries: close, stars: oscillations

1. Triple star systems

Recent works (e.g. Bear & Soker, 2016; Akashi 2016) have sparked interest in the subject of higher order multiples at the heart of planetary nebulae (PNe) and the role they might play in forming the most unusual and/or asymmetrical PNe, particularly given the large fraction observed in solar-like main sequence stars (Raghavan *et al.* 2010). However, to-date, there is no confirmed triple central star known.

1.1. *SuWt 2*

One of the best candidate triple central stars is NSV19992, the bright star near the apparent centre of the PN SuWt 2. Exter *et al.* (2010) reported that this star is, in fact, a binary system consisting of two A-type main sequence stars with an orbital period of 4.9 days. Neither of these stars could provide the ionising flux for the nebula, and Exter *et al.* (2010) hypothesise that the A-type binary may form part of a triple system along with the nebular progenitor. Combining data from multiple sources, they find some evidence for shifts in the systemic velocity of the binary that would be consistent with a possible third component. However, the morphology of SuWt 2 (Jones *et al.* 2010) is that of a rather canonical bipolar, seemingly inconsistent with the predicted "messy" structures for triple evolution (Bear & Soker 2016). In order to probe the possible triple system, Jones & Boffin (2016) acquired high-resolution VLT-UVES spectra of NSV19992 over the course of one-year, finding that there was no appreciable shift in the systemic velocity over the entire observing period. Furthermore, with the binary parameters (γ, K_1, K_2, T_0) they derive much of the data from Exter *et al.* (2010) phases well, with no need for a shift in systemic velocity (Jones & Boffin 2016). Collectively, this is strongly indicative that the A-type binary does not form a triple system with the nebular progenitor. Additionally, the systemic velocities of the nebula and A-type binary are found to differ by more than 15 km s^{-1}, it seems unlikely that the two are in any way connected and rather that the A-type binary is just a field star which, by chance, lies close to the projected centre of the nebula.

1.2. *Other chance alignments*

Somewhat surprisingly, Boffin *et al.* (2016) also find that the bright star near the projected centre of PN M3-2 is a binary system comprised of two A-type main sequence stars, this time with a ~2 day period. The chances of finding a chance alignment of a double A-type binary and a PN is slim (but not impossible), the probability of finding two such chance alignments is, however, exceptionally small. This might perhaps be taken as an indication that there may be some connection between the systems, for example, that one of the components of each binary system might be a born-again star. For SuWt 2, this interpretation seems extremely unlikely given that the stellar parameters derived by Exter *et al.* (2010) are not consistent with this evolution. Similarly for M3-2, a faint blue star, which is most likely the true central star, can be seen in deep imagery with good seeing (Boffin *et al.* 2016).

While not an A-type binary, it is worth mentioning that Méndez *et al.* (2016) recently showed that the bright A-type star, believed to form one half of a binary system, at the heart of NGC 1514 is, in fact, unrelated to the nebula and just another chance alignment. As such, chance alignments with field stars may not be as uncommon as predicted.

2. Long- and intermediate-period binary central stars

The connection between binary central stars and PN shaping is now beyond doubt (Hillwig *et al.* 2016). Given that other authors (e.g. Nordhaus 2016) will summarise the importance (and current state of our understanding) of post-common-envelope central stars as part of these proceedings, I will attempt to cover the longer-period systems, summarising our current knowledge of long- and intermediate-period binary central stars.

With more than 50 known binary central stars (see Figure 1) only a handful are known with periods greater than a few days. I will categorise these systems by the technique used in their discovery, and try to explain the inherent difficulties (and positive points) for each technique showing the difficulty faced in unveiling the hidden (but clearly very important) population of wide binaries in PNe.

2.1. *Radial velocity monitoring*

While the amplitude of photometric variability, due to irradiation or ellipsoidal modulation, falls off rapidly with longer periods, radial velocity variables (at favourable inclinations) remain detectable with modest resolution spectrographs out to periods of several days (up to several hundred days for high resolution spectrographs). However, this technique does have severe limitations in that it is intrinsically limited to brighter central stars (as high signal-to-noise is essential) and extremely costly in terms of observing time (as several different processes can contribute to radial velocity variability, it is essential to observe exhaustively in order to establish the periodic component associated with orbital motion from other sources of variability, for example, wind variability, De Marco *et al.* 2004).

Recently, radial velocity monitoring has been used to great effect by Van Winckel *et al.* (2014) in discovering the wide binaries at the heart of BD+33°2642 and LoTr 5 (which, coincidentally, was previously believed to be a triple system, a hypothesis not consistent with the observations of Van Winckel *et al.*) and by Manick *et al.* (2015) in the discovery of the 4 day period, post-common-envelope, Wolf-Rayet binary in NGC5189.

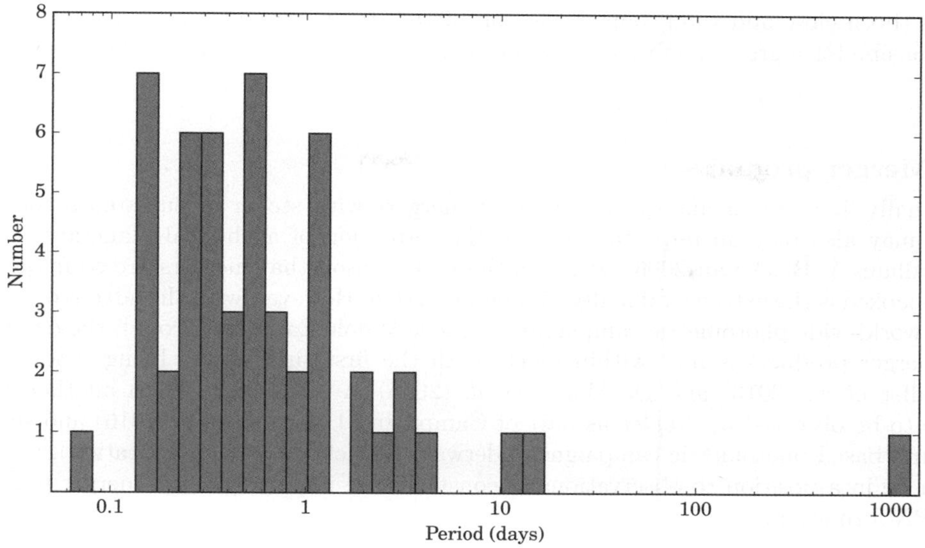

Figure 1. Period distribution of known binary central stars of PNe (data from the list maintained at http://drdjones.net/bCSPN).

2.2. *Giant and/or chemically polluted secondaries*

Several systems are known where the star at the apparent centre of the nebula cannot have produced the nebula but, as previously shown, this cannot be taken as evidence that the observed star forms part of a binary (or higher-order multiple) with the nebula progenitor (Méndez *et al.* 2015). There are cases where the star at the apparent nebular centre shows evidence of chemical pollution with AGB or post-AGB material (s-process elements, for example), this can be considered a strong indication that the star is associated with the nebula and that its progenitor was the source of chemical pollution (Miszalski *et al.* 2013). In these cases, while the radius of the polluted star and the level of pollution can place limits on possible periods, further radial velocity study is essential to constrain precisely the true period of the binary (e.g. LoTr 5 was known to host a G-type giant star before it's orbital period was found by Van Winckel *et al.* 2014). As the periods here are usually ~100 days, period determination therefore requires long-term monitoring with intermediate- to high-resolution spectrographs, which is often particularly difficult given the competitive nature of telescope time allocations.

2.3. *Infrared excesses*

The search for infrared excesses (Barker 2016), where the SED of the hot primary shows an additional component (often peaking in the red or infrared) due to a cooler companion, is perhaps the most promising methodology for revealing binary central stars as the technique does not depend on periodicity or inclination. Similarly, it is sensitive to a wide range of secondary masses (much more so than radial velocity monitoring or photometric variability monitoring). However, the technique is fraught with difficulty, requiring extremely high precision photometry in multiple bands often where the nebular contamination is appreciable and variable with wavelength, and unfortunately offers no clue as to the possible period of the binary (and, as such, central stars displaying IR excesses might be most appropriately used as a pre-selector for follow-up with other photometric or spectroscopic techniques, Aller 2016). Interestingly, the results from surveys for IR excess (De Marco *et al.* 2013, Douchin *et al.* 2015) indicate an extremely significant total (both close and wide) binary fraction further reinforcing that a binary pathway is responsibe for a great number of PNe, perhaps even crucial in a majority of cases.

3. Merger products

Finally, it is important to mention that mergers with stellar or sub-stellar companions may also play an important role in the formation of aspherical planetary nebula (Nordhaus & Blackman 2006). Perhaps the main reason that mergers are so frequently over-looked is the extreme difficulty in their detection. However, with the advent of *Kepler* and world-wide photometric campaigns (e.g. The Whole Earth Telescope), the detection of merger products is now within reach, with the first such object being reported by Handler *et al.* (2013) and De Marco *et al.* (2015) in NGC 6826. With another ~200 PNe to be observed by *Kepler* as part of Campaign 11 (Jacoby *et al.* 2016) and several ground-based photometric campaigns underway (Sowicka *et al.* in preparation), we may soon be in a position to observationally constrain the importance of a merger pathway for PN formation.

References

Aller, A. 2016, *These proceedings*

Akashi, M. 2016, *These proceedings*

Barker, H. 2016, *These proceedings*, arXiV:1611.02559

Bear, E. & Soker, N. 2016, arXiv:1606.08149

Boffin, H. M. J., Jones, D., Wesson, R., Beletsky, Y., Miszalski, B., Saviane, I., Monaco, L., Corradi, R. L. M., Santander-Garciá, M., & Rodríguez-Gil, P. 2016, *A&A*, submitted

De Marco, O., Passy, J.-C., Frew, D. J., Moe, M., & Jacoby, G. H. 2013, *MNRAS*, 428, 2118

De Marco, O., Long, J., Jacoby, G. H., Hillwig, T., Kronberger, M., Howell, S. B., Reindl, N., & Margheim, S. 2015, *MNRAS*, 448, 3587

De Marco, O., Bond, H. E., Harmer, D., & Fleming, A. J. 2004, *ApJ*, 602, 93

Douchin, D., De Marco, O., Frew, D. J., Jacoby, G. H., Jasniewicz, G., Fitzgerald, M., Passy, J.-C., Harmer, D., Hillwig, T., & Moe, M. 2015, *MNRAS*, 448, 3132

Exter, K., Bond, H. E., Stassun, K. G., Smalley, B., Maxted, P. F. L. & Pollacco, D. L. 2010 *AJ*, 140, 1414

Handler, G., Prinja, R. K., Urbaneja, M. A., Antoci, V., Twicken, J. D., & Barclay, T. 2013, *MNRAS*, 430, 2923

Hillwig, T. C., Jones, D., De Marco, O., Bond, H., Margheim, S., & Frew, D. 2016, *ApJ*, 832, 125, arXiV:1609.02185

Jacoby, G. H., Long, J., Kronberger, M., De Marco, O., & Hillwig, T. C. 2016, *AAS meeting abstracts*, 227, 238.01

Jones, D. & Boffin, H. M. J. 2016, *MNRAS*, submitted arXiV:1611.01819

Jones, D., Lloyd, M., Santander-García, M., López, J. A., Meaburn, J., Mitchell, D. L., Pollacco, D. L., O'Brien, T. J., & Vaytet, N. M. H. 2010, *MNRAS*, 408, 2312

Manick, R., Miszalski, B., & McBride, V. 2015, *MNRAS*, 448, 1789

Méndez, R. H., Kudritzki, R.-P., & Urbaneja, M. A. 2016, *ApJ*, 829, 73

Miszalski, B., Boffin, H. M. J., Jones, D., Karakas, A. I., Köppen, J., Tyndall, A. A., Mohamed, S. S., Rodríguez-Gil, P., & Santander-García, M. 2013, *MNRAS*, 436, 3068

Nordhaus, J. 2016, *These proceedings*

Nordhaus, J. & Blackman, E. 2006, *MNRAS*, 270, 2004

Raghavan, D., McAlister, H. A., Henry, T. J., Latham, D. W., Marcy, G. W., Mason, B. D., Gies, D. R., White, R. J., & ten Brummelaar, T. A. 2010, *ApJS*, 190, 1

Van Winckel, H., Jorissen, A., Exter, K., Raskin, G., Prins, S., Perez Padilla, J., Merges, F., & Pessemier, W. 2014, *A&A*, 563, 10

Discussion

FREW: The near by B-star to Su Wt 2 is too cool to explain the very high ionization of the nebula, so a hot ionizing source, hidden by the A-type central star is required.

TODT: Regarding the detection of CS companions via IR excess I wonder how you can tell this IR excess from that one caused by hot dust in PNe which already shows up in 2MASS (e.g. Phillips & Ramos-Larios 2006)?

JONES: The dust excess doesn't enter until redwards of the I-band and the searches primarily use I-band excess.

VAN WINCKEL: The A-type central stars (even binary + stars) are quite intriguing. Could they be rejuvenated A star from late thermal pulses?

JONES: At least in the case of Su Wt 2 this seems very unlikely based on the study by Exter *et al.* (2010). Smith *et al.* suggested near by B-type star could be source of ionizing radiation. David Frew says is too cold to explain high ionization.

MONTEX: I thought LoTr 5 was a suspected triple, have you considered following up this star?

JONES: As far as I know the only evidence for a triple there is the apparent inconsistency between the nebular and binary inclinations – which given the long period might not be unexpected.

Planetary Nebulae: Multi-Wavelength Probes of Stellar and Galactic Evolution
Proceedings IAU Symposium No. 323, 2016
X. Liu, L. Stanghellini & A. Karakas, eds.

doi:10.1017/S1743921317000540

[WN] central stars of planetary nebulae

H. Todt[1], B. Miszalski[2], J. A. Toalá[3,4] and M. A. Guerrero[4]

[1]Institute of Physics and Astronomy, University of Potsdam,
Karl-Liebknecht-Str. 24/25, 14476 Potsdam, Germany
email: htodt@astro.physik.uni-potsdam.de

[2]South African Astronomical Observatory,
PO Box 9, Observatory 7935, South Africa;
Southern African Large Telescope Foundation,
PO Box 9, Observatory 7935, South Africa

[3]Institute of Astronomy and Astrophysics, Academia Sinica (ASIAA),
10617 Taipei, Taiwan, Republic of China

[4]Instituto de Astrofísica de Andalucía, IAA-CSIC,
Glorieta de la Astronomía s/n, E-18008 Granada, Spain

Abstract. While most of the low-mass stars stay hydrogen-rich on their surface throughout their evolution, a considerable fraction of white dwarfs as well as central stars of planetary nebulae have a hydrogen-deficient surface composition. The majority of these H-deficient central stars exhibit spectra very similar to massive Wolf-Rayet stars of the carbon sequence, i.e. with broad emission lines of carbon, helium, and oxygen. In analogy to the massive Wolf-Rayet stars, they are classified as [WC] stars. Their formation, which is relatively well understood, is thought to be the result of a (very) late thermal pulse of the helium burning shell.

It is therefore surprising that some H-deficient central stars which have been found recently, e.g. IC 4663 and Abell 48, exhibit spectra that resemble those of the massive Wolf-Rayet stars of the nitrogen sequence, i.e. with strong emission lines of nitrogen instead of carbon. This new type of central stars is therefore labelled [WN]. We present spectral analyses of these objects and discuss the status of further candidates as well as the evolutionary status and origin of the [WN] stars.

Keywords. Stars: abundances – Stars: AGB and post-AGB – Stars: atmospheres – Stars: mass-loss – Stars: PN PB 8 – Stars: IC 4663 – Stars: Abell 48 – Stars: PMR 5 – Stars: Wolf-Rayet

1. Introduction

Wolf-Rayet central stars are hydrogen deficient central stars of planetary nebulae which exhibit in their spectra strong emission lines of helium, carbon and oxygen. Because their spectra resemble those of massive WC stars, they are called [WC] stars, with brackets to distinguish them from their massive counterparts. In spite of spectral similarities and comparable chemical composition, the formation of the low-mass [WC] stars is completely different from the formation of the massive WC stars. Stellar evolutionary models accounting for simultaneous burning and mixing explain the formation of a [WC] star by the occurrence of a thermal pulse (TP) at the very end or after the asymptotic giant branch (AGB) phase of a H-normal low-mass star. These models predict a hydrogen-deficient surface composition with carbon enriched up to $X_{\mathrm{C}} = 40\%$ after a late or very late TP (Althaus *et al.* 2005, Herwig 2001). Only in the case of a very late TP (VLTP) a supersolar nitrogen abundance of about $X_{\mathrm{N}} = 1\%$ is expected, but without any remaining hydrogen. Thus low-mass central stars with WN-like surface abundances are theoretically not expected.

2. The [WN] stars

2.1. *PB 8*

The CS PB 8 was classified as [WC 5-6] by Acker & Neiner (2003) based on a low-resolution spectrum. A spectral analysis based on a high resolution spectrum, Todt *et al.* (2010) revealed that this object has a stellar temperature of about 50 kK and an unusual composition with He:H:C:N:O=55:40:1:1:1 by mass and resembles spectroscopically a massive WN/C star.

Due to its unknown distance, also L is unknown, so whether this star is a true [WN/C] CS can only be inferred indirectly.

The nebula analysis by García-Rojas (2009) yield values for T_e and n_e which are typical for young PNe, as well as the small $v_{exp} \approx 30\,\mathrm{km/s}$. Regarding the luminosity distance, one finds that for a CSPN with $L/L_\odot = 6000$, PB 8 would be at a distance of 4.2 kpc and at a height of 300 pc above the Galactic plane, whereas a massive WR star would have at least $L/L_\odot = 2 \times 10^5$, and so at least a distance of 24 kpc and a height of 1.7 kpc above the Galactic plane. This is a rather untypical location for a massive WR star. So, PB 8 is indeed a CSPN.

We also considered the possibility that PB 8 may be a binary. Although this can not be excluded completely, it is rather unlikely as there were no shifts of radial velocities of spectral lines detected (Méndez 1989) and the nebula appears spherically symmetric, also in velocity space.

2.2. *IC 4663*

The first "pure" [WN] was discovered by Miszalski *et al.* (2012). Their spectral analysis based on an optical spectrum revealed that IC 4663 is an almost hydrogen-free ($< 2\,\%$ by mass) [WN] star, whose wind is 95% helium with only 0.8% nitrogen. Interestingly it is of early spectral subtype ([WN3]) with a temperature of about 140 kK and a relatively low mass-loss rate of about $2 \times 10^{-8}\,M_\odot\,\mathrm{yr}^{-1}$. Miszalski *et al.* (2012) showed that if IC 4663 were a massive WN star it would be at an implausible distance of 58 kpc and more than 8 kpc below the Galactic plane. Moreover, they discovered an AGB halo around IC 4663, proving it is a CSPN.

2.3. *Abell 48*

DePew *et al.* (2011) mentioned the CS of Abell 48, but it was unclear whether this is really a CSPN and not a massive WN star (Wachter *et al.* 2010 classified it as WN6). The CS Abell 48 was found to have a helium-rich (85%) wind with about 10% hydrogen, and a nitrogen abundance of about 5% (Todt *et al.* 2013, see Fig, 1; Frew *et al.* 2014). From the nebula analysis (line ratios of S II and Hα, as in Riesgo-Torado & López 2002) it was concluded that Abell 48 is indeed a CSPN. This is also consistent with the measured extinction and its significant proper motion - Abell 48 is a runaway. The CS has a mass-loss rate of $4 \times 10^{-7}\,M_\odot\,\mathrm{yr}^{-1}$ and a temperature of about 70 kK. Its spectral subtype is [WN5].

3. [WN] candidates

The earliest claim for the discovery of a [WN] central star was by Morgan *et al.* (2003), who observed the highly reddened ($E_{\mathrm{B-V}} = 3\,\mathrm{mag}$) object PMR 5, which shows a WN spectrum (subtype WN6) and a round nebula. However, it is not clear whether this is a PN or ring nebula around a massive WR star. While the electron density is consistent with a PN, the expansion velocity of $v_{exp} = 165\,\mathrm{km/s}$ is rather unusual for a PN, but

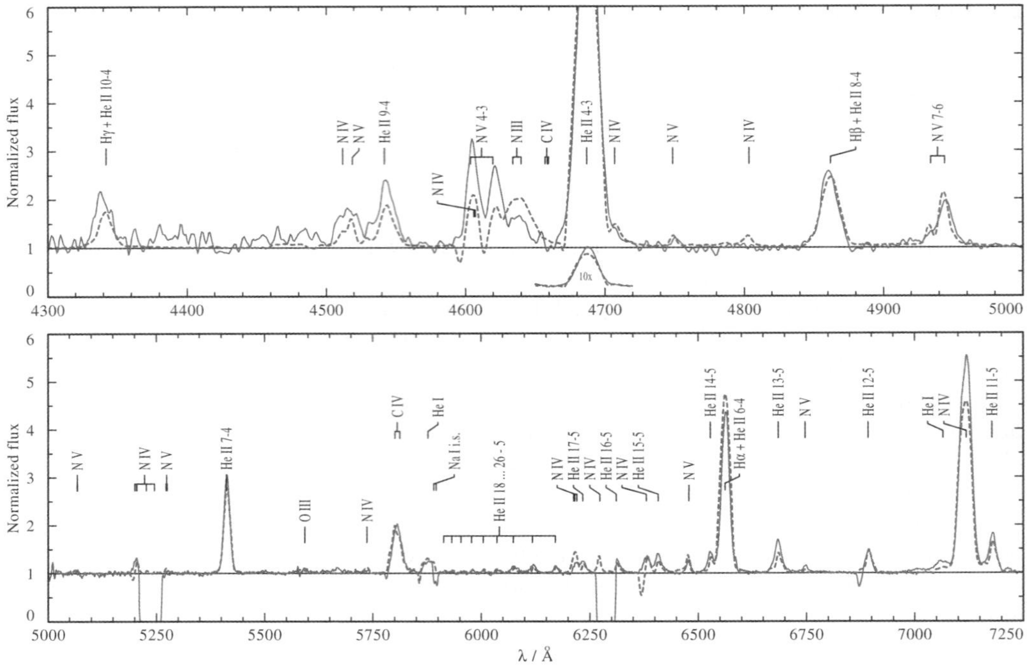

Figure 1. Optical spectrum of Abell 48: observation (blue, thin line) and best-fitting PoWR model (red, thick dotted line; see Todt *et al.* 2013), both normalized to the model continuum

typical for a ring nebula. Moreover, the luminosity distance is consistent with a massive WR star. For a typical CSPN luminosity of $L/L_{\odot} = 6000$ PMR 5 would be located at a distance of 0.5 kpc and at a height of 6 pc above the Galactic plane, while for a typical luminosity of a massive WN star PMR 5 would be at 2.9 kpc away and 35 pc above the Galactic plane. Hence, both locations are consistent with the corresponding type of star. However, a reddening of $E_{\mathrm{B-V}} = 3$ mag is rather untypical for a distance of only 500 pc. A spectral analysis based on the published spectrum revealed a chemical composition of a typical WNh star, but with a large mass fraction of N ($\approx 10\%$). Overall, chemistry and v_{exp} point to a massive WN star with a ring nebula.

So far, all of the confirmed [WN] stars are of earlier spectral subtypes. It may be that late [WN] stars have escaped detection because of their narrower stellar lines which can be confused with nebular emission lines, especially at low spectral resolution. Moreover, also some H-rich CSs show emission lines, e.g., NGC 6543, and are therefore classified as of WR/Of spectral type (Smith & Aller 1969). The stellar Hβ line, which is used for spectral classification of WN/Of stars (Crowther & Walborn 2011), is often blended with the nebular emission. Thus high spectral resolution is also needed to tell WN type central stars from Of type central stars.

4. Evolutionary status

The evolutionary channel by which [WN] stars are formed is unclear. None of the recent late TP scenarios predicts the observed surface compositions of PB 8, IC 4663, and Abell 48. It was speculated whether these objects belong to an alternative evolutionary sequence [WN] $\rightarrow$ O(He) (Miszalski *et al.* 2012, Werner 2012) or that [WN] stars are O(He) stars with higher masses (Reindl *et al.* 2014), evolving from RCB or sdO(He)

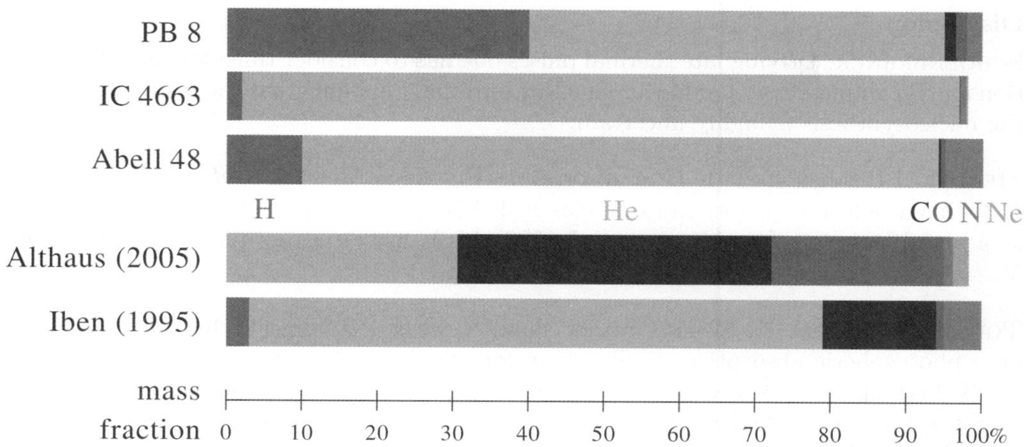

Figure 2. Observed abundance patterns of the [WN] stars vs. predicted abundances for H-deficient central stars in the VLTP / born-again scenario (Althaus *et al.* 2005, Iben & MacDonald 1995)

stars, which might be merger products of non-DA WDs. However, the merger scenario seems to be rather unlikely due to the long timescales involved, which are incompatible with the observed low ages of the PNe of the [WN] stars. Moreover, PB 8 and Abell 48 also contain hydrogen while the assumed progenitors are H-free.

Note that the "classical" born-again scenario by Iben & MacDonald 1995 yields abundances that are somehow similar to the observed ones, including residual hydrogen and and a supersolar nitrogen abundance (see Fig. 2).

References

Acker, A. & Neiner, C. 2003, *A&A*, 403, 659

Althaus, L. G., Serenelli, A. M., Panei, J. A., *et al.* 2005, *A&A*, 435, 631

Crowther, P. A. & Walborn, N. R. 2011, *MNRAS*, 416, 1311

DePew K., Parker Q. A., Miszalski B., De Marco O., Frew D. J., Acker A., Kovacevic A. V., Sharp R. G., 2011, *MNRAS*, 414, 2812

Frew, D. J., Bojičić, I. S., Parker, Q. A., *et al.* 2014, *MNRAS*, 440, 1345

García-Rojas, J., Peña, M., & Peimbert, A. 2009, *A&A*, 496, 139

Herwig, F. 2001, *Ap&SS*, 275, 15

Iben, Jr., I. & MacDonald, J. 1995, in Lecture Notes in Physics, Berlin Springer Verlag, Vol. 443, White Dwarfs, ed. D. Koester & K. Werner, 48

Méndez, R. H. 1989, in IAU Symposium, Vol. 131, Planetary Nebulae, ed. S. Torres-Peimbert, 261–272

Miszalski, B., Crowther, P. A., De Marco, O., *et al.* 2012, *MNRAS*, 423, 934

Morgan, D. H., Parker, Q. A., & Cohen, M. 2003, *MNRAS*, 346, 719

Reindl, N., Rauch, T., Werner, K., Kruk, J. W., & Todt, H. 2014, *A&A*, 566, A116

Riesgo-Tirado, H. & López, J. A. 2002, in RMxAC, ed. W. J. Henney, J. Franco, & M. Martos, Vol. 12, 174–174

Smith, L. F. & Aller, L. H. 1969, *ApJ*, 157, 1245

Todt, H., Peña, M., Hamann, W.-R., & Gräfener, G. 2010, *A&A*, 515, A83

Wachter S., Mauerhan J. C., Van Dyk S. D., Hoard D. W., Kafka S., Morris P. W., 2010, *AJ*, 139, 2330

Werner, K. 2012, in IAU Symposium, Vol. 283, IAU Symposium, 196–203

Discussion

SCHOENBERNER: Driving late thermal pulses one has to consider time-dependent convection and incomplete non-equilibrium nuclear burning. I imagine that the result concerning the nucleosynthesis is highly uncertain.

MENDEZ: PB 8 has 4541 He II in absorption (Fig. 2e in Mendez *et al.* 1988, A&A, 190, 113) that would not correspond to a classical WN spectral type, which show He II 4541 as a broad emission. (see Mendez *et al.* 1990, A&A, 229, 152, last paragraph of Section 3).

TODT: I think that He II 4541 rather shows a weak P Cygni-like line profile than a pure photospheric absorption profile. Even for massive WN stars exist examples where He II 4541 appears as weak P-Cygni line (e.g. WR 24, spectral sub-type WN6ha) or in absorption (SMC AB2 of spectral subtype WN5ha) due to a weaker wind. However, Crowther & Walborn (2011) define WN stars with help of the Hb profile. If it is in emission, as it is for PB 8 and WR 24, then it's a WN spectrum. If it shows a P-Cygni profile then they call it an Of/WN star.

DE MARCO: Have you ever analyzed the [WR] that is a member of a binary? Is it different in any way?

TODT: You may refer to NGC 5189. The optical and UV spectrum looks very similar to the spectra of other hot [WC] stars, e.g., PB 6, NGC 2867, NGC 6905, [S71d]3. Analyses by Keller *et al.* (2014) as well as my preliminary analysis didn't show any remarkable differences compared to other [WCE] stars.

MUTHUMARIAPPAN: The evolutionary sequences [WCL] $->$ [WCE] $->$ WC/PG1159 is likely, however, how do we account for the observed gap in the [WC5] - [WC8] subclasses of [WR] PN distribution?

TODT: The spectral subtypes of the [WC] stars correspond to different ionization stages of C, He, O and therefore reflect different stellar temperatures. It may be that these subtypes are not equally spaced with respect to the evolution timescale of the [WC] stars. So, perhaps, the subtypes [WC5] -[WC8] correspond to shorter phase in the evolution of [WC] stars making it more unlikely to observe them during this phase. Alternatively, one may think of [WCE] and [WCL] stars belonging to different populations. E.g. V605 Aql seemed to have evolved directly VLTP $->$ [WCE]. This seems to be corroborated by the different abundance patterns found for [WCL] and [WCE]. However, in this scenario it is not clear to what type of star the [WCL] stars eventually evolve if not to [WCE] stars.

STERLING: Can you expand on the models without convective overshoot, which maybe the best (so far) evolutionary channel for [WN] CSPN?

TODT: For the [WN] stars with significant amount of hydrogen (e.g. PB 8) we expect to get DA WDs. Even for those with very small amounts (e.g. IC 4663)this will be the case (Miller Bertolami *et al.* 2016). So far the parameter space for the VLTP/LTP/AFTP scenarios hasn't been explored enough to give distinct predictions, but the presence of hydrogen and nitrogen in the [WN] stars may point to an AFTP, where 14N could be the consequence of dredging up material only from the He-buffer zone or from hot-bottom burning in stars of M_i $\sim$ 4 M_Sun (Miller Bertolami *et al.*, priv. comm.)

Planetary Nebulae: Multi-Wavelength Probes of Stellar and Galactic Evolution
Proceedings IAU Symposium No. 323, 2016
X. Liu, L. Stanghellini & A. Karakas, eds.

doi:10.1017/S1743921317001533

New models for the evolution of central stars of planetary nebulae: Faster and Brighter

Marcelo M. Miller Bertolami

Instituto de Astrofisica de La Plata, CONICET-UNLP, Paseo del Bosque s/n, (B1900FWA) La Plata, Argentina. email: mmiller@fcaglp.unlp.edu.ar

Abstract. The post-asymptotic giant branch (AGB) phase is arguably one of the least understood phases of the evolution of low- and intermediate- mass stars. The recent post-AGB evolutionary sequences computed by Miller Bertolami (2016) are at least three to ten times faster than those previously published by Vassiliadis & Wood (1994) and Blöcker (1995) which have been used in a large number of studies. This is true for the whole mass and metallicity range. The new models are also ~0.1–0.3 dex brighter than the previous models with similar remnant masses. In this short article we comment on the main reasons behind these differences, and discuss possible implications for other studies of post-AGB stars or planetary nebulae.

Keywords. stars: AGB and post-AGB, planetary nebulae: general

1. Introduction

In the most simple picture, Planetary Nebulae (PNe) are formed by low- and intermediate-mass stars ($M_{\rm ZAMS} \sim 0.8$ –6 $M_\odot$) after the strong stellar winds at the end of the Asymptotic Giant Branch (AGB). At that point, stars contract and heat up, crossing the HR-diagram at constant luminosity before becoming sufficiently hot and bright to ionize the previously ejected material (Shklovsky 1957, Abell & Goldreich 1966, Paczyński 1970). The formation and detectability of PNe depends strongly on the interplay between two different timescales. The evolutionary timescale of the central star of the PN (CSPN), which provides the ionizing photons, and the dynamical timescale of the circumstellar material (Schönberner *et al.* 2007). In this manuscript we focus on the evolutionary timescales of the CSPNe. In particular, we will discuss the new results presented by Miller Bertolami (2016) from full stellar evolution computations of the post-AGB and CSPN phases. These new post-AGB models are based on state-of-the-art stellar evolution computations, they include an updated treatment of the AGB microphysics (radiative opacities and nuclear reaction rates) as well as of the macrophysics (convective boundary mixing and mass-loss rates). The models have been calibrated and tested to reproduce several observables during the post-AGB and previous evolutionary phases; e.g. solar radius and depth of the convective zone, width of the upper main sequence, mass range of carbon-rich AGB stars, oxygen abundances of PG1159 stars, semiempirical initial-final mass relationship, C/O ratios of AGB stars and CSPNe —see Miller Bertolami (2016) for details.

2. New post-AGB/CSPN evolutionary timescales

As shown in Fig.1, the post-AGB/CSPN crossing timescales ($\tau_{\rm cr}$) computed by Miller Bertolami (2016) are 3 to 10 times shorter those of the sequences of similar remnant mass ($M_{\rm CSPN}$) computed by Vassiliadis & Wood (1994) and Blöcker (1995). The new post-AGB models also evolve faster when compared at equal initial masses ($M_{\rm ZAMS}$; see Fig. 1) or

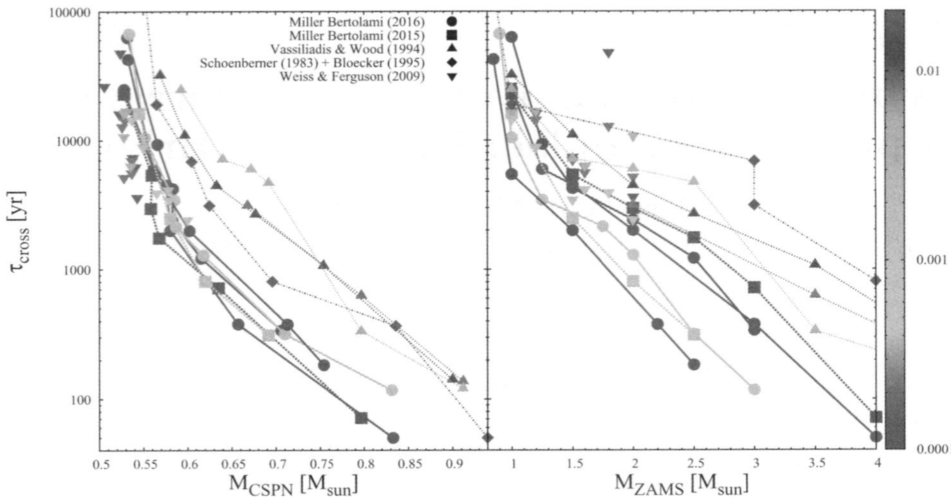

Figure 1. Post-AGB crossing timescales ($\tau_{\rm cr}$) of the sequences presented by different authors. Following Vassiliadis & Wood (1994) we have defined, and plotted, $\tau_{\rm cr}$ for all sequences as the time elapsed from $T_{\rm eff} = 10000\,K$ to the point of maximum $T_{\rm eff}$ in the HR diagram during the post-AGB evolution. *Left Panel:* $\tau_{\rm cr}$ as a function of the final remnant mass $M_{\rm CSPN}$. *Right Panel:* $\tau_{\rm cr}$ as a function of the initial progenitor mass $M_{\rm ZAMS}$. Color coding indicates the initial metallicities ($Z_{\rm ZAMS}$) of the sequences, as described in the color bar.

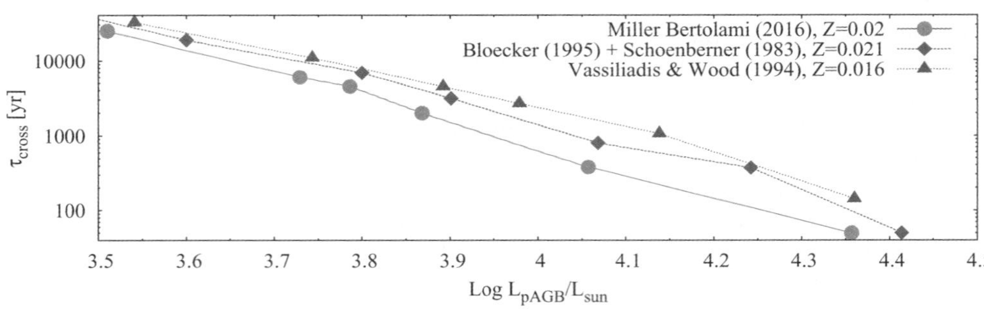

Figure 2. $\tau_{\rm cr}$ as a function of the post-AGB luminosity for sequences of similar metallicities.

post-AGB luminosities ($L_{\rm pAGB}$; see Fig. 2). The faster evolution of the new models can be traced back to the adoption of updated microphysics and the inclusion of convective boundary mixing during the thermal pulses on the AGB —see Miller Bertolami (2016) for a detailed discussion. The latter, in particular, leads to the occurrence of third dredge up for lower stellar masses and to carbon (C) pollution of the envelope. In fact, post-AGB timescales become even shorter in the case of the stellar models computed by Weiss & Ferguson (2009) and Miller Bertolami (2015), which assume a more intense convective boundary mixing during the thermal pulses on the AGB, and consequently have stronger third dredge-up episodes. C-pollution of the envelope increases the luminosity of the burning shell and decreases the critical envelope mass ($M_{\rm env}^{\rm crit}$) at which the models depart from the AGB. Then, as $\tau_{\rm cr}$ is mostly determined by the pace at which the hydrogen (H)-burning shell consumes the remaining H-rich envelope, an increase in the C-pollution of the envelope leads to a faster post-AGB evolution. It is worth emphasizing that the value of $M_{\rm env}^{\rm crit}$ cannot be arbitrarily set for a given sequence. $M_{\rm env}^{\rm crit}$ is mostly set by the H-burning shell luminosity as well as by the composition of the H-rich envelope. As noted

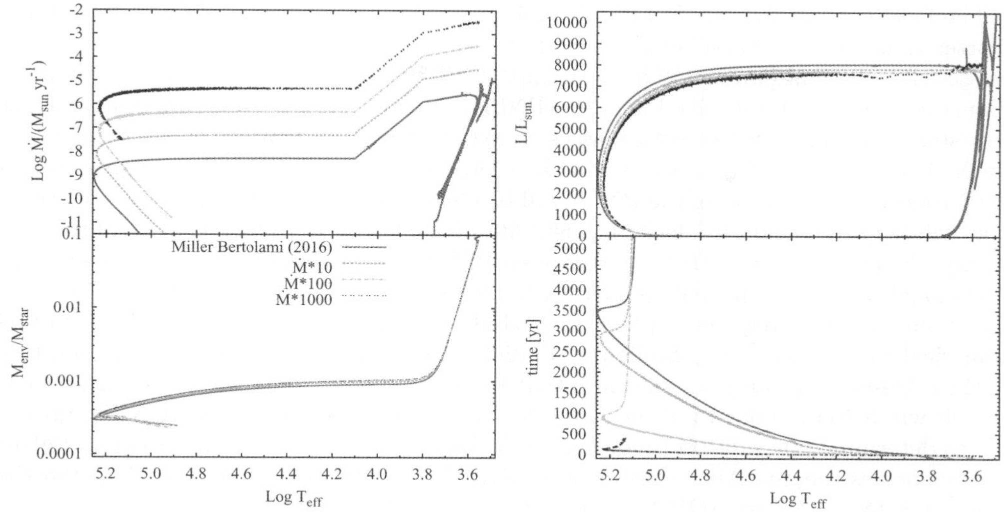

Figure 3. A numerical experiment. Comparison of the late AGB and post-AGB evolution of sequences computed with winds differing by 3 orders of magnitude. The reference sequence is the $M_{\rm ZAMS} = 1.25 M_\odot$, $Z = 0.001$ sequence from Miller Bertolami (2016).

by Paczyński(1971), H-burning post-AGB models have a very tight relationship between the effective temperature ($T_{\rm eff}$) and the envelope mass ($M_{\rm env}$). The departure from the AGB occurs as soon as $M_{\rm env}$ is reduced close to these corresponding post-AGB values. In particular, this means that the wind intensity at the end of the AGB plays no role in the determination of $M_{\rm env}^{\rm crit}$. This is fortunate, as the intensity of $\dot{M}$ at the end of the AGB, or the early post-AGB, is currently not known. In order to convince the reader of this counter-intuitive fact, we have recomputed the $M_{\rm ZAMS} = 1.25 M_\odot$, $Z_{\rm ZAMS} = 0.001$ sequence presented by Miller Bertolami (2016) artificially increasing the mass loss ($\dot{M}$) after the last thermal pulse by up to 3 orders of magnitude (Fig. 3, upper left panel). As seen in Fig. 3 (bottom left panel) the value of $M_{\rm env}^{\rm crit}$ is independent of $\dot{M}$ at the end of the AGB. Only the post-AGB timescales are affected (Fig. 3, bottom right panel) because at high values of $\dot{M}$ the H-rich envelope is mostly reduced by winds, speeding up the evolution. It should be noted, however, that $\dot{M}_{\rm AGB}$ can affect the post-AGB envelope mass in an indirect way. As $M_{\rm env}^{\rm crit}$ is strongly dependent on the luminosity of the sequence, and luminosity does not stay constant during the AGB phase, different $\dot{M}$-values during the whole AGB evolution can change the value of $M_{\rm env}^{\rm crit}$ for a given sequence.

3. Implications and final comments

Given the significantly shorter timescales, higher luminosities and different initial-final mass relations of the models presented by Miller Bertolami (2016), as compared with the usually adopted tracks from Vassiliadis & Wood (1994) and Blöcker (1995), one may wonder about the impact of these results for studies that rely on stellar evolution models as inputs. For example, Gesicki *et al.* (2016) studied 32 PNe from the bulge and obtained CSPNe-black-body temperatures and PNe expansion ages. They find that the CSPNe and progenitor masses derived from the post-AGB models of Miller Bertolami (2016) are in agreement with our current understanding of the stellar formation history of the Galactic Bulge and the white dwarf mass distribution. This was not the case when the tracks of Vassiliadis & Wood (1994) and Blöcker (1995) were used (Gesicki *et al.* 2014).

The fact that, for $M_{\rm ZAMS} \gtrsim 1.25\,M_\odot$, the post-AGB timescales of the new models are shorter than those of Vassiliadis & Wood (1994) and Blöcker (1995) for the same initial masses (Fig. 1, right panel) will have an impact in the understanding and modeling of the Planetary Nebula Luminosity Function (PNLF). Mendez (2016) concludes that with the new stellar models, CSPNe with masses as low as $0.58\,M_\odot$ are able to create relatively bright PNe —see also Zijlstra *et al.* in these proceedings. Consequently, the progenitors of PNe close to the cut-off of the PNLF will be older than previously expected, something that will help to explain the lack of sensitivity of the PNLF cut-off to the age of the harboring population. Due to their shorter post-AGB timescales and larger brightnesses, the new models are also expected to have an impact on the study of the diffuse X-ray emission from the inner regions of planetary nebulae by means of radiation-hydrodynamics numerical simulations —e.g. Steffen *et al.* (2008) and Toalá & Arthur(2014). In addition to these results, we can also speculate that the shorter post-AGB timescales of the new models will help to understand the formation of PNe around low mass post-AGB stars —e.g. as determined by asteroseismology, see Althaus *et al.* (2008), Calcaferro *et al.* (2016) and references therein. Also, shorter post-AGB timescales might help to understand the lack of CSPNe and post-AGB stars in M32 (Brown *et al.* 2008).

A word of caution, while the new models are to be preferred over the older grids, they are not devoid of uncertainties. In particular, the stellar models of Miller Bertolami (2016) fail to quantitatively reproduce the lifetimes of M- and C-type AGB stars, pointing to a need of a better calibration of convective boundary mixing and AGB-winds.

Acknowledgments

M3B thanks the IAU and the organizers for a travel grant and the waiving of the registration fee which allowed him to attend this wonderful Symposium. Part of this work was supported by ANPCyT through grant PICT-2014-2708 from FonCyT and grant PIP 112-200801-00940 from CONICET and by a Return Fellowship from the Alexander von Humboldt Foundation.

References

Abell, G. O. & Goldreich, P. 1966, *PASP*, 78, 232
Althaus, L. G., Córsico, A. H., Kepler, S. O., & Miller Bertolami, M. M. 2008, *A&A*, 478, 175
Blöcker, T. 1995, *A&A*, 299, 755
Brown, T. M., Smith, E., Ferguson, H. C., *et al.* 2008, *ApJ*, 682, 319-335
Calcaferro, L. M., Córsico, A. H., & Althaus, L. G. 2016, *A&A*, 589, A40
Gesicki, K. & Zijlstra, A. A. 2007, *A&A*, 467, L29
Gesicki, K., Zijlstra, A. A., Hajduk, M., & Szyszka, C. 2014, *A&A*, 566, A48
Gesicki, K., Zijlstra, A. A., & Miller Bertolami, M. M. 2016, `arXiv:1609.08680`
Mendez, R. H. 2016, `arXiv:1610.08625`
Miller Bertolami, M. M. 2016, *A&A*, 588, A25
Miller Bertolami, M. M. 2015, in: Dufour, P. and Bergeron, P. & Fontaine, G. (eds.), *19th European Workshop on White Dwarfs*, *ASP-CS* 493, p. 83
Paczyński, B. 1970, *AcA*, 20, 47
Paczyński, B. 1971, *AcA*, 21, 417
Schönberner, D., Jacob, R., Steffen, M., & Sandin, C. 2007, *A&A*, 473, 467
Shklovsky, I. S. 1957, in: G. H. Herbig (ed.), *Non-stable stars, proceedings from IAU Symposium no. 3, Cambridge University Press,*, p. 83
Steffen, M., Schönberner, D., & Warmuth, A. 2008, *A&A*, 489, 173
Toalá, J. A. & Arthur, S. J. 2014, *MNRAS*, 443, 3486
Vassiliadis, E. & Wood, P. R. 1994, *ApJS*, 92, 125
Weiss, A. & Ferguson, J. W. 2009, *A&A*, 508, 1343

Discussion

JONES: How applicable do you think these models are for post-CE systems?

MILLER BERTOLAMI: We don't know much about the ejection mechanisms in CE systems. But if the ejection of the common envelope is very fast and far from thermal equilibrium then the masses of the H-rich envelopes could be far from the predictions of the post-AGB models computed with steady winds prescriptions like mine. Then the time scale could be very different, so one should be very careful (and skeptical) when using this model for post-CE systems. In short, may be not very much depending on how does the envelope ejection happen.

DE MARCO: What is the new central star mass that makes a PN, judged as the CS that has a transition time of less than 50,000 yrs. What was it for the Vassiliadis+Wood 1994?

MILLER BERTOLAMI: "Crossing time" (i.e. the time from 7,000 K to maximum Teff in most of my H-burning sequences are below 50,000 yrs. It slightly depends on metallicity a bit but crossing times of less than 50,000 yrs are attained by H-burning sequences with masses M_CSPN $> 0.53\ M_\odot$ (M_ZAMS $> 1\ M_\odot$). Crossing times of less than 10,000 yrs correspond to models with M_CSPN $> 0.56 M_\odot$ (M_ZAMS $> 1.2\ M_\odot$). "Transition times" (the early post AGB evolution below Teff $<$ 7000 K) are ill defined and of a few kyr in my models, decreasing with increasing mass (but very uncertain). Vassiliadis+Wood crossing time scales shorter than 10,000 yrs are only obtained for models with M_CSPN $>$ 0.6 - 0.63 for H-burning sequences. And their crossing time scales are of a few tens of kyr but with no clear trend in mass or metallicity.

Q: Another crossing-time scale is found from the expansion ages of planetary nebulae. Have you investigated whether the stellar crossing time is consistent with kinematic ages?

ARNABOLDI: You were very quick to show the effects of metallicity of Tau_cross. What are the relevant parameters, in addition to the chemistry that drives this evolution?

STERLING: Do your AGB evolutionary models encounter convergence problems at low envelope masses? Does the envelope mass at the end of the AGB need to be parameterized/ estimated in your models?

MILLER BERTOLAMI: Yes, indeed. Also in our models we find a lot of convergence problems as we go to more massive and higher metallicity stars. That is why the models require a lot of baby sitting and are human time consuming to compute. No, the critical envelope mass at which the star departs from the AGB is natural result from the stellar structure computations. It is not artificially chosen. The value of the envelope at the departure from the AGB depends mainly on the structure (mass and radius) of the core and the composition of the envelope (and it is of course affected by the micro physics adopted, e.g. opacity).

Planetary Nebulae: Multi-Wavelength Probes of Stellar and Galactic Evolution
Proceedings IAU Symposium No. 323, 2016
X. Liu, L. Stanghellini & A. Karakas, eds.

doi:10.1017/S1743921317000989

The history of the Galactic bulge

A. A. Zijlstra[1], K. Gesicki[2] and M. M. Miller Bertolami[3]

[1]School of Physics & Astronomy, University of Manchester, Oxford Road, Manchester M13 9PL, UK

[2]Centre for Astronomy, Faculty of Physics, Astronomy and Informatics, Nicolaus Copernicus University, Grudziadzka 5, PL-87-100 Torun, Poland

[3]Instituto de Astrofísica de La Plata, UNLP-CONICET, Paseo del Bosque s/n, 1900 La Plata, Argentina

Abstract. Planetary nebulae form in stellar populations with ages from 1 to 10 Gyr, and can be used to trace their star formation histories. Here we apply this to the Galactic bulge, where there are indications both for an old origin and for younger stars. We use new stellar models, which have significant different evolutionary speeds during the post-AGB phase. We apply these new models to a sample of 32 planetary nebulae with HST imaging and VLT spectroscopy. The results show evidence for an old starburst, followed by continuous star formation until at least 2Gyr ago. This agrees very well with recent analysis of colour-magnitude diagrams of the bulge. We show that the new models can also explain the [OIII] luminosity functions, and predict the uniform luminosity cut-off both in spiral galaxies and old elliptical galaxies.

Keywords. planetary nebulae: general, Galaxy: bulge, Galaxy: evolution

1. Introduction

The origin of the Galactic bulge remains a controversial issue. For a long time, the central bulge was considered an old component of the Milky Way galaxy. Several studies found that the colour-magnitude diagram are consistent with an old population, of approximately 10 Gyr. But there are also indications for a much younger population, for instance from spectroscopy of micro-lensed main-sequence stars. The controversy is reviewed in Nataf (2016).

Planetary nebulae (PNe) form an important part of the puzzle of the bulge. There are currently 785 PNe known in the bulge; the total PNe population is estimated at 2000-3500. PNe form the end-phase of low- and intermediate mass stars. These stars lose their envelopes on the AGB, and the remaining stellar cores quickly heat up and ionise the ejecta, before becoming white dwarfs. If the masses of the stellar cores can be determined, the initial-final mass relations can be used to derive initial masses. The formation ages of the stars come from the stellar life time vs stellar mass relation. From this the star formation history can be derived.

This was done by Gesicki *et al.* (2014), but it resulted in an unusual age of order 1 Gyr. For older stars, with lower stellar masses, the stellar models (Bloecker 1995) evolve too slowly from the AGB to the PN phase, so that the ejecta are dispersed before ionization begins. Gesicki *et al.* suggest that these models must underestimate the speed of evolution during the post-AGB phase, and make the PN stars appear too young. A comparison with a few asteroseismologically-determined masses suggest that the models are a factor of 3 too slow. This adjustment gives bulge ages between 3 and 10 Gyr.

Recently, new stellar evolution models for the post-AGB phase were published. With these models, we revisit the age of the Galactic bulge as determined from PNe.

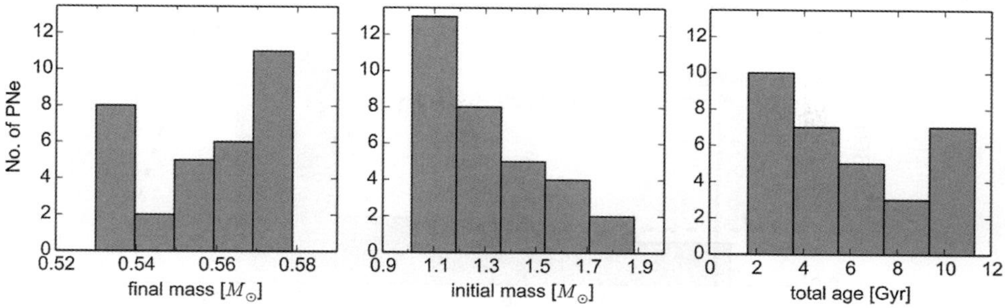

Figure 1. The core mass, initial mass, and stellar age distribution of the 32 Bulge PNe.

2. Stellar models and methods

The previous models for post-AGB evolution are based on the models of Schönberner (Blöcker 1995 and references therein) and Vassiliadis & Wood (1994). A new series of models was recently published by Miller Bertolami (2016) using updated opacities. For the same core mass, the new models are brighter and cross the HR diagram much faster. The two aspects are related, as the evolutionary speed is set by the rate at which the envelope mass is digested, which is determined by the rate of nuclear burning and thus the luminosity. Compared to the old tracks, the post-AGB evolution is accelerated by a factor of 3 to 8 (larger at lower masses). This factor is close to that predicted by Gesicki *et al.* (2014). We use the set of new tracks, interpolating between them to obtain a denser grid in core mass.

The so-called Torun method is used in the analysis (Gesicki *et al.* 214). A photo-ionization model is used to reproduce the emission-line spectrum, line fluxes and line images of a PN. Good resolution images and a known distance significantly improve this modeling. The photo-ionization model returns the ionization structure, the density profile and the electron temperature distribution. Subsequently, the line profiles are used to fit a model for the internal velocity field. This requires echelle spectra of lines of different ionization stages. The velocity field, density distribution and outer radius are used to derive an expansion age. This includes a correction factor to account for the acceleration of the outer regions, as derived from hydrodynamical models (Schönberner 2016, 2005).

The photo-ionization model also returns the temperature of the star. The PN expansion age and current stellar temperature are compared to the post-AGB model tracks, to derive the best-fitting stellar core mass. The heating rate is a very strong function of stellar mass, making this method more accurate than plotting the PNe on the HR diagram (Gesicki & Zijlstra 2007). The new stellar models have produced a systematic shift in the stellar mass vs heating rate relation.

Once the core masses are known, an initial-final-mass relation is used to obtain initial masses. Each initial mass corresponds to a different age of the star. The PN population yields the stellar death rates for that age. Combined with an initial mass function (a Salpeter function) this converts to a star formation rate for each age.

3. Results

We used a sample of 32 bulge PNe for which HST images and VLT/UVES echelle spectra are available. The original imaging program selected targets randomly from the Acker catalog, with the only constraints a position towards the Galactic bulge, and an angular diameter of less than 5 arcsec.

The final stellar masses, initial masses and age distribution of the sample are shown in Fig. 1. The stellar masses are significantly lower than found before, thanks to the

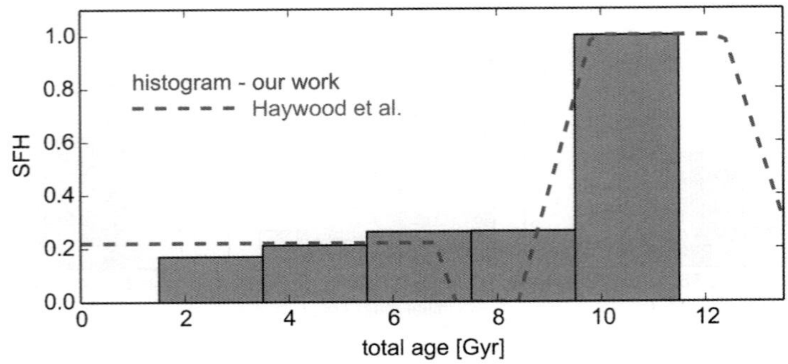

Figure 2. The normalized star formation history (SFH) of the Galactic bulge.

new stellar models. Stellar ages are between 2 and 12 Gyr, in better agreement with the expectations for the bulge than had previously been found. Still, the PN population cannot be solely due to an old population. A significant fraction is from younger stars.

To convert the age distribution to a star formation rate requires one final correction, regarding observational bias. This is the visibility time of the PN, described in detail in Gesicki *et al.* (2014): it takes into account that the time a PN remains visible is a function of stellar core mass, and is limited by how long it takes before the onset of ionization, and the duration of the phase before hydrogen burning ceases. The selection criterion on angular diameter also feeds into the visibility correction.

The resulting star formation history (SFH) is shown in Fig. 2. We find that the SFH shows a peak at ages appropriate for the traditional 'old' bulge, but there has been continuous star formation at a lower rate since. We cannot say whether the star formation continued over the past 2 Gyr.

This SFH is indicative of a mixed bulge, partly formed in a short burst during the early phase of the Milky Way, and partly consisting of younger stars. The latter may be related to the disk population, scattered into the bulge by the action of bar: this is the main explanation for a 'pseudo' bulge, a thickened inner region of a disk.

Fig. 2 also shows the SFH proposed by Haywood *et al.* (2016), based on modeling of colour-magnitude diagrams. This is in excellent agreement with the results derived from PNe. The results are a good validation of the new tracks, and show that PNe can be used as tracers of old and intermediate-age populations.

4. The O3 luminosity function

The new models are also important for extragalactic PN populations. The PN luminosity function (PNLF) in other galaxies shows a sharp cut-off at an [OIII] magnitude around −4: the cut-off magnitude is invariant across a range of galaxy types, from spirals to ellipticals (see also Mendez, these proceedings). With the older stellar models, this invariance was hard to understand. With the Blöcker tracks, the bright PNe come from young stars, and these are absent in elliptical galaxies (Ciardullo 2012). The new tracks can fit the PNLF with older stars.

Fig. 3 shows the PNLF as obtained from the new tracks. For each track, we have assumed a constant density, non evolving nebula, and calculated a photo-ionization model for a dense grid of stellar temperatures. For each model, we calculate the [OIII] flux. The method assumes that the nebula remains optically thick to ionizing photons and therefore will fail for the most evolved nebulae and possibly at the lowest masses where the nebulae are already evolved by the time ionization starts.

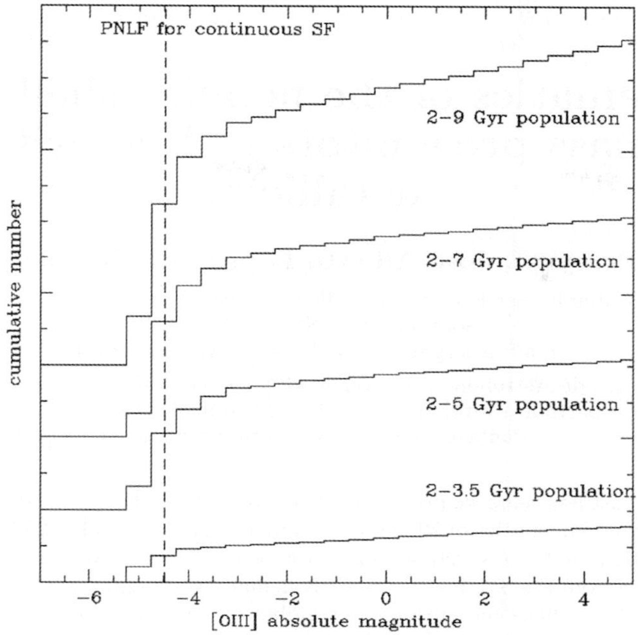

Figure 3. The [OIII] luminosity function of PNe, derived using the new post-AGB tracks.

The PNLF is shown for populations of a range of ages, assuming a constant star formation history. The dashed line shows the observed cut-off. The models reproduce this very well. The youngest stellar populations extend to brighter PNe, and populations older than 7 Gyr do not quite reach the cut-off luminosity.

For ages between 2 and 7 Gyr, the models confirm the presence of the cut-off at an invariant magnitude. This resolves a major problem in PN research. The new models show that the PNLF will give the same cut-off in all galaxy types, excluding those with exclusively stars older than 8 Gyr, and starburst galaxies which are dominated by very young stars. The results in this paper suggest the PNLF will also work for the Galactic bulge, based on the presence of a population younger than 8 Gyr.

References

Bloecker, T. 1995, A&A, 299, 755
Ciardullo, R. 2012, ApSS, 341, 151
Gesicki, K. & Zijlstra, A. A. 2007, A&A, 467, L29
Gesicki, K., Zijlstra, A. A., Hajduk, M., & Szyszka, C. 2014, A&A, 566, A48
Haywood, M., Di Matteo, P., Snaith, O., & Calamida, A. 2016, A&A, 593, A82
Miller Bertolami, M. M. 2016, A&A, 588, A25
Nataf, D. M. 2016, PASA, 33, e023
Schönberner, D., Jacob, R., & Steffen, M. 2005, A&A, 441, 573
Schönberner, D. 2016, Journal of Physics Conference Series, 728, 032001
Vassiliadis, E. & Wood, P. R. 1994, ApJS, 92, 125

Discussion

ARNABOLDI: Are you predicting the morphology of the PNLF as the population ages? Are you also computing the deviations from the Ciardullo's function when the population is rejuvinated?

Planetary Nebulae: Multi-Wavelength Probes of Stellar and Galactic Evolution
Proceedings IAU Symposium No. 323, 2016
X. Liu, L. Stanghellini & A. Karakas, eds.

doi:10.1017/S1743921317000886

The kinematics of the nebular shells around low mass progenitors of PNe with low metallicity

Margarita Pereyra[1], **José Alberto López**[2] **and Michael G. Richer**[2]

[1]Schlumberger Foundation Fellow, University of Southampton
Southampton, SO17 1BJ, UK
email: e.m.pereyra-talamantes@soton.ac.uk

[2]Instituto de Astronomía, Universidad Nacional Autónoma de México,
Apartado Postal 106, C.P. 22800 Ensenada, BC, México
email: jal@astrosen.unam.mx, richer@astrosen.unam.mx

Abstract. In the past few years we provided strong observational support for theoretical studies regarding the internal kinematics of Planetary Nebulae (PNe). A total of 257 objects segregated by different galactic populations were analized. Based upon spatially-resolved, long-slit, echelle spectroscopy drawn from the San Pedro Mártir Kinematic Catalogue of PNe †, we characterized the kinematics of PNe shells measuring their global expansion velocities. We present here a brief summary of these observational results, with a focus on our most recent study of about 26 PNe with low metallicity that appear to derive from progenitor stars of the lowest masses (including the halo PNe population). Low expansion velocities were found for these nebulae, less than $20\,\mathrm{km\,s^{-1}}$, which are most likely associated with a weak central star wind driving the kinematics of the nebular shell in this particular population.

Keywords. ISM: kinematics and dynamics, planetary nebulae: general, stars: evolution, low-mass, abundances

1. Introduction

Since the discovery of the existing link between the expansion velocity of PNe and the escape velocities of red giants by Abell & Goldreich (1966), the study of the internal kinematics of nebular envelopes has became crucial in understanding the final stages of intermediate and low-mass star's evolution. By analyzing the properties of the central stars of PNe and their influence on the kinematics of the nebular shells, we can investigate the mass-loss history of the AGB phase and the chemical evolution of the interstellar medium in their host galaxies. As a dynamically active system, the shells of planetary nebulae are perfect targets to study these processes and to test the predictions of stellar evolution models.

2. Overview

Current hydrodynamical simulations have successfully predicted the kinematic evolution of nebular shells through the entire planetary nebula phase, considering different masses and stellar populations for their progenitor stars (Mellema (1994), Villaver *et al.* (2002), Perinotto *et al.* (2004), Schönberner *et al.* (2005), García-Segura *et al.*

† Observations acquired at the Observatorio Astronómico Nacional in the Sierra San Pedro Mártir (OAN-SPM), B. C., Mexico. Available at *http://kincatpn.astrosen.unam.mx/*

(2006), Schönberner *et al.* (2010)). In the past few years, observational studies to support these predictions have also been performed (López *et al.* (2016)). By characterizing the internal kinematics of 131 objects from the Bulge population, Richer *et al.* (2010) have shown that the nebular shell starts expanding in a momentum driven phase with velocities of around 10-15 $\mathrm{km\,s^{-1}}$, very similar to the escape velocity of the AGB wind. This Bulge sample also provided strong evidence of the acceleration of the nebular shell predicted by hydrodynamical models at early stages of the central star's evolution, as the star evolves towards higher effective temperatures and the velocity of the stellar wind increases. According to models, at the very late stages, the nebular shell will lose its energy input power as the stellar wind weakens and the luminosity of the central star decreases. The effects of these changes have been observed in 100 PNe from the disc population analyzed by Pereyra *et al.* (2013). In this work, they found that the nebular shells with the largest velocities tend to have high luminosity central stars and a high degree of excitation. On the contrary, nebular shells with lower degree of excitation have lower velocities and correspond to the most evolved central stars.

3. Nebular shells around low mass progenitors of PNe

The sample under study includes 11 PNe from the Halo population and 15 seemingly bona-fide, low metallicity PNe (log(O/H) + 12 $\leqslant$ 8.0 dex) with kinematic data available from the SPM Catalogue (López *et al.* (2012)). The global expansion, or bulk flow velocity, of the nebular shell was determined from the matter with the highest emission within the spectrograph slit (i.e., *not* the post-shock velocity; Schönberner *et al.* (2005)). We obtained velocity measurements from [NII] λ6584, [OIII] λ5007, and Hα λ6563 emission lines, when available. Further details about data selection and velocity measurements can be found in Pereyra *et al.* (2016). The average velocity obtained for both groups, Halo and Low Metallicity PNe, is very similar in all of the emission lines considered (Table 1). The velocity distributions for the whole sample tend towards low values, with 62%, 70%, and 87% of the objects having expansion velocities below 20 $\mathrm{km\,s^{-1}}$ in the Hα λ6563, [OIII] λ5007, and [NII] λ6584 emission lines, respectively.

Table 1. Average expansion velocities for Halo and Low Metallicity PNe.

Group	$\overline{V}_{[\mathrm{N\,II}]}$ ± 2 ($\mathrm{km\,s^{-1}}$)	No. Objects considered	$\overline{V}_{\mathrm{H}\alpha}$ ± 2 ($\mathrm{km\,s^{-1}}$)	No. Objects considered	$\overline{V}_{[\mathrm{O\,III}]}$ ± 2 ($\mathrm{km\,s^{-1}}$)	No. Objects considered
Halo PNe	17	8	19	10	20	6
Low Metal PNe	17	7	19	14	18	11

By selecting Halo and low-metallicity PNe we bias our selected sample to a group of CSs of the lowest masses. Locating the central stars of these PNe in the Hertzsprung-Russell diagram, when temperatures and luminosities were available, we confirmed that they populate the low mass post-AGB tracks (1.5M$\odot$ or below, see Fig. 1). Although most of their nebular shells have low expansion velocities, which are among the lowest observed in PNe, their velocities do correlate with effective temperature. These results are not only in agreement with our previous findings but also with theory, which predicts that the stellar winds and the ionizing photon power output from the central star of a PN will be a consequence of the previous AGB evolution and both will increase during the post-AGB phase (Perinotto *et al.* (2004), Schönberner *et al.* (2005)). We conclude that

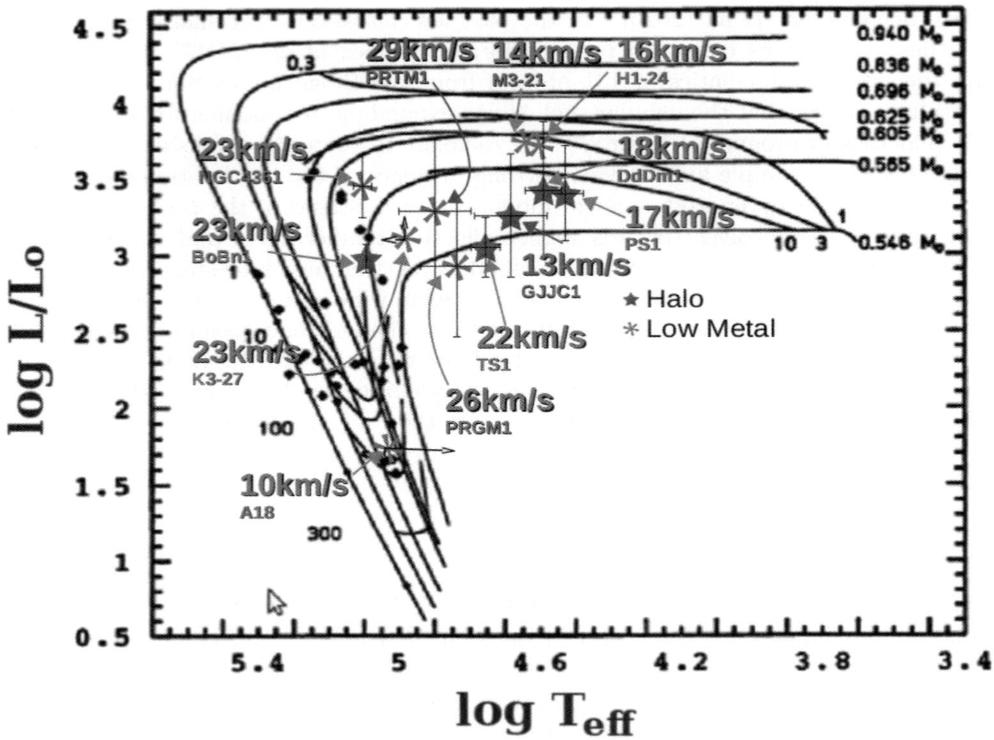

Figure 1. Coupling with the central star evolution. The central stars of these PNe with sub-solar metallicity seem to originate from low mass progenitors. Velocity measurements from Hα spectra are shown for most of the PNe (red label), except for GJJC1 and PRTM1 nebulae for which only [OIII] emission line spectrum was available (green label). The acceleration of the nebular shell is also observed in this sample, as expected from theory and previous observational results. Evolutionary tracks taken from Schönberner & Blöcker (1996).

the low expansion velocities found in this sample are directly related to the weaker stellar wind from the low-mass central stars driving the kinematics of their nebular shells.

4. Conclusions

By analyzing the internal kinematics of large samples of PNe segregated by stellar populations, we provided firm observational support for the theoretical predictions from hydrodynamical models. This work establishes a general description of the global kinematics of nebular shells and its time evolution throughout the entire planetary nebula phase (see Fig. 2). Both perspectives, theoretical and observational, are now consistent with each other but still there are some unanswered questions that need to be addressed. Our current understanding of the dynamics of PNe envelopes is mainly based on the assumption that all the nebular shells are produced by single star progenitors. However, a significant amount of binary central stars have been discovered over the past few years in the PNe Galactic population. These results suggest that the formation and evolution of planetary nebulae could be more complicated (De Marco (2009), Jones *et al.* (2014)). On the other hand, the newest theoretical models for the evolution of the post-AGB stars have pointed out that the evolutionary timescales for the central stars of planetary nebulae are at least 3 to 10 times faster than those computed by previous models (Miller

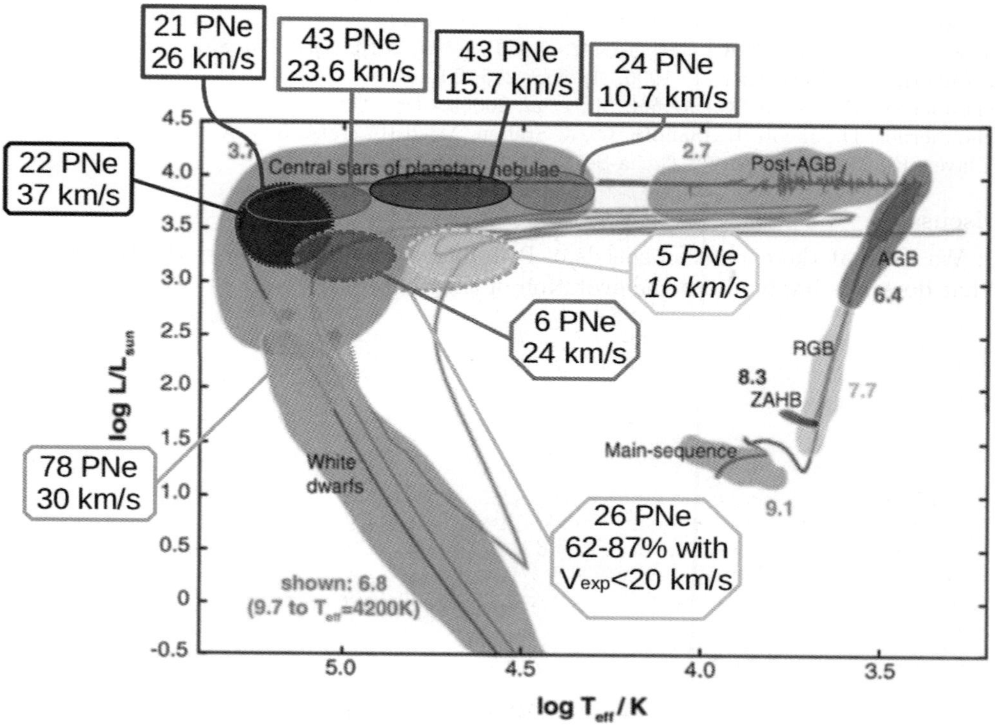

Figure 2. Observational framework for the time evolution of the internal kinematics of planetary nebulae. Hertzsprung-Russell diagram of a complete $2M_{\odot}$ evolutionary track from Herwig (2005). The labels indicate the average velocities obtained from the Bulge, Disc and Low-Mass samples. Our results clearly show that the nebular shells are accelerated at the early stages of the central star evolution in the planetary nebula phase, as the star becomes hotter and the stellar wind velocity increases. In the following evolution, when the inner parts of the shell lose their input energy from the central star, the expansion of the bulk of matter in the nebular shell slows down. For the PNe with low-mass progenitor stars we found nebular shells that expand slower that those around PNe originating from more massive AGB progenitors, which reflects the impact of the weaker stellar winds driving their kinematics.

Bertolami (2016)). These new timescales will have strong impact in the dynamics of the nebular shells of PNe, particularly for those nebulae with low-mass progenitors and should be taken into account in the radiation-hydrodynamic numerical simulations.

References

Abell G. O. & Goldreich, P. 1966, *PASP*, 78, 232
De Marco, O. 2009, *PASP*, 121, 316
García-Segura, G., López, J. A., Steffen, W., *et al.* 2006, *ApJ*, 646, L61
Herwig, Falk 2005, *Annu.Rev.A&A*, 43, 435
Jones, D., Santander-Garcia, M., Boffin, H. M. J., *et al.* 2014, *APNVI Conference*, 43
López, J. A., Richer, M. G., García-Díaz, Ma. T., *et al.* 2012, *RMxAA* 48, 3
López, J. A., Richer, M. G., Pereyra, M., & García-Díaz, M. T. 2016, *JPhCS*, 728
Mellema, G. 1994, *A&A*, 209, 915
Miller Bertolami, M. M. 2016, *A&A*, 588, A25
Pereyra M., Richer, M. G. & López, J. A. 2013, *ApJ*, 771, 114
Pereyra M., López, J. A. & Richer, M. G. 2016, *AJ*, 151, 53

Perinotto, M., Schönberner, D., Steffen, M., & Calonaci, C. 2004, *A&A*, 414, 993
Richer, M. G., López, J. A., García-Díaz, Ma. T., *et al.* 2010, *ApJ*, 716, 857
Schönberner, D. & Blöcker, T. 1996, *Ap&SS*, 245, 201
Schönberner, D., Jacob, R., Steffen, M., *et al.* 2005a, *A&A*, 431, 963
Schönberner, D., Jacob, R., Sandin, C., & Steffen, M. 2010, *A&A*, 523, A86
Villaver, E., Manchado, A., & Garía-Segura G. 2002, *ApJ*, 581, 1204

Discussion

Q: We see that the expansion speeds of PNe increase as they traverse the HR diagram. What does this imply about the evolution of the nebulae?

Planetary Nebulae: Multi-Wavelength Probes of Stellar and Galactic Evolution
Proceedings IAU Symposium No. 323, 2016
X. Liu, L. Stanghellini & A. Karakas, eds.

doi:10.1017/S1743921317000539

Planet destruction and the shaping of planetary nebulae

L. A. Boyle and M. P. Redman

Centre for Astronomy, National University of Ireland, Galway,
University Road, Galway, Ireland
email: `l.boyle2@nuigalway.ie`
email: `matt.redman@nuigalway.ie`

Abstract. The shaping of PNe as a result of an interaction with a planet is a hypothesis that has been suggested for nearly two decades. However, exploring the idea observationally is challenging due to the lack of capabilities needed to detect any evidence of such a scenario. Nonetheless, we propose that the hypothesis can be indirectly tested via a combination of exoplanet formation and evolution theories, the star and planet formation histories of the galaxy and the tidal evolution of star-planet systems. We present a calculation of the fraction of planetary nebulae in the galaxy today which have undergone an interaction with a planet, concluding that a significant number of visible planetary nebulae may have been shaped by a planet.

Keywords. Planetary Nebulae: general, Exoplanets

1. Introduction

One of the fundamental objectives still present in planetary nebulae (PNe) studies is to establish a characterisation of the mechanisms behind the array of morphologies observed. It is likely that central star binarity is the origin of the majority of bipolar planetary nebulae (Miszalski *et al.* 2009; Corradi *et al.* 2014; Jones *et al.* 2015). However, the period distribution of binary progenitor systems suggests that binary shaping is insufficient to account for the total population of non-spherical PNe (Raghavan *et al.* 2010). An encouraging idea is that a subset of them may be a result of the interaction of planets with single AGB progenitors (Soker 1996; De Marco & Soker 2011).

Orbital evolution calculations of planets around evolving stars have yielded crucial insights into the effects of mass-loss and stellar tides on a planets survival (Villaver & Livio 2007; Kunitomo *et al.* 2011; Nordhaus & Spiegel 2013; Madappatt *et al.* 2016). Given the correct combination of stellar mass, planet mass and initial semi-major axis, tidal forces can dominate over mass-loss and draw the planet into the envelope of its host. Whether the planet enters the envelope on the RGB or the AGB is crucial, since only an AGB interaction can result in the shaping of PNe. However, a planet entering the envelope of an RGB star can have significant consequences as it may impede the evolution of its host, ultimately preventing the formation of a PN at all (Nelemans 2010; Bear & Soker 2011). In the event that a planet enters the AGB envelope, its angular momentum and orbital energy will be deposited into the star leading to a possible in-spiral, destruction and shaping event (Nordhaus & Blackman 2006).

Unfortunately obtaining direct observational evidence for the planet shaping hypothesis is beyond the reach of current capabilities. The closest thing we have to observational evidence of a planet interaction in a PN is the suggestion that the presence of metal-rich components embedded within the normal nebular gas can provide a resolution to the long-standing *abundance discrepancy problem* (Wesson *et al.* 2005; Yuan *et al.* 2011).

The origin of such metal-rich components is still subject to debate, but one possibility put forward is that they may be the remnant of a destroyed planet (Liu *et al.* 2006; Corradi *et al.* 2015). In the absence of conclusive evidence we are still limited to a theoretical exploration, as obtaining an observational *planet fraction* to compliment the close binary fraction is challenging, if not yet impossible. Nonetheless, it is possible to make a prediction on the planet fraction based on our knowledge of the currently known exoplanets. We present a method to estimate the fraction of PNe which could have been shaped by the interaction of an AGB star with a planet.

2. Estimating the planet fraction of PNe: The known exoplanets

A primitive estimation for the planet-fraction can be sought by first looking to the distribution of currently known exoplanets. We downloaded information for the known exoplanets from the Exoplanet Orbit Database (exoplanets.org), and only included planetary systems which had been discovered, or followed up with, radial velocity measurements (thus where the star and planet masses and semi-major axes were available). This consisted in a total of 670 stars. The semi-major axis evolution of each planet system is modelled using the equations of orbital evolution around evolving stars given by

$$\dot{\mathrm{a}} = \mathrm{a}\frac{\dot{\mathrm{M}}_*}{\mathrm{M}_*} - \mathrm{f}\frac{\mathrm{a}}{\tau_{\mathrm{conv}}}\frac{\mathrm{M}_{\mathrm{env}}}{\mathrm{M}_*}\left(1+\frac{\mathrm{M}_{\mathrm{pl}}}{\mathrm{M}_*}\right)\frac{\mathrm{M}_{\mathrm{pl}}}{\mathrm{M}_*}\left(\frac{\mathrm{R}_*}{\mathrm{a}}\right)^8 \tag{2.1}$$

The first term in (2.1) represents the evolution due to mass-loss of the star, where M_{pl}, M_*, $\dot{\mathrm{M}}_*$ are the planet and stellar masses and the stellar mass-loss rate, respectively, and a is the semi-major axis of the planet. The second term in (2.1) is the evolution due to tides (Zahn 1977), where $\mathrm{M}_{\mathrm{env}}$ is the mass of the stellar envelope, R_* is the stellar radius and τ_{conv} is the convective timescale. The stellar parameters are calculated using the MESA stellar evolution code (Paxton *et al.* 2015). With the assumptions that all stars have a planet and that the stellar population is a good proxy for the PN progenitor population, we can make a simple estimation for the fraction of PN progenitors having interacted with a planet. To do this we categorise the hypothetical population as follows:

- **Systems which can shape a PN**
  - (*a*) Single stars interacting with a massive planet on the AGB
  - (*b*) Binary systems with a common envelope on the AGB
  - (*c*) Binary systems with no common envelope, but having a mild interaction
- **Non-interacting systems**
  - (*a*) Single stars with no interaction
  - (*b*) Binary systems with large separations
- **Systems which may prevent a PN**
  - (*a*) Single stars interacting with a massive planet on the RGB
  - (*b*) Binary systems with a common-envelope on the RGB

We do not model the orbital evolution of binary systems, instead we adopt the period distribution of Raghavan *et al.* (2010) who estimate that $\approx 50\%$ of stars form in binaries, where $\approx 30\%$ can undergo an interaction (14% on RGB, 16% on AGB).

The resulting population estimates are presented in Table 1. While it may appear that the majority of planet interactions occur on the RGB, we still do not know the true mass-period distribution of exoplanets. The radial velocity method is inherently biased towards the most close-in massive planets. This results in an apparent lack of planets with larger intitial separation capable of avoiding an interaction on the RGB, thus being available for an interaction on the AGB. Moreover, this analysis is based on the premise that the

Table 1. Population estimates based on exoplanet data.

	Single systems with planets					Binary systems			
Interaction		**RGB**	**AGB**	**None**			**RGB**	**AGB**	**None**
$N_{systems}$	670	196	50	424	$N_{systems}$	670	188	214	268
% single pop.	100	30	7	63	% binary pop.	100	28	32	40
% total pop.	50	15	3.5	31.5	% total pop.	50	14	16	20
% of all PNe		4					22		
% of non-spherical PNe		15					85		

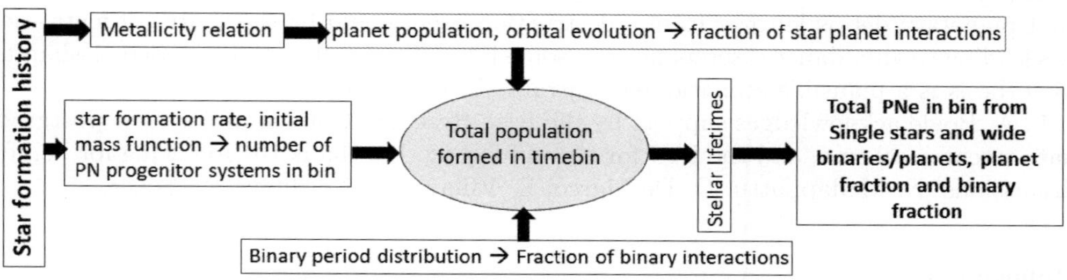

Figure 1. Simplified work-flow of a single time-bin in our model galaxy.

total population of PNe in the galaxy today evolved from a similar stellar population to that of the observed exoplanet hosts. For this assumption to hold true the galaxy would need to be a coeval population. In Section 3 we present a more robust approach.

3. Estimating the planet fraction of PNe: A galactic inventory

Since our aim is to predict the fraction of PNe in the galaxy whose progenitors would have suffered an interaction with a planet, we need to model the star-planet formation and evolution history. Our model galaxy is split into time-bins pre-determined by the star formation history provided by Behroozi *et al.* (2013). In each time-bin we calculate the total number of PNe progenitors which would have evolved to, and are currently still visible in, the present day galaxy by employing the star formation rate, the stellar initial mass function and stellar lifetimes. Combining the total number of PNe progenitor systems in each bin with a synthetic planet mass-period distribution (Ida & Lin 2005; Mordasini *et al.* 2015), our orbital evolution model described previously and the fraction of stars that have a planet (Gaidos & Mann 2014), we can determine the number of PNe from that time-bin that had a planet interaction. The criteria for systems resulting in non-spherical PNe for planets and binaries are the same as those outlined in Section 2. Finally, we sum the contributions from each bin allowing us to arrive at the fractions of PNe which have evolved via interactions with a planet, a binary or with no interaction. See Figure 1, and Boyle & Redman (2017, in prep).

4. Results and Implications

Our model predicts a total PN population of $\approx 6 \times 10^4$, of which $\approx 3.5 \times 10^3$(6%) would have evolved via an interaction with a massive planet, and $\approx 12 \times 10^3$(20%) via a binary interaction. This result suggests a total fraction of non-spherical PNe of $\approx$ 26%, translating to $\approx$23% and $\approx$77% of all non-spherical PNe having a planetary and binary origin, respectively. Moe & De Marco (2012) performed a thorough population synthesis of PNe from stellar binaries reporting a binary fraction of $\approx$70%, which provides us with a benchmark on which to compare our results. Our model predicts a total number of PNe discrepant with their result of 8,100 $\pm$ 2,300, however this is likely due to differing prescriptions for the star formation history, along with simplifications of our model. We anticipate that when appropriate refinements are made, and uncertainties taken into account that our resulting PN population and binary statistic will be in agreement.

While our model can still be regarded as preliminary, our result suggests that a significant fraction of PNe progenitors undergo interactions with orbiting massive planets. These interactions can potentially affect the morphology of the resulting PNe, complimenting the binary fraction. Additionally, this result provides support for the hypothesis that planetary debris leftover from a destruction event could be the source of metal-rich gas leading to abundance discrepancies in some PNe. We conclude that the planet-shaping hypothesis is a plausible scenario and warrants further investigation.

L. A. Boyle acknowledges support by the Irish Research Council under their postgraduate research scheme, and the IAU for travel funding. We thank the following for helpful discussions: N. Madappattu, O. De Marco, E. Villaver, J. Nordhaus, N. Soker.

References

Bear, E. & Soker, N. 2011, *MNRAS*, 411, 1792
Behroozi, P. S., Wechsler, R. H., & Conroy, C. 2013, *ApJ*, 770, 57
Corradi, R. L. M., Rodríguez-Gil, P., Jones, D., *et al.* 2014, *MNRAS*, 441, 2799
Corradi, R. L. M., García-Rojas, J., Jones, D., & Rodríguez-Gil, P. 2015, *ApJ*, 803, 99
De Marco, O. & Soker, N. 2011, *PASP*, 123, 402
Gaidos, E. & Mann, A. W. 2014, *ApJ*, 791, 54
Ida, S. & Lin, D. N. C. 2005, *ApJ*, 626, 1045
Jones, D., Boffin, H. M. J., Miszalski, B., *et al.* 2014, *A&AP*, 562, A89
Jones, D., Boffin, H. M. J., Rodríguez-Gil, P., *et al.* 2015, *A&AP*, 580, A19
Kunitomo, M., Ikoma, M., Sato, B., Katsuta, Y., & Ida, S. 2011, *ApJ*, 737, 66
Liu, X.-W., Barlow, M. J., Zhang, Y., Bastin, R. J., & Storey, P. J. 2006, *MNRAS*, 368, 1959
Madappatt, N., De Marco, O., & Villaver, E. 2016, *MNRAS*, 463, 1040
Miszalski, B., Acker, A., Moffat, A. F. J., Parker, Q. A., & Udalski, A. 2009, *A&AP*, 496, 813
Moe, M. & De Marco, O. 2012, *IAU Symposium*, 283, 111
Mordasini, C., Mollière, P., Dittkrist, K.-M., Jin, S., & Alibert, Y. 2015, *IJA*, 14, 201
Nelemans, G. 2010, *Ap&SS*, 329, 25
Nordhaus, J., & Blackman, E. G.2006, *MNRAS*, 370, 2004
Nordhaus, J. & Spiegel, D. S. 2013, *MNRAS*, 432, 500
Paxton, B., Marchant, P., Schwab, J., *et al.* 2015, *ApJS*, 220, 15
Raghavan, D., McAlister, H. A., Henry, T. J., *et al.* 2010, *ApJS*, 190, 1
Soker, N. 1996, *ApJL*, 460, L53
Villaver, E. & Livio, M. 2007, *ApJ*, 661, 1192
Wesson, R., Liu, X.-W., & Barlow, M. J. 2005, *MNRAS*, 362, 424
Yuan, H.-B., Liu, X.-W., Péquignot, D., *et al.* 2011, *MNRAS*, 411, 1035
Zahn, J.-P. 1977, *A&AP*, 57, 383

Session 5: The connection between binary evolution and PN phenomena

Chair: Roberto Mendez

Planetary Nebulae: Multi-Wavelength Probes of Stellar and Galactic Evolution
Proceedings IAU Symposium No. 323, 2016
X. Liu, L. Stanghellini & A. Karakas, eds.

doi:10.1017/S1743921317002022

Circumstellar spirals/shells/arcs: the message from binary stars

Hyosun Kim †

Academia Sinica Institute of Astronomy and Astrophysics, P.O. Box 23-141, Taipei 10617, Taiwan (email: hkim@asiaa.sinica.edu.tw)

Abstract. A consensus has grown in the past few decades that binarity is key to understanding the morphological diversities of the circumstellar envelopes (CSEs) surrounding stars in the Asymptotic Giant Branch (AGB) to Planetary Nebula (PN) phase. The possible roles of binaries in their shaping have, however, yet to be confirmed. Meanwhile, recurrent patterns are often found in the CSEs of AGB stars and the outer halos of PNe, providing a fossil record of the mass loss during the AGB phase. In this regard, recent molecular line observations using interferometric facilities have revealed the spatio-kinematics of such patterns. Numerical simulations of binary interactions producing spiral-shells have been extensively developed, revealing new probes for extracting the stellar and orbital properties from these patterns. I review recent theoretical and observational investigations on the circumstellar spiral-shell patterns and discuss their implications in linking binary properties to the asymmetric ejection events in the post-AGB phase.

Keywords. stars: AGB and post-AGB, binaries: general, circumstellar matter, stars: evolution, stars: fundamental parameters, stars: imaging, stars: late-type, stars: mass loss, stars: winds, outflows, submillimeter

1. Introduction

Since the launch of the Hubble Space Telescope (HST) in 1990, there have been many unexpected discoveries. One major finding among those is the recurrent patterns in the CSEs of sources in late stellar evolutionary phases (see Fig. 1), with AFGL 2688 being the archetypical target (Sahai *et al.* 1998). The HST image revealed its nearly concentric rings, which are interpreted as the limb-brightened surfaces of probable spherical shell-like structures. The pattern appear disconnected near the equatorial plane and brightest along the polar axis of the nebula due to the nonuniform illumination by direct starlight.

Su (2004) first summarized the properties of 10 sources showing the ring/arc patterns. Corradi *et al.* (2004) and Ramos-Larios *et al.* (2016) searched for the faint patterns in a systematic way by processing the HST and Spitzer archival imaging data, increasing the accumulated number of objects possessing the recurrent patterns to ~ 60. Recent molecular line observations using e.g., the Atacama Large Millimeter/submillimeter Array (ALMA) and Karl G. Jansky Very Large Array (JVLA) have been discovering the previously unknown spirals and shells. Table 1 shows some notable works in observations and theories on the recurrent patterns.

The kinematic timescales of the pattern intervals ($10^2 - 10^3$ years) are inconsistent with either stellar pulsation or thermal pulsation. Su (2004) showed that the timescales show no distinguishable trend along the evolutionary phase, implying that the ring/arc patterns are not a transient phenomenon. This is in conflict with one-dimensional time-dependent hydrodynamics calculations with ionization taken into account, resulting in

† EACOA fellow

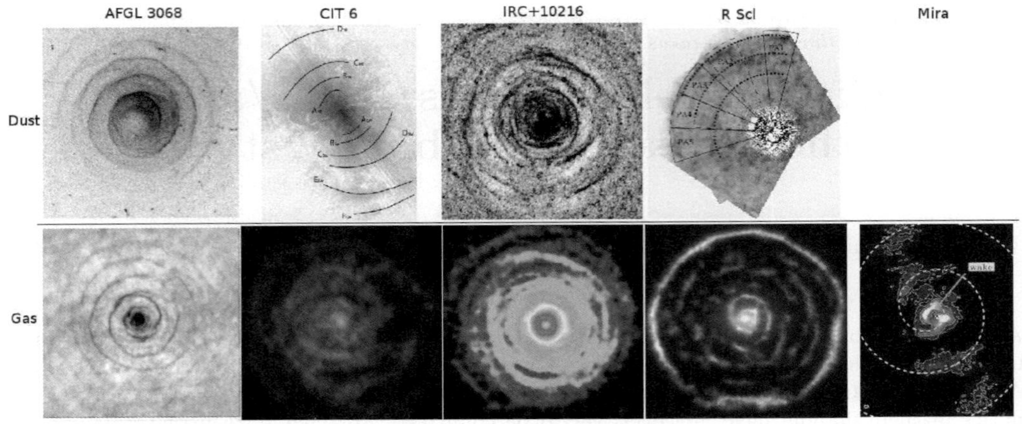

Figure 1. Upper panels: dust scattered light images of AFGL 3068 (Morris *et al. in prep.*), CIT 6 (Schmidt *et al.* 2002), IRC+10216 (Leão *et al.* 2006), and R Scl (Olofsson *et al.* 2010). **Lower panels:** CO center-frequency images of AFGL 3068 (Kim *et al.* 2017), CIT 6 (Kim *et al.* 2015b), IRC+10216 (Cernicharo *et al.* 2015), R Scl (Maercker *et al.* 2012), and Mira (Ramstedt *et al.* 2014). The sets of upper and lower panels are scaled in size.

1994	Soker – spiral due to wide binary motion
1997	Harpaz et al. – tidel interaction at periastron.
1998	Sahai (AFGL 2688) ... and many more from HST; long-term modulation of stellar pulsation
1999	Mastrodemos & Morris – binary
2004	Corral et al. (search for faint patterns)
2006	Mauron & Huggins (AFGL 3068 spiral, HST)
2011	Claussen (CIT6 rings, VLA)
2012	Kim & Taam (a, b=parameter study, c=AFGL3068 spiral); Maercker et al. (R Scl spiral)
2013	Kim et al. (CIT6 spiral, VLA)
2014	Ramstedt et al. (Mira, ALMA); Cernicharo et al. (IRC+10216 rose-window, IRAM)
2015	Decin et al. (IRC+10216 spiral, ALMA); Kim et al. (CIT6 spiral, SMA)
2016	Kim et al. (AFGL 3068 spiral, ALMA)

Table 1. Research highlights on the concentric ring-/arc-like patterns in the circumstellar envelopes of AGB stars and pPNe/PNe. In 1990s and 2000s, the HST observations and theories were rather independently investigated. In the current decade, thanks to the operations of high--sensitivity millimeter/submillimeter telescopes (e.g., ALMA, JVLA), the new detection of such patterns and their theoretical understanding have led to significant and synergistic progress.

the patterns disappearing within $\sim$ 3000 years after ionization starts (Meijerink *et al.* 2003). With the relevant model development (see §2) and the timescales consistent with the orbital periods of wide binaries, it is now commonly accepted that binary companions are responsible for modulating the mass distribution of CSEs in the recurrent form.

A remaining puzzle is their coexistence with bipolar structures in many pre-PNe (pPNe) and PNe. More than 80% of PNe are not spherical, but show a variety of

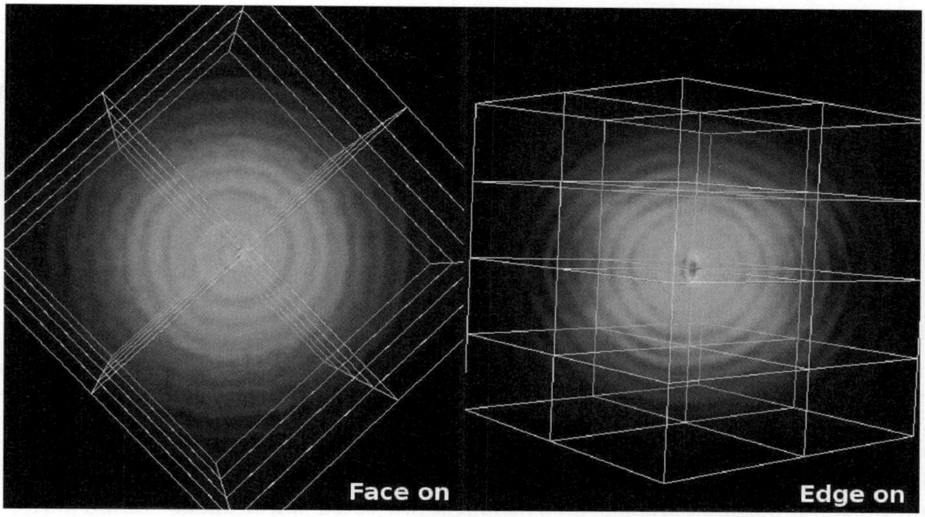

Figure 2. Circumstellar spiral-shell pattern induced by a binary in a circular orbit viewed at (*left*) face-on and (*right*) edge-on. The hydrodynamic simulation is performed using the FLASH3.0 code. The density levels vary from blue (low) to red (high). Notice that the edge-on pattern consists of bicentric half-circles. Also notice the overall flattened morphology. Credit: Hsieh, Kim, & Liu.

morphologies including highly bipolar or multipolar structures, and many of them also have outer faint ring-like patterns. The recurrent patterns have timescales of the order of a few hundred years, which correspond to the periods of *wide* binary stars. In contrast, the bipolar structures are the expected results of *close* binary interactions. I propose that a model involving a binary star system in an eccentric orbit can provide a plausible interpretation for the coexistence of these two geometrically distinct structures.

2. Binary-induced spiral-shell model

In a binary system, the orbital motion of the mass-losing AGB star introduces an anisotropy in the actual speeds of wind gusts (Soker 1994). The faster and slower wind gusts ejected over the orbit meet each other in the circumstellar space, forming a spiral pattern in the direction following the orbit. Mastrodemos & Morris (1999) showed, through smoothed-particle hydrodynamics simulations, the three dimensional spiral-shell nature of the circumstellar pattern induced by the binary orbital motion. The detailed structures are described in separate theoretical investigations for the patterns driven by the orbital motion of substellar-mass (e.g., planetary) companions (Kim & Taam 2012a) and by the reflex motion of a mass-losing star in a comparable-mass binary system (Kim & Taam 2012b). It is shown that the resulting geometrical details can be used to constrain the stellar masses and orbital properties (Kim & Taam 2012c; Kim *et al.* 2013).

The spiral-shell model provides a framework for understanding the various recurrent patterns, which include not only a perfect spiral but also circular rings depending on the orbit inclination angle (face-on vs. edge-on). In the edge-on view, the pattern shape is bicentric half-circles, in general; but when the orbital speed is sufficiently low compared to the wind speed, it takes the form of nearly concentric rings (see e.g., Fig. 2). The projected shape becomes an elongated spiral at intermediate inclination angles, from which the inclination angle can be derived (Fig. 3, and see Kim & Taam 2012c for details).

Using a simple geometric model, which is completely defined by five equations in Kim &

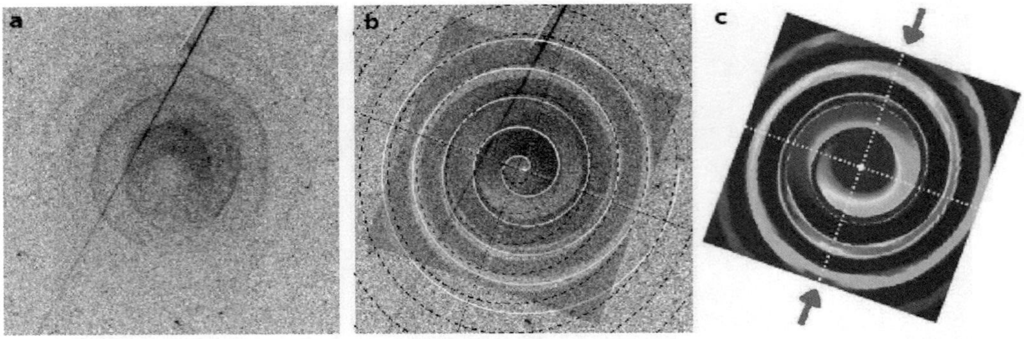

Figure 3. (a–b) AFGL 3068's HST image (25.″6 × 25.″6) is compared with (b–c) the density distribution in a hydrodynamic model (Kim & Taam 2012c). The arrows in (c) denotes the line of nodes of the binary orbital plane with the plane of the sky. The orbital period is almost free from the projection effect, and in this case, it is about 800 years.

Taam(2012c), a parameter space analysis for a projected spiral image can be performed. With four observables, mostly measured from the pattern shape, all binary orbital parameters can be derived. Once the parameter space is constrained, hydrodynamic models can be explored to fine tune the binary properties. Because the hydrodynamic modeling is expensive and time consuming, this pre-simulation analysis is very useful.

When an elongated spiral is projected onto a two-dimensional image, it could indicate an oblate ellipsoid halo viewed with a small inclination angle or an almost spherical halo with a very large inclination angle. This degeneracy is inherent in two-dimensional images such as that obtained from the HST. In the current era, high-resolution high-sensitivity interferometric data of molecular line emissions can be obtained providing kinematic information and, thus, the three-dimensional distribution of the pattern, lifting the degeneracy (see e.g., Kim *et al.* 2013 for details).

3. Eccentric long-period binaries

ALMA observations now reveal the detailed three-dimensional structure of the unambiguous and complete spiral pattern in AFGL 3068. In particular, the molecular line emission shows the innermost winding of the spiral pattern, which was absent in the HST dust scattered light image. It indicates that the innermost region of the CSE of AFGL 3068 is indeed extremely dense, thus, the penetration of interstellar UV photons required for scattering by the circumstellar dust is prohibited (see Fig. 1, leftmost panels). For the remaining windings, the dust and gas are well coupled.

The most exciting finding from the ALMA map of AFGL 3068 is the bifurcation of the spiral pattern, which is well reproduced by a highly eccentric-orbit binary model and cannot be mimicked by a circular-orbit model with any inclination (Fig. 4; Kim *et al.* 2017). This is the first solid piece of evidence for the existence of an eccentric binary system from a spiral pattern. We note that it is surprising that even the nearly perfect spiral source is not a circular-orbit binary. The similar and different patterns of various molecular line emissions are subject to detailed study of the radiative transfer and chemical compositional considerations to understand the physical conditions across the spiral shock.

An eccentric binary has also been suggested for another carbon star, CIT 6, based on the Submillimeter Array (SMA) observation (Fig. 5; see Kim *et al.* 2015): 1) the one-sided gaps of interarm emission, which can be reproduced by an eccentric binary model

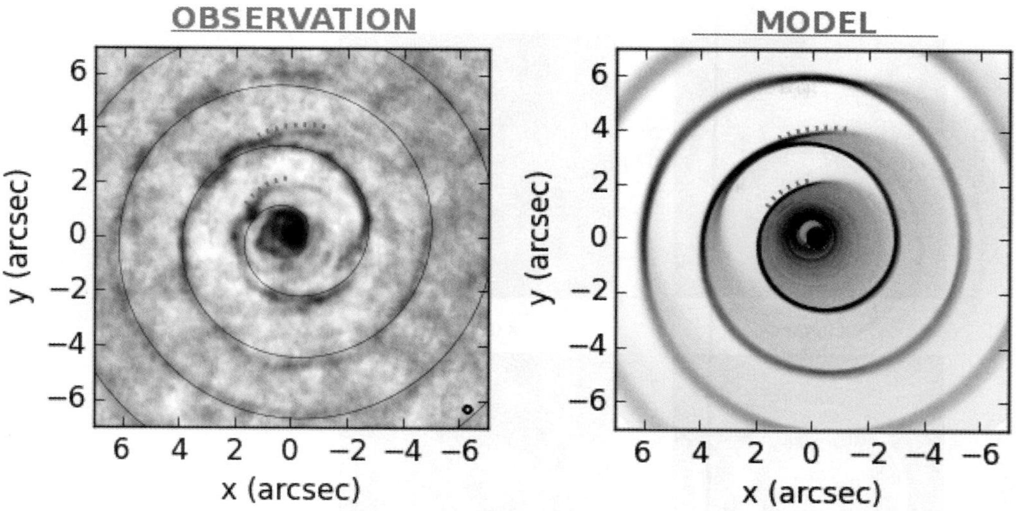

Figure 4. Bifurcation of the spiral pattern in AFGL 3068 revealing the eccentric binary nature. (*left*) ALMA observation versus (*right*) a hydrodynamic model of a binary with a large eccentricity of 0.8. The bifurcation is a distinctive characteristic of an eccentric binary and cannot be reproduced by a circular binary model (Kim *et al.* 2017).

(e.g. He 2007; Raga *et al.* 2011); 2) inner double spiral feature, which can be found in a relatively slow wind outflow (e.g., Mohamed & Podsiadlowski 2012) – the speculation is that a highly eccentric orbit leads to the penetration of the companion into the wind acceleration zone; 3) nascent bipolar (or multipolar) structure at the line edges, which would be a result of strong interaction between the binary stars at periastron passages.

In Table 2, these two carbon stars proposed as members of long-period eccentric binaries are compared to V Hydrae, a carbon star that is showing bullet-like mass ejection every 8.5 years, which is the possible orbital period of a binary (Sahai *et al.* 2016). Indeed, the derived orbital periods have a relation with the presence and geometry of the bipolar structures. For a binary in a circular orbit with these long periods, there is no mechanism to produce to the bipolar structure. Perhaps, strong interactions at the close periastron passages for a component in an eccentric binary can lead to a gradual decrease in the orbital separation, eventually leading to the formation of a disk around the companion star where bipolar outflows can be launched. This remains as a hypothesis and more theoretical investigations are needed. If true, the orbital eccentricity must be a key parameter linking the observed spiral-shell patterns to the bipolar structures.

Is it reasonable to expect such high eccentric orbits for these objects? Raghavan *et al.* (2010) showed that the distribution of orbital eccentricity in main-sequence binaries is random for systems with orbital periods greater than 10 days. Refer to the papers by D. Kamath and H. Van Winckel in this volume for the recent results for the eccentricity distribution of red giant binaries. The above three objects with orbital periods longer than several years are well within the regime that the circularization does not occur.

A triple system is another possible scenario for explaining the coexistence of outer long-period spiral-shell patterns and inner bipolar structures, although it should be very fine-tuned. The primary and secondary objects should be distant to explain the timescale of outer pattern spacing as observed to be of the order of a few hundred years, while the third body should be located near the companion star launching the bipolar structure. Among possible triple system configurations, only one such configuration works to explain the coexistence of two distinguished structures. Since the main-sequence fraction of triple

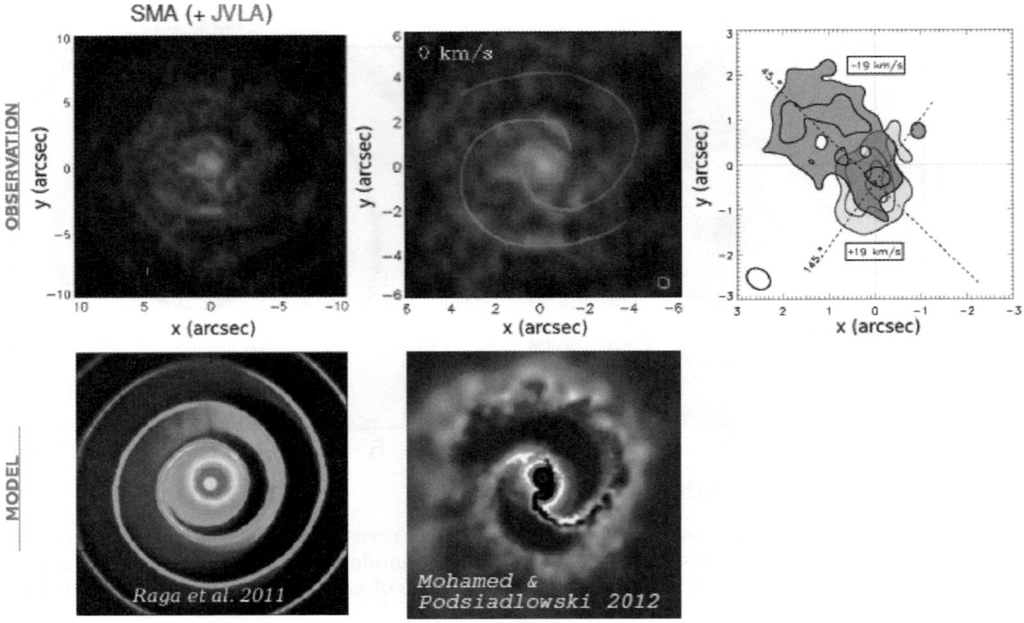

Figure 5. Hints of an eccentric binary in CIT 6 (Kim *et al.* 2015): (*left*) one-sided gaps of inter-arm emission, (*middle*) inner double spiral feature, and (*right*) nascent bipolar (or multipolar) structure. These features of CIT 6 revealed in the SMA observations are compared with the hydrodynamic model for a highly eccentric-orbit binary by Raga *et al.* (2011) and the slow wind model by Mohamed & Podsiadlowski(2012), respectively.

AFGL 3068	CIT 6	V Hydrae
$\sim$ 800 years no bipolar (Kim *et al.* 2017)	$\sim$ 300 years broadened bipolar (Kim *et al.* 2015)	8.5 years bullet-like ejection (Sahai *et al.* 2016)

Table 2. Comparison of three carbon stars AFGL 3068, CIT6, and V Hydrae tends to show a relation between the presence (and geometry) of bipolar structures and the orbital periods derived from either the spiral pattern or the mass ejection.

and higher-order multiple systems is less than 10%, it would be difficult to understand all known pPNe and PNe having recurrent patterns (> 60 objects, according to Ramos-Larios *et al.* 2016) by triple systems.

4. Implications

Spiral patterns are now frequently detected in the CSEs of stars in the AGB phase thanks to high-sensitivity high-resolution submillimeter/millimeter observations. This is a major step forward as many of them were not obvious in the dust scattered light images (Fig. 1). The complex rose-window pattern of IRC+10216 is understood in terms of a binary system, but requiring a mass loss variation (Cernicharo *et al.* 2015). For R Scl, a star that has a spiral and a thermal pulse shell, Maercker *et al.* (2012) used the spiral property as a timer to estimate the thermal pulsation period. In the case of Mira, the ALMA data reveal overall complex structures, but because it has a confirmed white dwarf companion, the partial arcs could be connected as a spiral (Ramstedt *et al.* 2014).

Many of these objects have a secondary point-like source as revealed in the near-infrared, optical, or radio continuum with apparent projected separations greater than 50 AU.

These five, best-known spiral/shell/arc objects are all in the AGB phase and are mostly carbon-rich. In order to determine whether the presence and evolution of circumstellar patterns depend on the stellar evolutionary phase and the chemical environment, molecular line surveys for AGB spirals in post-AGB sources are needed. Given such surveys, we can contemplate carrying out statistical studies of the known spiral/ring/arc sources. At the same time, analysis of the simplest objects, coupled with theoretical studies, will enable parameter estimation for a statistically meaningful sample. The golden era for studying the spiral-shell patterns is here, allowing us to progress in unravelling the nature and evolution of these evolved stars as encoded in their circumstellar patterns.

Acknowledgements

H.K. acknowlenges support for the travel and the work presented in this paper through EACOA Fellowship from the East Asia Core Observatories Association, which consists of the National Astronomical Observatories of China, the National Astronomical Observatory of Japan, the Korea Astronomy and Space Science Institute, and the Academia Sinica Institute of Astronomy and Astrophysics.

References

Cernicharo, J., Marcelino, N., Agúndez, M., & Guélin, M. 2015, A&A, 575, A91
Corradi, R. L. M., Sánchez-Blázquez, P., Mellema, G., Giammanco, C., & Schwarz, H. E. 2004, A&A, 417, 637
He, J. H. 2007, A&A, 467, 1081
Kim, H. & Taam, R. E. 2012a, ApJ, 744, 136
Kim, H. & Taam, R. E. 2012b, ApJ, 759, 59
Kim, H. & Taam, R. E. 2012c, ApJL, 759, L22
Kim, H., Hsieh, I.-T., Liu, S.-Y., & Taam, R. E. 2013, ApJ, 776, 86
Kim, H., Liu, S.-Y., Hirano, N., *et al.* 2015, ApJ, 814, 61
Kim, H., Trejo, A., Liu, S.-Y., *et al.* 2017, Nat. Astron., 1, 0060
Leão, I. C., de Laverny, P., Mékarnia, D., de Medeiros, J. R., & Vandame, B. 2006, A&A, 455, 187
Maercker, M., Mohamed, S., Vlemmings, W. H. T., *et al.* 2012, Nature, 490, 232
Mastrodemos, N. & Morris, M. 1999, ApJ, 523, 357
Meijerink, R., Mellema, G., & Simis, Y. 2003, A&A, 405, 1075
Mohamed, S. & Podsiadlowski, P. 2012, Baltic Astronomy, 21, 88
Olofsson, H., Maercker, M., Eriksson, K., Gustafsson, B., & Schöier, F. 2010, A&A, 515, A27
Raga, A. C., Cantó, J., Esquivel, A., Huggins, P. J., & Mauron, N. 2011, in Asymmetric Planetary Nebulae V Conference, ed. A. A. Zijlstra *et al.* (Jodrell Bank Centre for Astrophysics, Manchester), 185
Raghavan, D., McAlister, H. A., Henry, T. J., *et al.* 2010, ApJS, 190, 1
Ramos-Larios, G., Santamaría, E., Guerrero, M. A., *et al.* 2016, MNRAS, 462, 610
Ramstedt, S., Mohamed, S., Vlemmings, W. H. T., *et al.* 2014, A&A, 570, L14
Sahai, R., Trauger, J. T., Watson, A. M., *et al.* 1998, ApJ, 493, 301
Sahai, R., Scibelli, S., & Morris, M. R. 2016, ApJ, 827, 92
Schmidt, G. D., Hines, D. C., & Swift, S. 2002, ApJ, 576, 429
Soker, N. 1994, MNRAS, 270, 774
Su, K. Y. L. 2004, in ASP Conf. Ser. 313, Asymmetrical Planetary Nebulae III: Winds, Structure and the Thunderbird, ed. M. Meixner *et al.* (San Francisco, CA: ASP), 247

Discussion

CORRADI: Statistically, you should observe more circular and inclined spiral pattern rather than the spiral pattern that you expect when the systems are observed from "the pole". Is it what you see? Also, in PNe most patterns are clearly circular (e.g. NGC 6543), while in AGB stars you are mainly finding spirals.

KIM: The rings in PNe could be any of spirals or rings. Get a spiral model, block the central part by your hand, then you will find you cannot distinguish spiral/ring any more. Try it at home. Statistically, spirals are major in the binary model at the inclination angles 0 <= i <= 70 deg. except for the very edge-on case.

TORRES-PEIMBERT: Do you have any restrictions about the mass loss rate in your models?

KIM: The mass loss rate is used as an input parameter of the hydro dynamic model. With the full intensity data and through a radiative transfer calculation of the hydro dynamic model, the total mass distribution over the circumstellar envelope, and thus the rate, can be obtained. The constraint for the mass loss rate is possibly different for single vs. binary models, and for different inclination angles of the binary orbit.

DE MARCO: Have you determined what type of companion is implied by the ring structure of the Cat's Eye nebula? This PN is so close that we have put stringent limits on the companion mass and location ...

KIM: Not Yet. Let's do that.

LIU: It seems that observations hitherto of extrasolar planetary systems (proto-stellar disks) reveal spiral arms only in reflection light, not in e.g. dust emission. The case seems to be different in evolved stars (AGBs) or PNe. Would you like to comment on this?

MUTHUMARIAPPAN: Can one say something on the masses of and mass-loss rates from these spirals?

KWOK: You explain the outer structure as 2-dimensional spirals, but do you know that they are not projected 3-D shells?

Planetary Nebulae: Multi-Wavelength Probes of Stellar and Galactic Evolution
Proceedings IAU Symposium No. 323, 2016
X. Liu, L. Stanghellini & A. Karakas, eds.

doi:10.1017/S1743921317002216

Common Envelope Evolution: Implications for Post-AGB Stars and Planetary Nebulae

J. Nordhaus

National Technical Institute for the Deaf, Rochester Institute of Technology, One Bausch and Lomb Dr., Rochester, NY 14623, USA
email: nordhaus@astro.rit.edu

Center for Computational Relativity and Gravitation, Rochester Institute of Technology, One Bausch and Lomb Dr., Rochester, NY 14623, USA

Abstract. Common envelopes (CE) are of broad interest as they represent one method by which binaries with initially long-period orbits of a few years can be converted into short-period orbits of a few hours. Despite their importance, the brief lifetimes of CE phases make them difficult to directly observe. Nevertheless, CE interactions are potentially common, can produce a diverse array of nebular shapes, and can accommodate current post-AGB and planetary nebula outflow constraints. Here, I discuss ongoing theoretical and computational work on CEs and speculate on what lies ahead for determining accurate outcomes of this elusive phase of evolution.

Keywords. stars: AGB and post-AGB, (stars:) binaries (including multiple): close, stars: magnetic fields, stars: mass loss, (stars:) planetary systems, stars: evolution

---

1. Introduction

The extreme, non-spherical nebular morphologies observed in post-asymptotic-giant-branch (post-AGB) stars and their planetary nebulae (PNe) progeny are often attributed to interactions with orbiting companions. As low-mass stars ($\lesssim 8\ M_\odot$) evolve off the main sequence, expansion of the primary's radius is accompanied by strong mass loss, and affords new opportunities for powering outflows in binary systems. Potential shaping scenarios range from mild shocking of the outflow as the wind moves past the secondary to more extreme events such as Roche Lobe overflow and common envelope phases (Nordhaus *et al.* 2007, Chen *et al.* 2016).

The progenitor systems that are expected to incur such interactions are potentially common and span a range of binary types. For example, stellar, sub-stellar or compact objects initially orbiting $\lesssim$10 AU during the main sequence are expected to plunge into their host stars during the giant branches (Nordhaus *et al.* 2010, Nordhaus & Spiegel 2013). Rapid inspiral on fast timescales leads to a significant reduction in separation between the companion and the giant's core (Paczynski 1976). As the orbit shrinks, significant energy and angular momentum are transferred to the common envelope itself.

If the secondary is of sufficient mass, the CE can be ejected with the system emerging as a short-period binary (see Figure 1). If the companion is of insufficient mass to survive the interaction, it will be tidally disrupted as it approaches the giant's core (Nordhaus & Blackman 2006). The material from the disrupted companion forms an accretion disk inside the CE that is cold and dense compared to the ambient gas (Nordhaus *et al.* 2011). Even in these cases, accretion onto the proto-white dwarf core can eject the envelope resulting in a "single" star that, in effect, has been strongly influenced by the presence of a companion. Therefore, one complication in observationally determining the fraction of PN systems that evolved from binary systems is that:

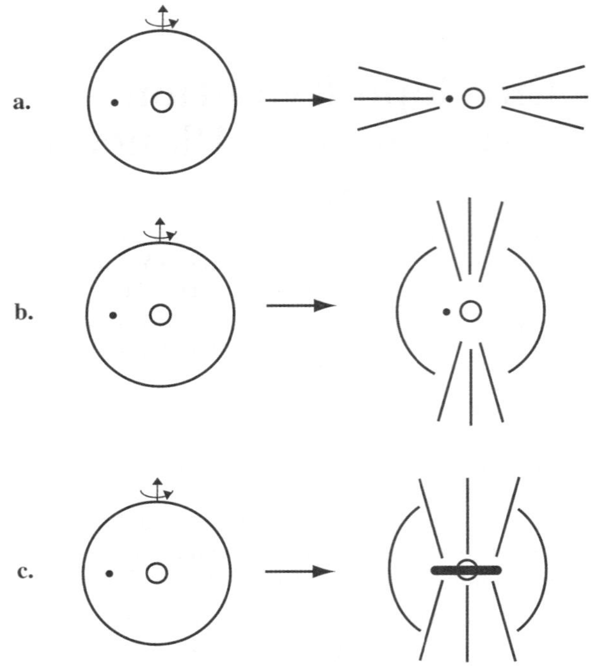

Figure 1. Possible outcomes of CE evolution. (a.) The companion directly ejects the envelope resulting, survives the CE and emerges as a post-CE binary (b.) Strong differential rotation in the CE induced by the plunging companion coupled with the presence of a deep convective zone, triggers a dynamo in the envelope resulting in a collimated bipolar outflow (c.) The companion is shredded into an accretion disk around the core which drives a collimated outflow and can lead to the formation of a highly-magnetized white dwarf (Nordhaus *et al.* 2011).

(i.) the absence of a detected companion does not imply that the progenitor system was a single star.

The converse is also highly applicable as common envelope phases are not the sole method for reducing orbital separation. In hierarchical three-body systems, Kozai-Lidov cycles exchange orbital inclination with eccentricity of the inner binary (Fabrycky & Tremaine 2007). In essence, the eccentricity of the inner binary can be pumped near unity at which point the orbit can circularize due to tidal dissipation. As the tertiary is distant, such systems would appear as post-CE binaries. Additionally, multi-body systems where $n > 3$ posses many methods that can generate close encounters. Therefore, we also have the added complication that:

(ii.) the detection of a short-period companion does not imply that the progenitor system evolved through a CE phase.

Since the white-dwarf mass function is strongly peaked, the population of short-period companions around low-mass white dwarfs may represent, at least in aggregate, a reasonable proxy for true descendants of CE phases. Disentangling the evolutionary outcome of a given system, let alone a population, requires a sufficient degree of fidelity from theoretical and numerical studies in conjunction with observational constraints.

2. Accretion during CEs satisfy observed post-AGB outflow constaints

One area which has recently had success in constraining formation scenarios for post-AGB systems comes from kinematic observations of large-scale outflows (Bujarrabal *et al.* 2001). From the measured outflow momenta, a minimum accretion rate, $\dot{M}_{\rm obs}$, necessary to power the outflow can be derived (Blackman & Lucchini 2014). The minimum accretion rates can be directly compared to theoretical models of engine-driving scenarios under the assumption of an accretor (main-sequence star or white dwarf) and the assumption that all accreted power goes into the outflow, $\dot{M}_{\rm max,theory}$. If $\dot{M}_{\rm max,theory} < \dot{M}_{\rm obs}$, that particular scenario can be confidently ruled out.

Recently, Blackman & Lucchini (2014) used observations from a sample of 19 post-AGB systems with kinematic outflow measurements. The authors considered both white dwarf and main-sequence accretors and four accretion paradigms: Bondi-Hoyle-Littleton (BHL) wind accretion, Roche-Lobe overflow consistent with the rate for the Red Rectangle, Wind-Roche-lobe overflow, and accretion during a common envelope phase. BHL wind accretion and Wind-Roche-lobe overflow were ruled out as explanations for the vast majority of objects. Roche-lobe overflow at the rate measured in the Red Rectangle could plausibly accommodate $\sim$1/3 of the sample. However, the only scenario that successfully powers all objects was accretion during a CE phase. This does not imply that common envelope accretion powers these systems, only that there is sufficient power available during the CE phase to explain the observed outflows.

3. CEs with Planetary and M-Dwarf Companions

Planets around main-sequence stars represent a reservoir of potential companions that might be expected to incur a CE phase and provide a source of accretion during the giant phases. Figure 2 shows the fates of known planetary systems with masses at least Jupiter's and host-star masses greater than or equal to the Sun's. Red dots indicate current semimajor axes of planets that will be engulfed, and tidally disrupted by their host stars. Blue circles indicate that the planet will avoid a common envelope phase with predicted final semimajor axes denoted by green dots. None of the systems shown will evolve to have circum-white dwarf planets within $\sim$6 AU. This implies that in the absence of scattering events, first-generation planets around white dwarfs are unlikely to be found in short-period orbits.

Nevertheless, these systems represent a source of accretion onto the white dwarf as they will tidally disrupt when the differential gravitational force from the proto-white dwarf exceeds the self gravity of the planet. Figure 3 shows the orbital period where tidal disruption (shredding) occurs (yellow dash-dot line) as a function of companion mass. Companions in the magenta region can survive the CE phase and emerge in short-period orbits to white dwarfs. Known low-mass-short-period companions to white dwarfs are denoted by blue circles and black diamonds. Grey-scale and marker size indicate white dwarf mass.

Note that for planetary masses, the allowable parameter space where planets survive the CE is very limited. This implies that most planets will tidally disrupt during the CE. Such disruption events result in initially geometrically-thick-hyper-critical accretion disks *within* the common envelope that are cold and dense compared to the hot stellar interior. These disks are prone to powerful outflows and may produce highly-magnetized white dwarfs via a dynamo (Nordhaus *et al.* 2011).

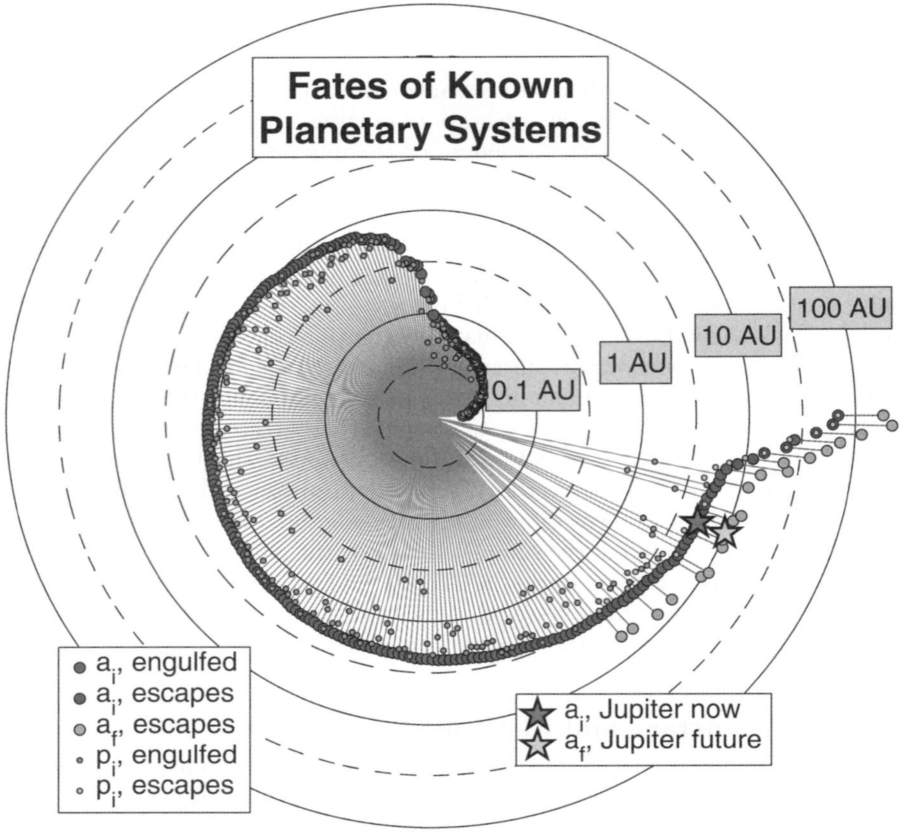

Figure 2. The fates of known planetary systems (Nordhaus & Spiegel 2013). Red and blue dots indicate current semimajor axes for exoplanets with masses at least Jupiter's and host-star masses greater than or equal to the Sun's. A red dot indicates that the planet will be engulfed while a blue dot indicates that the planet will escape engulfment; green dots indicate final circum-WD separations. Small light-red (light-blue) dots indicate current periastra of planets that will be (avoid being) engulfed. Current and final orbital semimajor axes of Jupiter are shown with magneta and yellow stars, respectively. Among the currently known planetary systems, none will evolve to have circum-WD jovian planets within ~6.4 AU.

4. Common Envelope Simulations

The common envelope phase occurs on dynamical timescales, is highly-non spherical and thus requires three-dimensional simulations with sufficient physical fidelity to properly model. In the last several years, grid-based attempts to simulate CE phases have been performed. Grid-based codes have an advantage over smooth-particle-hydrodynamics as they capture shocks (the companion's motion in the CE is supersonic) and utilize dynamical resolution if adaptive-mesh techniques are employed. Recent hydrodynamic grid-based simulations have been carried out with the ENZO, FLASH, and AREPO codes and have focused on near-equal mass binaries (see Table 3; Iaconi *et al.* 2016, Ricker & Taam 2012, Ohlmann *et al.* 2016). In these studies, the orbital energy available to unbind the CE should be sufficient even if radiative losses were severe. Under these conditions, none fully ejects the CE despite observing the dynamical inspiral phase. Additionally, all three predict final separations that are larger than those for these systems' observed post-CE counterparts. The authors conclude that processes on thermal timescales, unresolved

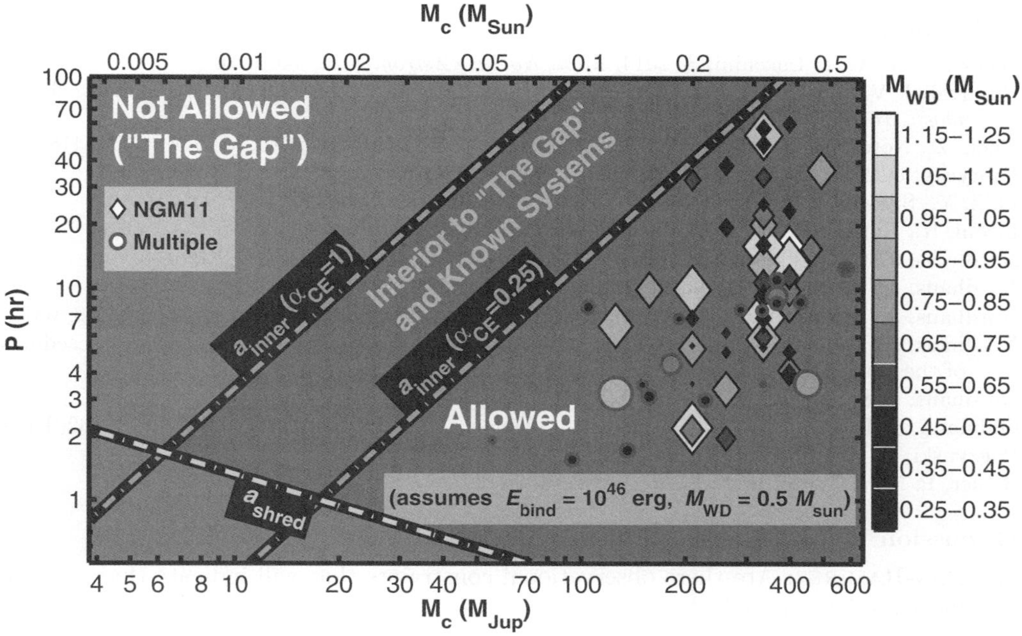

Figure 3. Known low-mass-short-period companions to white dwarfs (blue circles and black diamonds) compared to theoretical outcomes of CE evolution (Nordhaus & Spiegel 2013). Objects in the magenta region can survive the CE phase and emerge as short-period companions to WDs. The $a_{\rm inner}$ boundaries are determined by assuming that $\alpha_{\rm CE}$ of the liberated orbital energy during inspiral is used to unbind the CE. The $a_{\rm shred}$ boundary marks the location where companions tidally disrupt. Greyscale and marker size indicate white dwarf mass.

Table 1. Overview of current grid-based common envelope simulations.

Code	Primary $[M_\odot]$	Secondary $[M_\odot]$	Resolution* [cm]	Domain size [cm]
FLASH[1] (AMR)	1.05	0.6	2×10^{10}	4×10^{13}
ENZO[2] (AMR)	0.88	0.6	1.2×10^{11}	6×10^{13}
AREPO[3] (moving mesh)	1.98	0.99	7×10^{8}	3.3×10^{14}

Notes:
* Refers to smallest resolution present in the simulation.
[1] Ricker & Taam 2012
[2] Iaconi *et al.* 2016
[3] Ohlmann *et al.* 2016

microphysics, or other missing physical effects must be required to successfully eject the CE and produce post-CE systems that match observations (Ohlmann *et al.* 2016). Despite the initial challenges, these simulations represent the first step toward credible simulations of common envelopes. Future work must incorporate additional effects that are expected to be meaningfully important. These include, but are not limited to, incorporating magnetic fields, fully resolving convection in the giant's envelope, and including general relativistic effects for compact-object companions. Future work must also simulate a range of mass ratios for common envelope systems with various primaries.

References

Blackman, E. G. & Lucchini, S. 2014, *Mon. Not. R. Astron. Soc.*, 440, L16

Bujarrabal, V., Castro-Carrizo, A., Alcolea, J., & Sánchez Contreras, C. 2001, *Astron. Astrophys.*, 377, 868

Chen, Z., Nordhaus, J., Frank, A., Blackman, E. G., & Balick, B. 2016, MNRAS, 460, 4182

Fabrycky, D. & Tremaine, S. 2007, ApJ, 669, 1298

Fujii, Y., Spiegel, D. S., Mroczkowski, T., *et al.* 2016, ApJ, 820, 122

Iaconi, R., Reichardt, T., Staff, J., *et al.* 2016, MNRAS, in press.

Nordhaus, J. & Blackman, E. G. 2006, MNRAS, 370, 2004

Nordhaus, J., Blackman, E. G., & Frank, A. 2007, MNRAS, 376, 599

Nordhaus, J., Spiegel, D. S., Ibgui, L., Goodman, J., & Burrows, A. 2010, MNRAS, 408, 631

Nordhaus, J., Wellons, S., Spiegel, D. S., Metzger, B. D., & Blackman, E. G. 2011, Proceedings of the National Academy of Science, 108, 3135

Nordhaus, J. & Spiegel, D. S. 2013, MNRAS, 432, 500

Ohlmann, S. T., Röpke, F. K., Pakmor, R., Springel, V., & Müller, E. 2016, MNRAS, 462, L121

Paczynski, B. 1976, Structure and Evolution of Close Binary Systems, 73, 75

Ricker, P. M. & Taam, R. E. 2012, ApJ, 746, 74

Discussion

GUZMAN-RAMIREZ: Are there observational constraints that will indicate that a single star has experienced a CE?

NORDHAUS: For RGB/AGB stars, some possible indication might be isotopic abundances from extra mixing processes induced by inspiral of perhaps enhanced surface rotation rates.

DE MARCO: The Kozai-Lidov effect might be efficient, but only a very small combination of parameters could give you a compact binary in the middle of a PN that has avoided a common envelope phase.

NORDHAUS: The exact population calculation has not been worked out, but there might be more of these objects than one may imagine.

Planetary Nebulae: Multi-Wavelength Probes of Stellar and Galactic Evolution
Proceedings IAU Symposium No. 323, 2016
X. Liu, L. Stanghellini & A. Karakas, eds.

doi:10.1017/S1743921317002149

Post-common envelope PN, fundamental or irrelevant?

Orsola De Marco[1,2], T. Reichardt[1,2], R. Iaconi[1,2], T. Hillwig[3], G. H. Jacoby[4], D. Keller[5], R. G. Izzard[6], J. Nordhaus[7] and E. G. Blackman[8]

[1]Department of Physics & Astronomy, Macquarie University, Sydney, NSW 2109, Australia
email: orsola.demarco@mq.edu.au

[2]Astronomy, Astrophysics and Astrophotonics Research Centre, Macquarie University

[3]Department of Physics & Astronomy, Valparaiso University, Indiana, USA

[4]Lowell Observatory, Flagstaff, AZ, USA

[5]KU Leuven, Belgium

[6]Institute of Astronomy, Cambridge University, UK

[7]National Technical Institute for the Deaf, Rochester Institute of Technology, NY, USA

[8]University of Rochester, NY, USA

Abstract. One in 5 PN are ejected from common envelope binary interactions but *Kepler* results are already showing this proportion to be larger. Their properties, such as abundances can be starkly different from those of the general population, so they should be considered separately when using PN as chemical or population probes. Unfortunately post-common envelope PN cannot be discerned using only their morphologies, but this will change once we couple our new common envelope simulations with PN formation models.

Keywords. planetary nebulae: general, hydrodynamics, stars: AGB and post-AGB, binaries (including multiple): close, stars: evolution, stars: statistics

1. The fraction and chemistry of post-common envelope PN

At least one in 5 planetary nebulae (PN) derives from a common envelope (CE) binary interaction, where the PN is the ejected CE and the inner close binary is the result of the companion's in-spiral into the progenitor AGB star (Ivanova *et al.* 2013). This fraction is a lower limit because we cannot detect photometric variabilities $\lesssim$0.1 mag, due primarily to poor sampling of ground-based observations. A search carried out with the *Kepler Space Telescope* resulted in the detection of 4 variables out of 5 objects with data (De Marco *et al.* 2015). Of the 4 variables, one is a double-degenerate binary central star, and one is likely to be a pole-on binary. With variability amplitudes of 0.0007 and 0.0005 mag, neither would have been detected from the ground.

A binary search using data from *Kepler II* (K2), campaigns 0, 2, 7 and 11 is underway. Campaigns 0, 2 and 7 include a grand total of 15 PN, two of which have periodic variability consistent with binarity. The variability amplitudes, 0.05 and 0.02 mag, respectively, would not have been easily detected from the ground. However, the sensitivity of K2 is up to 5 times worse than for *Kepler* and neither of the 2 binaries detected with *Kepler* would have been detected by K2. In addition, 11 of the 15 K2 targets analysed are compact PN, while all the *Kepler* targets are extended: the variability detection threshold is therefore much worse for the bulk of the K2 targets. K2, Campaign 11 contains 139 viable PN, of which 22 are extended. Hopefully those data will return better statistics. The *Kepler* data show that low amplitude post-CE PN central binaries exist, as predicted

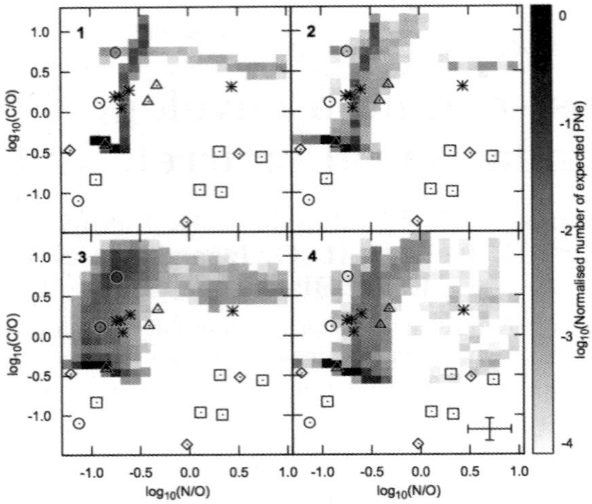

Figure 1. Observed vs. predicted C/O and N/O ratios for PN produced via different channels: (1) single stars, (2) wind mass-loss from binary star origin, (3) CE merger + wind mass loss, (4) CE ejection. Observation symbols: round: round PN; star: elliptical PN; square: bipolar PN; triangular: elliptical PN with bipolar core, rhomboid: quadrupolar PN. Observations: Leisy & Dennefeld (2006); Stanghellini *et al.* (2005). Figure from Keller, Izzard and Stanghellini, in prep.

and that the fraction of post-CE PN could be substantially larger than the lower limit of one in 5. This lower limit is already larger than the prediction from population synthesis of 9% for Solar metallicity of Izzard & Keller (2015, Fig. 1, which are consistent with previous work, e.g., Yungelson *et al.* (1993); Nie *et al.* (2012); Madappatt *et al.* (2016)). An over-abundance of post-CE PN could be explained either if some single stars make under-luminous PN (Soker & Subag 2005) or if some post-CE PN are mimics, such as for example Stroemgren spheres around hot stars (Frew & Parker 2010).

Expectations of the properties of post-CE PN are different from those of PN from other evolutionary channels. Post-CE PN have a curbed AGB evolution, resulting in reduced C/O ratios and lower s-process abundances. A secondary effect is that more massive stars are in binaries more often, something that could inflate the relative number of more massive stars in the post-CE PN population resulting in relatively larger N/O ratios. In Fig. 2 we can compare the top-left and bottom-right panels where we see the relatively broader distributions in the post-CE channel and also the larger, though still small, proportion of PN at high N/O and very low C/O not present in the single star channel. The observations are still insufficient for a proper comparison, but it is already clear that the observed bipolar PN at low C/O and mostly but not exclusively high N/O (more massive central stars) can only be (approximately) reproduced by the CE channel.

2. Observed and simulated post-common envelope PN

Unfortunately we cannot unambiguously recognise post-CE PN based on their morphology alone. Approximately 50 post-CE central stars are known to date and about two thirds of them are reasonably well imaged. As shown by Miszalski *et al.* (2009), many of these PN are approximately bipolar (e.g., A 41, K 1-2, Fig.2), but there is some diversity. Some PN display only a narrow waist (e.g., HaTr 4), sometimes with jets (A 63; jets not visible in Fig. 2). Ring nebulae (e.g., NGC 6337, Sp 1) are likely to be pole-on views of waists. More complex morphologies are also present: elliptical/bipolar PN with jets (NGC 6778, Fleming 1), more or less elliptical distributions of faint gas in which we see

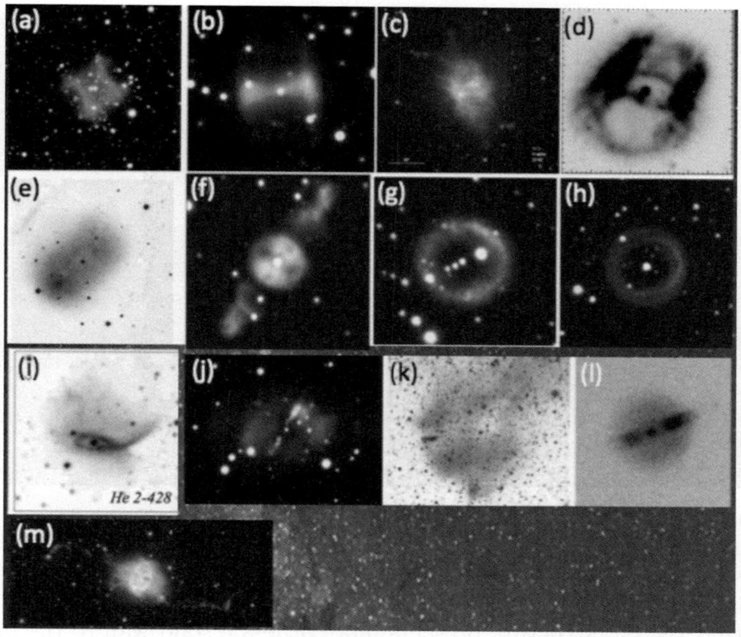

Figure 2. PN around post-CE central binaries; (a) A 63 (Mitchell *et al.* 2007); (b) HaTr 4 (Tyndall *et al.* 2012); (c) NGC 6778 (Guerrero & Miranda 2012); (d) A 41 (Jones *et al.* 2010); (e) A 65 (Huckvale *et al.* 2013); (f) ETHOS 1 (Miszalski *et al.* 2011); (g) NGC 6337 (Hillwig *et al.* 2010); (h) Sp 1 (Hillwig *et al.* 2016); (i) He 2-428 (Santander-García *et al.* 2015); (j) K 1-2 (Corradi *et al.* 1999); (k) HFG 1 (De Marco 2009); (l) PN G135.9+55.9 (Napiwotzki *et al.* 2005); (m) Fleming 1 (Boffin *et al.* 2012).

superimposed stark and very collimated double (PN G 135.9; the similar structures in K 1-2 may be filaments (Miszalski *et al.* 2009)) or single jet-like structures (HFG 1).

The only way to understand how these shapes result from CE ejections is via simulations, such as those by García-Segura *et al.* (1999) or Huarte-Espinosa *et al.* (2012). Nobody to date has simulated a post-CE PN *accounting for the common envelope ejection*. Simulations of the CE interaction have shown that the ejection of the envelope takes place primarily on the orbital plane (Sandquist *et al.* 1998; Passy *et al.* 2012). We have recently carried out a 1-million particle SPH simulation with the code PHANTOM (Iaconi *et al.* 2017) and the parameters of Passy *et al.* (2012), but starting the simulation at the time of Roche lobe overflow, something that allows the distribution of ejected gas to be more realistic. In addition, with SPH the entire gas distribution at the end of the simulation (~15 years after Roche lobe overflow) is known, albeit with varying resolution (from 0.3 $R_\odot$ near the binary to 80 $R_\odot$ near the outer parts of the distribution).

This simulation has not considered the fast wind, nor the ionising radiation. We could plausibly assume that the central star at the end of our simulation is hotter than ~20,000 K. If so, both ionising radiation and a fast tenuous wind would start blowing into the dense CE structure. While this has not been done yet in these simulations, using this image we can already make the following considerations. If a fast wind were to start at the end of the CE in-spiral it would meet with incredibly high densities, of the order of 10^{-8} g cm^{-3} in the vicinity of the stars (Fig. 3, left panel). Even the evacuated, funnel-like regions seen above and below the equatorial plane have high densities (~10^{-11} g cm^{-3}). By comparison typical PN simulations blow a fast wind into a much more tenuous medium with typical densities of 10^{-21} g cm^{-3}, e.g., García-Segura (2006).

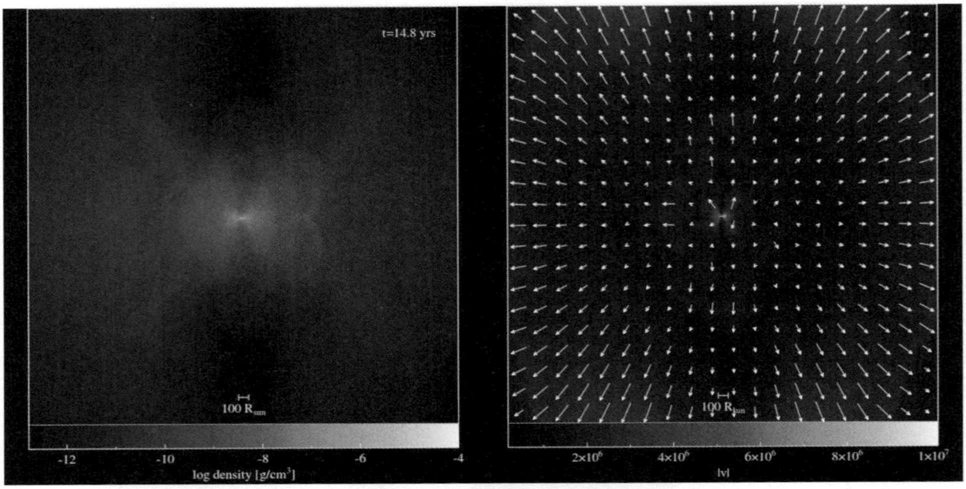

Figure 3. Density (left) and velocity (right, in cm s^{-1}; arrows indicate flow direction) slices perpendicular to the orbital plane at 14.8 years of an SPH CE simulation between a 0.88 M$_\odot$, RGB star and a 0.6 M$_\odot$ companion, carried out with the PHANTOM code (Iaconi *et al.* 2017).

This shows that the CE structure presents more or less a solid wall to the fast wind. On the other hand the entire CE is expanding at a range of speeds with the bulk of the volume expanding at 20 km s^{-1}. Hence, in 100-1000 years the densities will decline by 5 to 8 orders of magnitude and the wind will start sweeping and penetrating the CE.

References

Boffin, H. M. J., Miszalski, B., Rauch, T., *et al.* 2012, Science, 338, 773
Corradi, R. L. M., Perinotto, M., Villaver, E. 1999, *ApJ*, 523, 721
De Marco, O. 2009, *PASP*, 121, 316
De Marco, O., Long, J., Jacoby, G. H., *et al.* 2015, *MNRAS*, 448, 3587
Frew, D. J. & Parker, Q. A. 2010, *PASA*, 27, 129
García-Segura, G. 2006, in IAU Symp., 234, ed. Barlow & Méndez, 297
García-Segura, G., Langer, N., Różyczka, M., & Franco, J. 1999, *ApJ*, 517, 767
Guerrero, M. A. & Miranda, L. F. 2012, *A&A*, 539, A47
Hillwig, T. C., Bond, H. E., Afşar, M., & De Marco, O. 2010, *AJ*, 140, 319
Hillwig, T. C., Jones, D., De Marco, O., Bond, H. E., *et al.* 2016, *ApJ*, 832, 125
Huarte-Espinosa, M., Frank, A., Balick, B., *et al.* 2012, *MNRAS*, 424, 2055
Huckvale, L., Prouse, B., Jones, D., Lloyd, M., *et al.* 2013, *MNRAS*, 434, 1505
Iaconi, R., Reichardt, T., Staff, J., De Marco, O., *et al.* 2017, *MNRAS*, 464, 4028
Ivanova, N., Justham, S., Chen, X., De Marco, O., *et al.* 2013, *A&A Rev.*, 21, 59
Izzard, R. G. & Keller, D. 2015, in ASP Conf. Ser., 497, ed. Kerschbaum, Wing, & Hron, 161
Jones, D., Lloyd, M., Santander-García, M., *et al.* 2010, *MNRAS*, 408, 2312
Leisy, P. & Dennefeld, M. 2006, *A&A*, 456, 451
Madappatt, N., De Marco, O., & Villaver, E. 2016, *MNRAS*, 470, 317
Miszalski, B., Acker, A., Parker, Q. A., & Moffat, A. F. J. 2009, *A&A*, 505, 249
Miszalski, B., Corradi, R. L. M., Boffin, H. M. J., Jones, D., et al. 2011, *MNRAS*, 413, 1264
Mitchell, D. L., Pollacco, D., O'Brien, T. J., *et al.* 2007, *MNRAS*, 374, 1404
Napiwotzki, R., *et al.* 2005, in AIOP Conf. Ser., 804, ed. Szczerba, Stasinska, Gorny, 173
Nie, J. D., Wood, P. R., & Nicholls, C. P. 2012, *MNRAS*, 423, 2764
Passy, J.-C., De Marco, O., Fryer, C. L., Herwig, F., *et al.* 2012, *ApJ*, 744, 52
Sandquist, E. L., Taam, R. E., Chen, X., Bodenheimer, P., & Burkert, A. 1998, *ApJ*, 500, 909

Santander-García, M., Rodríguez-Gil, P., Corradi, R. L. M., *et al.* 2015, *Nature*, 519, 63
Soker, N. & Subag, E. 2005, *AJ*, 130, 2717
Stanghellini, L., Shaw, R. A., & Gilmore, D. 2005, *ApJ*, 622, 294
Tyndall, A. A., Jones, D., Lloyd, M., O'Brien, T. J., & Pollacco, D. 2012, *MNRAS*, 422, 1804
Yungelson, L. R., Tutukov, A. V., & Livio, M. 1993, *ApJ*, 418, 794

Discussion

Q: We have many wonderful common-envelope ejection simulations in the meeting. What are they teaching us about the formation of bipolar lobes? How can we utilize observations of lobe shapes and kinematics to constrain CE models?

LEAL-FERREIRA: You showed the case of this source in which we are not sure what is going on; maybe binary, maybe pulsation... Can a triple system fit that observation?

DE MARCO: I am not sure how that would work. We tried to think of various binary-related phenomena but nothing really stood out as a clear winner.

CORRADI: Given the important implications on the binary fraction of the results from Kepler, can you comment if there may be some selection bias in the sample? Could there be a selection bias in the sample of post CE PN in general?

DE MARCO: There should be no bias because the PN analyzed are all those found in the Kepler field. The only bias is the usual one affecting the entire PN sample (size, brightness, etc). In general we should consider whether any of the CE PN's and CE central stars' characteristics make them inherently easier to select out.

Planetary Nebulae: Multi-Wavelength Probes of Stellar and Galactic Evolution
Proceedings IAU Symposium No. 323, 2016
X. Liu, L. Stanghellini & A. Karakas, eds.

doi:10.1017/S1743921317001910

Where are the binaries in proto-planetary nebulae? Results of a long-term radial velocity study

Bruce J. Hrivnak[1] Hans Van Winckel[2], Griet Van de Steene[3], David Bohlender[4], Julius Sperauskas[5] and Wenxian Lu[1]

[1]Valparaiso University, Valparaiso, IN 46383, USA; email: bruce.hrivnak@valpo.edu

[2]Instituut voor Sterrenkunde, K. U. Leuven University, 3001 Leuven (Heverlee) Belgium; email: Hans.VanWinckel@ster.kuleuven.be

[3]Royal Observatory of Belgium, Ringlaan 3, 1180, Brussels, Belgium; email: gsteene@oma.be

[4]Dominion Astrophysical Observatory, Herzberg Institute of Astrophysics, 5071 West Saanich Road, Victoria, BC V9E 2E7, Canada; email: David.Bohlender@nrc-cnrc.gc.ca

[5]Vilnius University Observatory, Ciurlionio 29 Vilnius 2009, Lithuania; email: julius.sperauskas@ff.vu.lt

Abstract. We present the results of a long-term search (25 yrs) for radial velocity variability in a sample of seven bright proto-planetary nebulae showing axial symmetry. They all vary in velocity due to periodic pulsations. However, only marginal evidence is found for multi-year variations that might be due to a binary companion.

Keywords. binaries: spectroscopic – planetary nebulae: general – stars: AGB and post-AGB

1. Introduction

One of the outstanding problems in the study of planetary nebulae (PNe) concerns their shapes. While their AGB progenitors are basically spherical, PNe display a wide array of shapes, often axial or point-symmetric. Several mechanisms have been suggested to produce this morphology. These include the collimation of a fast wind by a torus produced by either (i) mass focused onto an orbital plan by a binary or (ii) mass lost preferentially in an equatorial plane by a rapidly rotating star. Another suggestion is (iii) a toroidal magnetic field that shapes the outflow of charged particles. While the first requires a binary explicitly, it is found in practice that the other two also require a binary to sustain them. Binaries companions to the central stars of PNe have been found in ∼15% of well-studied cases, primarily by photometric variability. Almost all have have short periods, less than a few days, and are assumed to be the result of common envelope evolution (see contributions by Nordhaus and by De Marco, this volume). A recent study demonstrated a tight correlation between the inclination of the binary and the inclination of the nebula (Hillwig *et al.* 2016). Periods longer than a few days are harder to find photometrically and fall into the realm of spectroscopic searches. See the review by De Marco (2009) of the evidence and role of binary central stars of PNe.

In this study, we have chosen to focus on a radial velocity search for binaries in proto-PNe (or pre-PNe; PPNe), the immediate precursors of PNe. They are distinguished by a large mid-IR excess due to cool dust from a detached circumstellar shell ("shell" sources). PPNe have the advantage of possessing F−G central stars, with more and narrower lines as compared to PNe. This allows the determination of more precise velocities. However,

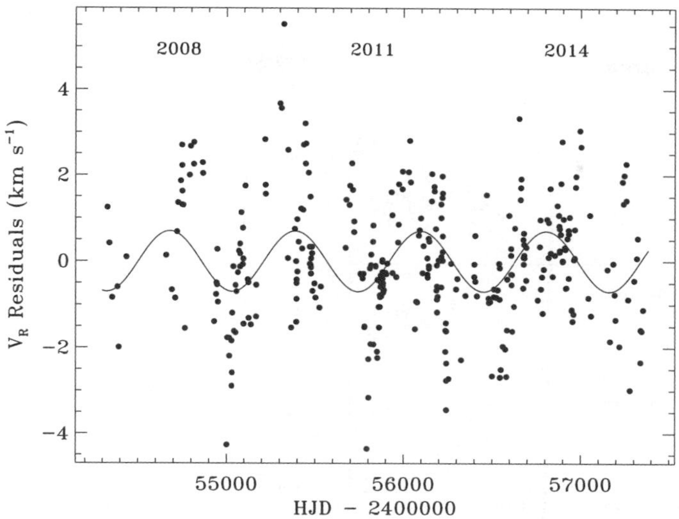

Figure 1. The observations of IRAS 22272+5435 following the removal of the periodic pulsations, fitted by a sine curve with P=710 days.

PPNe are known to vary due to pulsation, with periods in the range of 35−160 days (Hrivnak *et al.* 2010, Arkhipova *et al.* 2010), and this must be taken into account when searching for binary periods. These PPNe are to be distinguished from post-AGB F−G stars with near-IR excesses, all of which appear to be binaries (P∼120−1500 days) with circumbinary disks ("disc" sources; Van Winckel *et al.* 2009).

2. Observational Program

We have been carrying out a long-term observational study of the seven brightest (V<11) PPNe easily observable from the northern hemisphere using a 1−2 m telescope (Hrivnak *et al.* (2011)). They have been imaged in the visible (*HST*) and mid-IR and have bipolar or ellipsoidal shapes with sizes of a few arcsecs.

High-resolution spectroscopic observations were carried out with several telescopes and spectrographs over two different intervals of time. The initial observations were made in 1991−1995 at the Dominion Astrophysical Observatory (DAO) using a mechanical radial velocity spectrometer (RVS) attached to the 1.2m telescope. They revealed variations of ∼10 km s^{-1}, and in some cases periodic pulsation could be seen. Observations were re-initiated at the DAO in 2007, this time using a CCD detector. The program was enlarged to include observations made with the HERMES spectrograph (Raskin *et al.* 2011) on the 1.2-m Mercator Telescope in La Palma in 2009 and, for three of the objects, with a CORAVEL spectrometer on the 1.65-m telescope at the Moletai Observatory in Lithuania from 2008−2014.

3. Results

Multi-year studies. Pulsational variability was seen in all of the objects. In most of them, a clear periodicity could be seen with a timescale similar to what was found in their light curves (35−135 days). A search for longer, multi-year periodicity was unsuccessful in five of the objects, but some evidence was seen in two of them. It was revealed by variations in the mean seasonal velocities. For IRAS 22272+5435, the combined 2007−2015

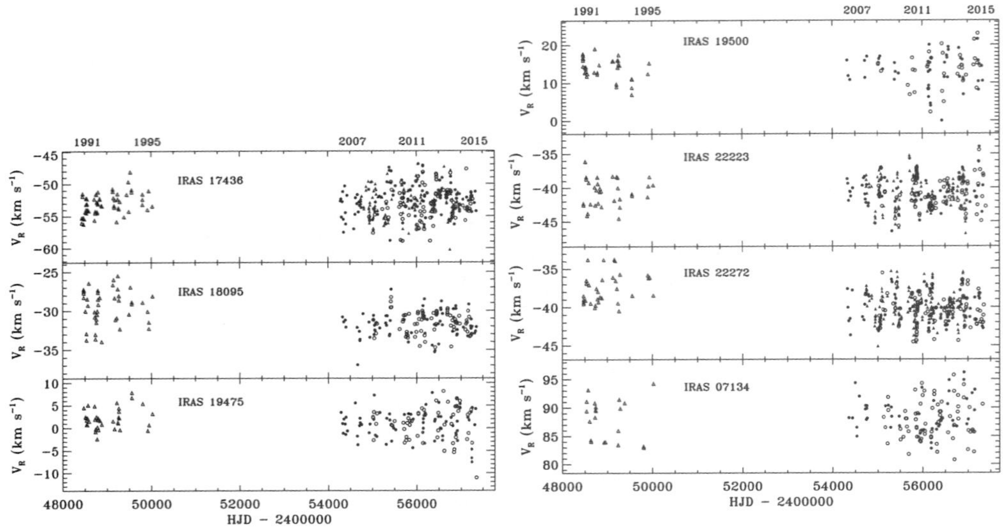

Figure 2. Comparison of the early (1991−1995) and recent (2007−2015) radial velocities.

data possessed a period of 710 days that just met our significance criteria (S/N=4.0 in frequency spectrum). Similarly, for IRAS 22223+4327, the 2009−2015 data suggested P≈770 days, but at a level just below our significance threshold. Searching the individual data sets supported periodic variability of this order in some cases but not in others. In neither of these cases is the result strong evidence for this longer, P∼ 2 year, period. In Figure 1 is shown for IRAS 22272+5435 a sine curve fit to the residuals of the radial velocities following the removal of the pulsational variations. While it fits the general trends in the data, the scatter is still large (σ∼1.3 km s^{-1}). Much of this, however, is due to the irregular nature of the pulsations, which can be seen in the varying amplitudes of the light and velocity curves.

Multi-decadal studies. The large total range of 25 years in the data of the two epochs (1991−1995 and 2007−2015) provides an opportunity to investigate even longer periodicities. However, this is complicated by inhomogeneities in the different radial velocity data sets. Comparison of velocities from the three sites showed systematic offsets in the data observed over the same time periods (2008−2015), even though they are each tied to IAU standard radial velocity stars. These offset values range from −1.1 to as large as +2.5 km s^{-1} in one case, with respect to the HERMES data, and these empirically-determined values were applied when combining the data sets. We attribute these systematic differences to several causes that manifest themselves in the measurements of the velocities of these complex spectra. These include the different spectral regions and ranges observed and different methods of fitting the profiles.

With these complications in mind, we compared the average values of the early DAO-RVS velocities (1991−1995) to the more recent data (2007−2015) adjusted to the HERMES values. These are shown in Figure 2. Differences between the two epochs were within the above offset values found among the different data sets (<± 2.0 km s^{-1}) except for two objects. For IRAS 18095+2704, the difference is 2.2 km s^{-1} and for IRAS 22272+5435, it is 3.0 km s^{-1}. These might suggest a long-term, multi-decade radial velocity variation, particularly in IRAS 22272+5435, but confirmation will require continued observations of the object with the same or similar instruments.

4. Discussion and Conclusions

Results thus far. The radial velocities of a sample of bright PPNe have been examined over several timescales, and several results have been found thus far.
1. Pulsation periods of 35–135 days have been found for each of them.
2. Evidence for multi-year periodic variations is weak, with only two of the objects showing suggestions of this (P≈2 yrs) at a level just at or below the significance criteria. And in these two cases, the results were seen in only some of the data sets.
3. Evidence for multi-decade variations is seen in two of the objects, being strongest in IRAS 22272+5435. However, as discussed above, the interpretations of the difference between the early and later data sets is tempered by the discovery of systematic differences (offsets) between the three recent contemporaneous data sets. The larger, multi-decade variation observed in IRAS 22272+5435 might be real, but will require continued long-term observations with the same or a similar device to confirm.

In our initial study of these seven PPNe, we discussed the possibility that our targets are biased toward low inclination as a result of our choice to observe those with bright central stars (Hrivnak *et al.* 2011). However, published modeling of their envelopes based on mid-IR observations indicates this not to be the case in general, and limiting values of P and M_2 were determined for undetected companions.

Ways ahead for progress. Several ways suggest themselves to make progress on the radial velocity search for binaries in PPNe.
1. Continue to observe these PPNe with the same or a similar device to investigate the suggested 2-year periodicity and the decadal shift in velocities.
2. Improve our analysis with a better selection of lines to probe deeper in the star where the effects of pulsation are lessened.
3. Observe edge-on PPNe, where one would see the full impact of the velocity along the line of site. We have begun such a study of three edge-on PPNe for which the central star is obscured in visible light but seen in the near-infrared. Initial results are suggestive, but only a few spectra have been obtained thus far and they require 8-m class telescopes.

Where are the binary companions? We conclude with the question that we began with. Where are the binaries?
1. They are present but hidden in long-period (>15 year) orbits – perhaps.
2. They are present but hidden due to the low mass of the secondaries (<0.2 $M_\odot$), which might even be planets (see contribution by Villaver, this volume) – perhaps.
3. They are present but hidden inside the atmosphere of the F–G star – unlikely due to timescale arguments for spiraling in.
4. They are merged with the central star – but there is no evidence to support this. They appear as normal post-AGB stars and are not rapid rotators.
5. They are absent and these PPNe are evolving as single stars.
Thus the present observational evidence, based on precise radial velocity measurements of PPNe, does not provide direct support for binary companions. To prove or exclude the presence of binaries is crucial to our understanding of the way(s) that bipolar PPNe and PNe form.

Acknowledgements

Financial support for the research of BJH has come from a series of NSF grants, most recently AST-1413660.

References

Arkhipova, V. P., Ikonnikova, N. P., & Komissarova, G. V. 2010, *Astron. Lett.*, 36, 269
De Marco, O. 2009, *PASP*, 316
Hillwig, T. H., Jones, D., & De Marco, O., *et al.* 2016, *ApJ*, 832, 125
Hrivnak, B. J., Lu, W., Maupin, R. E., & Spitzbart, B. D. 2010, *ApJ*, 709, 1042
Hrivnak, B. J., Lu, W., & Wefel, K. L., *et al.* 2011, *ApJ*, 734, 25
Raskin, G., Van Winckel, H., & Hensberge, H., *et al.* 2011, *AAp*, 526, 69
Van Winckel, H., Lloyd Evans, T., & Briquet, M., *et al.* 2009, *AAp*, 505, 1221

Discussion

Sloan: What dominates your error budget for radial velocity measurements in the optical spectra?

Hrivnak: The error budget is dominated by complex, variable line shapes due to pulsation. Also, S/N and aperture of telescopes available.

Q: Thanks for your extraordinary efforts over the years in collecting such as extraordinary set of observational data. I am particularly interested in the sources in you sample showing complex multi-polar morphologies. To me, this is very difficult to explain under the current assumption that these are formed by binary systems. Moreover, as they do not seem to display any particular photometric behaviour or signs of binarity in your data. Can you comment on this? Maybe a question to the experts on shaping of PNe ...?

Hrivnak: Thank you for your kind words of encouragement. One particular source, IRAS 19475+3119, is clearly multipolar. And I agree, such a morphology seems difficult to explained by a binary. Perhaps one could try to appeal to a model of a binary with a precessing disk (due to a 3rd star) and episodic mass out flows. But that is a question best addressed by the binary modelers.

Planetary Nebulae: Multi-Wavelength Probes of Stellar and Galactic Evolution
Proceedings IAU Symposium No. 323, 2016
X. Liu, L. Stanghellini & A. Karakas, eds.

doi:10.1017/S1743921317000643

Binary interactions on the RGB: Dusty post-RGB stars

Devika Kamath[1], Hans Van Winckel[1] and Peter Wood[2]

[1]Instituut voor Sterrenkunde, K.U.Leuven, Celestijnenlaan 200D bus 2401, B-3001 Leuven, Belgium
email: devika@ster.kuleuven.be

[2]Research School of Astronomy and Astrophysics, Mount Stromlo Observatory, Weston Creek, ACT 2611, Australia
email: wood@mso.anu.edu.au

Abstract. It is widely accepted that binary interactions are responsible for the shaping of planetary nebula. However, these binary interactions and evolutionary channels are poorly understood. Our recent study revealed a newly discovered population of low-luminosity, low-metallicity, likely binaries in the Magellanic Clouds: dusty post-RGB stars. They are likely to have evolved off the RGB via binary interaction. In this paper we present preliminary results of the first radial velocity monitoring of the post-RGB stars in the Small Magellanic Cloud (SMC) and the implications on stellar (binary) evolution. We also investigate their link, if any, to the planetary nebulae systems.

Keywords. techniques: radial velocities, stars: AGB and post-AGB, (stars:) binaries: general, stars: evolution, stars: Population II, (ISM:) planetary nebulae: general, Galaxy: general, (galaxies:) Magellanic Clouds.

1. Introduction

Dusty post-red giant branch (post-RGB) stars are a new class of evolved objects that lie below the tip of the RGB. These are dusty (with large infrared excesses $\geqslant 8\mu$m) low metallcity, evolved objects with luminosities below the tip of the RGB with an $M_{\rm bol}$(TRGB) = −3.6 based on studies by Frogel *et al.* (1983). These objects are very similar to the well known post-AGB stars, see Van Winckel (2003), in terms of their stellar parameters ($T_{\rm eff}$, $\log g$, [Fe/H] and E[$B-V$]) and circumstellar environment characteristics. However, they have luminosities less than the RGB tip (~ 2500 L$_\odot$). It is therefore likely that these objects are evolved, dusty post-RGB stars. These objects were first discovered in the Magellanic Clouds (MCs) by Kamath *et al.* (2016) during the search for post-AGB stars in the MCs (Kamath *et al.* 2014; Kamath *et al.* 2015).

During the RGB phase (where luminosities are <2500 L$_\odot$), single-star mass loss is insufficient to remove the H-rich envelope and produce a dusty post-RGB star (Vassiliadis & Wood 1993). The only way large amounts of mass loss and premature evolution off the RGB can occur is via binary interaction (Han *et al.* 1995). In the Galaxy, the binaries among the post-AGB stars all have a common property: their SEDs display a clear near-infrared excess indicating that circumstellar dust must be close to the central star, near sublimation temperature (Van Winckel 2003). It is now well established that this feature in the SED indicates the presence of a stable compact circumbinary disc (De Ruyter *et al.* 2006; Bujjarrabal *et al.* 2013; Hillen *et al.* 2013). These Galactic post-AGB binaries show orbital periods between 50 and 2000 days. It is possible that some of the Galactic post-AGB stars might be post-RGB stars but since we do not have accurate distances

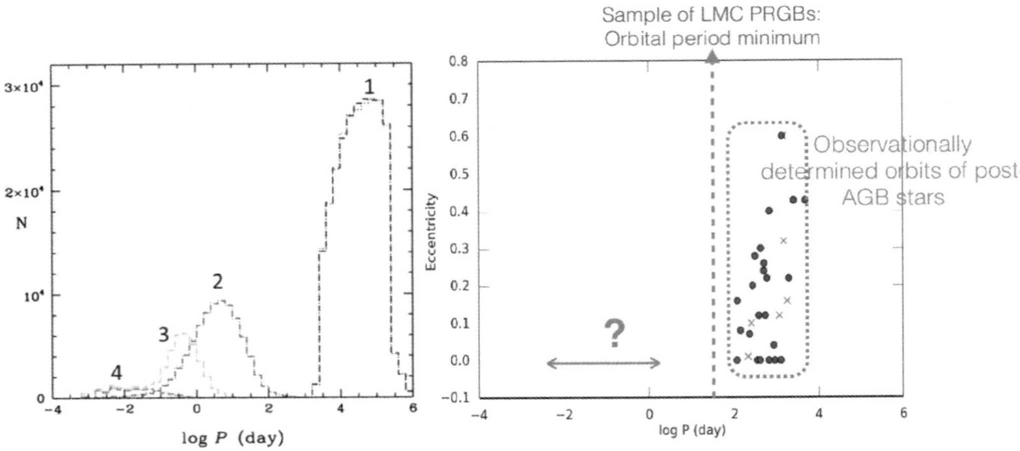

Figure 1. Left: The results of the population synthesis models which were normalised to the sequence E ellipsoidal variables of the Large Magellanic Cloud (LMC). The distribution of intermediate period binary post-AGB systems is given in black (curve1), the close binary post-AGB stars in blue (curve 2), the post-RGB and post-EAGB binaries in cyan/grey (curve 3) and the double degenerate secondaries in red (curve 4). Figure from Nie *et al.* (2012). Right: Our observed orbits for the Galactic systems (Van Winckel *et al.* 2009) and the SMC post-RGB stars. The red crosses indicate systems where the companion is certainly too massive to be a WD. The blue-filled circles represent systems where the companion can either be non-evolved or a cold WD. All orbits fall in the period range not predicted by the models. We note that some of these post-AGB stars could very well be binaries but the lack of accurate distances to these objects does not allow us to identify the possible post-RGBs amongst the post-AGB sample. The pink-dashed vertical line marks the minimum observed orbital period for the post-RGB stars in the SMC. As shown this is in stark contrast to that predicted by the models (denoted by the pink horizontal dashed double-headed arrow).

to the Galactic objects, identifying post-RGB systems in the Galaxy is not possible for now.

However, from the post-RGB stars in the MCs, we find that many of our new post-RGB candidates show similar disc-type SEDs to the post-AGB stars, pointing to binarity in these objects. Furthermore, in our recent study (Kamath *et al.* 2016) we were able to show that the likely precursors of these dusty post-RGB stars are the sequence-E variables, which are close binary red giants that show ellipsoidal light variations (Nicholls *et al.* 2012). In this paper, we present the preliminary results from the first radial velocity monitoring of the SMC post-RGB stars, their implications for binary evolution and we also aim to investigate the evolutionary link between these objects and planetary nebulae (PNe).

2. Radial Velocity Monitoring of post-RGB stars in the SMC

We recently initiated a radial velocity monitoring study of post-RGB stars in the SMC, to confirm their new channel of post-RGB (likely) binary evolution. We used X-shooter on UT2 at ESO, to obtain high-quality spectra (R = 10000 and a S/N per resolution element of ~70 - 80) of a carefully selected subset of our newly identified SMC post-RGB candidates, which covered the entire RGB luminosity space as well as showed diverse SED characteristics. We received 2 epochs of data, covering a period of 60 days, (ESO Proposal ID: 097.D-0800(A)).

On analysing the data, we were unable to find any radial velocity changes. This shows that these post-RGB objects, if they are binaries, are likely to have orbital periods greater than 60 days. This makes them, once again, similar to the post-AGB systems with intermediate to wide orbits (Van Winckel *et al.* 2007).

However, population synthesis models (Nie *et al.* 2012) normalised to the LMC binary red-giants (the likely progeny of these dusty post-RGB objects), predict that majority of our systems are likely to have orbital periods in the range of ~25 to 100 days, or up-to 500 days for the higher luminosity objects. We are continuing our radial velocity monitoring of these objects to cover the entire predicted orbital period range of 500 days. However, our preliminary results shows that these post-RGB systems are unlikely to evolve as the close-binaries or spiral-in systems, as predicted. In Fig 1, we show the discrepancy between the observed periods of the Galactic post-AGB and the SMC post-RGB binaries and the predictions from population synthesis models.

3. Do post-RGB stars evolve into planetary nebulae?

To establish a link between our newly identified post-RGB systems and the well known planetary nebula, we perform a simple exercise.

The lifetime of the post-RGB phase is determine by the rate of consumption of the H-envelope as well as the dissipation of the circumstellar environment (most likely a circumbinary disc). The observational lifetime of the post-RGB phase is $\sim 1.26 \times 10^5$ years. This agrees rather well, within the errors, with the predicted lifetime of $\sim$ 1.01×10^5 years (based on a stellar evolutionary model without mass-loss at $\log L/\mathrm{L}_\odot = 3.0$). The observed lifetime indicates a small amount of mass-accretion is required on the RGB. Post-RGB systems are expected to have discs which sustains the re-accretion. The discs around these stars may provide a reservoir of hydrogen that can be accreted back on to the high-luminosity star to fuel nuclear burning for an extended period, at the same time keeping the hydrogen envelope thick enough to maintain a relatively low Teff, which is insufficient to excite a PN. The lifetime of the disc (in the case of post-AGB systems but we expect this for post-RGB systems as well) is $\sim 10^4$ years (Bujarrabal *et al.* 2015) for a central stellar mass of $1.1\,\mathrm{M}_\odot$ and envelope mass of 4×10^{-3} $\mathrm{M}_\odot$. Therefore, taking the two lifetimes together, towards the end of the PRGB phase, when T_{eff} is high enough, there is probably no material left to ionise. This could imply that a common envelope interaction leading to some sort of stable mass transfer, as expected from post-RGB binaries may actually not produce a PNe. We note that the estimated and predicted birthrates are subject to several uncertainties.

4. Conclusions

A likely binary interaction on the RGB terminates the RGB evolution for some red giant binaries producing dusty post-RGB stars. Observationally, based on a very small sample, these systems have orbital periods greater than 60 days. This however is in contrast to the predicted orbital periods for these objects. This shows that we need to find out why our objects do not spiral-in or produce close binary systems. Furthermore, we need to identify what determines whether a system evolves into a spiral-in or a close binary or via stable mass transfer. To answer these questions we intend to carry out a systematic study of post-AGB and post-RGB binaries and compare their estimated orbital parameters to those predicted by binary evolution models. Our studies also show that a disc is an important ingredient to explain our system. However, it is still unclear as to how these discs evolve and the impact they have on the evolution of the post-AGB/post-RGB star. Based on the lifetime of post-RGB stars and the estimated lifetime of the circumbinary disk, we find that it is unlikely that these post-RGB stars would evolve into PNe. Likely progeny could be a white dwarf with a helium core, a cataclysmic variable or a sub-dwarf B star.

References

Bujarrabal, V., Alcolea, J., Van Winckel, H., Santander-García, M., & Castro-Carrizo, A. 2013, *A&A*, 557, A104

Bujarrabal, V., Castro-Carrizo, A., Alcolea, J., & Van Winckel, H. 2015, *A&A*, 575, arXiv:1502.01607

de Ruyter S., Van Winckel H., Maas T., Lloyd Evans T., Waters L. B. F. M., Dejonghe H., 2006, *A&A*, 448, 641

Frogel J. A., Cohen J. G., Persson S. E., 1983, *ApJ*, 275, 773

Han Z., Podsiadlowski P., Eggleton P. P., 1995, *MNRAS*, 272

Hillen M. *et al.*, 2013,*A&A*, 559, A111

Kamath, D., Wood, P. R., & Van Winckel, H. 2014, *MNRAS*, 439, 2211

Kamath, D., Wood, P. R., & Van Winckel, H. 2015, *MNRAS*, 454, 1468

Kamath, D., Wood, P. R., Van Winckel, H., & Nie, J. D. 2016, *A&A*, 586, L5

Nie, J. D., Wood, P. R., & Nicholls, C. P. 2012, *MNRAS*, 423, 2764

—. 2003, *ARAA*, 41, 391

Van Winckel, H., Lloyd Evans, T., Briquet, M., *et al.* 2009, *A&A*, 505, 1221

Vassiliadis, E. & Wood, P. R. 1993, *ApJ*, 413, 641

Nicholls, C. P., Wood, P. R., & Cioni, M.-R. L. 2010, *MNRAS*, 405, 1770

Discussion

Q: Now sdB stars relate to your systems. Do you find similar period distribution?

De Marco: The final separation does not just depend on $q = M_2/M_1$, but on the envelope binding energy. We reproduce quite a range of final separation in common envelope simulations. Another factor is how long, before the CE, the RLOF phase lasts and how much it depletes the envelope prior to the plunge.

Planetary Nebulae: Multi-Wavelength Probes of Stellar and Galactic Evolution
Proceedings IAU Symposium No. 323, 2016
X. Liu, L. Stanghellini & A. Karakas, eds.

doi:10.1017/S174392131700196X

Shaping planetary nebulae with jets in inclined triple stellar systems

Muhammad Akashi and Noam Soker

Department of Physics, Technion – Israel Institute of Technology, Haifa 32000, Israel;
akashi@ph.technion.ac.il, soker@physics.technion.ac.il

Abstract. We conduct three-dimensional hydrodynamical simulations of two opposite jets launched obliquely to the orbital plane around an asymptotic giant branch (AGB) star and within its dense wind, and demonstrate the formation of a 'messy' planetary nebula (PN), namely, a PN lacking any type of symmetry (highly irregular). In building the initial conditions we assume that a tight binary system orbits the AGB star, and that the orbital plane of the tight binary system is inclined to the orbital plane of binary system and the AGB star. We further assume that the accreted mass onto the tight binary system forms an accretion disk around one of the stars, and that the plane of the disk is in between the two orbital planes. The highly asymmetrical lobes that we obtain support the notion that messy PNe might be shaped by triple stellar systems.

Keywords. binaries: close – planetary nebulae – jets

1. Introduction

The notion that binary stellar systems or planetary systems shape planetary nebulae (PNe) has gained huge support over the years, both from theoretical considerations that point to the difficulties of models based on single asymptotic giant branch (AGB) stellar progenitors to shape PNe, and from a large number of observations and their analysis. In most cases of shaping by binary interaction, jets are involved. A relatively small number of papers raise the possibility that triple stellar systems are behind the morphologies of some PNe. In general there are two orbital planes. One is that of the more tight binary system, and the second one is that of the tight binary system motion around the center of mass with the third star. If the two orbital planes coincides, the interaction leads to a mass loss geometry that has a plane of symmetry, even if the mass loss geometry departs from axi-symmetry. Binary systems, e.g., if the interaction time scale is shorter than the orbital period or if the orbit is eccentric, can also cause departure from axisymmetry.

If the two orbital planes of the triple system are inclined to each other, the morphology of the descendant PN is likely to possess no symmetry plane. The descendant PN is actually likely to lack any symmetry; nor point-symmetry, nor axial-symmetry, and nor mirror symmetry. We term this a 'messy PN'.

Specifically, we consider the flow structure described by Soker (2004), where a tight binary system orbits an AGB star and accretes mass from the AGB wind. Because the orbital plane of the tight binary system is not parallel to the orbital plane around the AGB star, the jets' axis is not perpendicular to the orbital plane. We describe the initial setting and the numerical code in section 2. We then describe the numerical results in section 3. We summarize our results and call for the search for triple stellar systems in the center of messy PNe.

2. Numerical Set-up

2.1. *Initial conditions*

The two orbital planes, that of the tight binary system and that of the triple stellar system are inclined to each other. The tight binary system accretes mass from the wind. Due to the orbital motion of the two stars of the tight binary system around their mutual center of mass, the accreted mass possesses specific angular momentum that leads to the formation of an accretion disk around one (or two) of the stars. The plane of the accretion disk is close to being parallel to the orbital plane of the tight binary system. The accretion disk launches two opposite jets that are almost perpendicular to the orbital plane of the tight binary system.

We simulate two cases which differ in the inclination angle ϕ of the jets' axis to the line perpendicular to the orbital plane of the triple system. In the first case $\phi = 20^\circ$ and in the second case $\phi = 30^\circ$. The other parameters are identical for the two runs and are as follows. The orbital separation of the jets' origin (the location of the right binary system) to the center of the AGB star is $a_{123} = 30$ AU. The orbital period is $P_{123} = 67$ yrs corresponding to a total mass of the triple stellar system of $M_{123} = M_1 + M_2 + M_3 = 5 M_\odot$. (see Fig. 1).

At the beginning of each simulation the slow AGB wind fills the grid. The mass-loss rate and velocity of the slow AGB wind are $\dot{M}_w = 5 \times 10^{-5} M_\odot$ yr^{-1} and $v_w = 10$ km s^{-1}, respectively.

At $t = 0$ we start to launch two jets, along the $+z$ and $-z$ directions. The jets are conical, with a half opening angle of $\alpha_j = 30^\circ$. The jets opening angle influences the morphology of the nebula. However, we have limited numerical resources and cannot study more cases at this time. The initial velocity of the jets are $v_j = 500$ km s^{-1}, that is about equal to the escape speed from a main sequence star. The mass-loss rate into the two jets together is $\dot{M}_{2j} = 2 \times 10^{-7} M_\odot = 0.004 \dot{M}_w$. This corresponds to about 10% of the accretion rate from the wind.

2.2. *The numerical code*

Our simulations are performed by using version 4.0-beta of the FLASH code. We employ a full 3D adaptive mesh refinement (AMR) (10 levels; 2^{13}) using a Cartesian grid (x, y, z) with outflow boundary conditions at all boundary surfaces. The center of the grid is at (0,0,0). The AGB star is at a permanent place on the grid, it is placed at (0,0,-$a_{123} \times cos(\phi)$).

The slow wind and the ejected jets start with a temperature of 1000 K. The initial jets' temperature has no influence on the results (as long as it is highly supersonic) because the jets rapidly cool due to adiabatic expansion.

3. Results

We show three-dimensional density map for the case with inclination angle $\phi = 30$ (see Fig. 2).

4. Summary

As possible examples for PNe that might have been shaped by a triple stellar system Soker (2004) listed several PNe, including IC 2149, NGC 6210, and NGC 1514, that we also suggest here were shaped by a triple stellar progenitor.

N.S. is supported by the Charles Wolfson Academic Chair.

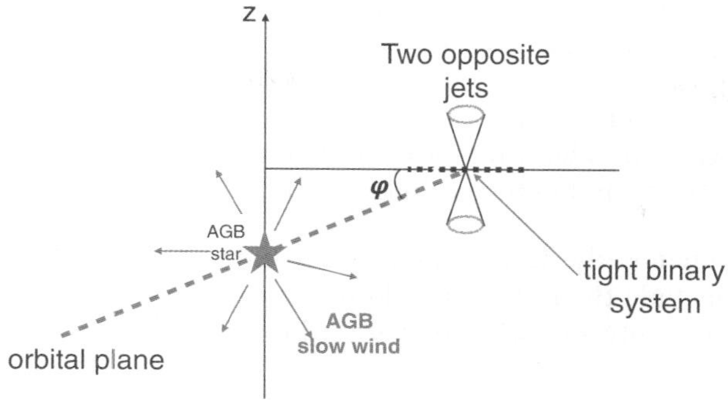

Figure 1. A schematic drawing of the outflow from the triple stellar system. The orbital plane of the tight binary system and the one of the triple system are inclined by an angle ϕ. The tight binary system accretes mass from the AGB wind and launches jets as indicated.

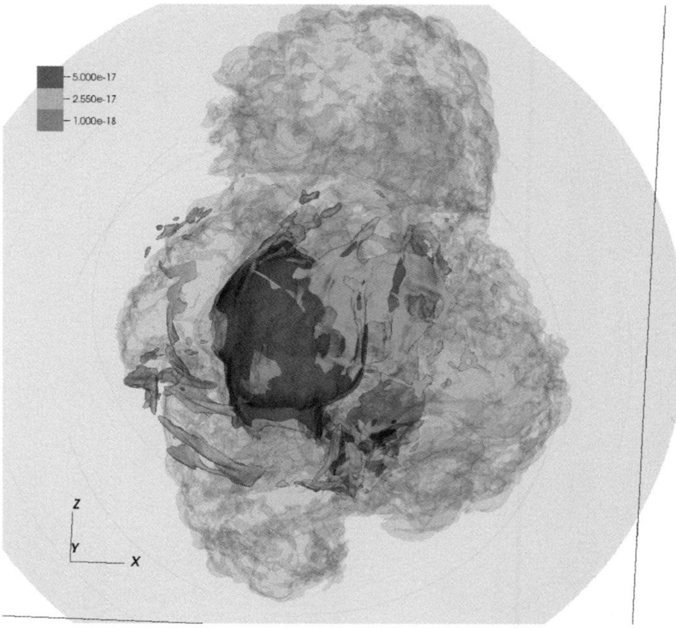

Figure 2. Three dimensional density map of the case $\phi = 30$.

References

Soker, N. 2004, MNRAS, 350, 1366
Soker, N. 2016, MNRAS, 455, 1584

Discussion

De Marco: What causes the misalignment of the plane of the accretion disk and the plane of the orbit (the AGN orbit)?

Akashi: The misalignment of the tight orbital plane is just an initial condition in our simulation. The accreted mass gains angular momentum from the tight binary. See Soker (2004).

Todt: Can you please comment on the plausibility of the assumed properties of the triple system, e.g. what are the eccentricities, how stable is such configuration over the stellar lifetime, what is the reason for the jet?

AKASHI: The properties and the parameters were chosen after an accurate calculation of the condition of accretion. The orbits are circular and stable. The reason for the jet is accretion from the AGB wind onto the tight binary, so you have disk and jet.

SCHOENBERNER: What is the time scale of these simulations? For applications to real objects one has to include the radiation field of the central source: The thermal pressure of ionized gas will certainly smooth the distorted envelope.

Planetary Nebulae: Multi-Wavelength Probes of Stellar and Galactic Evolution
Proceedings IAU Symposium No. 323, 2016
X. Liu, L. Stanghellini & A. Karakas, eds.

doi:10.1017/S1743921317002150

Post-AGB binaries as tracers of stellar evolution.

Hans Van Winckel

Institute for Astronomy
Department physics and astronomy KU Leuven
Celestijnenalan 200D, 3001 Leuven, Belgium
email: Hans.VanWinckel@kuleuven.be

Abstract. In this contribution the focus is on post-AGB binaries. It is now well established that these are often surrounded by stable long-lived circumbinary discs of gas and dust. Here we introduce our monitoring programme with our high-resolution spectrograph HERMES mounted on the 1.2m Mercator telescope. We illustrate the use of time-series high-resolution spectra and show that jets observed in many systems are launched at the location of the companion. The jet is likely originating from a circum-companion accretion disc. The link of these systems to some PNe relies on the detection of similar orbits and hence wide spectroscopic orbits among central stars of PNe. The conclusion is that Keplerian discs as well as circum-companion discs are fundamental to understanding the properties and evolution of these interacting evolved binaries.

Keywords. stars: AGB and post-AGB,(stars:) binaries: spectroscopic, techniques: radial velocities

1. Introduction

In the Galaxy some 80 optically bright post-Asymptotic Giant Branch (post-AGB) objects are identified with circumstellar dust near the sublimation temperature, while the central star is not in a dust-losing phase (e.g. Gezer *et al.* 2015 and references therein). In the Magellanic clouds, many more have been identified (Kamath *et al.* 2014; Kamath *et al.* 2015) It is now also well established that the circumstellar dust in these systems is trapped in a very compact stable disc (e.g. Hillen *et al.* 2016b and references therein) of gas and dust. The inner radius of the disc has been resolved with interferometric imaging by Hillen *et al.* 2016 and in some objects the kinematics were spatially resolved in CO and Keplerian rotation of the inner parts have been resolved (e.g. Bujarrabal *et al.* 2016a and references therein). The inferred luminosity of these objects is significant, pointing to a large scaleheight of the disc.

In the past decade it became clear that these object represent a peculiar phase in binary stellar evolution (Van Winckel 2003). The binaries are now not contact systems, but the orbits are now too short to accommodate an AGB star: these stars must have been subject to severe interaction in the past when the primary was at AGB dimensions. It is assumed that during the interaction, the disc, which seems now in hydrostatic equilibrium, was formed. Contrary to many central stars of PNe (see review of D. Jones these proceedings), the binaries did not suffer severe spiral-in and escaped the interaction phase on orbits between a few hundred and a few thousands of days (e.g. Van Winckel *et al.* 2009, Manick *et al.* 2016 and references therein).

We monitor northern bright (m(v) $<$ 12.0) objects with our high-resolution spectrograph HERMES (Raskin *et al.* 2011) mounted on our 1.2m Flemish Mercator telescope. We tuned our operational model such that regular monitoring can be obtained throughout the year. Our programme has been running since 2009 (e.g., Van Winckel *et al.*

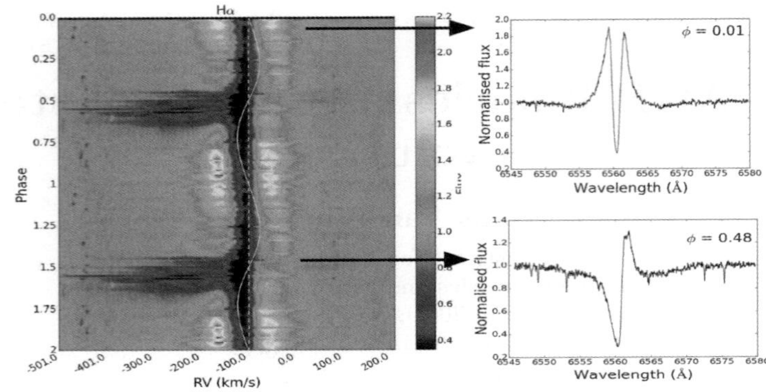

Figure 1. On the left panel we show the dynamic spectra around the H_α-line which we fold on the orbital period. On the right we show representative spectral line-profiles.

2010, Gorlova *et al.* 2014). Here we report on some newly discovered features in these time-series of high-resolution spectra.

2. Jet creation

The orbital elements are only the first observables deduced from our time-series (e.g. Van Winckel *et al.* 2009; Manick *et al.* 2016 and references therein). Also phased resolved line-profiles can be studied and here we focus on the H_α profile.

In Fig. 1 we show a two dimensional representation of the H_α profile of BD+46°442 (Gorlova *et al.* 2012, Bollen *et al.* 2017). The most apparent feature is the strong blue shifted absorption component which appears in the spectra during the phase when the companion is in lower conjunction. The interpretation of these P-Cygni profiles is that during this orbital phase, continuum photons of the evolved luminous primary are scattered out of the line-of-sight by line scattering on high-velocity hydrogen atoms. This high-velocity outflow originates around the companion, and is likely launched from a circumcompanion accretion disc (Bollen *et al.* 2017).

The modelling of these spectra (Fig. 2) is made complicated as the inclination is not very well constrained but the opening angle of the jet must be quite large. Doppler tomography, using the spectra obtained outside conjunction, allows us to reconstruct in velocity space the accretion disc of the companion which is the origin of the H_α shell profiles (Fig. 1, phase 0.1).

While not all objects have been studied in great detail, the phase resolved spectra shows that jet creation is very common with the launching side around the companion.

3. PNe connection

Ingredients which are deemed to be important also in the PNe phenomena are present in the post-AGB binaries with circumbinary discs. Binary interaction processes as well as jet creation processes are now resolvable in space and time. The evolutionary link between these post-AGB binaries and PNe is, however, not clear. The post-AGB systems may evolve too slow to ever become a PNe.

Moreover, while spiralled-in systems in PNe are common, wider spectroscopic orbits are largely unknown, with the noticeable exception of BD+33°2642 and LoTr5 (Van Winckel *et al.* 2014). For the latter we have obtained now data covering one full orbit of around 2800 days (Fig. 3).

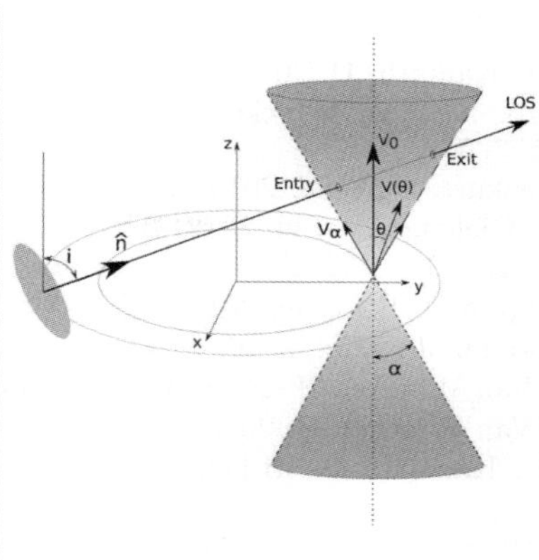

Figure 2. The geometric model. Fitting this model to the phase-resolved spectra indicate that the opening cone is wide (> 60 degrees) with a velocity law indicative of angle dependent outflow velocity which is higher on the symmetry axes and decreases towards the edge of the cone (Bollen *et al.* 2017).

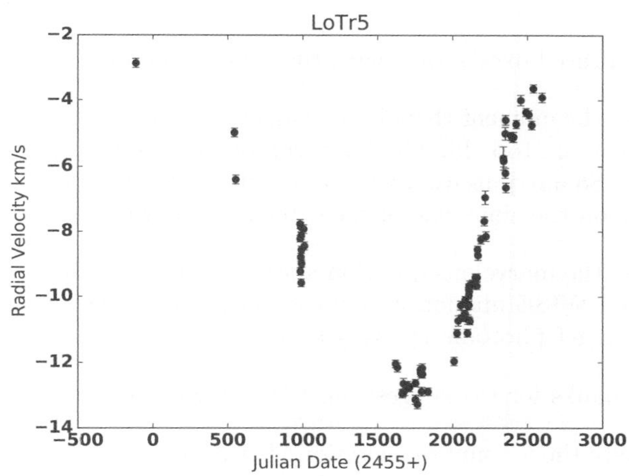

Figure 3. The radial velocity curve of LoTr5, one of the few central stars of PNe with wide spectroscopic orbits.

4. Conclusions

Binary post-AGB systems are characterised by rather large orbits (several hundreds of days) and show a common SED characteristic pointing to the presence of a stable circumbinary disc of gas and dust with a significant scale-height.

The systems are complex and we report here on the detection of high-velocity outflows at the location of the companion. The jets show wide opening angles and are likely launched in a circum-companion gas accretion disc.

The evolutionary connection to PNe is as yet unclear but more evolved counterparts of binary post-AGB stars are now possibly identified (Van Winckel *et al.* 2014). While spiralled-in systems are commonly observed among central stars of PNe (see review of D. Jones), this is not the case for post-AGB stars as these binaries have escaped a phase of strong interaction on rather wide and often eccentric systems (Van Winckel *et al.* 2009).

Disclosing the different binary evolutionary channels leading to the large variety of evolved systems is a challenge for the coming years.

References

Bollen, D., Van Winckel, H., & Kamath, D. 2017, *A&A*, in prep.

Bujarrabal, V., Castro-Carrizo, A., Van Winckel, H., Sanchez Contreras, C., & Santander-Garcia, M. 2017, *A&A*, 597, 5

Gezer, I., Van Winckel, H., Bozkurt, Z., *et al.* 2015, *MNRAS*, 453, 133

Gorlova, N., Van Winckel, H., Gielen, C., *et al.* 2012, *A&A*, 542, A27

Gorlova, N., Van Winckel, H., Vos, J., *et al.* 2014, ArXiv e-prints (1403.2287)

Hillen, M., Kluska, J., Le Bouquin, J.-B., *et al.* 2016, *A&A*, 588, L1

Hillen, M., Van Winckel, H., Menu, J., *et al.* 2016, ArXiv e-prints(1610.09930)

Kamath, D., Wood, P. R., & Van Winckel, H. 2014, *MNRAS*, 439, 2211

Kamath, D., Wood, P. R., & Van Winckel, H. 2015, *MNRAS*, 454, 1468

Manick, R., Van Winckel, H., Kamath, D., Hillen, M., & Escorza, A. 2016, ArXiv e-prints (1610.00506)

Raskin, G., van Winckel, H., Hensberge, H., *et al.* 2011, *A&A*, 526, A69

Van Winckel, H. 2003, *ARAA*, 41, 391

Van Winckel, H., Jorissen, A., Exter, K., *et al.* 2014, *A&A*, 563, L10

Van Winckel, H., Jorissen, A., Gorlova, N., *et al.* 2010,*MemSAI*, 81, 1022

Van Winckel, H., Lloyd Evans, T., Briquet, M., *et al.* 2009, *A&A*, 505, 1221

Discussion

CHEN: Does the infrared excess vary with the orbital motion?

VAN WINCKEL: For the moment there is no evidence for this. The stars are now not losing mass with dust: They are too old. The inner rim of the disc is Keplerian in rotation and located by the star. So any pulsation will be reflected into the rim. Whether the near-IR light curve changes on the timescale of the orbit is unknown as yet.

SLOAN: To address the marvelous question about infrared variability, it is good to recall that the deactivated WISE mission covers the entire sky every six months and in each visibility region, you get photometry every scan.

VAN WINCKEL: Thanks for the suggestion! We will certainly do so!

Q: Can you estimate the amount of accretion? Is the orbital plane of the binary alligned with the plane of the circumbinary plane? Do you have precession from the jet?

Q: Can you reconcile the large opening angle of the jet with the much narrower opening at the base of most bipolar lobes?

Session 6: PNe in nearby galaxies and chemical evolution

Chair: Romano Corradi

Planetary Nebulae: Multi-Wavelength Probes of Stellar and Galactic Evolution
Proceedings IAU Symposium No. 323, 2016
X. Liu, L. Stanghellini & A. Karakas, eds.

doi:10.1017/S1743921317000503

Abundances and gradients in nearby galaxies

Fabio Bresolin

Institute for Astronomy, University of Hawaii
2680 Woodlawn Drive, 96822 Honolulu, HI, USA
email: bresolin@ifa.hawaii.edu

Abstract. The study of chemical abundance gradients can provide essential information on the evolution of disk galaxies. Here I briefly review our current observational knowledge concerning the abundances of the ionized gas (H II regions and planetary nebulae) in nearby galaxies, and how they inform us about the time evolution of metallicity gradients.

Keywords. galaxies: spiral, galaxies: evolution, ISM: abundances

---

1. Introduction

The study of the chemical abundances in external galaxies continues to provide an essential testbed to probe the evolution of galaxies. In particular, for spiral galaxies the radial metallicity gradients and their temporal evolution encode the mechanisms of disk formation, and reflect the presence of inflows and outflows of gas, as well as radial gas flows along the disk, and react to the redistribution of stars (migration) and of metal-enriched gas during galaxy encounters and merger activity.

For observers of the gas-phase chemical abundances, oxygen is the most easily measured tracer of metallicity in galaxies. In this contribution I will therefore use the words 'metallicity' and 'oxygen abundance' (O/H) interchangeably, and will focus on a few of the main results obtained in nearby galaxies when trying to determine observationally the evolution of galactic radial abundance gradients. These measurements constitute essential ingredients for constraining models of galaxy formation and chemical evolution.

Since this is a volume on planetary nebulae (PNe), it is worth reminding ourselves of a few important issues that should be kept in mind when using them to trace the chemical abundances of galaxies and the evolution of gradients with time, and that we can disregard when dealing with chemical abundances derived from H II regions, the most commonly used tracers of α elements in galaxies:

(*a*) the O abundance we measure from the spectral lines of PNe could differ from the O abundance of the interstellar medium out of which the progenitor stars originated. Mixing and/or nucleosynthetic mechanisms could both produce or destroy oxygen in the stellar progenitors. Evidence that oxygen can be produced at low metallicity has been presented by Delgado-Inglada *et al.* (2015), who suggested that other elements, in particular argon and chlorine, should be considered more reliable metallicity tracers.

(*b*) the initial progenitor masses of the extragalactic PNe we observe are uncertain, and so are, as a consequence, their ages. The typical masses that are inferred are 1.5–2 $M_{\odot}$, with corresponding ages of 2–1 Gyr, but of course PNe with slightly lower masses would correspond to ages older by several Gyr.

(*c*) the effects of stellar radial migration are important when comparing the abundance gradients of old ($>$ 4 Gyr) and young populations (Roškar *et al.* 2008): the gradients would appear flatter at the present time than at the time of stellar birth. In addition, migration acts in the sense of increasing the chemical abundance dispersion with age at a given radius.

Keeping in mind these issues, we can attempt to use PNe to study the history of α elements, such as O, Ne, Ar, and S, in galactic disks. Hopefully, one can obtain a picture of the history of chemical enrichment via a comparison with tracers of the young population (H II regions and massive stars, for example). Since we obtain gas-phase abundances, we should correct, if possible, the derived values for the depletion of these elements onto dust grains. For oxygen this effect is on the order of 0.1 dex (Peimbert & Peimbert 2010).

2. Brief primer on abundance gradients in spiral galaxies

It has been known for decades that spiral disks display exponential decreases, from the center to their outskirts, of light, mass and O/H abundances. Thus, using the traditional units of 12+log(O/H) for the oxygen abundances, one obtains a linearly decreasing trend with radius. The units used to express this gradients are important. Using gradients expressed in dex kpc^{-1}, physically larger spiral galaxies display flatter gradients compared to smaller disks, which finds an explanation in models of the chemical evolution of galaxies as due to a faster evolution for the most massive systems. Such a trend virtually disappears if one normalizes the radial coordinate in terms of, say, the isophotal radius R_{25} or the effective radius r_e. Although with a significant scatter, gradients expressed using these units are similar for different galaxies (Sánchez *et al.* 2014, Ho *et al.* 2015). I would like to note, however, that the result obtained using R_{25} depends on the fact that we typically study spiral galaxies that have similar surface brightness properties. Bresolin & Kennicutt (2015) have shown that low surface brightness spiral galaxies display shallower slopes when these are expressed in dex R_{25}^{-1}, as a consequence of the fact that in these galaxies the typical reduction of O/H from the low surface brightness disk center to the isophotal radius is smaller than for galaxies with higher surface brightness disks. The use of the effective radius, obtained from the exponential distribution of the light in the disk, is not subject to this bias.

The homologous behavior of the gradients described above (similar normalized gradients for all galaxies) and the fact that flatter gradients in dex kpc^{-1} are found for more massive galaxies have been known for a long time, as is the increase of the characteristic O/H value with galaxy mass (or luminosity) and local surface mass density (the mass-metallicity and luminosity-metallicity relations). Investigations of abundance gradients carried out during the past decade have focused on a variety of additional issues, some of which are listed here:

- the use of massive stars, in alternative to the ionized gas (H II regions, PNe), to trace the distribution of metals
- the study of the effect of galaxy encounters, interactions and mergers on the radial metallicity distribution
- the derivation of abundance gradients for high-redshift systems
- the investigation of the metallicity in the extended disks of spiral galaxies, well beyond the isophotal radius
- the determination of metallicity gradients in low surface brightness spirals
- the azimuthal symmetry properties of metals in galactic disks

• the temporal evolution of the gradients, i.e. whether gradients become flatter or steeper with the evolution of the galactic disk

• the determination of the spatial chemical abundance mapping of large samples of spiral galaxies with the use of Integral Field Units in place of single- or multi-slit spectrographs.

I would also like to stress the importance of looking in regions within galaxies and at galactic systems where deviations from the pure exponential trend in the radial distribution of the metals have been found, since this informs us about interesting processes involving, for example, the interaction of spiral disks with their environment:

(*a*) in interacting galaxies the metallicity gradients can become quite shallow, a result that illustrates the radial redistribution of metals from the inner regions to the outer regions and the dilution of the inner regions as a consequence of gas flows triggered by the encounter. This effect can generate flat abundance gradients out to galactocentric distances of several tens of kpc (see, for example, Olave-Rojas *et al.* 2015).

(*b*) a number of extended disk galaxies display flat abundance gradients in their outskirts (e.g. Bresolin *et al.* 2012). The interpretation in this case involves a combination of low star formation efficiencies and the presence of an enriched gas inflow in the outer regions of the disks.

(*c*) some spiral galaxies show flatter gradients in their inner disks (e.g. Zinchenko *et al.* 2016). The mechanisms leading to this effect are poorly known, and could involve both external and secular processes.

2.1. *A word on nebular abundance diagnostics*

The determination of the chemical abundances of the ionized gas in galaxies relies on the ability of being able to characterize with sufficient accuracy the physical status of the gas, such as its temperature and density, their spatial distribution, and the properties of the ionizing field. In practice, we generally use a relatively simple picture concerning the physical status of the gas. The determination of the temperature, in particular, is critical, because the line emissivity of the forbidden, collisionally-excited lines that we typically use has an exponential dependence on it. The classical way of obtaining the electron temperature in the ionized gas involves the use of auroral lines, such as [O III]λ4363. This line is faint, and other faint lines that can also lead to a metallicity determination and have a shallower dependence on temperature, the metal recombination lines, lead to systematically higher (0.2-0.3 dex) oxygen abundances. In extragalactic H II region work one typically resorts to the use of brighter lines, and to a number of strong-line abundance diagnostics, that should be considered to have a statistical value for the metallicities they provide. A wide variety of such methods exists, with different calibration techniques, leading to very large systematic uncertainties. What is important for this review is to realize that even relative abundances can be affected, albeit to a lesser degree, and therefore the slope of the abundance gradients has a small dependence on the abundance diagnostic employed.

For these reasons it is important to i) use the same abundance diagnostic when trying to use both H II regions and PNe – this reduces the choice virtually to the auroral line method (or 'direct' method) only, since the recombination lines are too faint to be currently measured in PNe located in spiral galaxies beyond the Milky Way; ii) validate the gradients obtained using H II regions with the analysis of massive stars, since the two types of objects represent the same young galactic population. This kind of comparison has been carried out in a small number of nearby galaxies. In some cases the agreement between stellar and gaseous metallicity is excellent. For example, in NGC 300 the direct abundances obtained for H II regions by Bresolin *et al.* (2009) agree very well with the metallicity of both blue (Kudritzki *et al.* 2008) and red (Gazak *et al.* 2015) supergiant stars. Other galaxies, such as M33, show some level of disagreement, on the order of 0.2 dex. It is especially important to carry out this kind of comparison in metal-rich galaxies, since the direct method could fail, for a variety of causes, in high-metallicity environments. For this reason, recently Bresolin *et al.* (2016) have compared metallicities for blue supergiant and H II regions in the metal-rich galaxy M83. They showed that while direct abundances could underestimate the stellar abundances in the high-metallicity (solar and above) regime, there is in fact very little evidence for a systematic effect with metallicity, although one has to be careful in using strong-line methods empirically calibrated from direct abundance determinations. On the other hand, Bresolin *et al.* (2016) showed evidence that the metal recombination lines could provide chemical abundances that appear to be biased to large values in low-metallicity galaxies.

3. Time evolution of the gradients

The question of whether and how radial abundance gradient slopes evolve with time is crucial in understanding how galaxy disks evolve. Chemical evolution models predict changes that could go in either direction: gradients could become steeper or shallower with time. This result can depend on a number of factors, including the radial dependence of the star formation rate and gas infall, and the importance of feedback from supernovae and turbulence in the interstellar medium. Observationally, this issue has been addressed from two different points of view:

(*a*) study the metallicity of nearby galaxies using tracers that represent different population ages, such as young stars (OB stars, Cepheids, red and blue supergiants) or H II regions for the young age bin, and PNe and star clusters for older ages. This method is unfortunately affected by uncertainties in the ages of the old systems, such as PNe, and in their distances in the case of the Milky Way. Radial stellar migration impresses an additional level of uncertainty.

(*b*) obtain the metallicity distribution in high-redshift systems, presumably analogs or precursors of present-day spiral galaxies. In this case the limited spatial resolution has a negative impact on the technique. In addition, one needs to be sure that the young, high-redshift systems are directly comparable to the galaxies we study locally. In particular for the few lensed systems so far analyzed, a significantly steeper gradient than observed for nearby galaxies is found (Jones *et al.* 2015).

3.1. *Milky Way*

For the Miky Way the comparison between young (massive stars, H II regions) and old (PNe, open clusters) gradients during the course of the past several Gyrs has produced contradicting results, i.e. both a steepening (Stanghellini & Haywood 2010) and a flattening (Maciel & Costa 2009) of the gradient have been found to be consistent with the observational data, with additional analysis providing no evidence for either (Maciel & Costa 2013). This situation can be ascribed to uncertainties in distance and ages, and also to poor statistics.

Table 1. Studies of galaxies with abundance gradient information from different objects.

Galaxy	HII regions	PNe	BA stars
M33	Magrini *et al.* 2007, 2010 Bresolin *et al.* 2010	Magrini *et al.* 2009 Bresolin *et al.* 2010	Urbaneja *et al.* 2005 U *et al.* 2009
M31	Zurita & Bresolin 2012 Sanders *et al.* 2012	Sanders *et al.* 2012	Trundle *et al.* 2002 Venn *et al.* 2000 Przybilla *et al.* 2006
NGC 300	Bresolin *et al.* 2009	Stasinska *et al.* 2013	Kudritzki *et al.* 2008
M81	Stanghellini *et al.* 2010, 2014 Patterson *et al.* 2012	Stanghellini *et al.* 2010	Kudritzki *et al.* 2012

As mentioned before, chemical evolution models can produce both a flattening or a steepening of the gradient. An example of a model that produces a flattening is presented in the work by Kubryk *et al.* (2015), which also illustrates the effects of radial stellar migration on the gradient measured using old stars. This kind of flattening with time is in agreement with several other models present in the literature (e.g. Mollá 2014, Tissera *et al.* 2016). Recent CALIFA data also provide weak evidence for a steeper gradient in old (age > 6 Gyr) stellar populations compared to young (age < 2 Gyr) stars (Sánchez-Blázquez *et al.* 2014).

Recently Anders *et al.* (2016) (see also Xiang *et al.* 2015) used asteroseismology and spectroscopic data of field red giant stars to derive their distances and stellar parameters. They derived an evolution of the gradient with the age of the stars, which goes in the sense that the gradient appears *steeper* at the present time than it did Gyrs ago. This evolution is consistent with models in which the flattening at large lookback time is due to stellar radial mixing and migration.

3.2. *External galaxies*

The comparison betwen young and older gradients is in principle facilitated by examining an external galaxy, where distance uncertainties do not represent a problem, but unfortunately the issue of uncertain ages remains. At the same time, the change in the slope of the gradients predicted by models during the evolution of galaxies of the past few Gyrs is quite small, meaning that the empirical answer is understandably still rather uncertain.

As recently summarized by Magrini *et al.* (2016), there are only four spiral galaxies where the abundance gradients in the disk have been obtained for both H II regions and PNe using the direct method: M33, M31, NGC 300 and M81. Table 1, adapted from their paper, summarizes the references used for the ionized gas work. For the same galaxies we also have information on the massive star metallicities from different authors, as included in the last column of the table.

Fig. 1, also adapted from Magrini *et al.* (2016), but including also the results from the stellar work, shows the difference in the gradient slope, relative to the PN value and expressed in dex R_{25}^{-1}, for both H II regions (full circles) and massive stars (star symbols). The galaxies have been ordered by stellar mass, increasing from left to right. It can be seen that the data are consistent, in all cases, with a negative value of the slope difference, which means that the young population appears to provide a steeper gradient than the PNe. There are some issues, that cannot be dicussed in depth here. For example, the stellar metallicity gradient for M33 is significantly steeper than the abundance gradient from H II regions. Also, for the two most massive galaxies, M81 and M31, the stars indicate a flatter gradient than the H II regions, and in better agreement

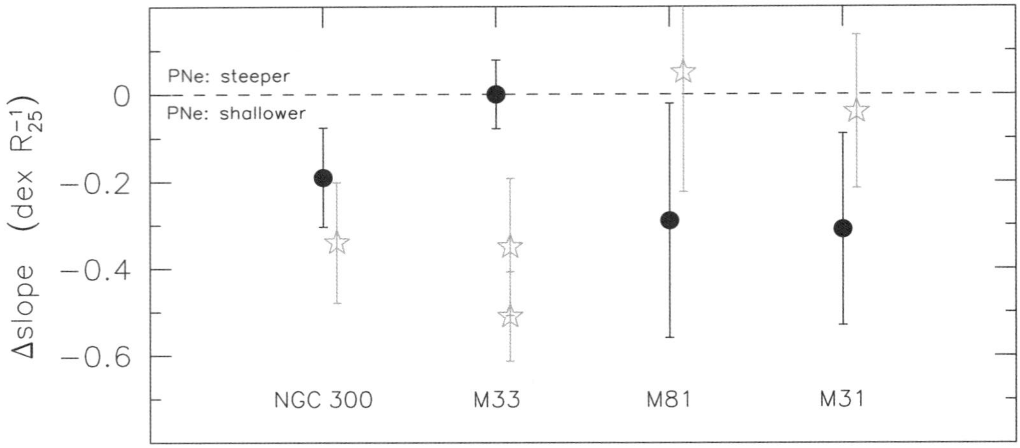

Figure 1. Difference in the abundance gradient slope, expressed in dex R_{25}^{-1}, as measured between H II regions and PNe (full circles) and young stars and PNe (star symbols), for the four galaxies of Table 1. The two stellar values for M33 refer to the works by Urbaneja *et al.* (2005) and U *et al.* (2009). For M31 the stellar-based slope comes from Trundle *et al.* (2002). (Partially adapted from Magrini *et al.* 2016).

with the PNe. In essence, however, the data presented in Fig. 1 shows that there is an indication, albeit weak, that the gradient has evolved into a steeper slope with age. Of course, the sample size is very limited. I should also add that it is possible that galaxies have evolved differently in relation to their environment, with different star formation rates and gas flow properties, and so it is conceivable that we should not expect this to be a general result, valid for all spiral galaxies.

Having shown the empirical result, it is interesting to consider what the galaxy evolution models predict as to the amount of slope change we could expect when comparing young populations with PNe. In order to try and answer this question one could take tailored models for some of the galaxies in Table 1. For example, Spitoni *et al.* (2013) produced a model of the evolution of the gradient with age for the galaxy M31, finding that the slope steepens with time. If we take a representative age for the PNe of 5 Gyrs, the estimated predicted change from these models is ∼0.15 dex R_{25}^{-1}, which is in fact smaller than the size of the errorbars in Fig. 1. Similarly for NGC 300, the models by Kang *et al.* (2016) predict that in the main disk of the galaxy there should be quite a small change in the gradient slope, this time flattening with time. In these two models and similar ones the chemical enrichment that occurred between a few Gyrs ago and the present time is on the order of 0.1 dex, which is roughly consistent with the change in the zeropoint of the gradients for the four galaxies examined by Magrini *et al.* (2016).

The small steepening of the gradients for the galaxies in Table 1 is also consistent with the cosmologically-motivated models presented by Gibson *et al.* (2013) and calculated with an enhanced prescription for the feedback from SNe, that predict a very small evolution of the gradients with time. These are, however, at variance with the much stronger evolution needed to explain, with a smaller feedback, the steep gradients measured for a few high-redshift systems. Such tension is exacerbated by observations of stellar clusters in M33 by Beasley *et al.* (2015), that also indicate a strong flattening of the metallicity gradient in this galaxy with time, in contrast with the results shown in Fig. 1.

In summary, different abundance and age tracers can produce opposite results when considering the evolution of the gradients. This can only partly be explained by age uncertainties and the small effect we are trying to measure.

4. PNe in the outer disk of M31

Although the PN population of M31 is discussed more in depth elsewhere in this volume, it is worth to briefly consider this system as an example of the diagnostic power of PNe to probe the α-element evolution in galaxies and the importance of the outskirts of galactic disks to constrain the evolutionary status of spirals. Recent work on the outer disk PN population in M31 has recently been published by Kwitter *et al.* (2012), Balick *et al.* (2013), Corradi *et al.* (2015) and Fang *et al.* (2015). These authors assume, for example based on kinematical information, that these PNe belong to the disk rather than the halo. While a few might actually not belong to the disk of the galaxy, these studies have shown an important number of PNe extending to galactocentric distances of 100 kpc or more, i.e. well beyond the isophotal radius. The abundance properties of these PNe are quite interesting, in that they deviate significantly from the radial trend described by young stars and H II regions in the inner disk, extrapolated to large radii. The radial abundance distribution of the PNe in the outer disk is virtually flat, with a mean O/H value only slightly ($\sim$0.2 dex) below solar. This behavior is in fact reminiscent of the cases presented at the end of Sect. 2, where deviations from a pure exponential decrease of the metallicity with radius were briefly introduced. The idea put forward by Balick *et al.* (2013) that a past galaxy encounter could explain the metallicity trend is consistent with that discussion.

References

Anders, F., Chiappini, C., Minchev, I., Miglio, A., Montalbán, J., Mosser, B., Rodrigues, T. S., Santiago, B. X., *et al.* 2016, ArXiv:1608.04951

Balick, B., Kwitter, K. B., Corradi, R. L. M., & Henry, R. B. C. 2013, *ApJ*, 774, 3

Beasley, M. A., San Roman, I., Gallart, C., Sarajedini, A., & Aparicio, A. 2015, *MNRAS*, 451, 3400

Bresolin, F. & Kennicutt, R. C. 2015, *MNRAS*, 454, 3664

Bresolin, F., Kennicutt, R. C., & Ryan-Weber, E. 2012, *ApJ*, 750, 122

Bresolin, F., Kudritzki, R.-P., Urbaneja, M. A., Gieren, W., Ho, I.-T., & Pietrzyński, G. 2016, *ApJ*, 830, 64

Bresolin, F., Ryan-Weber, E., Kennicutt, R. C., & Goddard, Q. 2009, *ApJ*, 695, 580

Bresolin, F., Stasińska, G., Vílchez, J. M., Simon, J. D., & Rosolowsky, E. 2010, *MNRAS*, 404, 1679

Corradi, R. L. M., Kwitter, K. B., Balick, B., Henry, R. B. C., & Hensley, K. 2015, *ApJ*, 807, 181

Delgado-Inglada, G., Rodríguez, M., Peimbert, M., Stasińska, G., & Morisset, C. 2015, *MNRAS*, 449, 1797

Fang, X., García-Benito, R., Guerrero, M. A., Liu, X., Yuan, H., Zhang, Y., & Zhang, B. 2015, *ApJ*, 815, 69

Gazak, J. Z., Kudritzki, R., Evans, C., Patrick, L., Davies, B., Bergemann, M., Plez, B., Bresolin, F., *et al.* 2015, *ApJ*, 805, 182

Gibson, B. K., Pilkington, K., Brook, C. B., Stinson, G. S., & Bailin, J. 2013, *A&A*, 554, A47

Ho, I.-T., Kudritzki, R.-P., Kewley, L. J., Zahid, H. J., Dopita, M. A., Bresolin, F., & Rupke, D. S. N. 2015, *MNRAS*, 448, 2030

Jones, T., Wang, X., Schmidt, K. B., Treu, T., Brammer, G. B., Bradač, M., Dressler, A., Henry, A. L., *et al.* 2015, *AJ*, 149, 107

Kang, X., Zhang, F., Chang, R., Wang, L., & Cheng, L. 2016, *A&A*, 585, A20

Kubryk, M., Prantzos, N., & Athanassoula, E. 2015, *A&A*, 580, A127

Kudritzki, R.-P., Urbaneja, M. A., Bresolin, F., Przybilla, N., Gieren, W., & Pietrzyński, G. 2008, *ApJ*, 681, 269

Kudritzki, R.-P., Urbaneja, M. A., Gazak, Z., Bresolin, F., Przybilla, N., Gieren, W., & Pietrzyński, G. 2012, *ApJ*, 747, 15

Kwitter, K. B., Lehman, E. M. M., Balick, B., & Henry, R. B. C. 2012, *ApJ*, 753, 12

Maciel, W. J. & Costa, R. D. D. 2009, in IAU Symposium, Vol. 254, The Galaxy Disk in Cosmological Context, ed. J. Andersen, Nordströara, B. m, & J. Bland-Hawthorn, 38

Maciel, W. J. & Costa, R. D. D. 2013, *Rev. Mexicana AyA*, 49, 333

Magrini, L., Coccato, L., Stanghellini, L., Casasola, V., & Galli, D. 2016, *A&A*, 588, A91

Magrini, L., Stanghellini, L., Corbelli, E., Galli, D., & Villaver, E. 2010, *A&A*, 512, A63

Magrini, L., Vílchez, J. M., Mampaso, A., Corradi, R. L. M., & Leisy, P. 2007, *A&A*, 470, 865

Mollá, M. 2014, Advances in Astronomy, 2014, 162949

Olave-Rojas, D., Torres-Flores, S., Carrasco, E. R., Mendes de Oliveira, C., de Mello, D. F., & Scarano, S. 2015, *MNRAS*, 453, 2808

Patterson, M. T., Walterbos, R. A. M., Kennicutt, R. C., Chiappini, C., & Thilker, D. A. 2012, *MNRAS*, 422, 401

Peimbert, A. & Peimbert, M. 2010, *ApJ*, 724, 791

Przybilla, N., Butler, K., Becker, S. R., & Kudritzki, R. P. 2006, *A&A*, 445, 1099

Roškar, R., Debattista, V. P., Stinson, G. S., Quinn, T. R., Kaufmann, T., & Wadsley, J. 2008, *ApJL*, 675, L65

Sánchez, S. F., Rosales-Ortega, F. F., Iglesias-Páramo, J., Mollá, M., Barrera-Ballesteros, J., Marino, R. A., Pérez, E., Sánchez-Blazquez, P., *et al.* 2014, *A&A*, 563, A49

Sánchez-Blázquez, P., Rosales-Ortega, F. F., Méndez-Abreu, J., Pérez, I., Sánchez, S. F., Zibetti, S., Aguerri, J. A. L., Bland-Hawthorn, J., *et al.* 2014, *A&A*, 570, A6

Sanders, N. E., Caldwell, N., McDowell, J., & Harding, P. 2012, *ApJ*, 758, 133

Spitoni, E., Matteucci, F., & Marcon-Uchida, M. M. 2013, *A&A*, 551, A123

Stanghellini, L. & Haywood, M. 2010, *ApJ*, 714, 1096

Stanghellini, L., Magrini, L., Casasola, V., & Villaver, E. 2014, *A&A*, 567, A88

Stasińska, G., Peña, M., Bresolin, F., & Tsamis, Y. G. 2013, *A&A*, 552, A12

Tissera, P. B., Pedrosa, S. E., Sillero, E., & Vilchez, J. M. 2016, *MNRAS*, 456, 2982

Trundle, C., Dufton, P. L., Lennon, D. J., Smartt, S. J., & Urbaneja, M. A. 2002, *A&A*, 395, 519

U, V., Urbaneja, M. A., Kudritzki, R.-P., Jacobs, B. A., Bresolin, F., & Przybilla, N. 2009, *ApJ*, 704, 1120

Urbaneja, M. A., Herrero, A., Kudritzki, R.-P., Najarro, F., Smartt, S. J., Puls, J., Lennon, D. J., & Corral, L. J. 2005, *ApJ*, 635, 311

Venn, K. A., McCarthy, J. K., Lennon, D. J., Przybilla, N., Kudritzki, R. P., & Lemke, M. 2000, *ApJ*, 541, 610

Xiang, M.-S., Liu, X.-W., Yuan, H.-B., Huang, Y., Wang, C., Ren, J.-J., Chen, B.-Q., Sun, N.-C., *et al.* 2015, Research in Astronomy and Astrophysics, 15, 1209

Zinchenko, I. A., Pilyugin, L. S., Grebel, E. K., Sánchez, S. F., & Vílchez, J. M. 2016, *MNRAS*, 462, 2715

Zurita, A. & Bresolin, F. 2012, *MNRAS*, 427, 1463

Planetary Nebulae: Multi-Wavelength Probes of Stellar and Galactic Evolution
Proceedings IAU Symposium No. 323, 2016
X. Liu, L. Stanghellini & A. Karakas, eds.

doi:10.1017/S1743921317001697

The evolution of the oxygen abundance radial gradient in the Milky Way Galaxy disk

Mercedes Mollá[1], Oscar Cavichia[2], Roberto D. D. Costa[3], Walter J. Maciel[3], Brad Gibson[4] and Angeles I Díaz[5,6]

[1]CIEMAT, Avda. Complutense 40, 28040 Madrid, Spain
email: mercedes.molla@ciemat.es

[2]Instituto de Física e Química, Universidade Federal de Itajubá, Av. BPS, 1303, 37500-903, Itajubá-MG, Brazil

[3]Instituto de Astronomia, Geofisica e Ciências Atmosféricas, Universidade de São Paulo, 05508-900, São Paulo-SP, Brazil

[4]E.A Milne Centre for Astrophysics, University of Hull, HU6 7RX, United Kingdom

[5]Universidad Autónoma de Madrid, 28049, Madrid, Spain

[6]Astro-UAM, Unidad Asociada CSIC, Universidad Autónoma de Madrid,28049, Madrid, Spain

Abstract. We review the state of our chemical evolution models for spiral and low mass galaxies. We analyze the consequences of using different stellar yields, infall rate laws and star formation prescriptions in the time/redshift evolution of the radial distributions of abundances, and other quantities as star formation rate or gas densities, in the Milky Way Galaxy; In particular we will study the evolution of the oxygen abundance radial gradient analyzing its relation with the ratio SFR/infall. We also compare the results with our old chemical evolution models, cosmological simulations and with the existing data, mainly with the planetary nebulae abundances.

Keywords. galaxies: evolution, galaxies: formation, galaxies: spiral, galaxies: abundances, Galaxy: abundances

1. Introduction

Chemical elements are mostly created by stellar nucleosynthesis, and eventually ejected and diluted in the interstellar medium when stars die. They are then incorporated into the successive generations of stars. This cycle of formation and death of stars and the subsequent ejection of elements may occur many times, if the star formation rate (SFR) is high, with elemental abundances increasing rapidly in the interstellar medium (ISM); If the SFR is low, the abundances will increase slowly, maintaining low during a long time. This way, elemental abundances give clues about when, how and with which rate stars form, existing a direct link between the evolutionary histories of galaxies or galaxy regions and the abundances measured within them. Chemical evolution models try to understand the metal enrichment process, predicting the elemental abundances expected in each region as depending on different inputs and hypotheses, as the Initial Mass Function (IMF), the star formation law, the existence or not of inflows or outflows of gas, or the stellar nucleosynthesis yields. Once computed a model for a galaxy, it is possible to compare with the observational data and deduce if our assumptions are valid or it is necessary to modify them.

Many numerical chemical evolution models have been presented in the literature from decades ago Lacey & Fall (1983), Lacey & Fall (1985), Díaz & Tosi (1984),Tosi & Díaz (1985), Matteucci & Francois (1989), Ferrini *et al.* (1994). All of them are usually computed to reproduce the Milky Way Galaxy (MWG) and other nearby spiral galaxies data as radial distributions of star formation, gas density, surface density stellar profile and the radial gradient of most common elements (C, N, O, Fe...), which define the present state of our Galaxy. The existence of a radial gradient of abundances in the spiral disks is well known (Henry & Worthey 1999). This gradient was first observed in the MWG (Shaver *et al.* 1983) and later in other external galaxies (McCall, Rybski & Shields 1985, Zaristky, Kennicutt & Huchra 1994), and it is now well characterized in our local Universe as shown by Sánchez *et al.* (2014)(S14). However, even fitting the present day data, differences among models appear for other evolutionary times. In particular, the time evolution of the radial gradient of abundances is not the same for all models. Thus, in Díaz & Tosi (1984), Tosi & Díaz (1985), Mollá, Díaz & Tosi (1990) or Chiappini, Matteucci & Gratton (1997) a flat radial gradient appears in the early times of evolution of a galaxy, steepening with time; on the contrary, in Ferrini *et al.* (1994), Prantzos & Aubert (1995), Mollá, Ferrini & Diaz (1997), Portinari & Chiosi (2000), and Hou, Prantzos & Boissier (2000), the gradient is steep, and then it flattens with time. In the first case, a disk is already formed in time zero with an infall of primordial gas which dilutes the enrichment of the disk due to the stars death. In the second type of models, the disk forms out from the gas infalling onto the equatorial plane. Therefore, the evolution of the metallicity gradient seems to be directly related on the formation process of the disk, stressing the importance of this point.

In Mollá, Ferrini & Diaz (1997) we analyzed the existing data for MWG about this question: planetary nebulae (PN) abundances, globular and open clusters metallicities and stellar abundances. Each data set has his own problems to determine the time evolution of the radial gradient. The PN are useful to estimate the radial gradient of oxygen, an element not modified by the nucleosynthesis in the progenitors stars of PN. The early data analyzing this question (Maciel, Costa & Uchida 2003) showed a radial gradient steeper than the one for the present time, although, due to error bars, the observational points mixed with the HII regions data in the plot of O/H *vs* galactocentric radius R. The open clusters problems are derived of the necessary classification in thin or thick disk or even halo populations before their use in a radial distribution of metallicity. Furthermore, the ages determination, coming from a fit of their spectra to stellar models, which depend on metallicity and on age simultaneously, added to differences depending on the use of LTE or NLTE models, may be uncertain. The estimate of age and metallicity of globular clusters (GC) seems less problematic since there was assumed that they form as a single stellar population, with a single burst of star formation† and, consequently, to obtain only an age and a metallicity for all the stars of a GC is more precise. Based on these GCs data, we concluded that the radial gradient of abundances in MWG was steeper in the past, and it has been flattening with time until arriving to the present value. However, these old objects may also be part of the thick disk or the halo populations. As for the open clusters, the kinematic information is essential to determine their group and, therefore, to know if the corresponding data may be included in the study of the time evolution of the radial distribution of **disk** abundances.

In Mollá & Díaz (2005)–hereinafter MD05–, we show 440 chemical evolution models, for 44 different galaxy masses and 10 possible star formation efficiencies in each one. From a MWG-like model, we obtained the evolution of the radial gradient of oxygen abundances, finding a gradient of ~ -0.2 dex kpc^{-1} for $z = 2$ which flattens with time until reaching the present time value $\sim -0.05 - 0.06$ dex kpc^{-1}. This behavior was in agreement with the cosmological simulations results from Pilkington *et al.* (2012) and also with the Yuan *et al.* (2011) value found at $z \sim 1.5$. However, it is necessary to take into account that this gradient was calculated with the O abundances of all radial regions

† Recent works conclude that there are at least two stellar generations in these old objects.

computed in our model (which arrived until 25 kpc). In fact, the optical radius of MWG is $\sim 13 - 15$ kpc, and, therefore, to include the regions out of a given radius might be considered as inadequate, since there thick or halo populations may being mixed with the thin disk. Taking into account that these outer regions of the galaxy are not evolved, this implies a steepening of the radial gradient, more strongly for the highest redshift since the parts not evolved of the disk will be more inner. If we only consider the regions within the optical radius to compute the radial gradient, the evolution of this last one along the redshift is smoother, with a value ~ -0.06 dex kpc^{-1} for all redshifts.

We have now revised our chemical evolution models to include more realistic and updated inputs in our code. In this work we analyze the corresponding results for our new MWG model, used to calibrating the model grid. We study in particular the evolution of the radial gradient of oxygen, searching for the possible differences with the old model from MD05 and checking if the new prescriptions related on stellar yields, star formation or infall rate have modified the behavior of this gradient. We also compare the results with the most recent data in the literature.

2. Abundances in the MWG

In the recent years, new sets of data have been published, as consequence of better instruments, techniques and telescopes. Anders, Chiappini & Minchev (2016) give a summary of the observations (see their Fig.5), referring to PN, open clusters, and stars from MWG in order to analyze the evolution of the radial gradient of metallicity along the time. The problem is that each type of objects shows a different evolution. Thus, PNs which gave a steep radial gradient for a time ∼8 Gyr compared with the one corresponding to the present time given by HII regions, give now, following the most recent findings, a flat evolution along the time; that is, the radial gradient for O is basically the same from 5-6 Gyr ago until now (Maciel & Costa 2013, M13), or even has steepen slightly on time (Stanghellini & Haywood 2010, S10). A radial gradient constant along the time is also obtained by Magrini *et al.* (2016) for M33, M31, M81 and NGC300, comparing their radial distributions of oxygen abundances given by HII regions and PN, finding no evidences of evolution of the radial gradient with time. The PN give precise measurements of abundances since they are very bright, allowing to observe them even being far away. However, to estimate their masses (an in consequence their ages) and their distances is not an easy task, as explained in Cavichia, Costa, & Maciel (2011), which makes the Magrini *et al.* (2016) constraint not totally certain. From open clusters data, practically all of works, as Friel *et al.* (2002), Chen, Hou, & Wang (2003), Magrini *et al.* (2009), Frinchaboy *et al.* (2013), Cunha *et al.* (2016), arrive to a similar conclusion: the radial gradient of metallicity was steeper in the past than now. The number of points of the complete sample in these works is usually small and, therefore, each defined bin of age is wide with a handful of points. Finally, stellar abundances, as given Nordström *et al.* (2004), Casagrande *et al.* (2011), Bergemann *et al.* (2014), Xiang *et al.* (2015), show a flattening of the radial gradient. However, the oldest age bin (around 12-13 Gyr old) has a radial gradient zero. It means that the radial gradient started flat, then it did steeper, and after it flattened again. Anders, Chiappini & Minchev (2016) do not agree with these results, obtaining a radial gradient steepening continuously with time. Looking at these data in detail, we find some problems for the precise determination of the radial gradient for the old ages bin. Bergemann *et al.* (2014) give the metallicity radial distribution as the Iron abundance dividing their sample in different stellar ages, with a radial range for the whole sample of ~ 4 kpc (actually not very wide), but the oldest stars are located in a reduced region around the Solar vicinity, with a radial range

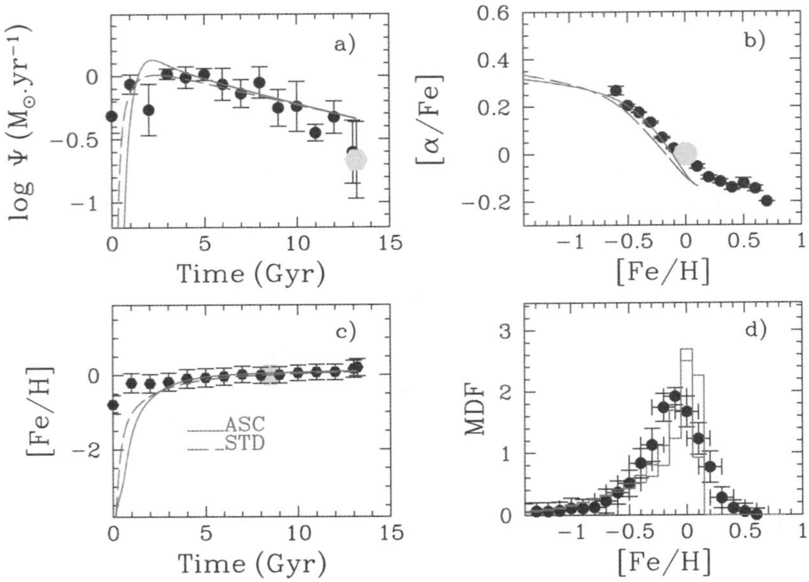

Figure 1. Solar Region evolution: a) SFR as a function of time; b) $[\alpha/Fe]$ as a function of $[Fe/H]$; c) $[Fe/H]$ as a function of time; and d) The metallicity distribution function. In all cases the blue and red lines represent the STD and ASC models, as labelled, while the black dots represent the Solar Region data as compiled in Mollá *et al.* (2015). The yellow large dot is the value of the SFR at the present time in a), the Solar abundance in b) and c), located in the time when the Sun formed in this last panel.

of $\sim$ 1 kpc. It is impossible to estimate a gradient with such a short radial range. This problem for the oldest age bin also occurs in Xiang *et al.* (2015), where the radial range reduces from the 3-4 kpc for the youngest bin to 1-1.5 kpc for the oldest one. For the CoRoGee sample from Anders, Chiappini & Minchev (2016) the total sample has a good radial range from 4 kpc to 14 kpc, but the oldest bin has a range much shorter, from 8 to 12 kpc. This produces a cloud of points around $\sim$ 9 kpc of galactocentric distance, which does not allow the computation of a precise radial gradient.

3. Multiphase Chemical Evolution models

Our basic model is well described in Ferrini *et al.* (1994), Mollá & Díaz (2005). We start with a spherical protogalaxy with a dynamical mass distribution given by the rotation curve. This initial gas falls onto the equatorial plane forming out the disk. In the halo the star formation follows a Schmid-Kennicutt law, while in the disk molecular clouds form first, and then stars are created from cloud-cloud collisions and from the interaction of massive stars with the surrounding clouds. Including the star mean lifetime and the stellar yields, we compute abundances for 15 elements until Fe, being this element mostly created by the Supernova type Ia yields from Iwamoto *et al.* (1999). In this standard model (STD), the star and molecular cloud formation are calculated with two efficiencies for each galaxy model, which we vary simultaneously with values among 0 and 1.

We have now an updated grid of models (Mollá *et al.* in preparation), where we use the stellar yields from Gavilán, Buell & Mollá (2005), Gavilán, Mollá & Buell (2006), Limongi & Chieffi (2003), Chieffi & Limongi (2004) and the IMF from Kroupa (2002), selected as the best combination of stellar yields and IMF from 144 models in Mollá *et al.* (2015). We have also revised the infall rates of gas, calculated to create disks as observed following

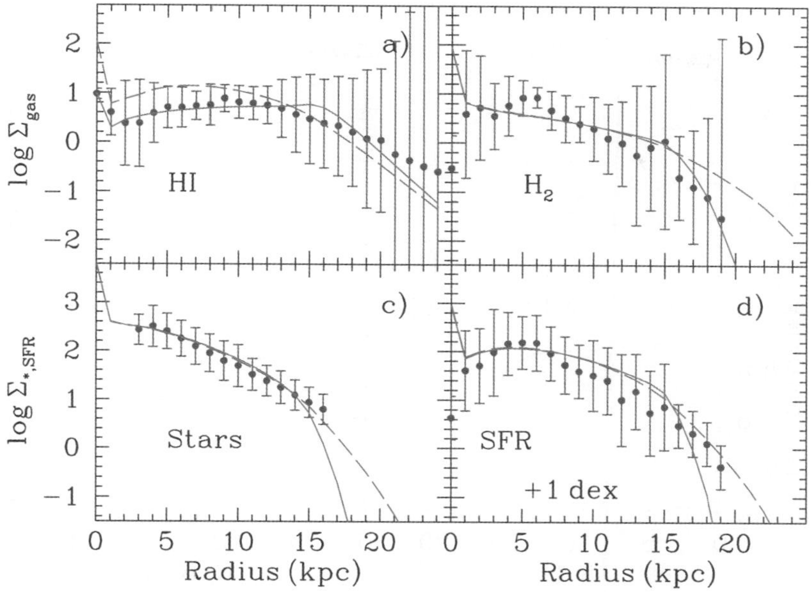

Figure 2. Radial distributions of surface density for: a) Diffuse gas; b) Molecular gas; c) Stellar profile; and d) SFR. In all cases the blue and red lines represent the STD and ASC models while the black dots represent the MWG data as compiled in the Appendix from Mollá *et al.* (2015).

Salucci *et al.* (2007). Details are given in Mollá *et al.* (2016). Moreover, we now follow the prescription by Ascasibar *et al.* (in preparation) to form molecular clouds from the diffuse gas. This model (ASC) is compared with our standard model (STD). The details are given in Mollá *et al.* (2017).

For a MWG-like model, we use a dynamical mass of 10^{12} M$_{\odot}$, able to create a disk of $\sim 7\,10^{10}$ M$_{\odot}$. In Fig. 1 we represent the evolution of the region located a $R = 8$ kpc as function of time in panels a) and c), and as function of the metallicity [Fe/H] in panels b) and d), compared with the Solar Region data. In Fig. 2 we draw the surface densities of HI, H_2, stars and SFR for the disk component, compared with the MWG data. We compare both STD and ASC models, finding that basically both fit the data within errors, with differences only in the outer regions with a more extended disk for STD.

4. Results: the time evolution of oxygen radial gradient

In Fig. 3 we show the radial distribution for O/H as given by STD (top panels) and ASC (bottom panels) prescriptions, for different redshifts as labelled in the top right panel. This way, we may check the effect of the star formation law on the evolution of the radial gradient. The distributions shows a sharp cut of the disk in ASC while this is smoother in STD, with a more extended disk, as said before. This is due to the effect of threshold of ASC, since the necessary density or metallicity must be reached to create molecular clouds. In left panels we show these distributions as a function of the galactocentric distance R (in kpc). In right panels we show the same distributions as a function of the normalized radius R/R_{eff}. It is clear that the slope of the distribution within the optical disk, defined as $R \sim 3\,R_{eff}$ (~ 15 kpc for the STD model at $z = 0$), is almost the same for all redshifts, as shown in right panels. We must remember that R_{eff} changes with redshift: the stellar disk increases their size when stars form in an inside-out scenario; at the early times stars form only in the inner disk. In the STD model, R_{eff} is

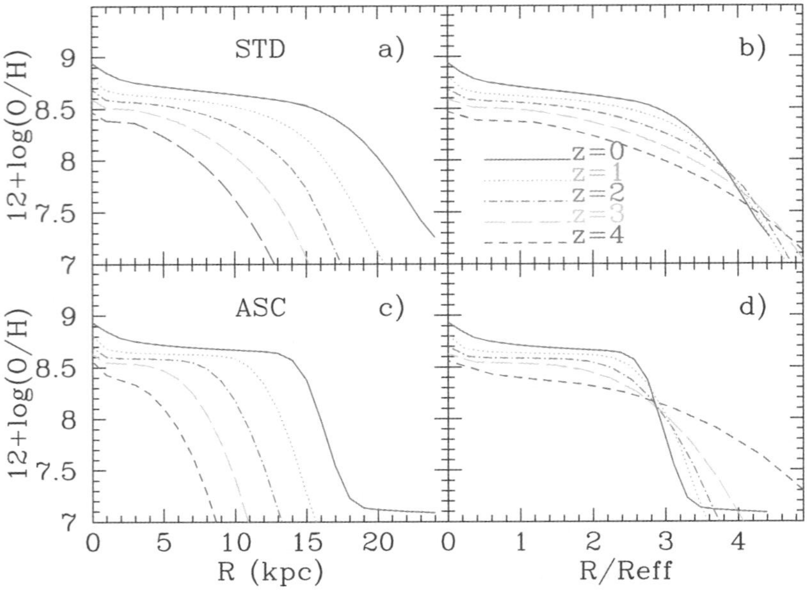

Figure 3. Radial distributions of the 12+log(O/H) abundances for different redshifts as labelled. Left panels as function of the galactocentric distance (in kpc) and right panels as function of the normalized radius R/R_{eff}. Top panels represent the evolution of STD, while the bottom ones refer to ASC.

around 5 kpc for the present time, but it is ~ 2.5 kpc at $z = 4$. However, R_{eff} increases very slowly in ASC model, being $R_{eff} \sim 1.6$ kpc at $z = 4$. Thus, at this redshift, the optical radius would be around ~ 7 kpc in STD model, but it only arrives to ~ 4.5 kpc in ASC. Therefore, it seems quite evident that the slope of the radial gradient is basically the same for all redshifts when is measured within the optical radius, where we must define a radial gradient of the disk oxygen abundance. If a radial gradient different than the one of the present time is obtained at other time/redshift, it implies that it is being measured with regions out of the optical disk, probably with the thick disk, or of the halo, not unexpected since the phases of the disk formation occur at $z > 2.5$.

We summarize our results in Fig. 4, with the evolution along redshift of the O/H radial gradient, measured as dex kpc^{-1} in the top panels and as dex R_{eff}^{-1} in the bottom ones. In left panels the gradient is calculated using the whole radial range that we run our MWG model, while in the right ones we use only regions within the optical radius. Top panel results are compared with our old model MD05, cosmological simulations by Gibson *et al.* (2013)(G13) and recent data as labelled, including M13 from PN and also Henry *et al.* (2010), Rupke, Kewley & Chiel (2010), Cresci *et al.* (2010), Queyrel *et al.* (2012), and Jones *et al.* (2013) as H10,R10, C10, Q12 and J13, respectively, for high redshift galaxies.

5. Conclusions

The main conclusions can be summarized as:

(*a*) The radial range where the least-squares straight line fit is done is crucial to determine the value of the abundance radial gradient.

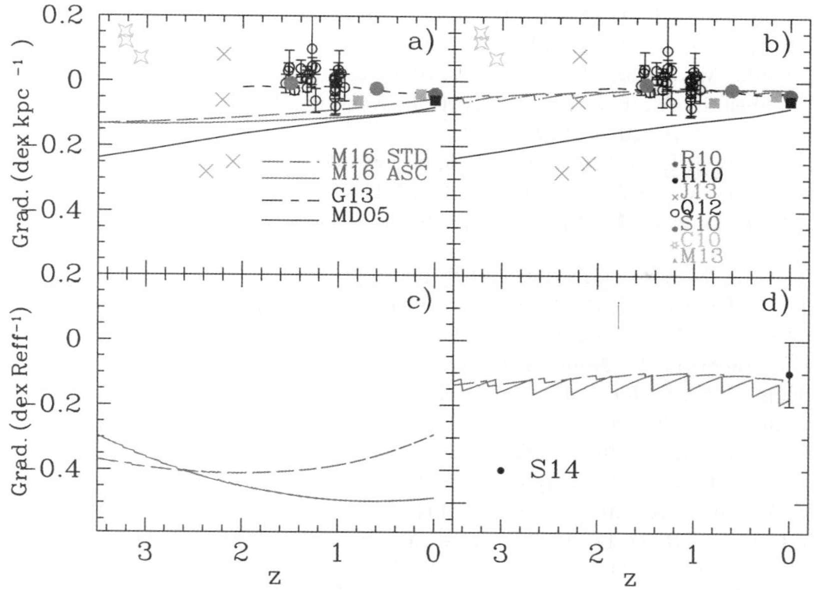

Figure 4. Evolution of the O/H radial gradient with the redshift z in dex kpc^{-1} (top panels) and dex R_{eff}^{-1} (bottom panels). In a) and c) calculated with the whole radial range. In b) and d) calculated with $R \leqslant 3\,R_{eff}$. In top panels, data are from H10, R10, S10, C10, Q12, J13 and M13. Models STD and ASC are our new models (M16) –this work and Mollá *et al.* (2017)–, our old model MD05 and simulation G13. In panel d) the present day value corresponds to the average gradient found by S14.

(*b*) Our MWG models produce a smooth evolution of the O abundances radial gradient with a value of -0.15 dex kpc^{-1} at $z = 3.5$ arriving to -0.05 dex kpc^{-1} at $z = 0$ when the whole radial range is used.

(*c*) The radial gradient has a value around ~ -0.05 dex kpc^{-1} for $z \leqslant 3.5$ when it is measured within the optical radius (variable with z).

(*d*) The radial gradient computed by our new models is flatter than the obtained in MD05 as consequence of the new infall rates, smoother than the old ones.

(*e*) If the radial gradient is measured as a function of a normalized radius (R/R_{eff}), differences arise between STD and ASC proceeding from the different grow of size of the stellar disk in both models (even with the same infall rates), if all radial regions are used.

(*f*) If the radial gradient is computed within the optical radius, as a function of a normalized radius (R/R_{eff}), it has a constant value of ~ -0.15 dex R_{eff}^{-1} for $z \leqslant 3.5$, which agrees with Sánchez *et al.* (2014) result for the CALIFA local galaxies.

References

Anders, F., Chiappini, C., Minchev, I., *et al.* 2016, *A&A*, submitted (arXiv:1608.04951)

Bergemann M., *et al.*, 2014, *A&A*, 565, A89

Casagrande, L., Schönrich R., Asplund M., Cassisi S., *et al.*, 2011, *A&A*, 530, A138

Cavichia O., Costa R. D. D., & Maciel W. J., 2011, *RMxAA*, 47, 49

Chen L., Hou J. L., & Wang J. J., 2003, *AJ*, 125, 1397

Chiappini, C., Matteucci, F., & Gratton, R. 1997, *ApJ*, 477, 765

Chieffi A. & Limongi M., 2004, *ApJ*, 608, 405

Cresci, G., Mannucci, F., Maiolino, R., *et al.* 2010, *Nature*, 467, 811 (C10)

Costa, R. D. D., Cavichia, O., & Maciel, W. J., 2013, *IAUS*, 289, 375

Cunha, K., Frinchaboy, P. M., Souto, D., *et al.* 2016, *Astronomische Nachrichten*, 337, 922
Diaz, A. I. & Tosi, M., 1984, *MNRAS*, 208, 365
Ferrini, F., Molla, M., Pardi, M. C., & Diaz, A. I., 1994, *ApJ*, 427, 745
Friel, E. D., Janes, K. A., Tavarez, M., Scott, J., Katsanis, R., *et al.*, 2002, *AJ*, 124, 2693
Frinchaboy, P. M., *et al.*, 2013, *ApJ*, 777, L1
Gavilán, M., Buell, J. F., & Mollá, M., 2005, *A&A*, 432, 861
Gavilán, M., Mollá, M., & Buell, J. F., 2006, *A&A*, 450, 509
Gibson, B. K., Pilkington, K., Brook, C. B., Stinson, G. S., & Bailin, J. 2013, *A&A*, 554, A47 (G13)
Henry, R. B. C. & Worthey, G. 1999, *PASP*, 111, 919
Henry, R. B. C., Kwitter, K. B., Jaskot, A. E., *et al.* 2010, *ApJ*, 724, 748 (H10)
Hou, J. L., Prantzos, N., & Boissier, S. 2000, *A&A*, 362, 921
Iwamoto, K., Brachwitz, F., Nomoto, K., *et al.* 1999, *ApJS*, 125, 439
Jones, T., Ellis, R. S., Richard, J., & Jullo, E. 2013, *ApJ*, 765, 48 (J13)
Kroupa, P., 2002, *Sci*, 295, 82
Lacey, C. G. & Fall, S. M. 1983, *MNRAS*, 204, 791
Lacey, C. G. & Fall, S. M. 1985, *ApJ*, 290, 154
Limongi, M. & Chieffi, A., 2003, *ApJ*, 592, 404
Maciel, W. J., Costa, R. D. D., & Uchida, M. M. M., 2003, *A&A*, 397, 667
Maciel, W. J. & Costa, R. D. D. 2013, *RMxA&Ap*, 49, 333 (M13)
Magrini, L., Sestito, P., Randich, S., & Galli, D. 2009, *A&A*, 494, 95
Magrini, L., Coccato, L., Stanghellini, L., Casasola, V., & Galli, D. 2016, *A&A*, 588, A91
Matteucci, F., & Francois, P. ,1989, *MNRAS*, 239, 885
McCall, M. L., Rybski, P. M., & Shields, G. A. 1985, *ApJS*, 57, 1
Mollá, M., Díaz, A. I., & Tosi, M., 1990, in Chemical and dynamical evolution of Galaxies, eds. F. Ferrini, J. Franco, & F. Matteucci, ETS Editrice (PISA), 577
Mollá, M., Ferrini, F., & Diaz, A. I. 1997, *ApJ*, 475, 519
Mollá, M. & Díaz, A. I. 2005, *MNRAS*, 358, 521 (MD05)
Mollá, M., Cavichia O., Gavilán M., & Gibson B. K. 2015, *MNRAS*, 451, 3693
Mollá, M., Díaz, A. I., Gibson, B. K., *et al.*, 2016, *MNRAS*, 462, 1329
Mollá, M., Ascasibar, Y., Díaz, A. I., & Gibson, B. K., 2017, *MNRAS*, submitted (M16)
Nordström B., *et al.*, 2004, A&A, 418, 989
Pilkington, K., Few, C. G., Gibson, B. K., *et al.* 2012, *A&A*, 540, A56
Portinari, L. & Chiosi, C. 2000, *A&A*, 355, 929
Prantzos, N. & Aubert, O. 1995, *A&A*, 302, 69
Queyrel, J., Contini, T., Kissler-Patig, M., *et al.* 2012, *A&A*, 539, A93 (Q12)
Rupke, D. S. N., Kewley, L. J., & Chien, L.-H. 2010, ApJ, 723, 1255 (R10)
Salucci, P., Lapi, A., Tonini, C., Gentile, G., *et al..*, 2007, *MNRAS*, 378, 41
Sánchez, S. F., Rosales-Ortega, F. F., Iglesias-Páramo, J., *et al.* 2014, *A&A*, 563, A49 (S14)
Shaver, P. A., McGee, R. X., Newton, L. M., Danks, A. C., *et al.* 1983, *MNRAS*, 204, 53
Stanghellini, L. & Haywood, M. 2010, *ApJ*, 714, 1096 (S10)
Tosi, M. & Díaz, A. I. 1985, *MNRAS* 217, 571
Xiang, M.-S., Liu, X.-W., Yuan, H.-B., *et al.* 2015, *Research in Astronomy and Astrophysics*, 15, 1209
Yuan, T.-T., Kewley, L. J., Swinbank, A. M., *et al.* . 2011, *ApJL*, 732, L14
Zaritsky, D., Kennicutt, R. C., Jr., & Huchra, J. P. 1994, *ApJ*, 420, 87

Discussion

Hensler: Part of the produced oxygen will be stored in the halo of galaxies (COS halo survey) or lost from it due to interactions with the environment. Its contribution to the oxygen enhancement depends on the time scale of this backflow cycle. How can you incorporate this effect into your models?

Q: UV absorption lines are used to measure the abundances of gas at large radii from spiral galaxies. Could you please summarize the results and how they compare to theoretical predictions of metal enrichment?

Planetary Nebulae: Multi-Wavelength Probes of Stellar and Galactic Evolution
Proceedings IAU Symposium No. 323, 2016
X. Liu, L. Stanghellini & A. Karakas, eds.

doi:10.1017/S1743921317001466

XSHOOTER spectroscopy of the enigmatic PN Lin49 in the SMC

Masaaki Otsuka[1], Francisca Kemper[1], Marcelo L. Leal-Ferreira[2], Isabel Aleman[2], Jeronimo Bernard-Salas[3], Jan Cami[4,5], Bram B. Ochsendorf[6], Els Peeters[4,5] and Peter Scicluna[1]

[1]ASIAA, 11F of Astronomy-Mathematics Building, AS/NTU. No.1, Sec. 4, Roosevelt Rd, Taipei 10617, Taiwan, R.O.C. email: otsuka@asiaa.sinica.edu.tw (MO)

[2]Leiden Observatory, University of Leiden, PO Box 9513, 2300 RA, Leiden, The Netherlands

[3]Department of Physical Sciences, The Open Univ., Milton Keynes, MK7 6AA, UK

[4]Department of Physics & Astronomy, The Univ. of Western Ontario, London, ON N6A 3K7, Canada

[5]SETI Institute, 189 Bernardo Ave, Suite 100, Mountain View, CA 94043, USA

[6]Space Telescope Science Institute, 3700 San Martin Drive, Baltimore, MD 21218, USA,

Abstract. We performed a detailed spectroscopic analysis of the fullerene C_{60}-containing planetary nebula (PN) Lin49 in the Small Magellanic Cloud (SMC). Lin49 is a C-rich and metal-deficient PN (Z~0.0006) and its nebular abundances are in agreement with the AGB model for the initially 1.25 $M_\odot$ stars with the metallicity $Z = 0.001$. By stellar absorption fitting with TLUSTY, we derived stellar abundances, effective temperature, and surface gravity. We constructed the photoionization model with CLOUDY in order to investigate physical conditions of Lin49. The model with the 0.005-0.1 μm radius graphite and a constant hydrogen density shell could not fit the ~1-5 μm spectral energy distribution (SED) owing to the strong near-IR excess. We propose that the near-IR excess indicates (1) the presence of extremely small carbon molecules or (2) the presence of high-density structure surrounding the central star.

Keywords. infrared: ISM, ISM: abundances, dust, molecules, planetary nebulae: individual (Lin49), stars: abundances

1. C_{60}-containing PN Lin49 in the SMC

Recently, *Spitzer*/IRS mid-IR spectrum revealed that Lin49 is a C-rich dust PN, showing strong C_{60} resonances at 17.4 and 18.9 μm and similar dust features such as the broad 11 μm and 30 μm features seen in the other C_{60}-containing Magellanic Cloud PNe Sloan *et al.* (2014). However, physical properties of the central star and the dusty nebula had been unknown. Therefore, we characterized Lin49 through abundance analysis, stellar absorption fitting, and photoionization modeling based on our own ESO/VLT X-SHOOTER UV to near-IR spectrum, the archived *Spitzer*/IRS spectrum, and the UV to mid-IR photometry data.

2. Nebular abundances and Stellar absorption line analysis

The resultant elemental abundances are listed in Table 1. With the XSHOOTER and *Spitzer*/IRS spectra, we determined the nebular abundances of nine elements. We found that the nebular abundances are in agreement with the AGB star nucleosynthesis model for the initially 1.25 $M_\odot$ and the metallicity $Z_\odot/20$ stars by Fishlock *et al.* (2014).

Table 1. Elemental abundances ($\log_{10}\epsilon(\mathrm{H}) = 12$). The values in the last two columns are the predictions in the AGB models by Fishlock *et al.* (2014) for initially $1.0\,M_\odot$ and $1.25\,M_\odot$ stars with $Z = 0.001$.

X	$\log_{10}\epsilon(\mathrm{X})$	$\log_{10}\epsilon(\mathrm{X}_\odot)$	[X/H]	$\log_{10}\epsilon(\mathrm{X}_{1.0\,M_\odot\mathrm{model}})$	$\log_{10}\epsilon(\mathrm{X}_{1.25\,M_\odot\mathrm{model}})$
He	10.80 - 11.01	10.93 ± 0.01	–0.13 - +0.08	10.99	11.01
C	8.46 ± 0.24	8.39 ± 0.04	+0.07 ± 0.25	8.06	8.56
N	6.93 ± 0.02	7.86 ± 0.12	–0.93 ± 0.12	7.15	7.26
O	8.11 ± 0.01	8.73 ± 0.07	–0.62 ± 0.07	7.58	7.68
Ne	7.18 ± 0.05	8.05 ± 0.10	–0.89 ± 0.11	6.89	7.37
S	6.02 ± 0.01	7.16 ± 0.02	–1.15 ± 0.02	5.99	6.00
Cl	4.03 ± 0.05	5.25 ± 0.06	–1.22 ± 0.08	4.07	4.08
Ar	5.48 ± 0.11	6.50 ± 0.10	–1.02 ± 0.15	5.27	5.28
Fe	4.55 ± 0.04	7.46 ± 0.08	–2.91 ± 0.09	6.37	6.38

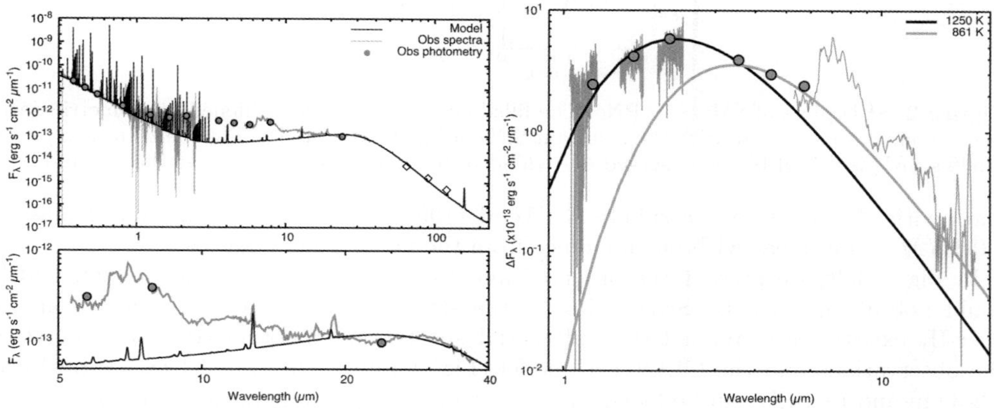

Figure 1. (*left panels*) (upper) Comparison between the CLOUDY model and observational data. The squares at 65, 90, and 120 μm are the *predicted* flux densities calculated by fitting the broad 30 μm feature. (lower) Closed-up plots for mid-IR wavelength. (*right panel*) Near-IR excess. The grey lines and the circles are the residual flux densities (ΔF_λ) between the observed X-SHOOTER and *Spitzer*/IRS spectra and photometry bands and the corresponding values obtained from the CLOUDY model.

We fit stellar absorption lines with the O-type star grid model OStar2002 by Lanz & Hubeny (2003) using the non-LTE model stellar atmospheres modeling code TLUSTY by Hubeny (1988). We determined the effective temperature ($T_{\rm eff}$) of 30 500 ± 500 K, the surface gravity ($\log g$) of 3.29 ± 0.06 cm s^{-2}, and ϵ(He,C,N,O,Si) of 10.9 ± 0.3, 9.0 ± 0.3, 7.6 ± 0.3, 8.6 ± 0.1, and 6.8 ± 0.3, respectively.

3. Photoionization modeling; finding of the near-IR excess

Using the photoionization code CLOUDY (Ferland *et al.* 2013), we investigated physical conditions of Lin49. As the SED of the ionization/heating source, we used the central star's spectrum synthesized by the TLUSTY and varied its intrinsic luminosity (L_*) to match the observed nebular elemental abundances, gas emission line fluxes, and broad band fluxes/flux densities. We supposed that the infrared continuum is due to graphite grains; we adopted the radius a = 0.005-0.1 μm and the $a^{-3.5}$ size distribution. We derived the gas and dust masses of 0.11 $M_\odot$ and $4.3\times10^{-5}\,M_\odot$, respectively. The dust temperature is in the range from 79 K to 825 K. We determined the core-mass of 0.53-0.57 $M_\odot$. The location of the derived L_* and $T_{\rm eff}$ on evolutionary tracks of post AGB

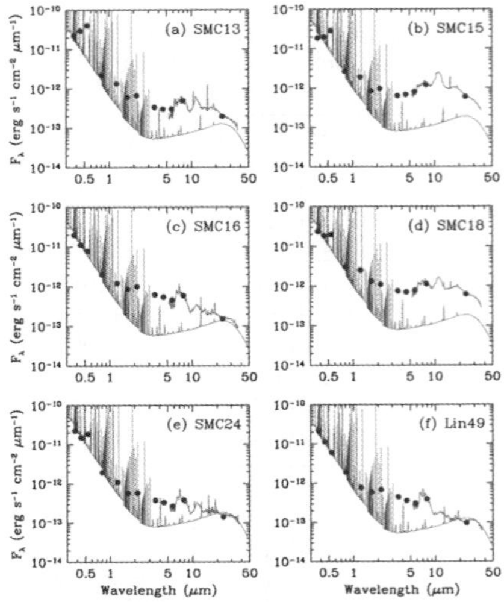

Figure 2. SED plots of SMC C_{60} PNe. The filled circles are the de-reddened photometric data. In each panel, we compare with the resultant SED of Lin49 presented in Fig 1 left. In each panel, Lin49's SED is scaled to the observed de-reddened flux density of each PN at the I_c band.

stars with $Z = 0.001$ of Vassiliadis & Wood (1994) indicates that the initial mass is 1.0-1.5 $M_\odot$, consistent with our inference from the nebular abundances.

In Fig. 1 left, we present the observed and the modeled SEDs. The predicted SED could not fit the near-IR SED owing to the strong near-IR excess. To focus on the near-IR excess, we examined the differential spectra and photometry points between the observations and the CLOUDY model (Fig. 1 right); the near-IR excess peaks at ~2-3 μm and its SED can be fitted by a single blackbody temperature of 1250 K.

4. Interpretations for the near infrared excess

In Fig. 2, we plot the observed data of SMC C_{60} PNe and compare their SED plots with the scaled Lin49's model SED. The near-IR excess is a common feature in SMC C_{60} PNe. The near-IR excess might hold the important information to disclose the C_{60} formation process in PNe. The near-IR excess in these PNe might indicate case 1: the presence of small carbon clusters and also fullerene precursors (in the XSHOOTER K-band spectrum of Lin49, we did not detect any H_2 and CO lines) or case 2: the presence of a sub-structure surrounding the central star.

4.1. *Stochastic heating by extremely small particles*

For small molecules, a single photon energy E_{ph} produces the difference between the maximum and minimum temperatures $\Delta T = E_{ph}/3Nk$, where N is the number of the atoms in a molecule and k is the Boltzmann constant. We obtained a N of ~39. As a pragmatic problem, with grains composing of 39 C-atoms only, it is difficult to reproduce the observed broad continuous near-IR excess feature seen in Lin49; to get a continuum like behavior, enough interacting vibrational modes are necessary.

4.2. *The presence of a sub-structure surrounding the central star*

Therefore, the near-IR excess would be due to the presence of a sub-structure surrounding the central star such a disc. We can fit the near-IR excess in Lin49 (Fig. 3 right), supposing the two density shells as explained in Fig. 3 left. Since the temperature in the disc

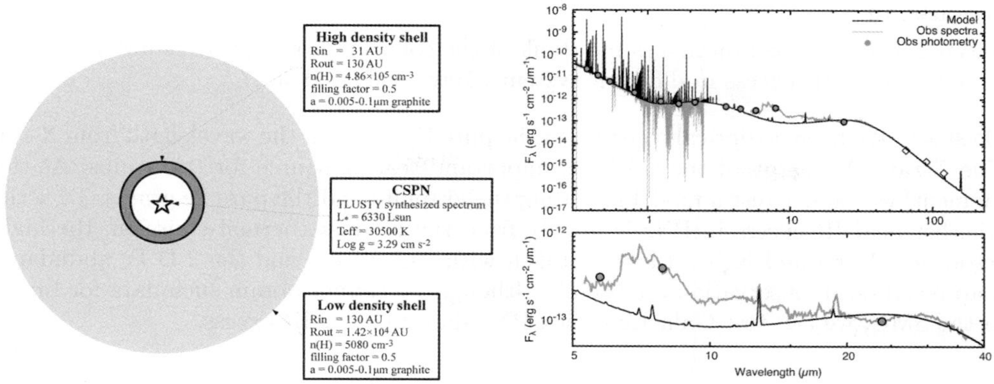

Figure 3. (*left panel*) The explanation of the two density shell model. The central star is located in the geometrical center of two spherical nebulae, so called *high density shell* and *low density shell*. As the ionization and heating source, we used the spectrum of the central star (CSPN) synthesized by the TLUSTY fitting. (*right two panels*) (upper) Comparison between the observed SED plots and the predicted SED by the two density shell model. (lower) Closed-up plot for mid-IR wavelength. The dust temperature was in the range of 79 K to 1453 K. The lines and symbols in both of panels are as defined in Fig. 1 left.

decreases with increasing radius, at the inner radius of the disc only the elements such Fe with a high condensation temperature are caught in grains. Indeed, the Fe abundance of Lin49 is extremely deficient ([Fe/H] = −2.9) compared to heavy α-elements (e.g., [Ar/H] = −1.0). If Lin49 has a disc, the large carbon molecules such as fullerene were relatively easily formed. Detailed discussions are found in Otsuka *et al.* (2016).

5. Future works

The near-IR excess is a common feature in SMC C_{60} PNe. However, since there are no available near-IR spectra of these SMC C_{60} PNe to confirm the near-IR excess spectroscopically, it is still far from the firm conclusion on the C_{60} formation in these SMC PNe. It is worth investigating the SMC C_{60} PNe in order to understand the C_{60} formation in the circumstellar environment; the well determined distance to the SMC allows us to accurately determine the luminosity of the central star, the size of the nebula, and physical conditions of the gas, dust, and C_{60} in the nebula, and then clarify the current evolutionary stage of the central star and estimate the initial mass. Based on the larger sample, we can conclude that the near-IR excess indicates the presence of a sub-structure such as a disc surrounding the central star and the fullerene was efficiently made within this disc. By investigating the near-IR excess of the SMC C_{60} PNe, we can find what conditions are essential in the C_{60} formation in PNe.

References

Ferland, G. J., Porter, R. L., van Hoof, P. A. M., *et al.* 2013, *RMxAA*, 49, 137
Fishlock, C. K., Karakas, A. I., Lugaro, M., & Yong, D. 2014, *ApJ*, 797, 44
Hubeny, I. 1988, *Computer Physics Communications*, 52, 103
Lanz, T. & Hubeny, I. 2003, *ApJS*, 146, 417
Otsuka, M., Kemper, F., Leal-Ferreira, M. L., *et al.* 2016, *MNRAS*, 462, 12
Sloan, G. C., Lagadec, E., Zijlstra, A. A., *et al.* 2014, *ApJ*, 791, 28
Vassiliadis, E. & Wood, P. R. 1994, *ApJS*, 92, 125

Discussion

SLOAN: The Fe depletion makes me think of the solid Fe proposed by IainMcDonald to explain a near-IR excess in different systems. Have you tried this?

OTSUKA: We have no optical constant of the pure Fe grains in the wavelength from X-ray. The X-ray-UV energy of the CSPN is important heating source for the grains. At this moment, we cannot test a model including the graphite and the pure Fe grains. To verify whether near-IR excess in PNe is mainly from the Fe dust thermal emission, the high-resolution J or I and K band images taken with AO-system and the 2-D Fe abundance map by slit-scan observation are useful. Although we cannot obtain such data for Lin 49 in the SMC, we can test Galactic C-rich PNe showing near-IR excess.

Planetary Nebulae: Multi-Wavelength Probes of Stellar and Galactic Evolution
Proceedings IAU Symposium No. 323, 2016
X. Liu, L. Stanghellini & A. Karakas, eds.

doi:10.1017/S1743921317000679

Abundances and Gradients in M31 – A Chemical Study of Planetary Nebulae in the Substructures

Xuan Fang[1,2], **Rubén García-Benito**[3], **Martín A. Guerrero**[3], **Xiaowei Liu**[4,5], **Yong Zhang**[1,6] **and Haibo Yuan**[7]

[1]Laboratory for Space Research, University of Hong Kong, Pokfulam Road, Hong Kong, China
email: fangx@hku.hk

[2]Department of Earth Sciences, University of Hong Kong, Pokfulam Road, Hong Kong, China

[3]Instituto de Astrofísica de Andalucía, Glorieta de la Astronomía s/n, 18008, Granada, Spain

[4]Department of Astronomy, School of Physics, Peking University, Beijing 100871, China

[5]Kavli Institute for Astronomy and Astrophysics, Peking University, Beijing 100871, China

[6]Department of Physics, University of Hong Kong, Pokfulam Road, Hong Kong, China

[7]Department of Astronomy, Beijing Normal University, Beijing 100875, China

Abstract. We present deep optical spectroscopy of seven planetary nebulae (PNe) in the substructures of M31, three in the Northern Spur and four associated with the Giant Stream. The spectra were obtained with the OSIRIS spectrograph on the 10.4 m GTC. The detection of the [O III] λ4363 auroral line in all PNe of our sample enables reliable abundance determinations. Our targets have low N/O (<0.5) and He/H ratios, indicating that they are probably Type II PNe. The PNe in our sample have rather homogeneous oxygen abundances, with an average value of 8.56±0.10. Based on the abundances as well as the spatial and kinematical information of our targets, we speculate that the Northern Spur and the Giant Stream might have the same origin. We raise a hypothesis that the dwarf satellite M32 might be responsible for these two substructures. New observations have recently been made to assess this hypothesis.

Keywords. galaxies: abundances, galaxies: evolution, galaxies: individual (M31), ISM: abundances, planetary nebulae: general, stars: evolution

1. Introduction

Planetary nebulae (PNe) are excellent tools to study the chemistry, kinematics, and stellar populations of host galaxies. Being bright in emission lines (e.g., [O III] λ5007), PNe are easily detected in the Andromeda Galaxy (M31), the nearest (780 kpc) large spiral system. Abundances of PNe in the disk and bulge of M31 have been studied for decades (e.g., Jacoby & Ford 1986; Jacoby & Ciardullo 1999; Kwitter et al. 2012) and abundance gradients were derived, which help to understand chemical evolution of M31. Recent observations using the 10 m class telescope have pushed abundance determination of M31 PNe to higher levels of accuracy and invoked novel ideas of formation and evolution of M31 (Balick *et al.* 2013; Corradi *et al.* 2015; Fang *et al.* 2015).

Although chemical abundances of disk PNe in M31 have been extensively studied, those associated with the substructures was largely ignored. It has been widely accepted that large galaxies evolve from merging of smaller galaxies (White 1978). The relics of galaxy interaction can be seen in the outskirts of galaxies in the form of various substructures such as tidal tails or stellar streams. Numerous substructures have been discovered in the outskirts of M31 (e.g., Ibata *et al.* 2001; McConnachie *et al.* 2009). The exact origin

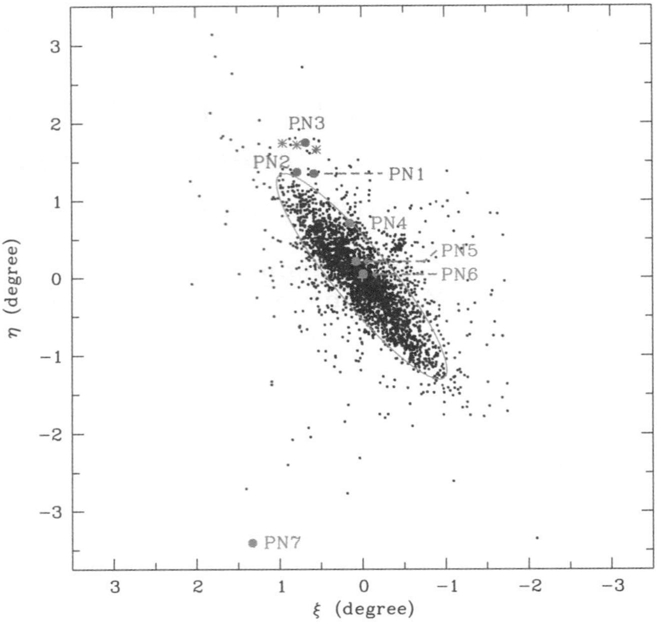

Figure 1. Distribution of PNe in M31. Objects are mostly from the catalogs of Merrett *et al.* (2006) and Yuan *et al.* (2010). Red asterisks are the three Northern Spur PNe observed by Fang *et al.* (2013). The PNe observed by Fang *et al.* (2015), three in the Northern Spur (blue filled circles, named PN1–3) and four associated with the Giant Stream (magenta fill circles, named PN4–7) are highlighted. The green ellipse is a 2°-radius (27.4 kpc) round disk centered on M31 with an inclination angle of 77.7° and a position angle of 37.7°.

of the southern stellar stream (also known as the Giant Stream) and the Northern Spur, the two most prominent stellar substructures of M31, are still unclear, although possible connection of these two has been proposed (Merrett *et al.* 2003). Detailed studies of PNe in the substructures may provide important clues to the assemblage history of the galaxy.

We have tried to obtain deep optical spectra of PNe in the substructures of M31. Our first effort was made with the 5.1 m Hale Telescope to observe three PNe in the Northern Spur, but with limited data quality (Fang *et al.* 2013). The follow-up observations were carried out using the 10.4 m Gran Telescopio Canarias at Observatorio del Roque de los Muchachos (ORM, La Palma). The observations and results are presented in the following sections.

2. Observations

Target Selection. Our targets were selected according to kinematics (see Figure 7 in Fang *et al.* 2015): six were selected from the catalog of Merrett *et al.* (2006), and the other one was identified by the Large Sky Area Multi-Object Fiber Spectroscopic Telescope (LAMOST; Yuan *et al.* 2010). Three are in the Northern Spur and the other four are kinematically associated with the Giant Stream (see Figure 1). The sky-projected galactocentric distances range from 0.7 to 50.2 kpc, and magnitudes $m(\lambda5007)\sim20.48$–21.51.
Observations and Data Reduction. Observations of the seven PNe were made with the Optical System for Imaging and low-intermediate-Resolution Integrated Spectroscopy (OSIRIS) at the 10.4 m Gran Telescopio Canarias (GTC), from 2014 August 23 to 29 (2014B) in service mode. The seeing is 0.5–1.0 arcsec during the observations. The grism R1000B and a 1 arcsec-wide long slit were used. The detector of OSIRIS consists of a mosaic of two Marconi CCDs with 2048×4096 pixels each. The pixel size is 15 μm

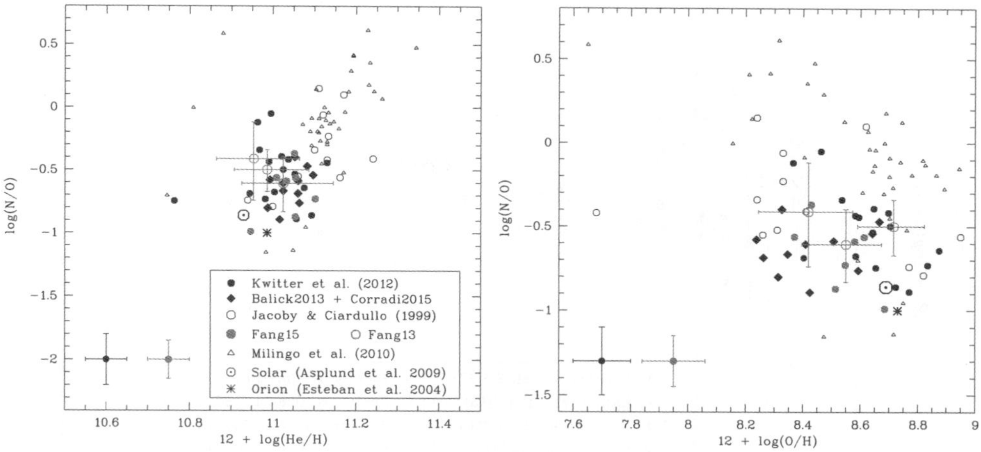

Figure 2. Left: log(N/O) vs. 12 + log(He/H). Right: log(N/O) vs. 12 + log(O/H). Different symbols represent different PN samples (see legend). Explanation of data sets is given in the text. Red open circles (along with error bars) are the three Northern Spur PNe observed by Fang *et al.* (2013, Fang13), and red filled symbols are the seven PNe observed by Fang *et al.* (2015, Fang15). Representative error bars of our targets and those of Kwitter *et al.* (2012, also Balick *et al.* 2013 and Corradi *et al.* 2015) are given in the lower-left corner.

(~0.127 arcsec). CCD data were 2×2 binned in the standard observing mode. This instrument setup gives a wavelength coverage 3630–7800 Å with a spectral resolution of FWHM~6 Å at 2.06 Å pixel^{-1}. For each PN, four 20-40 min exposures were made. Exposures of the spectrophotometric standard stars were also made for each target. Arc lines of neon and HgAr lamps were obtained for wavelength calibration. All data were then reduced following standard procedures using IRAF (see details in Fang *et al.* 2015). The 1D spectrum of each target PN was finally extracted for analysis.

3. Results

Abundance Correlations. Elemental abundances of He, O, N, Ne, S and Ar relative to hydrogen were derived for our target PNe. Figure 2 show the N/O versus He/H and O/H. Our sample are compared with the disk PNe observed by Jacoby & Ciardullo (1999), Kwitter et al. (2012), Balick *et al.* (2013) and Corradi *et al.* (2015). The Galactic Type I PNe of Milingo *et al.* (2010) are also presented for purpose of comparison. Our sample have low N/O (<0.5) and He/H ratios, indicating they are probably Type II (i.e., their progenitor stars may have low masses). Not only our substructure PNe, but also those recently observed M31 disk PNe seem to be Type II.

Abundance Gradient. The oxygen gradient of M31 is presented in Figure 3. Our substructure PNe generally have consistent oxygen abundances within the errors, and have an average value of 8.56±0.12. The oxygen abundance of target PN7, whose projection-corrected galactocentric distance might be ⩾100 kpc, is 0.11 dex lower than the solar value (8.69, Asplund *et al.* 2009). Abundances of PN5 and PN6 are higher than the disk and bulge samples of Jacoby & Ciardullo (1999) and Sanders et al. (2012) at similar distances. The homogeneous distribution of oxygen in our sample indicates that the Northern Spur and the Giant Stream may have the same origin. We speculate that these two substructures might have been tidally stripped from the dwarf satellite M32, given its spatial position. For the disk PNe, oxygen gradient has a negative value (−0.011±0.004 dex kpc^{-1}; Kwitter *et al.* 2012) within 50 kpc, but tends to be flattened further out. Although PN6

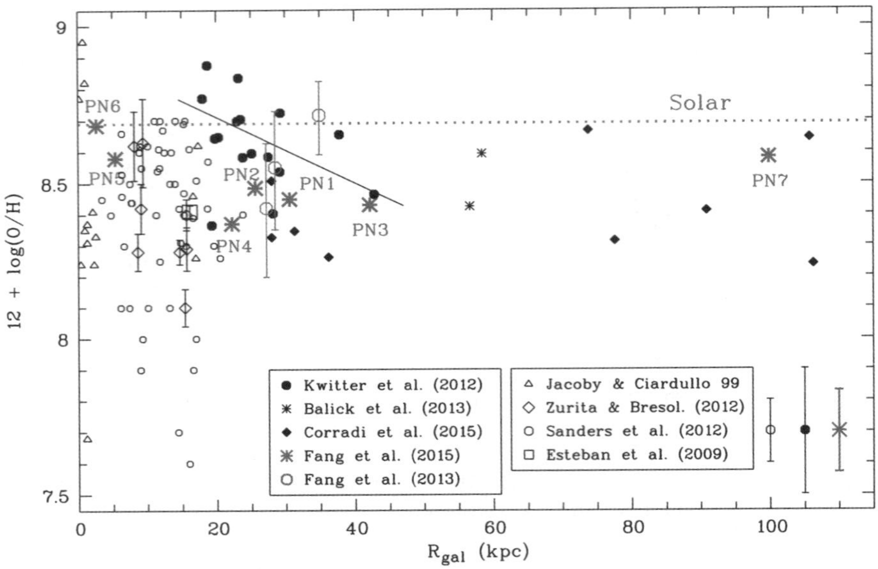

Figure 3. Oxygen abundance gradient of PNe and H II regions in M31. Our targets are red symbols, and their ID PN1–7 are the same as in Figure 1. Open diamonds are H II regions observed by Zurita & Bresolin (2012), and open square is the H II region K932 studied by Esteban *et al.* (2009). The other symbols are PNe. Explanation of the other data sets is given in the text. The black straight line is a linear fit to the disk sample of Kwitter *et al.* (2012). The horizontal red dotted line is the solar value of Asplund *et al.* (2009).

seems to have higher O/H than the other targets in our sample, its abundance uncertainty could be relatively larger due to its proximity to the galaxy center.

4. New Observations

We have obtained new GTC optical spectra of another 11 M31 PNe in 2016B: three in the Northern Spur, five associated with the Giant Stream, two in the dwarf satellite M32, and one in the outer halo. We have now 18 substructure PNe observed with the GTC. A systematic study of nebular abundances and properties of progenitor stars using photoionization modeling is underway (X. Fang *et al.*, in preparation). We well also assess whether M32 is responsible for the two substructures.

References

Asplund, M., Grevesse, N., Sauval, A.-J., & Scott, P. 2009, *ARA&A*, 47, 481
Balick, B., Kwitter, K. B., Corradi, R. L. M., & Henry, R. B. C. 2013, *ApJ*, 774, 3
Corradi, R. L. M., Kwitter, K. B., Balick, B., *et al.* 2015, *ApJ*, 807, 181
Esteban, C., Bresolin, F., Peimbert, M., *et al.* 2009, *ApJ*, 700, 654
Fang, X., García-Benito, R., Guerrero, M. A., *et al.* 2015, *ApJ*, 915, 69
Fang, X., Zhang, Y., García-Benito, R., Liu, X.-W., & Yuan, H.-B. 2013, *ApJ*, 774, 138
Ibata, R. A., Irwin, M. J., Lewis, G. F., *et al.* 2001, *Natur*, 412, 49
Jacoby, G. H. & Ciardullo, R. 1999, *ApJ*, 515, 169
Jacoby, G. H. & Ford, H. C. 1986, *ApJ*, 304, 490
Kwitter, K. B., Lehman, E. M. M., Balick, B., & Henry, R. B. C. 2012 *ApJ*, 753, 12
McConnachie, A. W., Irwin, M. J., Ibata, R. A., *et al.* 2009, *Natur*, 461, 66
Merrett, H. R., Kuijken, K., Merrifield, M. R., *et al.* 2003, *MNRAS*, 346, L62
Merrett, H. R., Merrifield, M. R., Douglas, N. G., *et al.* 2006, *MNRAS*, 369, 120
Milingo, J. B., Kwitter, K. B., Henry, R. B. C., & Souza, S. P. 2010, *ApJ*, 711, 619
Sanders, N. E., Caldwell, N., McDowell, J., & Harding, P. 2012, *ApJ*, 758, 133

White, S. D. M. 1978, *MNRAS*, 184, 185
Yuan, H.-B., Liu, X.-W., Huo, Z.-Y., *et al.* 2010, *RAA*, 10, 599
Zurita, A. & Bresolin, F. 2012, *MNRAS*, 427, 1463

Discussion

ARNABOLDI: As PI of the PN.S, I congratulate for the project to follow up PNs from Merrett's survey. Do you have plans to follow up discovery of PNs in the halo of M31?

LIU: LAMOST has hitherto surveyed a few thousand M31/M33 PNe and candidates, and obtained positive detections for over a thousand of them, including some new identifications (Yuan *et al.* 2010, RAA, 10, 599; Xiang *et al.*, this volume). Some of the GTC targets reported in this talk were selected from the LAMOST new findings.

STANGHELLINI: You found that your targets do not have enhanced He/H nor N/O, and you defined them as Type II. Could they be Type III, belonging to a thick disk/high velocity sample?

STERLING: Two questions: How many magnitudes below the bright PNLF cutoff are the PNe in your sample? Secondly, you detected C II 4267 in some targets. Did you also detect O II ORLs so that you could determine C/O ratios in some objects in your sample?

FANG: Most PNe in our sample are below the PNLF cutoff; 2) We have detected some weak features whose wavelengths agree with the O II ORLs, but low resolution (R $\sim$ 1000) of the spectra prevents a further study of the C/O ORL ratios.

Planetary Nebulae: Multi-Wavelength Probes of Stellar and Galactic Evolution
Proceedings IAU Symposium No. 323, 2016
X. Liu, L. Stanghellini & A. Karakas, eds.

doi:10.1017/S1743921317003222

The Origins of Planetary Nebulae in the Inner and Outer Disks of M 31

Bruce Balick[1], Karen Kwitter[2], Romano Corradi[3,4], Rebeca Galera Rosillo[4,5] and Richard Henry[6]

[1]Astronomy Department, University of Washington, Seattle WA, 98195-1580, USA

[2]Department of Astronomy, Williams College, Williamstown MA, 01267-2565, USA

[3]Gran Telescopio de Canarias, Center for Astrophysics in La Palma, S/C de Tenerife, E38712, Spain

[4]Departamento de Astrofisica, Universidad de La Laguna, La Laguna Tenerife, E-38206, Spain

[5]Instituto de Astrofisica de Canarias, La Laguna, Tenerife, E-38205, Spain

[6]Richard Henry, H. L. Dodge Department of Physics & Astronomy, University of Oklahoma, Norman, OK 73019, USA

Abstract. The planetary nebulae (PNe) of M 31 are receiving considerable attention as probes of its structure and chemical evolution in a galactic environment that is putatively similar to the Milky Way. We have obtained deep spectra for about 30 luminous PNe in M 31's inner disk and beyond (R_{gal} < 105 kpc). The entire ensemble of PNe exhibit O/H ~ 2/3 solar with no discernible radial gradient, in stark contrast to the H II regions of M 31. This suggests that the outer PNe in M 31 formed from a common O-rich ISM at least 5 GY ago. We infer that the outer PNe and the underlying stellar population have little common history in M 31, and that the formation of the O-rich PNe preceded any putative encounter with M 33 ~2–3 Gy ago.

Keywords. planetary nebulae: general, galaxies: abundances, galaxies: halos, ISM: evolution

1. Introduction

Undisturbed spiral galaxies exhibit a standard form consisting of stellar ensembles with distinct morphologies (bulges, disks, and halos). Presumably each of these ensembles shares distinct histories across the class of spirals. Their differences in shapes are echoed by their distinct kinematic patterns: stars in disks rotate as a whole about the symmetry axis of the disk whereas halo stars follow elongated orbits with random orientations. The stars of the nearest regular spiral galaxy, M 31, also follow this pattern, though some faint streams of stars extend in projection out to about 100 kpc (e.g., Ibata *et al.* 2001, Majewski *et al.* 2003, Ibata *et al.* 2014).

Thus we imagine that M 31 is a fairly normal spiral galaxy with a typical formation history. Its proximity facilitates observational studies of M 31's present stellar distribution and kinematics. Kinematic studies, coupled with large-scale abundance studies, allow us to glimpse the evolutionary history of M 31 in detail. Like nearly all spirals, the substantial differences in metallicity of stars in the disk and the halo of M 31 support the idea that the formation and evolutionary histories of both stellar ensembles are distinct and fairly characteristic outcomes of the evolutions of all spirals. More to the point here, large-scale studies of M 31 are especially valuable since the origins of thick disks and halos of spirals are still somewhat unclear (cf. Yoachim & Dalcanton 2005, Bernard *et al.* 2015).

Gaseous tracers are often more easily observable probes of galactic structure kinematics and abundances than are stars, especially in the inner disk of M 31 where H II regions are numerous. Most planetary nebulae ("PNe") and virtually all H II regions lie inside its photometric radius $R_{25} = 20.6$ kpc. The H II regions in M 31 display roughly solar metal abundances with mild radial gradients (Zurita & Bresolin 2012), much like other spirals of comparable size and mass (cf. Magrini *et al.* 2016 and various references within).

PNe are the only gaseous tracers of alpha-element abundances (such as O and Ne) beyond R_{25} in M 31. Their luminous emission lines ($\sim 10^4$ $L_\odot$) render them easily identifiable, useful for tracing bulk kinematics, and viable for measuring metal abundances and their large-scale gradients in spirals. In this paper we use PNe throughout M 31, including the region of the halo—and wherever stars younger than $\sim 10^{10}$y are found—to trace the abundance of O/H and surmise its chemical history.

An extensive PN identification survey of PNe in M 31 using [O III] fluxes was published by Merrett *et al.* (2006, "M+06"). In their analysis they found over 2700 PNe in M 31 whose kinematics are distinct from the stellar streams, M32, and dwarf galaxies in their fields. As expected, the PNe in this sample follow the distribution of red giants. Also the luminosity functions of the PNe follow a universal pattern at all radii where statistics can be compiled. Somewhat oddly, however, M+06 showed that most of the PNe beyond R_{25} display an ensemble pattern of kinematics similar to that of the PNe in inner disk and the HI rotation curve. A few PNe are found in this "exodisk" at even larger radii. Almost none of the entire sample have the kinematics expected of a halo population.

2. Observations & Methodology

PNe selected from the M+06 catalogue of PNe in M 31 have a complete set of observable optical nebular oxygen line fluxes that have long been used for determining O/H in the PNe of the Milky Way and the Magellanic Clouds. Accurate O/H abundances require good flux measurements of the faint [O III]λ4363 line. To this end we used 8–10-m class telescopes for PNe lying within about two magnitudes of the most luminous [O III] emitters (aka "M_* PNe"). The O/H abundances of all targets have been consistently derived using well-understood methodology (e.g., Kwitter & Henry 2001, Henry *et al.* 2004). Ionization models argue that selecting very [O III]-luminous PNe does not induce a strong bias in the outcome of O/H values.

Most of the data in this study were obtained using the 8.4-m Gemini-North Telescope (see Kwitter *et al.* 2012 for details) and the 10.4-m Gran Telescopio Canarias (see Balick *et al.* 2013 and Corradi *et al.* 2015). Nearly all of the targets are kinematic members of M 31's exodisk. However, last year we obtained spectra of an additional six PN spectra of luminosity-selected PNe in M 31 that are mostly within R_{25}. These are included here (Galera-Rosillo, 2016, this volume).

Values of O/H derived from our measurements are shown in Fig. 1. The most recent results are shown as large orange dots and earlier measurements as red and blue dots. Blue dots identify PNe that do not have disk- or exodisk-like kinematics. The solar abundance is indicated as a horizontal line. A linear gradient fitted to the red and orange dots is not robust: that is, the sign of its slope varies when any third of the data points are removed. (Even the sign of the slope may well change with new data.)

One straightforward conclusion is that the ensembles of disk and exodisk PNe in our study share an unexpectedly large O/H abundance and an insignificant radial gradient throughout M 31. Indeed, $<$[O/H]$>\sim -0.2$, far in excess of the metallicity measured in the underlying stellar population beyond R_{25}, $<$[Fe/H]$>\sim -0.7$. Given this and the fact that the kinematics of genuine halo stars and PNe in the exodisk show very different aggregate kinematic properties, we conclude that the histories of these stellar populations have little in common. This may also apply within the disk of M 31 also (Magrini *et al.* .

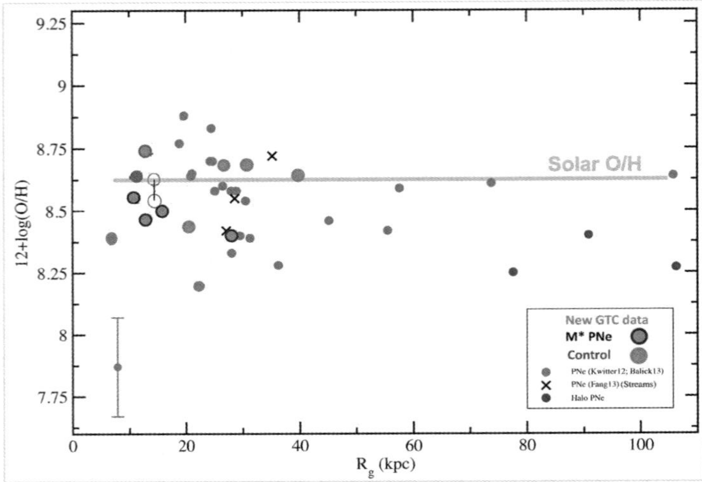

Figure 1. Our O/H measurements of PNe in M 31 as a function of deprojected disk radius (dots). See text for details.

The exodisk PNe are otherwise chemically ordinary. Their Ne/O and He/H ratios are typical of those in the Milky Way and the Magellanic clouds. One the other hand, initial masses of the progenitor stars lie well below the $\sim$4-M$_\odot$) limit for enriching nitrogen or helium (Karakas & Lattanzio). Hence the highly unusual values of O/H and Ne/H must have originated in the ISM from which the progenitor stars of the PNe formed.

We note that two PNe in the sample appear to have a high N/O abundances ($\gtrsim$0.7). However, their [NII]6584/6548 flux ratios depart substantially from 2.9 so the anomalous N/O ratios need independent confirmation.

Finally, our result of near-solar O/H for the PNe of M 31's exodisk are probably not replicated in other large spiral galaxies (Magrini *et al.* . Accordingly, the historical origin of these PNe in M 31 is likely to be different from those in other undisturbed spiral galaxies. The kinematics of PNe beyond R_{25} in other spirals is a good test this assertion.

3. Interpretation & Conclusions

To summarize, our abundance results coupled with aggregate kinematic data suggest that (with only minor exceptions) the entire ensemble of PNe presently located at $\sim 5 < R_{gal} < 100$ kpc formed from a large-scale ISM characterized by nearly solar alpha-element abundances. No trace of this ISM remains; it exists today only in the inner 10 kpc of M 31's inner disk (e.g., Zurita & Bresolin, but see also Olsen *et al.* 2006).

Hence the PN formation process may not have been steady throughout M 31's lifetime but rather occurred during some sort of an unusual global enrichment event. On the other hand, as pointed out to us privately by Harriet Dinerstein, there are other nuclear pathways for enriching oxygen. (However, such mechanisms don't apply to the PNe in other spiral galaxies, including in the Milky Way where PNe exhibit values of O/H that are similar to H II regions and red giant stars.)

We have used our ionic abundances to constrain CLOUDY models that make it possible to estimate the luminosities and temperatures of the central stars of the PNe, L_* and T_*, respectively. Some of this work is still underway. We have plotted the preliminary results on an H–R diagram with evolutionary tracks that indicate the final masses and ages of the central stars. When comparing our points to new nearby tracks of the post-AGB evolution of central stars by Miller Bertolami (2016) we find characteristic initial stellar masses lie in the range ~1.2–1.3 $M_\odot$. Hence the PNe in our study have ages $\geqslant 5$ Gy. So the ISM from which the central stars formed was probably in place well before a putative encounter between M 33 and M 31 2–3 Gy ago (McConnachie *et al.* 2009) that initiated the stellar streams threading the halo of M 31 (cf. Ibata *et al.* 2014).

The same exodisk kinematic pattern applies an extended stellar disk of red giant stars at $R_{gal} > 40$ kpc by Ibata *et al.* 2005, Chapman *et al.* 2006. Thus the population PNe in our study has a stellar counterpart, most of which exhibit [Fe/H] values closer to that expected of old halo stars: $\leqslant -0.7$ (Chapman *et al.* Richardson *et al.* Hammer *et al.* 2010).

In the preceding talk Dr. Xuan Fang reviewed O/H abundances of the M 31's PNe located in its outer stellar streams. Dr. Fang independently finds very similar O/H abundances and gradients as we find in M 31's exodisk. We surmise that although the PNe that he studied may have formed from the same metal-rich ISM as did ours, they could have acquired the kinematics of the streams during a subsequent encounter with M 33–M 31.

While the putative M 31–M 33 encounter is a convenient way to account for the kinematics of the ensemble of PN beyond R_{25}, it may not actually be relevant. Note that all but three of the exodisk PNe have relaxed into disk-like orbits. The relaxation time of a star at an orbital radius of 100 kpc is approximately its orbital time, 2 GY. That is, the kinematics of PNe were probably established prior to an encounter (and any remnants of the metal-rich ISM from which they formed may well have been stripped and ejected).

References

Balick, B., Kwitter, K. B., Corradi, R. L. M., & Henry, R. B. C. 2013, *ApJ*, 774, 3

Bernard, E. J., Ferguson, A. M. N., Richardson, J. C., *et al.* 2015, *MNRAS*, 446, 2789

Chapman, S. C., Ibata, R., Lewis, G. F., *et al.* 2006,*ApJ*, 653, 255

Corradi, R. L. M., Kwitter, K. B., Balick, B., Henry, R. B. C., & Hensley, K. 2015, *ApJ* 807, 181

Hammer, F., Yang, Y. B., Wang, J. L., *et al.* 2010, *ApJ*, 725, 542

Henry, R. B. C., Kwitter, K. B., & Balick, B. 2004, *AJ*, 127, 2284

Ibata, R., Chapman, S., Ferguson, A. M. N., *et al.* 2005, *ApJ*, 634, 287

Ibata, R., Irwin, M., Lewis, G. F., & Stolte, A. 2001, *ApJ* (Letters), 547, L133

Ibata, R. A., Lewis, G. F., McConnachie, A. W., *et al.* 2014, *ApJ*, 780, 128

Kwitter, K. B. & Henry, R. B. C. 2001, *ApJ* 562, 804

Karakas, A. I. & Lattanzio, J. C. 2014, *PASA*, 31, e030

Kwitter, K. B., Lehman, E. M. M., Balick, B., & Henry, R. B. C. 2012, *ApJ*, 753, 12

Majewski, S. R., Skrutskie, M. F., Weinberg, M. D., & Ostheimer, J. C. 2003, *ApJ*, 599, 1082

Magrini, L., Coccato, L., Stanghellini, L., Casasola, V., & Galli, D. 2016, *A&A*, 588, A91

McConnachie, A. W., Irwin, M. J., Ibata, R. A., *et al.* 2009, *Nature* 461, 66

Merrett, H. R., Merrifield, M. R., Douglas, N. G., *et al.* 2006, *MNRAS*, 369, 120

Miller Bertolami, M. M. 2016, *A&A*, 588, A25

Olsen, K. A. G., Blum, R. D., Stephens, A. W., *et al.* 2006, *AJ*, 132, 271

Richardson, J. C., Ferguson, A. M. N., Mackey, A. D., *et al.* 2009, *MNRAS*, 396, 1842

Vassiliadis, E. & Wood, P. R. 1994, *ApJS*, 92, 125

Yoachim, P. & Dalcanton, J. J. 2005, *ApJ*, 624, 701

Zurita, A. & Bresolin, F. 2012, *MNRAS*, 427, 1463

Discussion

OTSUKA: I think that the estimation of the current luminosity and core-mass of the central star and the initial mass based on the post-AGB evolution tracks depends on the shape of the adopting SED of the central star. Did you adopt the theoretical stellar atmosphere model? or blackbody SED?

BALICK: We adopt SED models from the options available within CLOUDY. Since I don't run the models I will guess that we adopt blackbody models for the emergent UV spectrum and then vary the stellar luminosity until the CLOUDY-estimated line fluxes and ratios agree with our observations.

Q: Could the halo have formed from the interaction with the dwarf galaxy which Kniazev *et al.* (2014) claim left a remnant in the halo, at M31 NE?

BALICK: Possible, but unlikely. Small (dwarf) galaxies have small metal abundances ([O/H] $\sim<$ 0.3). What's more, I would guess that the numbers of assimilated dwarfs would be unreasonably large given the numbers of known dwarfs today. In addition, these encounters might not explain the disk-likekinematics of the ensemble of PNe at large radii from M31's nucleus. Finally, it would be nice if an evolutionary scenario accounted for the presence of an [O/H] gradient that is flat and continuous between projected nuclear distances from the inner disk ($\sim$10 kpc) to the survey limit ($\sim$100 kpc).

STERLING: You mentioned that the PNe observed appear to be too old to be from the starburst 3 Gyr ago. Is this based on the estimated central star masses? If so, how did you estimate the luminosities and temperatures of the central stars for mass determinations?

BALICK: Yes, the estimate is based on estimates of L* and T* based on (1) our spectra; (2) emission-line luminosities; and (3) most of all CLOUDY models that successfully predict the observables using mass and temperature models of central star ionizing radiation.

Session 7: PNe outside of the local group
Chair: Letizia Stanghellini

Planetary Nebulae: Multi-Wavelength Probes of Stellar and Galactic Evolution
Proceedings IAU Symposium No. 323, 2016
X. Liu, L. Stanghellini & A. Karakas, eds.

doi:10.1017/S1743921317000564

Planetary nebulae as kinematic tracers of galaxy stellar halos

Lodovico Coccato

European Southern Observatory, Karl-Schwarzschild Strasse 2, D-85748 Garching, Germany.
email: lcoccato@eso.org

Abstract. The kinematic and dynamical properties of galaxy stellar halos are difficult to measure because of the faint surface brightness that characterizes these regions. Spiral galaxies can be probed using the radio HI emission; on the contrary, early-type galaxies contain less gas, therefore alternative kinematic tracers need to be used. Planetary nebulae (PNe) can be easily detected far out in the halo thanks to their bright emission lines. It is therefore possible to map the halo kinematics also in early-type galaxies, typically out to 5 effective radii or beyond. Thanks to the recent spectroscopic surveys targeting extra-galactic PNe, we can now rely on a few tens of galaxies where the kinematics of the stellar halos are measured. Here, I will review the main results obtained in this field in the last decades.

Keywords. galaxies: general; galaxies: haloes; galaxies: kinematics and dynamics; ISM Planetary Nebuale: general.

1. Introduction

The study of kinematics and chemical content in the outer regions of galaxies is extremely important to understand dynamics, dark matter content and distribution, and to get constrain on formation processes of the galaxies themselves. Indeed, the dynamical timescale of the stellar halos is comparable to the age of the galaxy (e.g., Binney & Tremaine 1987), therefore the imprints of formation processes can be still be preserved in the mass and orbital distribution and stellar population properties.

The challenge of these studies resides in the faint surface brightness level of the stellar halos. Indeed, either traditional long-slit or integral field absorption line spectroscopy require long integration times to go beyond 2 effective radii (e.g., Coccato *et al.* 2010a,b, 2011; Greene *et al.* 2012, 2015). This is particularly challenging for early-type galaxies because, contrarily to their spiral counterparts, they do not contain as much gas to be used as kinematic tracer in their outskirts.

Using planetary nebulae (PNe) as alternative tracer for the kinematics of the underlying stellar population revealed to be a winning strategy. Indeed, PNe can be detected far out in galaxy halos, thanks to their bright [O III] λ5007 emission line that outshines the stellar continuum background. The idea has been exploited by many groups, targeting a large variety of nearby galaxies including our own. The basic strategy includes the identification of the PNe candidate (e.g., by means of multi band photometry) and their spectroscopic follow-up to get the radial velocity and, eventually, emission line ratios for chemical studies. Because this two-stages approach can include a number of systematic effects (e.g. change of observing conditions between exposures), a dedicated instrument was build: Planetary Nebulae Spectrograph (PN.S, Douglas *et al.* 2002). The PN.S exploits the simultaneous counter-dispersed imaging technique, which allows to identify and obtain the radial velocity of a PNe in a single exposure.

The purpose of this review is to highlight the main results in obtained by using PNe as kinematic tracers in the last decades.

2. Measuring the distribution and kinematics of the PNe population

The first thing that needs to be done to establish if PNe are indeed useful to extend the information of stars to larger radii is to check whether or not the PNe represent the properties of the underlying stellar population in the regions where their observations overlap. This is done in two ways: by comparing the number density of PNe with the stellar surface brightness profiles (Section 2.1), and by comparing the stellar kinematics with the one derived from absorption line spectroscopy (Section 2.2).

2.1. *Spatial distribution of PNe*

The most common way to compare the PNe density distribution and stellar surface brightness is divide the galaxy into concentric annuli, count the PNe within each annuls, and compare the number density (counts divided by area) to the mean stellar surface brightness in that area. One important aspect to consider is the completeness of the sample, which can be established by populating the original image with simulated sources and re-run the PNe detection procedure. The number density of PNe needs to be corrected for incompleteness before being compared to the stellar surface brightness. The scaling factor between light and PNe counts is related to the PNe specific frequency (the so-called α parameter, Jacoby 1980); see for example Arnaboldi (2012) for a review.

2.2. *Kinematics*

Unlike absorption line spectroscopy, in which the kinematics measured at a given position *is* the luminosity weighted kinematics of all the stars along that line of sight, the velocity of a given PN represents that object only, and does not necessarily reflect the mean behaviour of the population. In other words, a single PNe velocity determination is like a "random number" extracted from the line-of-sight velocity distribution. Thus, it is necessary to proceed on a statistical way, and use the combined information of many PNe around a given area to "reconstruct" the mean kinematics (velocity and velocity dispersion) of that area. Then, one can compare the kinematics obtained from stars and PNe in the regions where both tracers are available.

Several techniques have been developed either to reconstruct the 2D velocity kinematics or to use the individual velocities in dynamical models (Romanowsky & Kochanek 2001; Amorisco & Evans 2012; Foster *et al.* 2013; Arnold *et al.* 2014). Here we focus on the "Gaussian Kernel smoothing" technique developed in Peng *et al.* (2004), with "Adaptive Kernel" as introudced in Coccato *et al.* (2009). Basically, the kinematics at each point in the field of view is obtained computing weighted moments of the velocity distribution of the entire PNe population. Each PN contributes with a weight that depends on its distance to that point. The weights dependency from distance is a Gaussian function, where its dispersion (kernel amplitude) depends on the local density of PNe. Regions with high PNe density have smaller kernel amplitude, regions with low PNe density have larger kernel amplitude. Figure 1 compares the two-dimensional map of individual PN velocities with the two-dimensional maps of velocity and velocity dispersions reconstructed using this technique.

2.3. *Comparison between stellar and PNe properties*

The comparison between stars and PNe revealed that the distribution and kinematics of PNe is in good agreement with the stellar surface brightness and kinematics in the

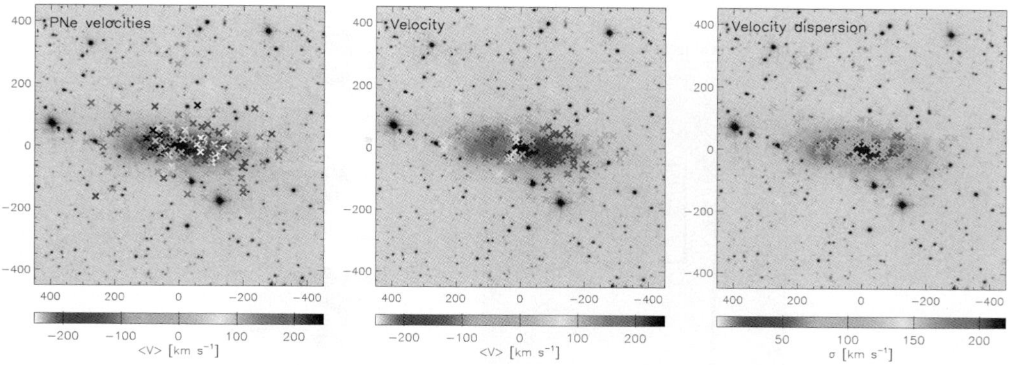

Figure 1. Recovering the kinematics of the PNe population from individual radial velocities. The left panel shows the position of the PNe (crosses) detected in NGC 1023, superimposed to the image of the field of view (data from Noordermeer *et al.* 2008). The color of each cross is proportional to the radial velocity of the PNe (minus the galaxy systemic velocity), as indicated by the color bar. The central and right panels show the values of the reconstructed velocity and velocity dispersion fields at each PNe position, using the adaptive Gaussian kernel smoothing. The spatial scale is in arcseconds.

regions where these tracers overlap. Moreover, for early-type galaxies, the PNe spatial distribution is consistent with the extrapolation of the Sérsic fit to the stellar surface brightness, rather than a de Vaucouleurs plus an exponential disk, in line with the view of "spheroidal" halos around early-type galaxies, rather than "disk-like" structures (e.g. Coccato *et al.* 2009, and references therein).

These finding have an important practical application: the combined information of the kinematics and distribution of stars (that typically probe the central effective radius) and PNe (in the outer regions) can be used to derive the entire dynamic picture of a galaxy and its halo (see Section 3.2.1).

There are, however, few exceptions to this picture. The most outstanding example is given by E6 galaxy NGC 821 (Figure 2). Its PNe system (Coccato *et al.* 2009) shows a rotation axis which is $\sim 100°$ misaligned from that of the stars (Proctor *et al.* 2009; Foster *et al.* 2016). These exceptions and, more in general, the differences between the kinematics obtained from different tracers, are seen as an indication of recent satellite accretion in the halos (Coccato *et al.* 2013; Foster *et al.* 2014; Romanowsky *et al.* 2014; Longobardi *et al.* 2015).

3. Results

3.1. *Early works*

3.1.1. *The Milky-Way*

The use of PNe radial velocities to measure the kinematic properties of a galaxy was first applied to our own Milky Way. In their pioneering work, Schneider & Terzian (1983) exploited the measured radial velocities of 252 galactic PNe to derive the galactic rotation curve beyond the solar circle. Their results showed a rising rotation curve out to ~ 15 kpc, in agreement and extending further out, the previous results obtained with O-B stars and CO clouds.

The study was extended also to the MW bulge by Durand *et al.* (1998) using 867 galactic PNe and later by Beaulieu *et al.* (2000) using 373 bulge PNe. At odd with previous results, there was no evidence for a rising disk rotation curve in these studies.

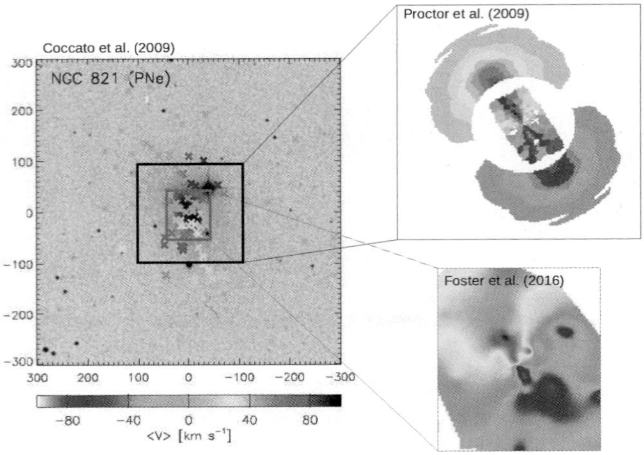

Figure 2. Comparison between the velocity field of the PNe population (left panel) compared with the velocity field of the stars (right panels). PNe data are from Coccato *et al.* (2009), stellar data are from Proctor *et al.* (2009, for the inner $\sim 200'' \times 200''$) and Foster *et al.* (2016, for the inner $\sim 100'' \times 120''$). The direction of rotation of PNe is $\sim 100°$ misaligned with respect to the direction of rotation of the and stars. The spatial scale on the left panel is in arcseconds.

The kinematic of the bulge was found to be consistent with a linearly increasing rotation velocity (Durand *et al.* 1998). The comparison with N-body models showed that the kinematics of the bulge is consistent with models where the MW bulge and bar formed from the disk via bar-forming instabilities (Beaulieu *et al.* 2000).

Nowadays, the census of galactic PNe with radial velocities includes about 2500 objects (e.g., the HASH database, Parker *et al.* 2016).

3.1.2. *The Andromeda Galaxy M31*

The next obvious candidate for these studies was our neighbor galaxy M31. Nolthenius & Ford (1987) detected and measured the radial velocities of 37 PNe within 15 and 30 kpc along the photometric major axis of M31. Despite the scatter in their measurements, they showed that the rotation curve of M31 flattens at ~ 220 km s^{-1} outside 20 kpc from the galaxy center. The first quantum step in the number of PNe detection in M31 and its satellites was done in Merrett *et al.* (2006), with the publication of a catalog of ~ 2700 sources. This catalog allowed to identify a new satellite (Andromeda VIII) and to probe the stellar distribution in M31 out to large distance from the galaxy center: they found is no indication of a cut-off in M31's disk out to four scale lengths, and no signs of a spheroidal halo population in excess of the bulge out to 10 effective bulge radii.

3.2. *Early-Types*

The use of PNe radial velocity measurements has revealed the following kinematic properties for the halos of early-type galaxies (e.g., Douglas *et al.* 2007; Coccato *et al.* 2009; Teodorescu *et al.* 2010; McNeil *et al.* 2010, and references therein):

- The velocity dispersion and $V_{\rm RMS}$ have a large variety of radial profiles, including flat profiles (generally associated to more massive galaxies), declining profiles (generally associated to less massive ellipticals), and even rising profiles (generally associated to lenticulars with a prominent stellar halo, or disturbed systems).
- Kinematic twists and misalignment in the velocity fields are more frequent at large radii than what observed in the inner regions. Twists are also observed in some fast rotator galaxies.

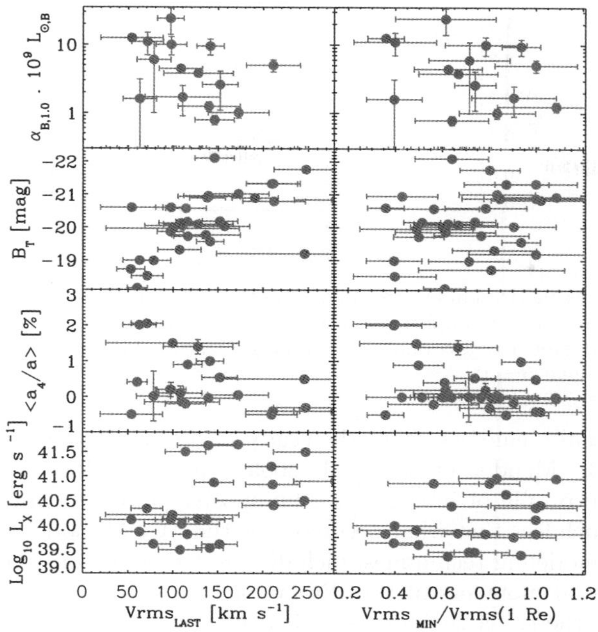

Figure 3. Correlations between the halo kinematics and other galaxy properties (updated from Coccato *et al.* 2009). The halo kinematics is parametrized with the outermost value of $V_{\rm rms}$ (left) and its normalization at 1 effective radius (right), as obtained from PNe radial velocities. Each point represents a galaxy. From top to bottom, the panels indicate: PNe specific frequency, total B-band luminosity, shape parameter, and total X-ray luminosity. Galaxies that are classified as fast rotators are shown in blue, slow rotators in red.

• Outer halos are characterized by more complex radial profiles of the specific angular momentum-related $\lambda(R)$ parameter than observed within one effective radius. Some galaxies are more rotational dominated at large radii than in their central parts (i.e., the $\lambda(R)$ profile increase in the halo), some other are pressure supported at large radii (i.e., the $\lambda(R)$ profile decreases in the halo).

• The halo kinematics are correlated with other galaxy properties, such as total B-band and X-ray luminosity, isophotal shape, total stellar mass, V/σ and PNe specific frequency, with a clear separation between fast and slow rotators (Figure 3).

• That the halos of massive ellipticals sometimes host kinematic substructures, interpreted as relics of accretion events. Kinematic substructures are also found in the halos of spiral galaxies (e.g., Herrmann *et al.* 2009).

3.2.1. *Mass distribution in Early-Type galaxies: "lost and found" Dark Matter*

The main science driver for studying the kinematics of PNe in Early-Type galaxies is to use them as tracer for the gravitational potential in order to infer the mass distribution and the fraction of dark matter.

Initial works that included a significant amount of PNe in the analysis were carried on massive nearby ellipticals, such as NGC 1316 (Arnaboldi *et al.* 1998), NGC 5128 (Hui *et al.* 1995; Peng *et al.* 2004), and more recently on NGC 4649 (Teodorescu *et al.* 2011) - just to cite few examples. The signature of the presence of dark matter was clearly present in the kinematic profile at large radii. Other works led by several team that combined the information of other tracers to the PNe, reached similar results (e.g., Napolitano *et al.* 2005; Deason *et al.* 2012; Forbes *et al.* 2016).

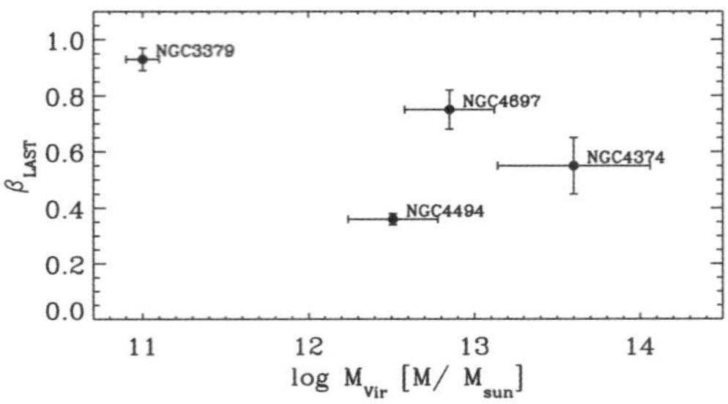

Figure 4. Trend between virial mass (from Forbes *et al.* 2017, Table 1) and anisotropy parameter β (from Fig. 23 in Morganti *et al.* 2013, and from Fig. A2 in Napolitano *et al.* 2011). The value of β refers to the outermost determination. Radially biased orbits are those with high values of β; isotropic models have $\beta = 0$.

Controversial results emerged from the study of ordinary ellipticals. Studies by Ciardullo *et al.* (1993); Méndez *et al.* (2001, 2009); Romanowsky *et al.* (2003) indicated that the kinematic profile of intermediate luminosity ellipticals are consistent with no dark matter, or much less that what predicted by the ΛCDM model. The different teams adopted different modeling techniques, including isotropic or constant anisotropic Jeans models fitting several moments of the PNe velocity distribution, which lead to results consistent to each others. These findings challenged not only the view that *all* galaxies are embedded in massive halos, but also alternative theories such as MOND (Milgrom 1983), that were not able to reconcile the luminous mass in these galaxies with the observed decreasing velocity dispersion profiles.

The solution was found by assuming that the stellar orbits are anisotropic, and that their anisotropy varies with radius (becoming more radial outward). This reconciled the observed dispersion profiles with MOND (Milgrom & Sanders 2003). Moreover the increase of radial anisotropy at large radii is consistent with the numerical simulations of disk-galaxy mergers in a ΛCDM universe (Dekel *et al.* 2005).

Following this idea, more elaborated dynamical models of PNe datasets revealed that the inclusion of radial variation in the anisotropy profile was indeed able to reconcile the dark matter content in low-luminous ellipticals with the predictions of ΛCDM, altough a certan amount of degeneracy between mass, shape, and anisotropy exists for nearly spherical systems (e.g., de Lorenzi *et al.* 2009). In addition, as indicated in Figure 4, the new models suggest that lower mass halos have stars in orbits that have higher radial anisotropy than the stars in high mass halos (Morganti *et al.* 2013, Napolitano *et al.* 2011, Napolitano, priv. comm.), in line with what found also using other tracers (e.g., Zhu *et al.* 2016).

3.3. *Lenticulars*

Lenticular galaxies are mainly composed by a spheroid and a disk without spiral arms. Therefore, it is reasonable to assume that some PNe are associated to the spheroid and some to the disk. The association of a PNe to one component or the other can be done by combining the surface brightness spheroid/disk decomposition with PNe kinematic data, assuming a non rotating model for the spheroid component (Cortesi *et al.* 2011). One application is to recover the disk kinematics by removing the contamination from the PNe associated to the spheroid in order to study scaling relations such as the Tully-Fisher at large radii. Cortesi *et al.* (2013) found that: i) the disks of S0s have higher velocity dispersion than the spirals with the same rotational velocity; and that ii) S0s lie around one magnitude below the Tully-Fisher relation for spiral galaxies, i.e., disks of S0s have higher circular velocity that those of spirals of the same luminosity. This was interpreted by the authors as an indication of a mild harassment process that that heated disks of spirals and transformed them into S0s.

3.4. *Spirals*

Altough spiral galaxies have gas that can be used to probe their kinematics and mass distribution at large radii, PNe are still useful to probe the vertical stellar velocity dispersion (σ_Z) in nearly face-on systems. Herrmann & Ciardullo (2009) studied a sample of five nearby, low-inclination spiral galaxies. They found that: i) 4/5 spirals have constant mass-to-light ratio out to ~ 3 optical scale lengths (h_Z); ii) outside $\sim 3h_Z$, σ_Z becomes flat with radius; iii) the disks of early type spirals have higher values of mass-to-light ratio and are closer to maximal than the disks of later-type spirals; iv) the inner halos of spirals are better fit by pseudo-isothermal laws than by NFW models.

The PNe in spiral galaxies were also used to measure the stellar migration in stellar disks to investigate its influence on the variation of metallicity gradients in stellar disks with time (Magrini *et al.* 2016).

4. Summary

In this review I have illustrated the main findings obtained by using planetary nebulae as kinematic tracers in galaxy halos, sometimes persuing an historical approach. The main lessons that we can learn in these last decades on the topics are:

• In the regions were PNe and absorption line data overlap, PNe well trace the surface brightness and kinematics of stars (with few exceptions). Therefore, PNe can be used to measure the properties of galaxy halos by extending the information of stellar kinematics to large radii, and to unveil signatures of accretions events also in galaxies with a relatively undisturbed morphology.

• Variable radial anisotropy needs to be considered in dynamical models in order to obtain a reliable estimate of the dark matter content and distribution in the halos. There is the indication lower mass halos have stars in orbits that have higher radial anisotropy than the stars in high mass halos.

• The properties of the PNe population in spiral galaxies suggest that the disks of early type spirals have higher values of mass-to-light ratio and are closer to maximal than the disks of later-type spirals.

• The study of PNe kinematics in S0 galaxies revealed that diks of S0s have higher velocity dispersion that those of spirals, and that S0s lie one magnitude below the Tully-Fisher relation with respect to spiral galaxies of the same luminosity.

References

Amorisco, N. C. & Evans, N. W. 2012, *MNRAS*, 424, 1899
Arnaboldi, M. 2012, in *IAU Symposium*, Vol. 283, IAU Symposium, 267–274
Arnaboldi, M., Freeman, K. C., Gerhard, O., *et al.* 1998, *ApJ*, 507, 759
Arnold, J. A., Romanowsky, A. J., Brodie, J. P., *et al.* 2014, *ApJ*, 791, 80
Beaulieu, S. F., Freeman, K. C., Kalnajs, A. J., Saha, P., & Zhao, H. 2000, *AJ*, 120, 855
Binney, J. & Tremaine, S. 1987, Galactic dynamics
Ciardullo, R., Jacoby, G. H., & Dejonghe, H. B. 1993, *ApJ*, 414, 454
Coccato, L., Arnaboldi, M., & Gerhard, O. 2013, *MNRAS*, 436, 1322
Coccato, L., Arnaboldi, M., Gerhard, O., *et al.* 2010a, *A&A*, 519, A95
Coccato, L., Gerhard, O., & Arnaboldi, M. 2010b, *MNRAS*, 407, L26
Coccato, L., Gerhard, O., Arnaboldi, M., *et al.* 2009, *MNRAS*, 394, 1249

Coccato, L., Gerhard, O., Arnaboldi, M., & Ventimiglia, G. 2011, *A&A*, 533, A138
Cortesi, A., Merrifield, M. R., Arnaboldi, M., *et al.* 2011, *MNRAS*, 414, 642
Cortesi, A., Merrifield, M. R., Coccato, L., *et al.* 2013, *MNRAS*, 432, 1010
de Lorenzi, F., Gerhard, O., Coccato, L., *et al.* 2009, *MNRAS*, 395, 76
Deason, A. J., Belokurov, V., Evans, N. W., & McCarthy, I. G. 2012, *ApJ*, 748, 2
Dekel, A., Stoehr, F., Mamon, G. A., *et al.* 2005, *Nature*, 437, 707
Douglas, N. G., Arnaboldi, M., Freeman, K. C., *et al.* 2002, *PASP*, 114, 1234
Douglas, N. G., Napolitano, N. R., Romanowsky, A. J., *et al.* 2007, *ApJ*, 664, 257
Durand, S., Acker, A., & Zijlstra, A. 1998, *A&AS*, 132, 13
Forbes, D. A., Alabi, A., Romanowsky, A. J., *et al.* 2016, *MNRAS*, 458, L44
Forbes, D. A., Alabi, A., Romanowsky, A. J., *et al.* 2017, *MNRAS*, 464, L26
Foster, C., Arnold, J. A., Forbes, D. A., *et al.* 2013, *MNRAS*, 435, 3587
Foster, C., Lux, H., Romanowsky, A. J., *et al.* 2014, *MNRAS*, 442, 3544
Foster, C., Pastorello, N., Roediger, J., *et al.* 2016, *MNRAS*, 457, 147
Greene, J. E., Janish, R., Ma, C.-P., *et al.* 2015, *ApJ*, 807, 11
Greene, J. E., Murphy, J. D., Comerford, J. M., Gebhardt, K., & Adams, J. J. 2012, *ApJ*, 750, 32
Herrmann, K. A. & Ciardullo, R. 2009, *ApJ*, 705, 1686
Herrmann, K. A., Ciardullo, R., & Sigurdsson, S. 2009, *ApJ*, 693, L19
Hui, X., Ford, H. C., Freeman, K. C., & Dopita, M. A. 1995, *ApJ*, 449, 592
Jacoby, G. H. 1980, *ApJS*, 42, 1
Longobardi, A., Arnaboldi, M., Gerhard, O., & Mihos, J. C. 2015, *A&A*, 579, L3
Magrini, L., Coccato, L., Stanghellini, L., Casasola, V., & Galli, D. 2016, *A&A*, 588, A91
McNeil, E. K., Arnaboldi, M., Freeman, K. C., *et al.* 2010, *A&A*, 518, A44
Méndez, R. H., Riffeser, A., Kudritzki, R.-P., *et al.* 2001, *ApJ*, 563, 135
Méndez, R. H., Teodorescu, A. M., Kudritzki, R.-P., & Burkert, A. 2009, *ApJ*, 691, 228
Merrett, H. R., Merrifield, M. R., Douglas, N. G., *et al.* 2006, *MNRAS*, 369, 120
Milgrom, M. 1983, *ApJ*, 270, 365
Milgrom, M. & Sanders, R. H. 2003, *ApJ*, 599, L25
Morganti, L., Gerhard, O., Coccato, L., Martinez-Valpuesta, I., & Arnaboldi, M. 2013, *MNRAS*, 431, 3570
Napolitano, N. R., Capaccioli, M., Romanowsky, A. J., *et al.* 2005, *MNRAS*, 357, 691
Napolitano, N. R., Romanowsky, A. J., Capaccioli, M., *et al.* 2011, *MNRAS*, 411, 2035
Nolthenius, R. & Ford, H. C. 1987, *ApJ*, 317, 62
Noordermeer, E., Merrifield, M. R., Coccato, L., *et al.* 2008, *MNRAS*, 384, 943
Parker, Q. A., Bojičić, I. S., & Frew, D. J. 2016, *Journal of Physics Conference Series*, 728, 032008
Peng, E. W., Ford, H. C., & Freeman, K. C. 2004, *ApJ*, 602, 685
Proctor, R. N., Forbes, D. A., Romanowsky, A. J., *et al.* 2009, *MNRAS*, 398, 91
Romanowsky, A. J., Arnold, J. A., Brodie, J. P., *et al.* 2014, in Astronomical Society of the Pacific Conference Series, Vol. 486, Multi-Spin Galaxies, ASP Conference Series, ed. E. Iodice & E. M. Corsini, 169
Romanowsky, A. J., Douglas, N. G., Arnaboldi, M., *et al.* 2003, *Science*, 301, 1696
Romanowsky, A. J. & Kochanek, C. S. 2001, *ApJ*, 553, 722
Schneider, S. E. & Terzian, Y. 1983, *ApJ*, 274, L61
Teodorescu, A. M., Méndez, R. H., Bernardi, F., Riffeser, A., & Kudritzki, R. P. 2010, *ApJ*, 721, 369
Teodorescu, A. M., Méndez, R. H., Bernardi, F., *et al.* 2011, *ApJ*, 736, 65
Zhu, L., Romanowsky, A. J., van de Ven, G., *et al.* 2016, *MNRAS*, 462, 4001

Discussion

HUANG: Did you correct the assymetric drift when deriving the rotation curve using PNe as tracers?

Planetary Nebulae: Multi-Wavelength Probes of Stellar and Galactic Evolution
Proceedings IAU Symposium No. 323, 2016
X. Liu, L. Stanghellini & A. Karakas, eds.

doi:10.1017/S1743921317000771

The PN.S ETG survey

M. Arnaboldi[1,2], C. Pulsoni[3], O. Gerhard[3] and PN.S Consortium

[1]European Southern Observatory, K. Schwarzschild str. 2, 85748 Garching, Germany
email: marnabol@eso.org

[2]INAF - Astrophysical Observatory of Turin, Strada Osservatorio 20, 10025, Pino Torinese, Italy

[3]Max-Planck-Institut für Extraterrestrische Physik, Giessenbachstrasse, 85748 Garching, Germany
email: pulsoni@mpe.mpg.de; gerhard@mpe.mpg.de

Abstract. We have analyzed the velocity fields in the halos of thirty-three early-type galaxies (ETGs) with planetary nebulae (PNs) as tracers, reaching radii of up to $\sim 8R_e$. The sample comprises twenty-five galaxies from the Planetary Nebulae Spectrograph (PN.S) ETG survey and eight further galaxies with extended PN velocity data from the literature and new Counter-Dispersed imaging observations. The catalogues from these thirty-three ETGs provide astrometric positions and heliocentric line-of-sight velocities for a total of 8354 extragalactic PNs. All these catalogues are treated homogeneously for the identification of kinematic outliers and the separation between main galaxy/satellites in each galaxy field-of-view. We discuss the rotation velocities, velocity dispersion profiles and radial trends of the specific angular momentum, separating between slow rotators and fast rotators. We compare the properties of the $V/\sigma(R)$ and $\sigma(R)$ profiles with predictions of 2D velocity fields from hydro-dynamical cosmological simulations. We briefly mention the possible origin of the quasi-Keplerian steeply decreasing profile galaxies which encompasses about one fourth of the current sample.

Keywords. ISM: planetary nebulae: general. Galaxies: elliptical and lenticular, cD; evolution; formation, halos, kinematics and dynamics

---

1. Introduction

The outer halos of bright early-type galaxies are important sites for testing the late mass assembly process predicted by the ΛCDM hierarchical model. According to the two-phase formation scenario, an initial fast assembly stage is dominated by dissipative processes and rapid star formation, fueled either by infall of cold gas or by major mergers. It is then followed by minor merger episodes that increase the size of the galaxy halos in an efficient way (Oser *et al.* 2010, Hirschmann *et al.* 2012). The late assembly of the outer halos is imprinted in distinctive signatures of the spatial distributions of the accreted halo stars, and their kinematics. The numerical simulations from Abadi *et al.* (2006) show that while the dark matter particles have a mild radial anisotropy, the orbits of accreted stars are strongly radially biased, with radial anisotropy monotonically increasing with radius. Recent hydro-dynamical simulations of galaxy assembly show that the relative fraction of *in-situ* stars and accreted stars in a galaxy varies with the galaxy mass (Rodriguez-Gomez *et al.* 2016), and that these two components have steeper/shallower radial gradients of their surface brightness profiles (Abadi *et al.* 2006, Pillepich *et al.* 2014). Thus the radial trends of the spatial distribution their substructures are valuable observables that can be used to constrain the accretion histories of galaxies. Furthermore, kinematics observations of halo tracers can be used to constrain the orbital distribution and dynamical mass in the systems.

Probing the spatial distribution of stars and their motions at distances larger than two effective radii (the effective radius, or R_e, is the radius that contains half of the total light) is challenging as the surface brightness of the halos becomes only a few percent of the sky brightness. This is particularly difficult for absorption line measurements, given the required high signal-to-noise ratio for the determination of the high order moments of the line-of-sight velocity distribution (LOSVD) that are sensitive to the stellar orbits (e.g. Gerhard 1993). This is resolved by using stellar probes that overcome the limit of steeply decreasing surface brightness profiles, such as globular clusters (GCs, e.g. Brodie *et al.* 2014) and Planetary Nebulae (PNs).

PNs are the late phases of Sun-like stars in the mass range $8-1$ $M_\odot$. In old stellar populations like those found in the halos of early-type galaxies, one star every $10^6-10^7 L_\odot$ is expected to be in such phase (Buzzoni *et al.* 2006). PNs are established probes of their parent stars because their number scales with the total bolometric luminosity of the associated stellar population, via the luminosity specific PN number, or α parameter for short, whereby $N_{PN} = \alpha L_{Bol,gal}$. In normal ETGs, the number of PNs trace the light associated with the stellar component in a galaxy (Coccato *et al.* 2009; Cortesi *et al.* 2013; Coccato 2017, this volume). PNs are easily identified in narrow band imaging surveys (Arnaboldi *et al.* 2002; Longobardi *et al.* 2013), as the nebula surrounding the central core star (a hot white dwarf) is very efficient in re-emitting up to 15% of the UV-continuum in the optical green line of the [OIII] doublet, at 5007Å (Dopita *et al.* 1992). Once the PN candidates are identified, the follow-up spectroscopy provides the line-of-sight velocities (LOSVs) of these individual stars with accuracy down to a few kms^{-1} (Arnaboldi *et al.* 2008, Longobardi *et al.* 2015). The PN LOSVDs as function of radius, the 2D velocity fields and the point-symmetry properties of the velocity fields, as well as the specific angular momentum $\lambda(R)$ profiles, can be measured for stellar halos for a wide range of galaxy luminosities and masses, whereas other kinematic probes (such as globular clusters X-rays and lensing measurements) are biased towards the massive galaxies.

A recent development is that in cD galaxies, at radii larger than 70 kpc, a transition is seen from galaxy to intracluster light, signalled by the measurement of a *PN-scarce* vs. a *PN-rich* outer stellar population (e.g M87, Longobardi *et al.* 2013) and M49, Hartke *et al.* 2017, this volume).

Because the detailed properties of individual stellar halos are the results of a stochastic process (Rodriguez-Gomez *et al.* 2016), information on the accretion is to be looked for in the different spatial and velocity distributions of the discrete tracers. In the case of GCs and PNs, their specific luminosity frequencies depend on total luminosities, absolute magnitudes and morphological types of their galaxy hosts, such that GCs frequencies are largest for the smallest and most massive galaxies, while the PN frequency is largest for L^* galaxies (Coccato *et al.* 2013). A vivid example is provided by the halo of M31, the giant Andromeda galaxy. The map of globular clusters in the M31 halo overlaid on the surface density map of metal poor stars from the PANDAS survey shows that GCs are prominent on streams at very large radii ($R > 50$ kpc) but do not trace the bulk of the light in the brighter inner halo, see Veljianoski *et al.* (2014). This complementarity is reminiscent of spatial segregation that arises in hydro-dynamical simulations, see for example Murante *et al.* (2007) and also recent computations from Amorisco (2017). In these accretion histories, the star particles that come from merging of massive progenitors are found preferentially at smaller radii, while star particles that come into the system from stripping events of low mass progenitors are found at larger radii. Thus different accretion histories (i.e. stochasticity) and progenitors may lead to different observed velocity fields and spatial distribution of tracers for galaxies of the same stellar mass.

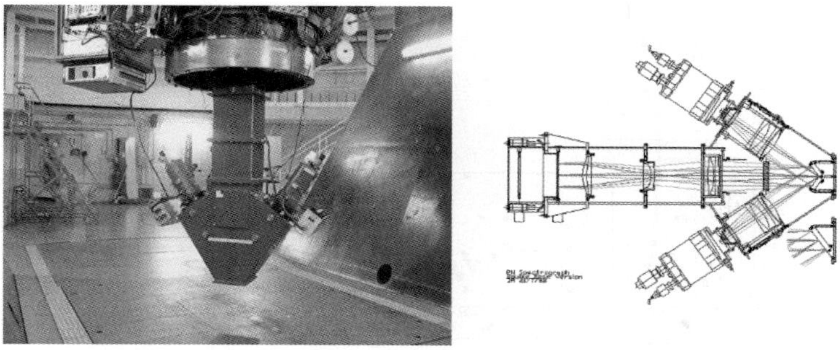

Figure 1. Left: the PN.S mounted at the Cassegrain Focus of the WHT on La Palma Island, Spain. Right: optical layout of the PN.S, indicating the dichroic (left), the narrow band filter (middle) and the concave grating (top right) that splits the pupil image into two - Right and Left - arms.

The purpose of the Planetary Nebulae Spectrograph (PN.S) ETG project is to map the spatial, velocity and angular momentum distribution for stellar halos in ETGs, using PNs as tracers. In what follows we present a short summary of this survey which is now the most extended survey of extragalactic PNs in ETGs to date.

2. The PN.S survey of ETGs

The PN.S ETGs survey is based on data collected with the PN.S mounted at the William Herschel Telescope (WHT) on La Palma, see an image of the PN.S instrument in Fig. 1. The PN.S is a custom built instrument designed for counter dispersed imaging (Douglas *et al.* 2002), that allows the identification of the [OIII] λ 5007 Å monochromatic emission and the measurement of its Doppler shifted observed wavelength in a single observing run. The PN.S works with a pair of images through a single narrow-band filter and dispersed in opposite directions by a concave grating element, as shown by the optical layout in Fig. 1. Despite a slightly lower optical efficiency than direct imaging (because of the use of gratings) the PN.S achieves equally good results because the images can be obtained simultaneously and are thus less influenced by changes in photometric conditions. A PN identification consists of two 3σ events in the same row of both detectors, and there is more signal redundancy in a pair of matching peaks than in a single peak of the same significance. All the observing time spent on the galaxy field-of-view contributes both to the identification of PNs in a given galaxy field and to the spectroscopy to measure the Doppler shift of the [OIII] λ 5007 Å emission.

The goal of the PN.S ETG survey (P.I. Magda Arnaboldi) is to measure the kinematics, dynamics, angular momentum, and mass in ETG halos in the nearby Universe, within a distance of $D \leqslant 25$ Mpc. The range of structural parameters, such as luminosity, central velocity dispersion, ellipticity and isophotes shapes are shown in Fig. 2. Twenty-five ETGs were observed with the PN.S; additional two galaxies were observed with new Counter Dispersed Imaging (CDI) with FORS2@VLT, and six PN catalogs were collected from the literature. All catalogs were uniformly validated for completeness, removal of background galaxy emissions and kinematical outliers (Arnaboldi *et al.* in prep.). The catalogs were then processed to identify the contribution from nearby satellites using the probability membership method from McNeil-Moylan *et al.* (2012). In summary, we were able to measure astrometry and LOSVs for nearly 80 to 700 PNs in each sample galaxy, typically covering radii out to 5 to 8 R_e, giving a total detected sample of 8354 extragalactic PNs.

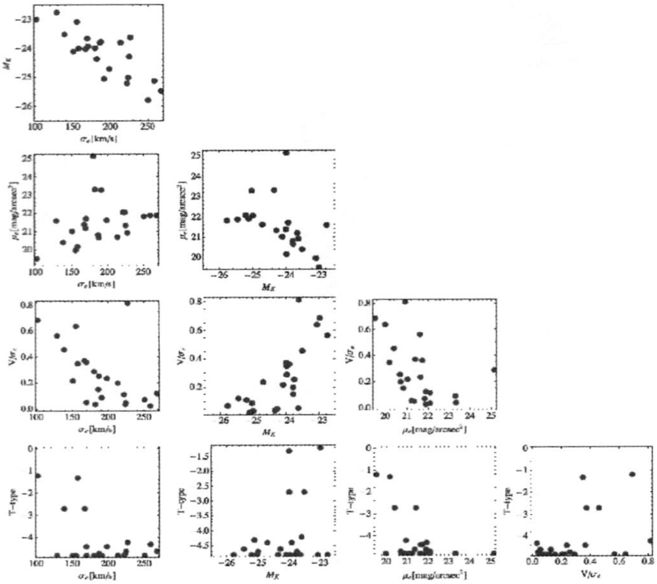

Figure 2. Structural parameters, i.e. luminosity, central velocity dispersion, ellipticity and isophote shapes, for the magnitude limited sample of galaxies covered by the PN.S ETG survey.

In Pulsoni *et al.*, in prep., we obtained the extended velocity fields and assessed the symmetry properties of the LOSVDs in the outer halos. A few galaxies for which the most radially extended and statistical significant samples are available show 2D velocity fields that are not point-symmetric, e.g. NGC 5128. Such properties indicate that the stellar kinematics in the outer regions are out of dynamical equilibrium, because of recent merger events. When the 2D velocity field is compatible with point symmetry, the PN catalog can be folded, thus doubling the number of tracers. The smoothed 2D maps of the first (i.e. average velocity $v(R)$) and second moment (i.e. velocity dispersion $\sigma(R)$) of the LOSVD are then compared with the simulated maps of $v/\sigma(R)$ and $\sigma(R)$ out to $5R_e$ for ETGs from Wu *et al.* (2014). We briefly summarize the results of the comparison in turn.

• The measured halo v/σ at $5R_e$ for the ETGs in the PN.S survey correlate with those within one R_e for most of the sample galaxies. Thus the division between slow and fast rotators in the outer halos is similar as in the cores, as measured by the ATLAS^{3D} project (Cappellari *et al.* 2011). There are some more complicated cases though, where single galaxies cross the separating line in specific angular momentum λ_R (Emsellem *et al.* 2007), as we go from the cores to the outer halos (Pulsoni *et al.* 2017).

• The $\sigma(R)/\sigma(R_e)$ vs. R/R_e profiles for the ETGs in the PN.S survey show quasi-Keplerian steeply falling profiles for about one fourth of the galaxies in the sample. Interestingly these profiles are not measured for the simulated halos of ETGs in Wu *et al.* (2014). The open questions about the nature of quasi-Keplerian ETGs are: are they face-on disks of fast rotators? Is the quasi-Keplerian behavior signaling low-mass halos or strong radial anisotropy? As shown by de Lorenzi *et al.* (2008), de Lorenzi *et al.* (2009) and Morganti *et al.* (2013), quasi-Keplerian profiles are compatible with ΛCDM dark matter halos, and a range of dark matter fractions and anisotropy profiles. For galaxies with decreasing $\sigma(R)$ profiles, there is a strong degeneracy in dynamical models for the mass determination. Current work is in progress for a Jeans-Kurtosis analysis of the spherical ETGs (Napolitano *et al.* 2017, in prep.).

3. Summary

In this contribution we briefly covered the following observational results:

• GCs and PNe trace a complementary mix of progenitor systems.

• Outer halos are dynamically young structures. Some may even be out of equilibrium and violate point symmetry.

• Angular momentum in the halo and within one R_e are correlated, but some transition cases exist, which cross the fast-rotator/slow-rotator division.

• Halo orbits are mildly to strongly radially biased. This may be a record of radial infall orbits.

• Velocity-dispersion dichotomy: what is the nature of the quasi- Keplerian ETGs?
Future projects for PN studies include using PN tracers for measuring Oxygen abundance, PNLF morphologies and their relations to parent stellar populations, in addition to the kinematical mapping of the outer halos.

4. Acknowledgments

The authors would like to acknowledge the contribution and support from Dr Nigel Douglas, first P.I. of the PN.S. We would like to thank the staff at the WHT for support during the observations, and the SOC of the IAU symposium 323 for the opportunity of presenting the results from the completed PN.S ETGs survey.

References

Abadi, M., Navarro, J. F., & Steinmetz, M. 2006, *MNRAS*, 365, 747
Amorisco, N. 2017, *MNRAS*, 464, 2882
Arnaboldi, M. *et al.* 2002, *AJ*, 123, 760
Arnaboldi, M. *et al.* 2008, *ApJL*, 674, 17
Brodie, J. *et al.* 2014, *ApJ*, 796, 52
Buzzoni, A., Arnaboldi, M., & Corradi, R. 2006, *MNRAS*, 368, 877
Cappellari, M. *et al.* 2011, *MNRAS*, 413, 813
Coccato, L. *et al.* 2009, *MNRAS*, 394, 1249
Coccato, L. *et al.* 2013, *MNRAS*, 436, 1322
Cortesi, A. *et al.* 2013, *A&A*, 549, 115
de Lorenzi, F. *et al.* 2008 *MNRAS*, 385, 1729
de Lorenzi, F. *et al.* 2009 *MNRAS*, 395, 76
Dopita, M. A., Jacoby, G. H., & Vassiliadis, E. 1992, *ApJ*, 389, 27
Douglas, N. *et al.* 2002, *PASP*, 114, 1234
Emsellem, E. *et al.* 2007, *MNRAS*, 379, 401
Gerhard, O. 1993, *MNRAS*, 265, 213
Hirschmann, M. *et al.* 2012, *MNRAS*, 419, 3200
Longobardi, A. *et al.* 2013, *A&A*, 558, 42
Longobardi, A. *et al.* 2015, *A&A*, 579, 135
McNeil-Moylan, E. K. *et al.* 2012 *A&A*, 539, 11
Morganti, L. *et al.* 2013, *MNRAS*, 431, 3570
Murante, G. *et al.* 2007, *MNRAS* 377, 2
Oser, L. *et al.* 2010, *ApJ*, 725, 2312
Pillepich, A. *et al.* 2014, *ApJ*, 784, 161
Rodriguez-Gomez, V. *et al.* 2016, *MNRAS*, 458, 2371
Veljianoski, J. *et al.* 2014, *MNRAS*, 441, 2929
Wu, X. *et al.* 2014, *MNRAS*, 438, 2701

Planetary Nebulae: Multi-Wavelength Probes of Stellar and Galactic Evolution
Proceedings IAU Symposium No. 323, 2016
X. Liu, L. Stanghellini & A. Karakas, eds.

doi:10.1017/S1743921317001144

Resolving the Disk-Halo Degeneracy using Planetary Nebulae

S. Aniyan[1], K. C. Freeman[1], M. Arnaboldi[2], O. Gerhard[3], L. Coccato[2], M. Fabricius[3], K. Kuijken[4] and M. Merrifield[5]

[1]Research School of Astronomy & Astrophysics, Australian National University
email: suryashree.aniyan@anu.edu.au

[2]European Southern Observatory, Garching
[3]Max-Planck-Institut fur Extraterrestrische Physik, Garching
[4]Leiden Observatory, Leiden University
[5]School of Physics and Astronomy, University of Nottingham

Abstract. The decomposition of the 21 cm rotation curve of galaxies into contribution from the disk and dark halo depends on the adopted mass to light ratio (M/L) of the disk. Given the vertical velocity dispersion (σ_z) of stars in the disk and its scale height (h_z), the disk surface density and hence the M/L can be estimated. Earlier works have used this technique to conclude that galaxy disks are submaximal. Here we address an important conceptual problem: star-forming spirals have an old (kinematically hot) disk population and a young cold disk population. Both of these populations contribute to the integrated light spectra from which σ_z is measured. The measured scale height h_z is for the old disk population. In the Jeans equation, σ_z and h_z must pertain to the same population. We have developed techniques to extract the velocity dispersion of the old disk from integrated light spectra and from samples of planetary nebulae. We present the analysis of the disk kinematics of the galaxy NGC 628 using IFU data in the inner regions and planetary nebulae as tracers in the outer regions of the disk. We demonstrate that using the scale height of the old thin disk with the vertical velocity dispersion of the same population, traced by PNe, results in a maximal disk for NGC 628. Our analysis concludes that previous studies underestimate the disk surface mass density by ~ 2, sufficient to make a maximal disk for NGC 628 appear like a submaximal disk.

Keywords. galaxies: kinematics and dynamics, galaxies: spiral, galaxies: halos

1. Introduction

Decomposing the rotation curve of spiral galaxies into contribution from the disk and the dark halo provides a direct technique for studying the structure of the dark halos. The decomposition depends however on the adopted M/L of the stellar disk (van Albada *et al.* 1985) which is still uncertain. Adopting different M/L values can result in a maximal† or a submaximal disk, both of which can fit the observed rotation curves equally well (but with very different dark halos). The M/L value is therefore critical to obtaining the parameters of the dark halos, such as their scale densities and scale lengths. These quantities are cosmologically significant, because the densities and scale radii of dark halos follow well-defined scaling laws and can be used to measure the redshift of assembly of halos of different masses (Kormendy and Freeman 2016).

Several techniques have been used to break the disk-halo degeneracy but they all present challenges. One of the more direct methods to break this degeneracy uses the

† A maximal disk has the maximum M/L value consistent with the observed rotation curve and a non-hollow dark halo. Typically, the disk of a maximal disk provides about 85% of the rotational velocity at the peak of the rotation curve (Sackett 1997).

vertical velocity dispersion of tracers in the disks to measure the surface mass density of the disk (e.g. Bottema 1997). Using the 1D Jeans equation in the vertical direction, the vertical luminosity-weighted velocity dispersion σ_z and the vertical exponential disk scale height h_z together give the surface mass density Σ of the disk via the relation:

$$\Sigma = f\sigma_z^2/Gh_z \tag{1.1}$$

where G is the gravitational constant and f is a geometric factor, known as the vertical structure constant, that depends weakly on the adopted vertical structure of the disk. The surface brightness of the disk and the surface mass density together give the M/L of the disk, which is needed to break the disk-halo degeneracy.

The scale height h_z of the thin disk cannot be measured directly for face-on galaxies. Studies of edge-on disk galaxies show a correlation between the scale height and indicators of the galaxies' mass scale, such as the absolute magnitude and the circular velocity (eg Yoachim & Dalcanton 2006). We can, therefore, estimate the scale height statistically using other known features of the galaxy.

The other parameter, σ_z of the disk, can be measured either from: (1) spectra of the integrated light of the disk in relatively face-on galaxies, or (2) the velocity distribution of a population of stellar tracers (such as planetary nebulae).

The early dynamical estimates of vertical velocity dispersion in external galaxies (e.g. Bottema 1997) indicated that the disks are submaximal, consistent with more recent estimates. The DiskMass Survey (Bershady *et al.* 2011) used integral-field spectroscopy to measure the stellar kinematics of the disks of near face-on galaxies and find that the dynamical stellar M/L obtained from the surface mass density is about 3 times lower than the M/L from maximum disk hypothesis and conclude that disks are submaximal. Work by Herrmann & Ciardullo (2008, 2009a, 2009b) looked at 5 near-face-on spirals (including our target galaxy) using PNe as tracers and found that in the later-type (Scd) systems, the disks are clearly sub-maximal. In earlier (Sc) galaxies, this discrepancy is smaller but still present; only the very-early Sab system M94 has evidence for a maximal disk.

An important conceptual problem has, however, been overlooked in the earlier studies described above. Eqn. 1.1 comes from the vertical Jeans equation for an equilibrium disk. It is therefore essential that the vertical disk scale height h_z and the vertical velocity dispersion σ_z should refer to the same population of stars.

The red and near-infrared measurements of h_z of edge-on disk galaxies are dominated by the red giants of the older, kinematically hotter population, away from the dust plane. On the other hand, σ_z is usually measured from integrated light spectra near the Mg b lines ($\sim 5150 - 5200$ Å), since this region has many absorption lines and the sky is relatively dark. The disks of star-forming spirals include a population of young (ages < 2 Gyr), kinematically cold stars among a population of older, kinematically hotter stars. The red giants of this mixed young + old population provide most of the absorption line signal that is used for deriving velocity dispersions from the integrated light spectra of galactic disks.

Therefore, in Eqn. 1.1, we should be using the σ_z of the *older* disk stars in combination with the h_z of this same population for an accurate determination of the surface mass density. However, because of limited signal-to-noise ratio (SNR) of previous studies, a single kinematical population is adopted for the velocity dispersion whereas, ideally, the dispersion of the older stars should be extracted from the composite observed spectrum of the younger and older stars.

Adopting a single kinematical population for a composite kinematical population gives a σ_z that is smaller than the dispersion of the old disk giants (for which the h_z was measured), and hence underestimates the surface density of the disk. A maximal disk will then appear submaximal. A recent study of the K-giants in the solar neighbourhood by Aniyan *et al.* (2016) showed that the young stars contribute significantly to the total light and that the velocity dispersion derived assuming a single population of tracers leads to the disk surface mass density being underestimated by a factor ~ 2.

We present the analysis of the disk kinematics of the galaxy NGC 628. We demonstrate that using the scale height of the old thin disk with the vertical velocity dispersion of the same population, will result in a maximal disk for NGC 628.

2. Observations and Data Analysis

We look at the face-on galaxy (inclination $\sim 6.5°$) NGC 628. Two complementary analyses were carried out: (1) integral light spectroscopy using the VIRUS-W instrument at McDonald observatory, and (2) using the Planetary nebula spectrograph (PN.S) instrument at La Palma to identify PNe in the outer regions of the disk.

The VIRUS-W is an optical-fibre-based Integral Field Unit (IFU) spectrograph at the 2.7m Harlan J. Smith Telescope at McDonald Observatory in Texas. The IFU has 267 fibres with a spectral resolution of R $\sim$ 8700. The spectral coverage is from 4802 – 5470 Å. The VIRUS-W observations were carried out in October 2014. The IFU was placed at several positions around a radius of $\sim$ 1 scale length. The spectrum from each fibre was summed up to build the SNR of the final spectrum. We thus end up with a spectrum that is representative of the kinematics of the galaxy at a mean luminosity weighted radius of 78″. We used the pPXF code (Cappellari & Emsellem 2004) to retrieve the line of sight dispersions (σ_{LOS}) of the kinematically hotter and colder stellar components of the thin disk. We then calculated the σ_z from the σ_{LOS} for both components, with conventional assumptions about the anisotropy of the velocity dispersion.

The planetary nebula spectrograph (PN.S) is an imaging spectrograph designed for efficient observation of extragalactic PNe, that operates on the 4.2m William Herschel Telescope at La Palma. The PN.S has a 'left' and 'right' arm in which the light is dispersed in opposite directions. Combining these two counter-dispersed images allows the PNe to be detected and their radial velocities to be measured in a single observation. The data for NGC 628 were acquired in September 2014. We obtained 14 images centred on the centre of the galaxy, each with an exposure time of 1800s. These 14 left and right images were stacked to create the final galaxy image with high SNR. For this study, `sextractor` was used to identify [OIII] emitters in the stacked left and right image. As an added check, we also performed a visual inspection of the identified sources, to reject any obvious extended sources and add any faint point-like objects that the code may have missed. This resulted in a catalogue of 716 point sources identified in the left and right arm of NGC 628. Figure 1 shows a region of identified sources on the image from the left arm. However, this sample is a mixture of HII regions and PNe, since both can have strong [OIII] emissions. We calculated the bright luminosity cut-off for this galaxy and only include the objects fainter than this cut-off in our analysis. This excludes most of the obvious, bright resolved HII regions. The objects were then put into radial bins with 155 objects in each bin. The dispersions for the hot and cold components in each bin were then calculated using a maximum likelihood estimator code written in python.

3. Results and Discussion

We used integrated light data from the VIRUS-W spectrograph and PNe velocities from the PN.S instrument to extract the velocity dispersion of the disk. We extract two σ_z values: one representing the kinematically cold disk and another representing the kinematically hot disk of NGC 628. We compared our values of the hot and cold component

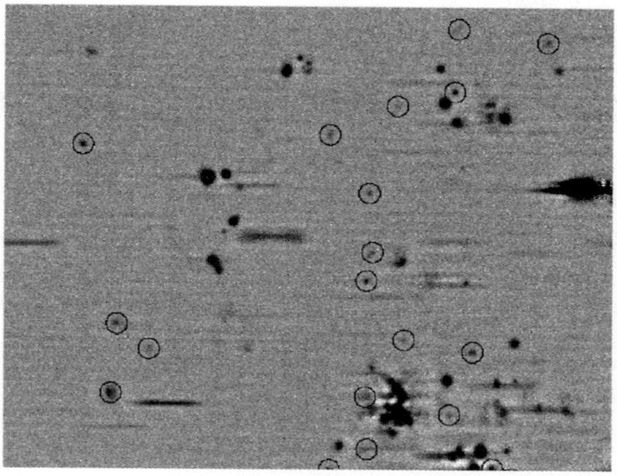

Figure 1. A part of the field in the image from the left arm of the planetary nebula spectrograph showing the identified planetary nebulae. The horizontal streaks are the spectra of stars in the field, limited in wavelength by the bandpass of the narrow band [OIII] filter.

vertical velocity dispersion with the results from Herrmann & Ciardullo (2009b), who assumed a single population of PNe and did not separate out the contribution of the cold and hot components. The one-component fit to our data agrees fairly well with the results from Herrmann & Ciardullo (2009b). Our two-component fit shows a hotter old disk population, with a velocity dispersion that follows the expected radial exponential decline with twice the galaxy's radial scale length, indicating that NGC 628 has an M/L ratio that is approximately constant with radius.

We use the calculated σ_z values to determine the disk surface mass density at each radial bin. Our central disk surface mass density is a factor of ~ 2 times larger than the value obtained by Herrmann & Ciardullo (2009b). We use BVRI photometry from the literature to calculate the M/L in different bands in this galaxy. We do, in fact, get a constant M/L value for NGC 628, and our M/L values are about twice those found by Herrmann & Ciardullo (2009b). This galaxy appears to have a near-maximal disk.

This is the first report from a larger study of nearby face-on spirals, using a combination of integrated light and PNe tracers. The analysis of these galaxies is currently underway.

Acknowledgements

The authors would like to thank the staff at McDonald Observatory and WHT for providing technical support during observations. SA would like to thank the SOC for granting an oral contribution to present this work, and for awarding a prize for this talk.

References

Aniyan, S. *et al.* 2016, *MNRAS*, 456, 1484
Bershady, M. A. *et al.* 2011, *ApJ*, 739, L47
Bottema, R. 1997, *A & A* 328, 517
Cappellari, M. & Emsellem, E. 2004, *PASP*, 116, 138
Herrmann, K. A., Ciardullo, R., Feldmeier, J. J., & Vinciguerra, M. 2008, *ApJ*, 683, 630
Herrmann, K. A. & Ciardullo, R. 2009, *ApJ*, 703, 894
Herrmann, K. A. & Ciardullo, R. 2009, *ApJ*, 705, 1686
Kormendy, J. & Freeman, K. C. 2016, *ApJ* 817, 84
Sackett, P. D. 1997, *ApJ* 483, 103
van Albada, T. S., Bahcall, J. N., Begeman, K., & Sancisi, R. 1985, *ApJ*, 295, 305
Yoachim, P. & Dalcanton, J. J., 2006, *AJ*, 131, 226

Planetary Nebulae: Multi-Wavelength Probes of Stellar and Galactic Evolution
Proceedings IAU Symposium No. 323, 2016
X. Liu, L. Stanghellini & A. Karakas, eds.

doi:10.1017/S1743921317001946

The VIRGO PN population and the mass assembly in M87

Alessia Longobardi[1,2], Magda Arnaboldi[3] and Ortwin Gerhard[2]

[1]Kavli Institute for Astronomy and Astrophysics, Peking University,
5 Yiheyuan Road, Haidian District, Beijing 100871, P. R. China,
email: alongobardi@pku.edu.cn
[2]MPI für extraterrestrische Physik, Giessenbachstrasse 1, D-85741 Garching, Germany
email: alongobardi@mpe.mpg.de, gerhard@mpe.mpg.de
[3]European Southern Observatory, Karl-Schwarzschild-Strasse 2, D-85748 Garching, Germany
email: marnabol@eso.org

Abstract. Cosmological simulations allow us to study in detail the evolution of galaxy halos in cluster environments, but the extremely low surface brightness of such components makes it difficult to gather observational constraints. Planetary nebulas (PNs) offer a unique tool to investigate these environments owing to their strong [OIII] emission line. We study the light and kinematics of the Virgo cluster and its central galaxy, M87, prime targets to address the topic of galaxy formation and evolution in dense environments. We make use of a deep and extended PN sample (~300 objects) that extends out to 150 kpc in radius from M87's centre. We show that at all distance the galaxy halo overlaps with the Virgo intracluster light (ICL). Halo and ICL are dynamically distinct components with different parent stellar populations, consistent with the halo of M87 being redder and more metal rich than the ICL. The synergy between PN kinematic information and deep V/B-band photometry made it possible to unravel an ongoing accretion process in the outskirt of M87. This accretion event represents a non-negligible perturbation of the halo light, showing that this galaxy is still growing by accretion of smaller systems.

Keywords. galaxies: clusters: individual (Virgo cluster), galaxies: halos, galaxies: individual (M87), galaxies: kinematics and dynamics, planetary nebulae: general

1. Introduction

Galaxy halos are faint stellar components made of stars gravitationally bound to the individual galaxies. In galaxy clusters these halos may be surrounded by IC stars (Gonzales *et al.* 2005). The formation of ICL and extended halos around central cluster galaxies (cD) is closely related to the transformation of galaxies in clusters (De Lucia *et al.* 2012) and, hence, their physical properties reflect the formation of the galaxies and their hosting clusters. When studying cDs we might wonder whether any photometrical and dynamical difference does exist between the central galaxy halo and the ICL. Usually no differentiation between them is possible and their physical properties are determined by treating them as a continuum. But is this really the case?

The Virgo cluster, the nearest large scale structure in the Universe, and its central galaxy M87 are prime targets to address the topic of galaxy/cluster formation and evolution, and to answer the question raised above. Virgo is a dynamically young cluster with a complex network of extended tidal features, suggesting that its core is not completely in equilibrium (Mihos *et al.*2005). At its centre lies the giant elliptical galaxy M87 with a cD halo that extends out to $\sim$ 150 kpc in radius, and down to 27 mag arcsec^{-2} in the V-band. The observed blue colour gradients towards its outskirts (Rudick *et al.* 2010), hence the inferred metallicity and age gradients (Montes *et al.* 2014), are consistent with a late build up of its halo, suggesting that here we can find a great deal of information on the physical mechanisms governing the mass assembly in the Universe.

However, the most significant challenge to investigate galaxy halos far away from their centres is that these regions have a very low surface brightness, making it difficult for most of the traditional techniques to be successful. PNs, offer a unique tool to investigate these environments owing to their strong [OIII] emission line, whereby $\sim 15\%$ of the luminosity of the central star is re-emitted in the forbidden [OIII] line at $\lambda 5007$ Å(Dopita *et al.*1992). Hence they make it possible to study M87 beyond its brightest regions, where the halo blends into the ICL.

2. PN surveys in the Virgo core and M87

Previous PNs studies in Virgo and M87. Several large-scale [OIII]$\lambda 5007$ Å surveys of intergalactic fields have been carried out with the goal of mapping the distribution, kinematics, and luminosity function of PNs in Virgo, starting from the discovery of ICPNs by Arnaboldi *et al.* (1996). The study of ICPNs constrained the fraction of the ICL to 10% in the Virgo core (Feldmeier *et al.*2003). This percentage drops in lower density regions suggesting that the ICL is poorly mixed and that Virgo is a dynamically young cluster (Castro *et al.* 2009, Aguerri *et al.* 2005). Moreover, using PN kinematics and spatial distribution, Doherty *et al.* (2009) gave evidence for the truncation of the luminous halo of M87 due to an earlier mass interaction, and they showed that PNs might be separable into components that are bound to either the potential of the galaxy or of that of the cluster.

The large Suprime-Cam@Subaru and FLAMES@VLT PN surveys. To further investigate these findings our group decided to carry out deep photometric and spectroscopic PN surveys, and gather a uniformly distributed PN sample containing ~15 times more data than any other PN sample in these regions. They are described in what follows.

We first carried out a photometric survey with Suprime-Cam@Subaru that covered a $0.43\,\mathrm{deg}^2$ region centred on M87 (Longobardi *et al.*2013), obtaining images with a (on-band) narrow-band filter centred on the redshifted [OIII]$\lambda 5007$Å emission line (at the Virgo cluster distance) and a (off-band) broadband V-filter. As consequence of their bright [OIII] and very faint continuum, extragalactic PNs can be identified based on their colour (on-off band mag) vs on-band mag properties. This led to a large and deep photometric sample of PNs, that we spectroscopically followed-up with the FLAMES@VLT. The 287 sources classified as PNs were identified from their narrow and symmetric redshifted $\lambda 5007$Å [OIII] emission line, the presence of the second $\lambda 4959$Å [OIII] emission line, and the absence of significant continuum. Data were acquired with a spectral resolution of R = 22500. With this setup, the measured line-of-sight (LOS) velocities were obtained with an estimated median accuracy of $4.2\,\mathrm{kms}^{-1}$.

3. PNs as tracers of light and stellar populations

The different dynamics of Virgo ICL and M87 halo. The study of the PN kinematics by Longobardi *et al.* (2015) (L15a) showed that the M87 halo and the Virgo ICL are dynamically distinct components with different velocity and density distributions. Using Gaussian fitting and robust sigma clipping procedures they showed that the line of sight velocity distribution (LOSVD) is bimodal and splits into two Gaussian components , one that feels the potential of the galaxy halo and one that feels the potential of the cluster.

The halo-ICL bimodality is shown in Fig. 1 where we plot the projected phase-space distribution for both halo (red asterisks) and IC (blue asterisks) PNs. We also show

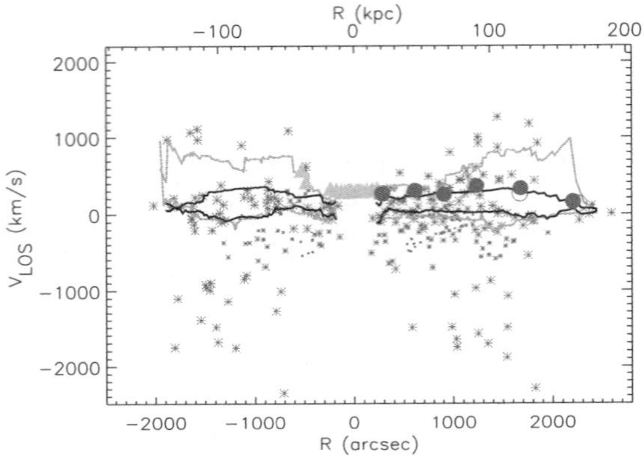

Figure 1. Velocity phase-space diagram V_{LOS} vs. major axis distance R for both M87 halo (red asterisks) and IC component (blue asterisks) from L15a. Different point sizes represent different probability to belong to either the halo or the ICL. Black dashed-dot and black continuous lines depict the running average, and running dispersion computed from the 254 halo PNs weighted by their probability. Red full dots represent the robust estimate of the halo dispersion from L15, while red circle show the velocity dispersion when the 'crown' is statistically subtracted (see text). Green dashed-dot and green continuous lines represent the same quantities for the halo+IC component system. It is clear that including the ICL component leads to overestimate the velocity dispersion. Cyan triangles are IFU velocity dispersion data from Murphy *et al.* (2011, 2014) consistent with contamination from ICL (Longobardi *et al.* 2017).

the LOSVD 'running average', and 'running dispersion', i.e the first and second order moments of sub-sequences of 30 terms of the full data-set, for both the only halo (black lines) and the total halo+IC (green lines) system. While the overall mean velocity is not affected by the IC component, this population has an impact on the estimate of the dispersion. If halo and IC populations are not separated, the dispersion is overestimated, reaching, at its maximum, the Virgo sub-cluster A velocity dispersion $\sim 800\,\mathrm{kms}^{-1}$.

The PN population of the Virgo ICL and M87 stellar halo. Virgo ICL and M87 galaxy halo not only do not share the same dynamics but they also differ in density profiles and underlying stellar populations. This is shown by the different properties of the PN population they are associated with, such as the α-parameter, i.e. the number of PNs per bolometric luminosity, and the planetary nebula luminosity function (PNLF). The M87 halo is more centrally concentrated and it is characterised by a smaller α-parameter than the ICL, typical of older and more massive systems. Moreover the ICL PNLF shows an overall shallower gradient than the halo PNLF and has a dip present at ~1-1.5 mag fainter than the bright cutoff. This is an evolutionary feature that is observed only in luminosity functions of younger and star forming systems. All these evidences lead to the conclusion that M87's halo and Virgo's ICL have two different parent stellar populations, consistent with the halo being redder and more metal rich than the ICL (L15a).

PNs as tracers of galaxy evolution: the 'crown' of M87 It is now known that the M87 halo shows kinematic signatures of accretion. Using PNs Longobardi *et al.* (2015) (L15b) showed that the M87 velocity phase-space is complex with a kinematic substructure that can be seen as cold features in a space dominated by random motion as consequence of a not complete phase-mixing of a disrupted satellite. In the light this substructure can be seen as a 'crown' shaped features, 70 kpc north the M87's centre, causing an increase in light by $\gtrsim 60\%$. The crown represents the densest region of the entire event that extends $\sim$ 50 kpc along the major axis, where we also observe a variation in the colour: along the minor axis where no crown PNs are observed the galaxy is redder (Fig 2).

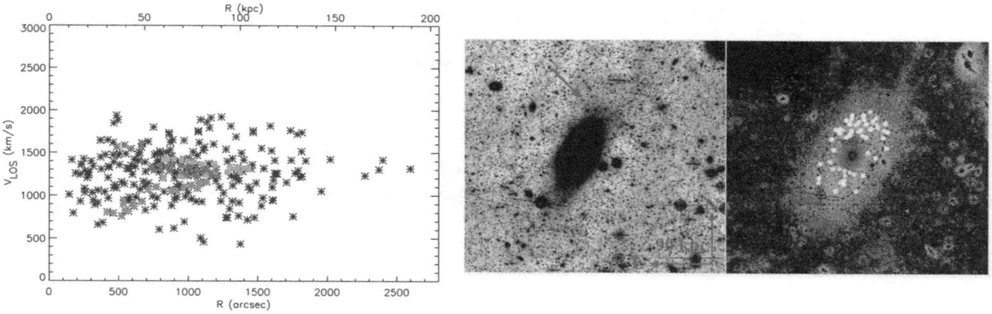

Figure 2. Left panel Velocity phase-space for the PNs (black asterisks) in the halo of M87. The green circles and magenta diamonds show the PNs associated with the crown substructure (L+15b). Orange squares show a GC substructure from Romanowsky *et al.* (2012). **Central panel**. Unsharped masked image of M87. The *crown* is visible at distance 60-90 kpc along the major axis, NW of M87. **Right panel**: (B-V) colour image of M87 (Mihos *et al.* 2016) with substructure PNs overplotted (white dots). The ellipse indicates the isophote at 90 kpc. The crown is found in a region where (B-V)$\sim$0.8, bluer than on the minor axis.

4. Conclusions

PNs represent a great tool to investigate regions in space characterised by very low surface brightnesses and where we can find signatures of the physical processes involved in the formation and evolution of astronomical objects. At the Virgo cluster core, PNs allowed us to trace the transition region between M87 galaxy halo and Virgo ICL, showing us that the M87 stellar halo is distinct from the surrounding ICL in its kinematics, density profile, and parent stellar population. PNs also traced an accretion event in M87 that has caused an important modification of its outer halo in the last Gyr. This accretion event, first revealed as kinematic substructure and then revealed as feature in the light, shows us that beyond $\sim$ 60 kpc M87's halo is still assembling, and the PN kinematics reflects the dynamical complexity of a system still growing.

References

Aguerri, J. A. L., Gerhard, O. E., Arnaboldi, M., *et al.* , 2005 *AJ*, 129, 2585

Arnaboldi, M., Freeman, K. C., Mendez, R. H., *et al.* , 1996 *ApJ*, 472, 145

Castro-Rodríguez, N., Arnaboldi, M., & Aguerri, J. A. L. , 2009 *A&A*, 507, 621

De Lucia, G., Weinmann, S., Poggianti, B. M., Aragon-Salamanca, *et al.* , 2012 *MNRAS*, 423, 1277

Doherty, M., Arnaboldi, M., Das, P., *et al.* , 2009 *A&A*, 502, 771

Dopita, M. A., Jacoby, G. H., & Vassiliadis, E. , 1992 *ApJ*, 389, 27

Feldmeier, J. J., Ciardullo, R., Jacoby, G. H., & Durrell, P. R. , 2003 *ApJS*, 145, 65

Gonzalez, A. H., Zabludoff, A. I., & Zaritsky, D. 2005, *ApJ*, 618, 195

Longobardi, A., Arnaboldi, M., Gerhard, O., *et al.* , 2013 *A&A*, 558, A42

Longobardi, A., Arnaboldi, M., Gerhard, O., & Hanuschik, R. , 2015a *A&A*, 579A, 135L

Longobardi, A., Arnaboldi, M., Gerhard, O., & Mihos, J., C. , 2015b *A&A*, 579L, 3L

Longobardi, A., Arnaboldi, M., Gerhard, O. & C. Pulsoni , 2017 *A&A*, in prep.

Mihos, J. C., Harding, P., Feldmeier, J., & Morrison, H. , 2005 *ApJ*, 631, L41

Mihos, J. C., Harding, P., Feldmeier, J., *et al.* , 2016 *arXiv e-prints*, 1611.04435

Montes, M., Trujillo, I., Prieto, M. A., & Acosta-Pulido, J. A. , 2014 *MNRAS*, 439, 990

Murphy, J. D., Gebhardt, K., & Adams, J. J. , 2011 *ApJ*, 729, 129

Murphy, J. D., Gebhardt, K., & Cradit, M. , 2014 *ApJ*,785,143

Romanowsky, A. J., Strader, J., Brodie, J. P., *et al.* , 2012 *ApJ*, 748, 29

Rudick, C. S., Mihos, J. C., & Harding, P. , 2010 *ApJ*, 720, 569

Discussion

Q: Do globular clusters trace the same kinematics and substructures as PNe?

Q: The chevron that you show is like the tail of a comet. Is there any evidence for the remains of the shredded galaxy? Might a wavelet analysis help find theremnant?

Planetary Nebulae: Multi-Wavelength Probes of Stellar and Galactic Evolution
Proceedings IAU Symposium No. 323, 2016
X. Liu, L. Stanghellini & A. Karakas, eds.

doi:10.1017/S1743921317000709

The halo of M49 and its environment as traced by planetary nebulae

Johanna Hartke[1], Magda Arnaboldi[1], Alessia Longobardi[2,3], Ortwin Gerhard[3], Ken Freeman[4] and Sadanori Okamura[5]

[1]European Southern Observatory, Karl-Schwarzschild-Str. 2, 85748 Garching, Germany, email: [jhartke;marnabol]@eso.org

[2]Kavli Institute for Astronomy and Astrophysics, Peking University, 5 Yiheyuan Road, Beijing 100871, P. R. China, email: alongobardi@pku.edu.cn

[3]Max-Planck-Institut für Extraterrestrische Physik, Giessenbachstrasse, 85748 Garching, Germany, email: gerhard@mpe.mpg.de

[4]RSAA, Mt. Stromlo Observatory, 2611, Canberra, Australia, email: Kenneth.Freeman@anu.edu.au

[5]Department of Advanced Sciences, Faculty of Science and Engineering, Hosei University, 184-8584 Tokyo, Japan, email: sadanori.okamura@hosei.ac.jp

Abstract. We investigate the stellar halo of the nearby elliptical Virgo-cluster galaxy M49 using Planetary Nebulae (PNe). M49 is the second-brightest galaxy of the Virgo cluster and is at the center of the Virgo subcluster B. We present an extended catalogue extracted from a narrow-band survey carried out with Subaru's Suprime Cam, consisting of 735 PNe down to a limiting magnitude of $m_{5007} = 29.3$. This PNe population traces the halo out to 155 kpc from the galaxy's center, which provides accurate measurement of the luminosity-specific PN-number (α-parameter) in the inner and outer regions of M49's halo. We are also able to determine the morphological variation of the planetary nebulae luminosity function (PNLF), that may trace different parent stellar populations. This enables us to identify the transition from the PN-scarce, possibly metal-rich, galaxy halo to the PN-rich, metal-poor, outer component.

Keywords. galaxies: clusters: individual (Virgo), galaxies: elliptical and lenticular, cD, galaxies: individual (M49), planetary nebulae: general

1. Introduction

In the current paradigm of hierarchical structure formation, galaxies build-up to their current size by a two-phase process consisting of in-situ star formation and then further growth through mergers and accretion of smaller structures (e.g. Oser *et al.*, 2012). These accretion events leave long-lasting signatures in the galaxies' halos which can be identified in the kinematic phase-space or in the density distribution of stars. Numerical simulations predict that a significant fraction of halo stars have been accreted in the most massive galaxies (e.g. Cooper *et al.*, 2013). In order to better understand the late stages of galaxy formation, we need to identify accretion events and their time scales.

M49 (NGC 4472) is the brightest galaxy in the Virgo Cluster. It is surrounded by a diffuse halo that contains a layered system of shells that is suggestive of a hierarchical build-up (Mihos *et al.*, 2005, Janowiecki *et al.*, 2010). Furthermore, there is ongoing accretion of the dwarf irregular galaxy VCC 1249 (Arrigoni Battaia *et al.*, 2012). Extended multi-band photometry of the halo provides evidence for an extended, blue halo (Mihos *et al.*, 2013). Single stellar population models predict corresponding metallicities from [Fe/H] $\simeq -1.0$ to -1.5 (Bruzual & Charlot, 2003). This low metallicity and the large size of this halo are in conflict with predictions from conventional hierarchical assembly:

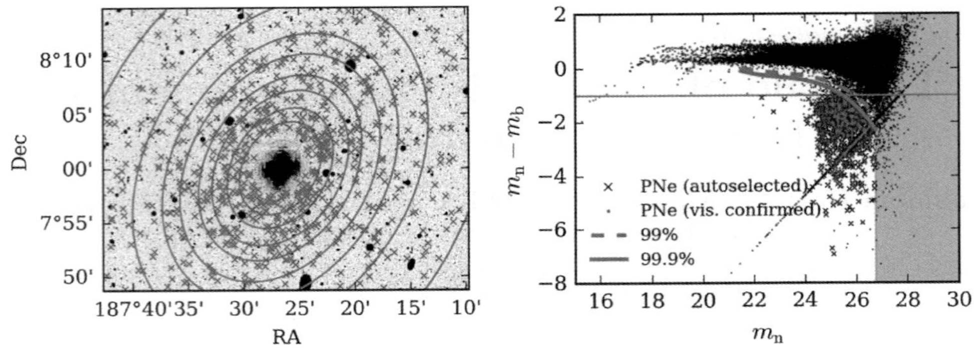

Figure 1. *Left:* [OIII] image of M49 with PN candidates indicated by red crosses. Elliptical bins with $PA = -31^\circ$ and $\epsilon = 0.28$ (Kormendy *et al.*, 2009) are spaced such that they contain ~ 70 objects in each bin. *Right:* CMD of automatically detected sources. Visually confirmed PN candidates are plotted with red dots. The solid blue line indicates the region where 99.9% of a simulated continuum population with no colour excess would lie.

progenitors which are expected to have built a M49-sized halo would have contributed higher metallicities than the one measured in M49's outer halo (see e.g. Tremonti *et al.* (2004) for the mass-metallicity relation for galaxies).

Planetary nebulae (PNe) are the only single *stellar* tracers which can be used to study the halos of galaxies at the distance of the Virgo Cluster. In the following we present a new, homogeneous and statistically complete sample of PNe in M49's extended halo.

2. PNe selection

We have obtained deep [OIII] and V-band images with the Suprime-Cam at the 8.2m Subaru telescope. The mosaic camera covers an area of $34' \times 27'$, thus M49's halo can be observed out to a major-axis distance of 155 kpc. We select PNe using the on-off band technique, making use of two characteristic features of the PN spectrum: the bright [OIII]5007 Å emission and the absence of a continuum. PNe can thus be identified through their on-band [OIII] emission and absence of emission on the broad-band V-image. We use an automatic selection procedure, which selects point-like candidates with [OIII] - V colour excess (Arnaboldi *et al.*, 2002, 2003). The left panel of Fig. 1 shows the selected 735 PN candidates which have been detected within the limiting magnitude of $m_{[\mathrm{OIII,lim}]} = 29.3$. The limiting magnitude of this survey is defined as the magnitude where the recovery fraction of a simulated PN population drops below 50%. We also use this simulated population in order to evaluate the photometric and detection completeness of our survey. The right panel of Fig. 1 illustrates the colour-magnitude diagram (CMD) and the location where the PN candidates are selected.

3. The PN number density distribution and α-parameters

Previous studies have shown that the PN number density follows stellar light in early-type galaxies (Coccato *et al.*, 2009, Cortesi *et al.*, 2013, Longobardi *et al.*, 2013). The stellar light is characterized by a Sérsic profile out to $r \sim 915''$ (Kormendy *et al.*, 2009) after which it starts to flatten (Capaccioli *et al.*, 2015). The PN number density profile follows the stellar light closely, and it flattens already beyond a radius $r \sim 730''$ (Hartke *et al.*, 2016). The flattening is likely due to the presence of an additional component

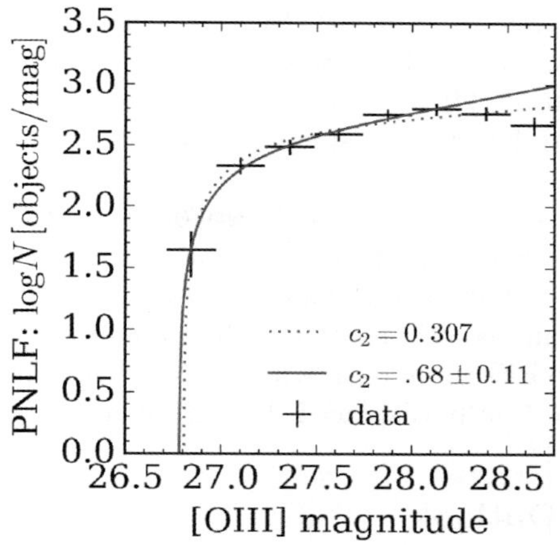

Figure 2. PNLF of M49 corrected for completeness and Ly-α contamination.

associated with the outer halo, e.g a contribution from intragroup light (IGL) from the Subcluster B at whose center M49 resides.

The total number of PNe N_{PN} and the total bolometric luminosity L_{bol} of the underlying stellar population are related through the luminosity-specific PN number, also known as α-parameter: $N_{\mathrm{PN}} = \alpha L_{\mathrm{bol}}$. We determine the α-parameter in the two radial regimes by fitting a two-component model for the PN number density; one population follows a Sérsic-like radial profile and the contribution from the IGL is constant with radius. The best-fit model is found to have a PN-yield that varies, with the outer halo contributing roughly three times more PN than the inner halo. The higher abundance of PNe in the outer regions is likely due to the presence of PNe associated with the surrounding IGL. A higher PN-yield was also observed in M87, where Virgo's intracluster PNe were kinematically identified in Longobardi *et al.* (2015) after having observed a flattened number density profile (Longobardi *et al.*, 2013).

4. The PNLF of M49

We also characterize M49's PN sample by its luminosity function (PNLF) and we use the analytical formula that was introduced by Longobardi *et al.* (2013) to fit the PNLF of M87:

$$N(M) \propto e^{M c_2}(1 - e^{3(M^\star - M)}). \tag{4.1}$$

This is the generalized form of the PNLF that allows for variation of the faint-end slope by having c_2 as a free parameter. The Ciardullo *et al.* (1989) formula is then a particular case with $c_2 = 0.307$. Because of the empirical evidence for a bright cut-off consistent with a absolute magnitude $M^\star = -4.51$ (Ciardullo *et al.*, 2002), the PNLF can be used as a secondary distance indicator. Using the generalized formula, we determine a distance modulus of $\mu_{\mathrm{PNLF}} = 31.29^{+0.07}_{-0.08}$ for M49, which agrees with the distance determined from Surface Brightness Fluctuations (SBF, $\mu = 31.17 \pm 0.07$, Mei *et al.*, 2007). Several authors have debated about the discrepancy between SBF and PNLF distances (e.g Mendez *et al.*, 2016 at the IAUS 323, Ciardullo *et al.*, 2002, Ciardullo, 2012). We believe that this discrepancy may arise because of a contamination of the sample by Ly-α emitters at redshift $z = 3.1$, which is statistically corrected for in the M49 sample (see Hartke *et al.* (subm.) for further details). The Ly-α-corrected PNLF is shown in Fig. 2.

Another characteristic of the analytic PNLF is its slope, which is set by the parameter c_2. Steeper slopes are measured for PN populations with older parent stars (e.g Ciardullo *et al.*, 2004, Longobardi *et al.*, 2013). We fit a PNLF to the two radial regimes identified in the previous section and determine a steeper PNLF slope in the outer region compared to the inner one.

We then investigate the PNLF in four quadrants in the inner halo in order to trace local variation of the PNLF. We find that one of the quadrants has a significantly flatter PNLF slope than the other three. The flatter slope is likely due to a young, bright population of PNe, which points to an accretion event. While no substructure is prominent in the light at the location of the PN-population with flatter PNLF, it is reasonable to associate them with stars that were tidally displaced during the accretion of VCC 1249.

5. Conclusions and Outlook

We have presented a PN-survey with unprecedented coverage of the halo of M49. We are able to trace the transition between galaxy and intragroup light based on the flattening of the PN number density profile compared to the light. Our fit of a two-component model points towards an IGL population with a three-times higher α-parameter compared to the main halo. We have accurately determined the distance to M49 using the bright cut-off of the PNLF and have resolved the PNLF-SBF distance discrepancy at least for this object. The morphology of the PNLF shows variation that points us to a change in the physical properties (like age, metallicity) of the stellar populations, which is likely to arise from the accretion of VCC 1249. Next, we want to obtain high-resolution spectra of the PNe in order to kinematically map M49's extended halo. We will then be able to disentangle intragroup and halo PNe and identify substructures from accretion events in the PN-velocity phase-space.

References

Arnaboldi, M., Aguerri, J. A. L., Napolitano, N. R., *et al.* 2002, *AJ*, 123, 760
Arnaboldi, M., Freeman, K. C., Okamura, S., *et al.* 2003, *AJ*, 125, 514
Arrigoni Battaia, F., Gavazzi, G., Fumagalli, M., *et al.* 2012, *A&A*, 543, A112
Bruzual, G. & Charlot, S. 2003, *MNRAS*, 344, 1000
Capaccioli, M., Spavone, M., Grado, A., *et al.* 2015, *A&A*, 581, A10
Ciardullo, R., Jacoby, G. H., Ford, H. C., & Neill, J. D. 1989, *ApJ*, 339, 53
Ciardullo, R., Feldmeier, J. J., Jacoby, G. H., *et al.* 2002, *ApJ*, 577, 31
Ciardullo, R., Durrell, P. R., Laychak, M. B., *et al.* 2004, *ApJ*, 614, 167
Ciardullo, R. 2012, *ApSS*, 341, 151
Coccato, L., Gerhard, O., Arnaboldi, M., *et al.* 2009, *MNRAS*, 394, 1249
Cooper, A. P., D'Souza, R., Kauffmann, G., *et al.* 2013, *MNRAS*, 434, 3348
Cortesi, A., Merrifield, M. R., Coccato, L., *et al.* 2013, *MNRAS*, 432, 1010
Hartke, J., Arnaboldi, M., Longobardi, A., *et al.* 2016, *A&A*, subm.
Janowiecki, S., Mihos, J. C., Harding, P., *et al.* 2010, *ApJ*, 715, 972
Kormendy, J., Fisher, D. B., Cornell, M. E., & Bender, R. 2009, *ApJS*, 182, 216
Longobardi, A., Arnaboldi, M., Gerhard, O., *et al.* 2013, *A&A*, 558, A42
Longobardi, A., Arnaboldi, M., Gerhard, O., & Hanuschik, R. 2015, *A&A*, 579, A135
Mei, S., Blakeslee, J. P., Côté, P., *et al.* 2007, *ApJ*, 655, 144
Mendez, R. H. 2016, ArXiv e-prints 1610.08625
Mihos, J. C., Harding, P., Feldmeier, J., & Morrison, H. 2005, *ApJL*, 631, L41
Mihos, J. C., Harding, P., Rudick, C. S., & Feldmeier, J. J. 2013, *ApJL*, 764, L20
Oser, L., Naab, T., Ostriker, J. P., & Johansson, P. H. 2012, *ApJ*, 744, 63
Tremonti, C. A. *et al.* 2004, *ApJ*, 613, 898

Discussion

MENDEZ: Jacoby *et al.* (1990) got a smaller PNLF distance to M49. Did you check your magnitudes against theirs for any objects you may have in common?

HARTKE: We have not cross-match the individual catalogues. We wish to commenton the different sample sizes ∼30 PNe vs. ∼730 in our sample. The difference in the bright cut-off is probably due to different ways of accounting for contamination from $z = 3.1$ Ly a emitters, which also have strong 5007 Aemission.

TORRES-PEIMBERT: How did you correct for the Ly alpha galaxy population?

HARTKE: We statistically correct for Ly-a emitters at $z = 3.1$ based on their luminosity function as determined by Gronwall *et al.* (2007). In every mag-binwe subtract the number of expected Ly-a emitters from the completeness-corrected number of PNe.

Planetary Nebulae: Multi-Wavelength Probes of Stellar and Galactic Evolution
Proceedings IAU Symposium No. 323, 2016
X. Liu, L. Stanghellini & A. Karakas, eds.

doi:10.1017/S1743921317000461

The Planetary Nebulae Luminosity Function (PNLF): current perspectives

Roberto H. Méndez

Institute for Astronomy, University of Hawaii,
2680 Woodlawn Drive, Honolulu, HI 96822, USA
email: mendez@ifa.hawaii.edu

Abstract. This paper starts with a brief historical review about the PNLF and its use as a distance indicator. Then the PNLF distances are compared with Surface Brightness Fluctuations (SBF) distances and Tip of the Red Giant Branch (TRGB) distances. A Monte Carlo method to generate simulated PNLFs is described, leading to the last subject: recent progress in reproducing the expected maximum final mass in old stellar populations, a stellar astrophysics enigma that has been challenging us for quite some time.

Keywords. galaxies: distances and redshifts, planetary nebulae: general, stars: AGB and post-AGB, evolution

1. Historical introduction

The field of extragalactic PNs as tools for distance determination was pioneered by Holland Ford and George Jacoby, later joined by Robin Ciardullo. They started detecting and measuring PNs in nearby galaxies in the late 70's and early 80's. The measured fluxes in [O III] 5007 can be transformed into "Jacoby magnitudes" $m(5007)$ (Jacoby 1989). The PNLF then indicates how many PNs are found at each apparent magnitude $m(5007)$. It was noticed that the absolute magnitudes $M(5007)$ of the brightest PNs in galaxies of known distance always had a similar value. On this mostly empirical basis, they proposed to assume that the bright end of the PNLF is universal, and use it as a standard candle. An analytical representation of the PNLF was introduced by Ciardullo *et al.* (1989). The absolute magnitude M^* of the bright end needs to be calibrated using a galaxy at a known distance, for example the bulge of M 31. After testing the method on other nearby galaxies, Jacoby *et al.* (1990) applied it to several galaxies in the Virgo cluster, obtaining similar distances to all those galaxies, and an average distance of 15 Mpc, implying a high value of the Hubble constant that was later essentially confirmed by the HST Key project on cepheid distances. As of today, the PNLF method has given distances to about 60 galaxies.

2. The discrepancy with SBF distances

In the meantime, other methods of distance determination have become popular, among them the SBF method (Tonry and Schneider 1988, Tonry *et al.* 2001, Blakeslee *et al.* 2009). Around the year 2000, it became apparent that there is a systematic difference between PNLF and SBF distances, with PNLF distance moduli being smaller by about 0.3-0.4 mag at distances in excess of 10 Mpc (Méndez 1999; Ciardullo *et al.* 2002). A more recent comparison adding a few more galaxies is shown in Fig. 16 of Teodorescu *et al.* (2010).

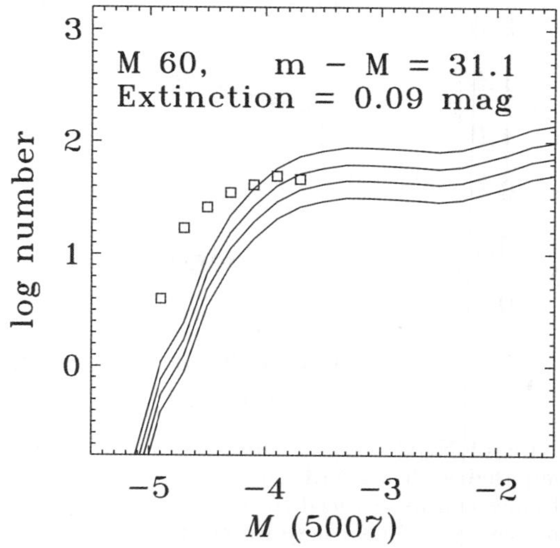

Figure 1. Observed [O III] λ 5007 PNLF of M 60 (squares). The PNLF distance to M 60 corresponds to a distance modulus 30.7 (Teodorescu *et al.* 2011). Here I have forced this observed PNLF to be at the SBF distance modulus 31.1 (Blakeslee *et al.* 2009). Therefore all the PNs have become 0.4 mag brighter. The simulated PNLF (solid lines) is the same one used by Teodorescu *et al.* (2011).

I have no room to describe ideas about the reason for the discrepancy (see e.g. Ciardullo *et al.* 2002, and particularly the careful review by Ciardullo 2012); but it is important to show what happens if we assume that the SBF distances are correct. Figure 1 shows how the increased distance forces the observed PNLF to become more luminous. The brightest PNs become too bright. Why don't we see such superbright PNs in nearby galaxies? If SBF distances are confirmed, this will require an explanation.

3. TRGB distances as a tie-breaker

We need a third method to decide if it is the PNLF distances or the SBF distances, or even both, that are systematically wrong. I believe our best choice for this test is TRGB distances. The method is described by Makarov *et al.* (2006); it has been applied to many galaxies up to distances of about 10 Mpc. The agreement with PNLF distances is good (Ciardullo 2012). Unfortunately, right now we have only one TRGB distance to a galaxy more distant than 14 Mpc: M 87 (Bird *et al.* 2010). This single TRGB distance agrees with the SBF distance. Ideally, we want at least ten TRGB distances in the Virgo and Fornax clusters to yield a convincing result and decide once and for all what is the truth. This will be possible with the James Webb Space Telescope (JWST), which is expected to allow TRGB distance determinations up to at least 30 Mpc.

4. Modeling the PNLF

It is not necessary to rely exclusively on the analytical PNLF representation. An alternative is to produce simulated PNLFs, using Monte Carlo techniques. This approach is described by Méndez *et al.* (1993), Méndez & Soffner (1997), and Méndez *et al.* (2008). The idea is to generate a set of PN central stars with random post-AGB ages and masses. Using a family of post-AGB evolutionary tracks, for every pair (age, mass) the corresponding central star luminosity and surface temperature can be derived. Using

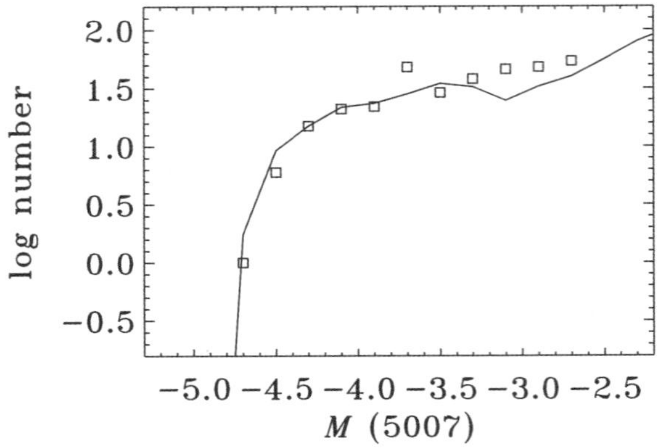

Figure 2. Observed PNLF of NGC 4697 (squares; distance modulus 30.1; from Fig. 14 of Méndez *et al.* 2001), compared with a PNLF simulation produced as described in Section 6. Now the maximum final mass can be reduced to 0.58 $M_\odot$, in much better agreement with the initial-final mass relation (see Cummings, these Proceedings).

recombination theory and empirical information about leaking of H-ionizing photons, the nebular $H\beta$ luminosities are calculated. Finally, generating suitable ratios $I(5007)/I(H\beta)$ for different regions of the HR diagram, the luminosities in 5007 are obtained, and the PNLF is built.

5. The mass distribution and the maximum final mass enigma

In such PNLF simulation work we have always used a mass distribution similar to that of white dwarfs. For a population without recent star formation, it is necessary to cut the mass distribution at a certain maximum final mass, because all the more massive stars have already evolved into white dwarfs. Here comes the problem: since we expect elliptical galaxies to have old populations, the initial mass is more or less 1 $M_\odot$, and therefore the maximum final mass is expected to be at most 0.55 $M_\odot$. The PNLF should become very faint. However, the observed PNLF bright end stays constant. From the observed central star luminosities, using traditional post-AGB evolutionary tracks, the maximum final mass is predicted to be 0.63 $M_\odot$; a severe inconsistency.

6. New evolutionary tracks and their consequences

The old tracks required a central star mass of 0.63 $M_\odot$ to produce log $L/L_\odot = 3.9$ as the star abandons the AGB. New evolutionary tracks by Miller Bertolami (2016) have considerably reduced the required central star mass. Figure 2 shows a fit to the PNLF of the elliptical galaxy NGC 4697 (Méndez *et al.* 2001), produced with a simulation based on an analytical approximation to Miller Bertolami's evolutionary tracks. Now the mass distribution can be truncated at 0.58 $M_\odot$. This is perhaps not a full solution to the maximum final mass enigma, but it is certainly a substantial step in the right direction. With a little help from the initial-final mass relation as described by Cummings in these Proceedings, we may be close to solving this persistent problem.

7. PN formation rates

There is another important consequence of the new Miller Bertolami tracks. In addition to being more luminous at any given mass, these more luminous central stars move faster along their tracks; there is less available fuel, and it is consumed more quickly. The times required to cross the HR diagram at high luminosity are shorter.

It is customary to define a formation rate $B_{\rm PN}$, in PNs per year per $L_\odot$; a PN visibility duration $t_{\rm PN}$ in years (e.g. 30,000); the total luminosity $L_{\rm T}$ of the sampled population in $L_\odot$; and the total PN population $n_{\rm PN}$ (only a fraction of these -the brightest- are actually observed). Then we can write $n_{\rm PN} = B_{\rm PN} L_{\rm T} t_{\rm PN}$. The luminosity-specific PN formation rate α in PNs per $L_\odot$ is given by $\alpha = B_{\rm PN} t_{\rm PN} = n_{\rm PN}/L_{\rm T}$.

Now introduce the new evolutionary tracks, with their shorter crossing times. The percentage of visible PNs goes down, because we have not modified $t_{\rm PN}$. So from a given number of observed PNs we derive a larger $n_{\rm PN}$. Consequently $B_{\rm PN}$ and α increase. The new values of $B_{\rm PN}$ and α may turn out to be in better agreement with the predictions of simple stellar population (SSP) theory (Renzini & Buzzoni 1986; Buzzoni *et al.* 2006).

8. A view to the future

1. We may be on our way to finally solving the maximum final mass enigma (Miller Bertolami's new evolutionary tracks, plus a little help from the initial-final mass relation).
2. The PNLF cannot be expected to play a dominant role in future extragalactic distance determination; other methods have advantages. For example, the PNLF method cannot migrate toward the infrared, like the SBF and TRGB methods.
3. On the other hand, once the PNLF-SBF distance problem is fixed, we will gain confidence to use the PNLF and consequent limits on PN formation rates to learn about post-AGB evolution in a variety of stellar populations.

References

Bird, S., Harris, W. E., Blakeslee, J. P., & Flynn, C. 2010, *A&A*, 524, A71
Blakeslee, J. P., Jordan, A., Mei, S., *et al.* 2009, *ApJ*, 694, 556
Buzzoni, A., Arnaboldi, M., & Corradi, R. L. M. 2006, *MNRAS*, 368, 877
Ciardullo, R. 2012, *Ap&SS*, 341, 151
Ciardullo, R., Feldmeier, J. J., Jacoby, G. H., *et al.* 2002, *ApJ*, 577, 31
Ciardullo, R., Jacoby, G. H., Ford, H. C., & Neill, J. D. 1989, *ApJ*, 339, 53
Jacoby, G. H. 1989, *ApJ*, 339, 39
Jacoby, G. H., Ciardullo, R., & Ford, H. C. 1990, *ApJ*, 356, 332
Makarov, D., Makarova, L., Rizzi, L., *et al.* 2006, *AJ*, 132, 2729
Méndez, R. H. 1999, in: A. Heck & F. Caputo (eds.), "Post-Hipparcos Cosmic Candles", *Ap. & Space Sci. Library*, 237, 161 (Kluwer)
Méndez, R. H., Kudritzki, R. P., Ciardullo, R., & Jacoby, G. H. 1993, *A&A*, 275, 534
Méndez, R. H., Riffeser, A., Kudritzki, R. P., *et al.* 2001, *ApJ*, 563, 135
Méndez, R. H. & Soffner, T. 1997, *A&A*, 321, 898
Méndez, R. H., Teodorescu, A. M., Schönberner, D., *et al.* 2008, *ApJ*, 681, 325
Miller Bertolami, M. M. 2016, *A&A*, 588, A25
Renzini, A. & Buzzoni, A. 1986, in: C. Chiosi & A. Renzini (eds.), "Spectral Evolution of Galaxies", *Ap. & Space Sci. Library*, 122, 195 (Reidel)
Teodorescu, A. M., Méndez, R. H., Bernardi, F., *et al.* 2010, *ApJ*, 721, 369
Teodorescu, A. M., Méndez, R. H., Bernardi, F., *et al.* 2011, *ApJ*, 736, 65
Tonry, J., Dressler, A., Blakeslee, J. P., *et al.* 2001, *ApJ*, 546, 681
Tonry, J. & Schneider, D. P. 1988, *AJ*, 96, 807

Discussion

SCHOENBERNER: According to our hydro models, the visibility time is controlledby nebular expansion speed and how small surface brightness or line fluxes onecan observe. For

Frew's local sample, the visibility time is about 20,000years see Paper VIII(?) of our series on dynamical evolution of PNe).

De Marco: Self extinction of the most massive planetaries explains the absenceof a sprinkling of super bright PNe, brighter than even the cut-off. If this iscorrect, how can it be that dust self obscuration is so precise as to eliminateexactly and only those PNe brighter than the cut-off?

Mendez: The extinction would affect preferentially those nebulae around themost massive stars, because the star evolves more quickly and the nebula isdenser.

Arnaboldi: Have you considered whether systematic differences between PNLF and SBFdistances at $D > 10$ Mpc can be caused by contamination of Ly alpha emittersat $z = 3.14$? In M49, M87 we have reached consistency – see Longobardi+2013and Hartke+2016 in prep. – this conference –

Mendez: In several cases we have concluded that Ly a emitters cannot be animportance source of contamination (Teodorescu et al. 2010, ApJ, 721, 369; 2011ApJ, 736, 65). I cannot rule out an isolated case, but Ly a emitters do notseem to me as a plausible solution for the whole PNLF-SBF discrepancy.

Session 8: First results from Gaia and looking to the future

Chair: Albert Zijlstra

Planetary Nebulae: Multi-Wavelength Probes of Stellar and Galactic Evolution
Proceedings IAU Symposium No. 323, 2016
X. Liu, L. Stanghellini & A. Karakas, eds.

doi:10.1017/S1743921317002137

Gaia and the Planetary Nebulae

Minia Manteiga[1], Carlos Dafonte[1], Marco A. Álvarez[1], Daniel Garabato[1], Ana Ulla[2] and Carme Jordi[3]

[1]Dept. of Information and Communication Technologies, University of A Coruña, E-15071 A Coruña, Spain

[2]Dept. of Applied Physics, University of Vigo, E-32004 Vigo, Spain
[3]Dept. of Astronomy and Meteorology, E-08028 Barcelona, Spain

Abstract. About three years after Gaia was set into orbit, the first release of Gaia data, DR1, has been published by ESA and DPAC. Gaia's first archive contains the results from the analysis of the initial 14 months of mission data. Outstandingly, it includes TGAS, Gaia-Tycho2 astrometric solution for about 2 million stars up to visible magnitude around 11.5.

In addition to the five parameter astrometric solution for the sky (positions, parallaxes and proper motions), future Gaia data releases will provide the spectral energy distributions from 330-1050 nm, together with radial velocities, for sources brighter than visible magnitude 16. Within this context, the relevance of Gaia data for the study of PN is be briefly presented.

Keywords. Gaia, distances, astrometry, parallax

1. Parallaxes and distances in DR1/TGAS

Gaia DR1 (Brown *et al.* 2016 and Prusti *et al.* 2016) consists of three parts: 1) an astrometric data set which contains the positions, parallaxes, and mean proper motions for about 2 million of the brightest stars in common with the Hipparcos and Tycho-2 catalogues (TGAS); 2) the positions and G-band magnitudes for an additional 1.1 billion sources (the secondary astrometric data set); and 3) the G-band light curves of ~3000 Cepheid and RR Lyrae stars. As a consequence of the very preliminary nature of this first Gaia data release, there is a number of important limitations to the data quality. These limitations are documented in the A&A papers that followed Gaia DR1, with further information provided on-line (gaia.esac.esa.int). TGAS astrometric uncertainties as a function of Gaia passband magnitude G are summarized in Table 1. It should be noted that an extra systematic error as large as 0.3 mas has been reported by DPAC, a fact not to be ignored in DR1.

A sample of 22 bright PN and proto-PN is contained in TGAS. However, only 7 of those appear with relative parallax errors (1-sigma) below 43%. Table 2 shows such data, and some of the derived distances in comparison with other non model-dependant previously published distances. This information is presented and analysed in detail by Stanghellini *et al.* in these proceedings. It is worth noting that deriving distances by

G (mag)	σ_π (mas)
9.3	0.24
11	0.32
12	0.64

Table 1. TGAS uncertainties in parallax.

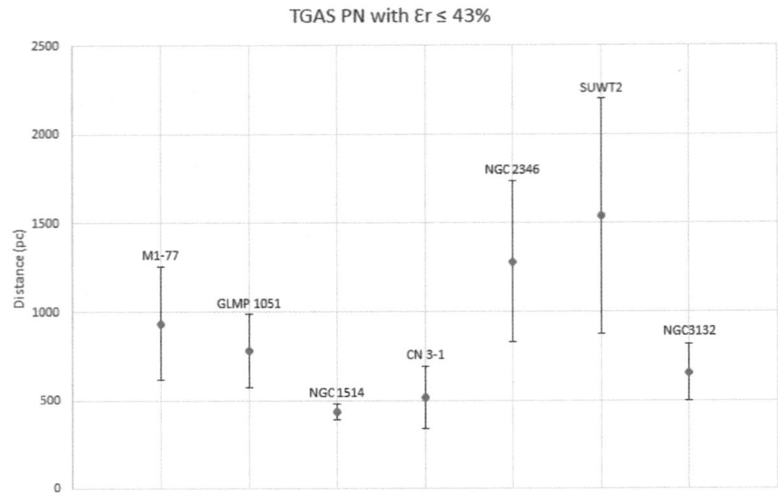

Figure 1. Distances to PN in DR1.

largerly uncertain parallaxes is not straithforward and some prior information regarding the star distribution should be taken into account as for instance in Astraatmadja & Bailer-Jones (2016). Figure 1 shows TGAS distances for PN with parallax relative errors less than 43%. In future releases, as more Gaia epoch measurements will be added, most accurate parallaxes will be derived.

DPAC end-of-mission (eom) expectation is that Gaia parallaxes will improve by a factor of 20-30. This will imply that PN brighter than magnitude 15 can be expected to have their eom parallaxes derived with accuracies of the order of 25 microarcsec, which turns into relative errors around 10% at 5 kpc and 20% at 10 kpc.

2. Spectral Energy Distributions from Gaia BP/RP photometers

Observations from the photometric instrument will be specially interesting to planetary nebula research. Gaia blue and red spectrophotometers (BP/RP) cover, respectively, the $330 - 660$ nm and $650 - 1050$ nm wavelength regions, with a dispersion ranging from $4 - 32$ nm per pixel in the blue to $7 - 15$ nm in the red. These spectra have the potential to discover a significant number of young compact PN and postAGB objects with emission lines. In fact, the most intense emission lines are visible in BP/RP simulated spectrophotometry. This enables DPAC to define spectral indices to diagnose the physical nature of the sources and to disentangle PN from other emission line objects. DPAC has developed an algorithm for unsupervised classification of Gaia spectrophotometric observations based on Self Organizing Maps (SOM), that is expected to facilitate the segmentation of astronomical object types (Fustes, Manteiga *et al.* (2013)), in order to make a full use of Gaia potential to discover new objects.

3. PN morphological information from the astrometric instrument and the Sky Mapper detections

Gaia will observe not only point sources but many objects with a structure (extended objects). These are expected to be essentially galaxies, but may also be star forming regions or planetary nebulae. The morphological information of the structure of these

PN G name	other name	π(mas)	$1/\pi$(pc)	$\sigma(\pi)$	$\epsilon_r(\pi)$	G(mag)	other distance
PN G104.8-06.7	M 2-54	-0.33		0.35		12.147	
PN G076.6-05.7 proto	V* V1853 Cyg	0.27		0.38	141	11.030	
PN G89.3-02.2	M1-77	1.07	935	0.36	34	11.825	2500 ± 500
PN G099.3 -01.9 proto	GLMP 1051	1.28	781	0.35	27	12.273	
PN G029.1-21.2 proto	LSIV-12 111	-0.64		0.56		11.312	
PN G165.5-15.2	NGC 1514	2.29	437	0.24	10	9.220	550 ± 190
PN G64.7+05.0	BD+30 3639	0.28		0.43	154	10.242	2689
IRAS 18371-3159 proto	PM 1-248	0.63		0.87	138	11.992	
PN G038.2+12.0	Cn 3-1	1.93	518	0.66	34	12.304	4680 ± 1620
PN G166.1+10.4	IC 2149	0.25		0.4	160	11.216	1950 ± 450
PN G345.2-08.8	IC1266	0.22		0.49	223	11.246	2670 ± 760
PN G332.9-09.9	He 3-1333	0.19		0.65	342	11.094	9840 ± 3360
PN G334.8-07.4	Hen 3-1312	0.31		0.24	77	11.057	4400
PN G315.1-13.0	He 2-131	0.09		0.35	389	10.676	590 ± 180
PN G331.1-05.7	Hen 2-172	0.74		0.68	92	11.648	
PN G215.6+03.6	NGC 2346	0.78	1282	0.27	35	11.147	1060 ± 340
PN G321.0+03.9	He 2-113	0.36		0.62	172	11.219	7600 ± 2170
PN G052.7+50.7 proto	BD+33 2642	0.1		0.3	300	10.709	
PN G311.0+02.4	PN SuWt 2	0.65	1538	0.28	43	11.911	2300 ± 200
PN G316.1+08.4	He 2-108	-0.01		0.87		12.635	3630 ± 1040
PN G291.3+08.4	PHR J1134-5243	0.19		0.41	216	12.961	5080 ± 1010
PN G272.1+12.3	NGC 3132	1.52	658	0.36	24	10.040	540 ± 140

Table 2. Parallaxes and errors for PN in DR1. Column 6 contains relative errors in parallaxes. See text for details.

objects is present in Gaia observations and can, therefore, be retrieved from them. Current DPAC software is prepared to analyse extended sources using the astrometric field (AF) CCD scans. The windows that will be transmitted to Earth for image reconstruction have pixel angular resolutions of 59 mas AL (along scan) and 177 mas AC (across scan), respectively. Windows sizes vary depending on the source magnitude, from 1.06x2.12 arcsec (ALxAC) for sources brighter than $G = 13$, to 0.71x2.12 arcsec for fainter ones (Krone-Martins, Ducourant *et al.* (2013)). For each image, 2D maps will be measured and described by parameters such as concentration of light, asymmetry, smoothness and momentum. Then, a morphological classification will be performed via a Support Vector Machine algorithm. This morphological information will be published in Gaia final DR, around year 2023. Compact PN will be included in this analysis.

Gaia Sky Mapper (SM) CCD windows have angular sizes of 4.72x2.12 arcsec (ALxAC) and a pixel resolution $\geqslant$ 118 mas and they can also be retrieved for further analysis. SM is used to detect and select objects for higher resolution in the AF. The detection criteria of pointlike sources had to de relaxed for the detection of solar system objects. As a result, the filamentary structures of ionized gas around PN are being recovered. Figure 2 (from Castañeda-Pons (2015)), shows such detections around NGC 6302, the Butterfly Nebula, whose central star (CS) was not detected in these Gaia observations. An astrometric

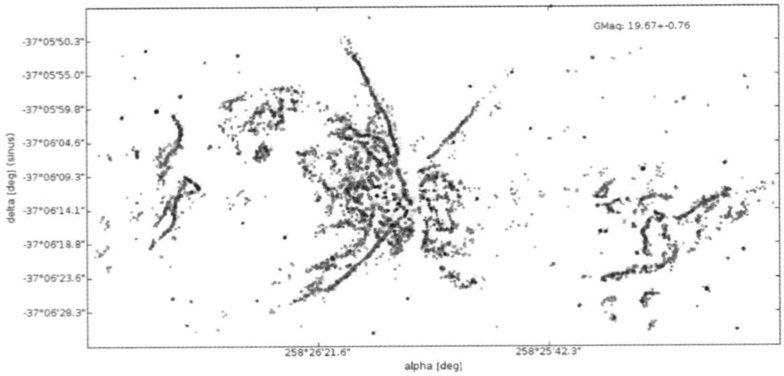

Figure 2. Ionized gas detections around NGC 6302 as scanned by Gaia Sky Mapper. Credits: Castañeda-Pons (2015)

accuracy of around 70 mas, for such SM detections along different observation epochs, has been found (Mignard (2016), DPAC). This means that both SM and AF detections around PN can be used for kinematics studies when, for instance, combined with HST images. This opens a whole new area of studies.

4. Conclusions

Gaia is certainly providing useful measurements for the study of PN. Some of the main possibilities offered by the mission data can be summarized as follows:

- TGAS parallaxes are, by now, very limited by source brightness and accuracy. We need to wait for the next data releases for accurate distance derivations
- Gaia AF instrument detections can provide parallaxes for nebulae with no CS detected
- We expect that BP/RP SEDs can allow to identify new PN (not to forget that Gaia is dealing with big statistics)
- Images of compact nebulae (sizes $\leqslant 4.72$ arcsecs) will be treated and classified
- Morphological information from SM detections can be obtained for extended PN
- SM and AF detections around PN can be used for kinematics (combined with HST) studies

References

Brown, A. G. A., Vallenari, A., *et al.* 2016, *A&A*, doi:10.1051/0004-6361/201629512
Prusti, T., de Bruijne, J. H. J., *et al.* 2016, *A&A*, doi:10.1051/0004-6361/201629272
Astraatmadja, T. L. & Bailer-Jones, C. A. L. 2016, *ApJ*, arXiv:1609.07369, in press
Krone-Martins, A., Ducourant, C., *et al.* 2013, *A&A*, Volume 556, id.A102, 11 pp
Fustes, D., Manteiga, M., *et al.* 2013, *A&A*, Volume 559, id.A7, 10 pp
Castañeda-Pons, J. B. 2015, PhD Thesis, University of Barcelona

Planetary Nebulae: Multi-Wavelength Probes of Stellar and Galactic Evolution
Proceedings IAU Symposium No. 323, 2016
X. Liu, L. Stanghellini & A. Karakas, eds.

doi:10.1017/S1743921317002125

Galactic planetary nebulae as absolute probes: The view from Gaia

Letizia Stanghellini[1], **Beatrice Bucciarelli**[2], **Mario G. Lattanzi**[2] **and Roberto Morbidelli**[2]

[1]National Optical Astronomy Observatory
950 N. Cherry Avenue, Tucson (AZ) 85719, USA
email: lstanghellini@noao.edu

[2]INAF-Osservatorio Astronomico di Torino (Italy)
email: lattanzi@oato.inaf.it, bucciarelli@oato.inaf.it, morbidelli@oato.inaf.it

Abstract. We searched the first Gaia data release for Galactic central stars of planetary nebulae (CSPNe) for parallaxes in order to determine the distances of the hosting PNe. For the small sample of PNe for which a comparison is available, we show that distances derived from Gaia parallaxes agree, within the uncertainties, with the individual PN distances derived by other reliable methods. While Gaia parallaxes available for Galactic CSPNe are still few, and with high uncertainties, we studied the possibility of building a PN distance scale by using the Gaia distances as calibrators. We found that a scale built on the relation between the linear nebular radius and its surface brightness has promising future applications.

Keywords. planetary nebulae: general; stars: distances

1. Motivation

Distances are among the most important basic parameters needed to study the physics of planetary nebulae (PNe) and their central stars (CSs). Large sample of Galactic PNe with reliable individual distances are sought for in a variety of applications. Distances to PNe have been derived from spectroscopic binary CSs, cluster membership, reddening, and nebular expansion, in ~ 40 Galactic PNe (see Stanghellini *et al.* 2008, and references therein), which represents a small fraction of the sample of several hundred, spectroscopy-confirmed Galactic PNe (Frew *et al.* 2016). Statistical distances, calibrated on these known distances, are commonly used. The problem with some of the statistical distance calibrators is that they rely at some level on modeling and assumptions. For example, reddening distances assume that the interstellar absorption toward a PN is similar to that of nearby stars, and does not account for possible patchiness in the ISM, while expansion distances assume that PN ejecta evolve homogeneously without acceleration. As a consequence, both the individual distances and the distance scale derived based on those can be severely misleading. The best individual distances are those derived for spectroscopic binary CSs, but only few of those are available to date.

2. Parallaxes of CSPNe in the first Gaia data release

The first Gaia data release (DR1, Gaia Collaboration *et al.* 2016) a few weeks ago has allowed us to retrieve the CSPN parallaxes that have been measured by Gaia with sufficiently low uncertainties to allow a distance determination. By searching DR1 against the

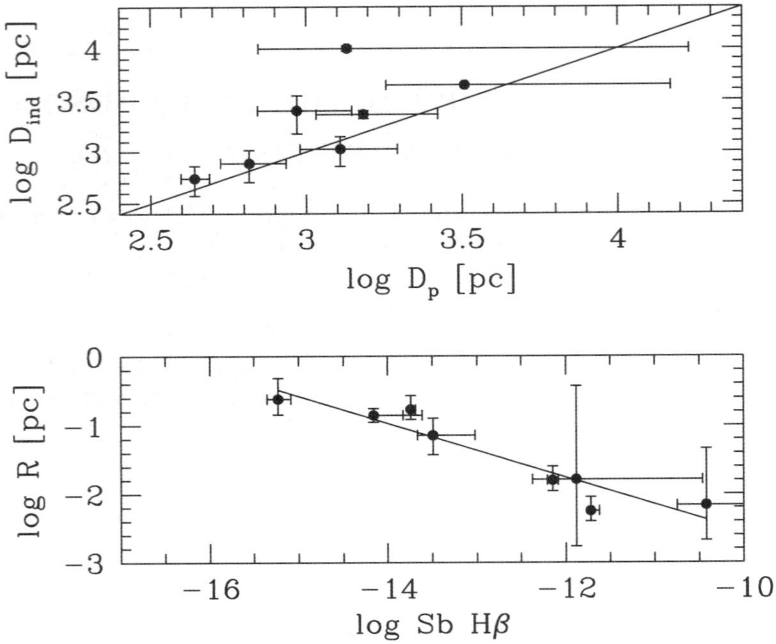

Figure 1. Top panel: Comparison between Gaia parallax distances and independent distances, as given in Table 1. Bottom panel: The physical radius – surface brightness relation obtained by calibrating with Gaia parallaxes. The resulting linear fit of the data, log R [pc] = -0.4 log Sb Hβ - 6.5, is an initial guess of the Galactic PN distance scale derived from the first Gaia release. Excluding Sa St 2-12 from the fit does not vary the correlation significantly.

astrometric positions corresponding to spectroscopically confirmed Galactic PNe (Acker *et al.* 1992, and updates) we found parallaxes for 8 CSPNe whose uncertainties are below 100%, thus usable for a positive distance determination. In Table 1 we list these targets (Cols. 1 and 2), the measured parallaxes of their CSs (3), their distances derived from Gaia parallaxes (4), and, if available in the literature, the independent distances (5), with the reference coded in Column (6). Independent distances are from spectroscopic parallaxes, except for M 1-77 and NGC 2346, whose distances are derived with the extinction method. Note that the distance to Su Wt 2 may refer to a companion star rather than the CS. It is also worth noting that the PN nature of Sa St 2-12 may be uncertain. The high relative error of trigonometric parallaxes and the non-linear relations among the physical parameters forbid a direct application of the first-order error propagation formula; left and right (asymmetric) error bars are therefore estimated by inserting $x \pm \sigma(x)$ into the corresponding functional relation, $\sigma(x)$ being the known standard deviation of the observed quantity x. Gaia parallax distances of CSPN span a broad range, and they agree with other independent distances within the errors, as shown in Fig. 1, where we plotted the available independent distances for the Gaia PN set against the best individual distances available in the literature. The best match is for nearby PNe, as expected, but even for PC 11, for which CS spectroscopy sets it at 10,000 pc, the two distances agree within the large parallax uncertainties. Note that all errorbars of the independent distances do not account for the intrinsic method uncertainties,

Table 1. Distance of Galactic PNe from Gaia parallaxes and from other methods.

Name	**Alias**	**p±σ(p)** [mas]	**log(D$_p$)** [pc]	**log(D$_{ind}$)** [pc]	ref
PN G038.2+12.0	Cn 3-1	1.932±0.656	$2.71^{+0.18}_{-0.13}$	...	...
PN G089.3-02.2	M 1-77	1.072±0.357	$2.97^{+0.18}_{-0.13}$	$3.025^{+0.15}_{-0.22}$	HW88
PN G165.5-15.2	NGC 1514	2.286±0.239	$2.64^{+0.05}_{-0.04}$	$2.74^{+0.12}_{-0.17}$	A15
PN G215.6+03.6	NGC 2346	0.778±0.269	$3.11^{+0.18}_{-0.13}$	$3.025^{+0.12}_{-0.17}$	G86
PN G272.1+12.3	NGC 3132	1.524±0.364	$2.82^{+0.12}_{-0.09}$	$2.89^{+0.13}_{-0.18}$	C99
PN G311.0+02.4	Su Wt 2	0.655±0.277	$3.18^{+0.24}_{-0.15}$	3.36±0.04	E10
PN G331.1-05.7	PC 11	0.741±0.682	$3.13^{+1.10}_{-0.28}$	4.00±0.03	P10
PN G334.8-07.4	Sa St 2-12	0.310±0.243	$3.51^{+0.66}_{-0.25}$	3.64	P04

which are typically very hard to quantify, as discussed in the original references listed in Table 1.

3. The Galactic PN distance scale from Gaia parallaxes

While there are still too few calibrators, and their uncertainties remain relatively high, the relation between physical radius and surface brightness is well defined (see Fig. 1, bottom panel) at all surface brightness in our range. We use the surface brightness of PNe measured in Hβ, after correcting for the interstellar extinction. The linear correlation coefficient of the two sets of parameters of Fig. 1 (bottom panel) is 0.95, showing a very tight correlation between the distance-dependent vs. the distance-independent parameter, make it a very promising distance scale. It is worth recalling that future data releases from Gaia will provide parallaxes of stars with V<15 mag with a precision of 0.03 mas (Lindegren *et al.* 2016). From Acker *et al.* (1992) we find that there are ~50 Galactic PNe in this magnitude range, and whose statistical distances are estimated to be smaller than 3000 pc (Stanghellini *et al.*, 2008). For this group of PNe, Gaia will provide final parallaxes of better than 10% relative uncertainties. For another ~40 PNe the estimated parallax relative uncertainties will be of the order of 20%. This wealth of new data will definitely help to constraint the scale. The present work is preliminary to set the stage for these future data sets.

References

Acker, A., Marcout, J., Ochsenbein, F., *et al.* 1992, *The Strasbourg-ESO Catalogue of Galactic Planetary Nebulae. Parts I*, II., by Acker, A.; Marcout, J.; Ochsenbein, F.; Stenholm, B.; Tylenda, R.; Schohn, C. European Southern Observatory, & Garching (Germany), 1992, 1047 p., ISBN 3-923524-41-2,

Aller, A., Montesinos, B., Miranda, L. F., Solano, E., & Ulla, A. 2015, *MNRAS*, 448, 2822 (A15)

Ciardullo, R., Bond, H. E., Sipior, M. S., *et al.* 1999, *AJ*, 118, 488 (C99)

Frew, D. J., Parker, Q. A., & Bojičić, I. S. 2016, *MNRAS* 455, 1459

Gaia Collaboration, Brown, A. G. A., Vallenari, A., *et al.* 2016, arXiv:1609.04172

Gathier, R., Pottasch, S. R., & Pel, J. W. 1986, *A&A*, 157, 171 (G86)

Huemer, G. & Weinberger, R. 1988, *A&AS*, 72, 383 (HW88)

Lindegren, L., Lammers, U., Bastian, U., *et al.* 2016, arXiv:1609.04303
Pereira, C. B. 2004, *A&A*, 413, 1009 (P04)
Pereira, C. B., Baella, N. O., Daflon, S., & Miranda, L. F. 2010, *A&A*, 509, A13 (P10)
Stanghellini, L., Shaw, R. A., & Villaver, E. 2008, *ApJ*, 689, 194-202

Planetary Nebulae: Multi-Wavelength Probes of Stellar and Galactic Evolution
Proceedings IAU Symposium No. 323, 2016
X. Liu, L. Stanghellini & A. Karakas, eds.

doi:10.1017/S1743921317002174

Gaia: mapping the dynamics and morphologies of Galactic Planetary Nebulae: First results from *Gaia* Data Release 1

Nicholas A. Walton[1], Marina Rejkuba[2], Jeremy R. Walsh[2] and Albert A. Zijstra[3]

[1]Institute of Astronomy, University of Cambridge,
Madingley Road, Cambridge CB3, OHJ
email: naw@ast.cam.ac.uk

[2]European Southern Observatory,
Karl-Schwarzschild Strasse 2, 85748, Garching, Germany
email: mrejkuba@eso.org and jwalsh@eso.org

[3]Jodrell Bank Centre for Astrophysics, Alan Turing Building,
Manchester, M13 9PL, UK
email: a.zijlstra@manchester.ac.uk

Abstract. This paper comments on the use *Gaia* in studying the internal dynamics and morphologies of Planetary Nebulae (PN). It is noted that the second and subsequent releases of *Gaia* data, will have significant potential in unravelling PN morphologies.

Keywords. planetary nebulae, distances, surveys, astrometry.

1. Introduction

Planetary Nebulae (PN) are a brief evolutionary stage through which low and intermediate mass stars pass towards the end of their evolution, between red giant and white dwarf. They play an important role in the processing of a number of elements into the surrounding interstellar medium. They can be used as powerful probes of the kinematical structure of the Milky Way, and can provide insights into the chemical evolution history of the Galaxy. Understanding the global role of PN is limited due to large uncertainties in individual distances and to a detailed knowledge of the dynamics of their nebulae. These factors in turn constrain the absolute parameters of PN, such as their sizes, luminosities, masses, lifetimes and determination of the overall Galactic PN population.

We briefly discuss the impact of the ESA *Gaia* mission data in supporting detailed studies of the expansion histories and morphological evolution of planetary nebulae.

2. *Gaia* and the *Gaia* Data Release 1

The ESA *Gaia* satellite was launched at the end of 2013. From the completion of its flight commissioning in July 2014, it collects data over its five year nominal mission to allow the determination of highly accurate positions, parallaxes, and proper motions for >1 billion sources brighter than magnitude 20.7 in the white-light photometric band G of *Gaia*. Its astrometry is complemented by multi-colour photometry, measured for all sources observed by *Gaia*, and radial velocities which are collected for stars brighter than G∼17. The mission and science goals are summarised in Gaia Collaboration *et al.* (2016b).

The first major data release of *Gaia* data (*Gaia* DR1) was made 14th September 2016 (Gaia Collaboration *et al.* 2016a). The *Gaia* DR1 contains astrometry, G-band photometry, and a small number of variable star light curves, for a total of ~1.14 billion sources. The two main astrometric components of *Gaia* DR1 consist of a primary astrometric data set with positions, parallaxes, and mean proper motions for ~2 million stars in common between the *Gaia* DR1, Hipparcos and Tycho-2 catalogues and a secondary astrometric data set containing the positions for an additional ~1.14 billion sources. The primary data set is the realisation of the Tycho-Gaia astrometric solution (TGAS), the principles of this are outlined in Michalik *et al.* (2015). The typical uncertainty is about 0.3 mas for the positions, and about 1 mas yr^{-1} for the proper motions. For the secondary data set the typical uncertainty on the positions is ~10 mas. Photometric data is released for all sources in the primary and secondary astrometric data sets, containing the mean *Gaia* G-band magnitudes. The brightest source in *Gaia* DR1 has a magnitude G = 3.2, while the majority of the sources (99.7%) are in the range $11.2 \leqslant G \leqslant 21$ (van Leeuwen *et al.* 2016).

3. *Gaia* and Planetary Nebulae

Gaia will detect the central stars of many planetary nebulae down to a G magnitude of 20.7. The only parallaxes available in *Gaia* DR1 are those of the TGAS for stars brighter than a visual magnitude of ~11.5 (Lindegren *et al.* 2016). Thus only the brightest PN are likely to have observable parallaxes from the *Gaia* DR1 release. This indeed is the case, with only a small number (<20) of the brightest central stars (Stanghellini *et al.* 2016) with a measured *Gaia* DR1 parallax. However, the situation will greatly improve with much more extensive set of parallaxes available with upcoming *Gaia* data releases, from *Gaia* DR2 onwards†.

Gaia is optimised for the detection of point sources, and in general is not sensitive to extended objects (with sizes >0.5 arcsec) (de Bruijne *et al.* 2015). The left side of Fig. 1 shows part of a *Gaia* Sky Mapper (SM) CCD image overlaid with detected sources. Red markers are for the brightest stars (G<13 mag) which are downlinked as small images. Yellow markers are for stars 13<G<16 mag; these are binned in across-scan direction. The cyan markers are the faintest stars with 16<G<20 mag where across-scan binning is done as well. Square shape markers, indicate those stars that will be used for *Gaia* attitude control.

However, as noted in de Bruijne *et al.* (2015), *Gaia* is able to resolve point like and somewhat extended structure. The exact performance of *Gaia* in detecting non-point like objects is complex, and is a function of object morphology, brightness and at which scan angle *Gaia* observes the source. In this respect, many features will only be observed at certain *Gaia* scan angles.

The sub-structure within extended objects such as PN is thus potentially detectable as point-like sources by *Gaia*. Sampling of these points may enable a detailed view of extended structure within the nebula. The capability of *Gaia* in this area was demonstrated by the early Ecliptic Pole Scanning observations of the large PN NGC 6543‡, where the complex nebula is decomposed by *Gaia* into thousands of individual mapping points. The right hand image of Fig. 1 shows the ~84 000 *Gaia* detections that were made in the area around NGC 6543 from 25 July to 21 August 2014.

† The *Gaia* data release scenario is available at `http://www.cosmos.esa.int/web/gaia/release`
‡ The *Gaia* Image of the Week (IoW_20141205) explores this in more detail, see `http://www.cosmos.esa.int/web/gaia/iow_20141205`

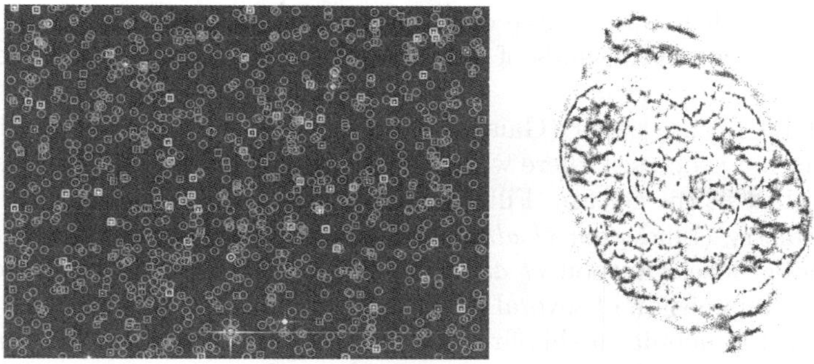

Figure 1. The left hand image shows a *Gaia* Sky Mapper image of a field obtained during commissioning (Image: ESA/Gaia/DPAC/Airbus DS). The right hand image shows the ~84,000 *Gaia* detections that were made in the area around NGC 6543 from 25 July to 21 August 2014 (Gaia image credit: ESA/Gaia/DPAC/UB/IEEC)

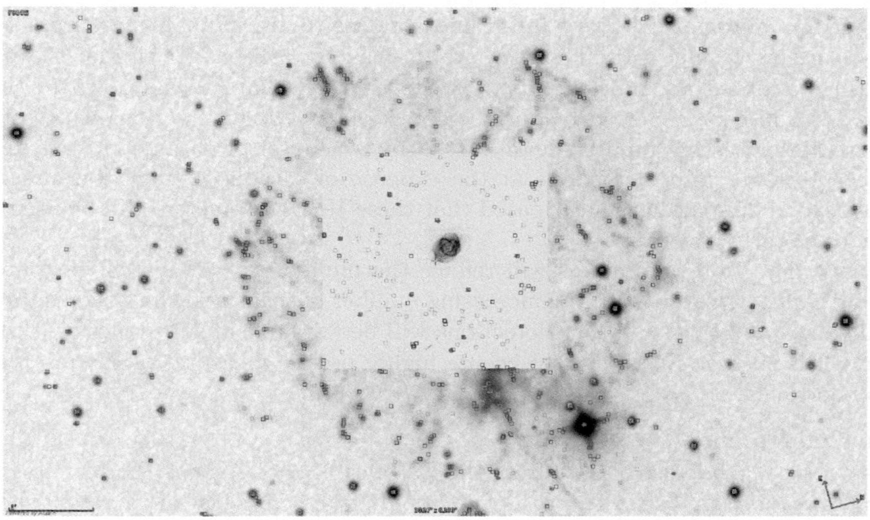

Figure 2. NGC 6543 as seen in Gaia DR1. The red dots are Gaia DR1 source detections, whilst the blue dots are USNO B1.0, and green dots GSC 2.3 sources. The background image is from DSS, whilst the central region is HST, both taken from CDS/Aladin.

With the five year nominal lifetime *Gaia* astrometry, it should be possible to accurately map the evolution of the plane of sky expansion of PN such as NGC 6543 and from line of sight velocities, the complex dynamical structure of the expanding nebula.

4. The Morphologies of Planetary Nebulae as Revealed by *Gaia*

Many PN have complex structure as described in a number of talks elsewhere in these proceedings. The PN morphologies can help infer the presence of, for instance, binary central stars.

Initial searches of the *Gaia* DR1 catalogues were made around a number of the brighter PN. One example is searching for source detections around NGC 6543. Fig. 2 shows the DSS image around NGC 6543, together with a high resolution HST image of the central region. Overlaid are the detections of stars as given in the USNO B1.0 and GSC 2.3 catalogues, along with detections in the *Gaia* DR1 catalogue. It is immediately apparent that there are very few *Gaia* source detections in and around NGC 6543. The question is where have the many thousands of detections that were indicated after commissioning gone?

The *Gaia* DR1 release paper (Gaia Collaboration *et al.* 2016a) notes that, as part of the data validation processes, there was a significant filtering of sources detected by *Gaia* but not published in the release. Filtering occurred at the processing sub-system level, thus for astrometry (Lindegren *et al.* 2016), for photometry (van Leeuwen *et al.* 2016), and at the global level (Arenou *et al.* 2016) where all data was combined for the data release. It is a combination of several of these filters that has resulted in the low number of detections in the nebula in this first data release. For instance, the requirement that each source should have been observed by *Gaia* at least 5 times meeting a threshold uncertainty limit.

The NGC 6543 observations were obtained in the early operational stages of *Gaia* with the at launch video processing unit (VPU) parameters active. As noted in de Bruijne *et al.* (2015), these govern the detection of sources in the SM *Gaia* CCDs. The detection algorithm automatically uses full-frame SM data to discriminate stars from spurious objects, such as cosmic rays and solar protons, using PSF-based criteria; the parameter settings adopted at launch ensured a high level of completeness down to the faint limit at G = 20.7 mag, but with spurious detections in the (diffraction) wings of bright stars essentially passing unfiltered. In May 2016, a new set of parameters was uploaded that accepts fewer false detections at the expense of a reduced detection efficiency of objects beyond 20 mag. It is anticipated that these VPU parameters will be used for the remainder of the mission.

The new detection parameters will reduce the number of faint extended sources detected by *Gaia*. However, with the increasing number of scans and thus *Gaia* data of PN, it is fully expected that sufficient detections will be available of sub-structure within the nebula of PN such as NGC 6543 to ensure highly precise measures of PN morphologies to be undertaken.

Acknowledgements

This work has made use of data from the European Space Agency (ESA) mission *Gaia* (`http://www.cosmos.esa.int/gaia`), processed by the *Gaia* Data Processing and Analysis Consortium (DPAC, `http://www.cosmos.esa.int/web/gaia/dpac/consortium`). Funding for the DPAC has been provided by national institutions, in particular the institutions participating in the *Gaia* Multilateral Agreement.

The Digitized Sky Surveys were produced at the Space Telescope Science Institute under U.S. Government grant NAG W-2166.

This research has made use of "Aladin sky atlas" developed at CDS, Strasbourg Observatory, France.

References

Arenou, F., Luri, X., Babusiaux, C., *et al.* 2016, *A&A* (special Gaia issue)
J. H. J. de Bruijne, J. H. J., Allen, M., Azaz, S., *et al.* 2015, *A&A*, 576, A74
Evans, D. W., Riello, M., De Angeli, F., *et al.* 2016, *A&A* (special Gaia issue)
Gaia Collaboration, Brown, A. G. A., Vallenari, A., *et al.* 2016, *A&A*, 595, A1
Gaia Collaboration, Prusti, T., de Bruijne, J. H. J., *et al.* 2016, *A&A*, 595, A1
van Leeuwen, F., Evans, D. W., De Angeli, F., *et al.* 2016, *A&A* (arXiv:1612.02952)
Lindegren, L., Lammers, U., Bastian, U., *et al.* 2016, *A&A*, 595, A4
Michalik, D., Lindegren, L., & Hobbs, D. 2015 *A&A*, 574, A115
Stanghellini, L., Bucciarelli, B., Lattanzi, M. G., & Morbidelli, R. 2016 (arXiv:1609.08840)

Planetary Nebulae: Multi-Wavelength Probes of Stellar and Galactic Evolution
Proceedings IAU Symposium No. 323, 2016
X. Liu, L. Stanghellini & A. Karakas, eds.

doi:10.1017/S1743921317000758

Conference Summary

Sun Kwok[1,2]

[1]Laboratory for Space Research,
The University of Hong Kong, Hong Kong, China
email: sunkwok@hku.hk

[2]Dept. of Earth Sciences,
The University of Hong Kong, Hong Kong, China

Abstract. The progress in planetary nebulae (PN) research reported in this symposium is reviewed in the context of our current understanding of the PN phenomenon.

Keywords. stars: AGB and post-AGB, (stars:) binaries: symbiotic, stars: evolution, (ISM:) planetary nebulae: general

1. Evolution of the subject

Research in planetary nebulae (PN) has undergone significant evolution over the years. Before the IAU PN Symposium in Ithaca in 1977, observational studies of PN were primarily focused on optical spectroscopy of atomic lines resulting from recombination and collisional excitation, and the determination of elemental abundance from these observations. After 1977, it was realized that PN consist of more than ionized gas, and observations have extended to infrared observations of continuum emissions from dust and millimeter-wave observations of line emissions from molecules. After the introduction of CCD detectors and the availability of observation platforms from space, detailed morphological structures of PN as well as exterior haloes were discovered through high-dynamic-range imaging (Jewitt *et al.* 1986, Corradi *et al.* 2003).

On the theoretical side, we knew that PN are progenitors of white dwarfs and descendents of red giants (Shklovsky 1956) and the PN shells are ejected outer envelopes of red giant stars (Abell and Goldreich 1966). However, the details of central star evolution and shell ejection mechanism were not understood. The identification of H-shell burning on an electronic-degenerate core as the energy source of central stars of PN (Paczyński 1971) made possible the construction of realistic evolutionary tracks (Schönberner 1979). This also put PN as a post-asymptotic-giant-branch (post-AGB) phenomenon. The realization that PN are formed through an interacting winds process solved many of the problems faced by sudden-ejection models (Kwok *et al.* 1978). New interacting-winds dynamical models coupled with time-dependent photoionization models opened new ways to interpret the complex morphological structures observed (Schönberner *et al.* 2014).

The ability to observe PN in extragalactic systems by large telescopes has also allowed PN to be used as kinematic tracers (Meatheringham *et al.* 1988) and standard candles for distance determination (Jacoby 1989). A summary of our modern understanding of PN is given by Kwok (2000).

2. Interaction with other fields

The field of PN research did not develop in isolation. This discipline of research has benefited from developments in other fields and progress in the field has influenced many

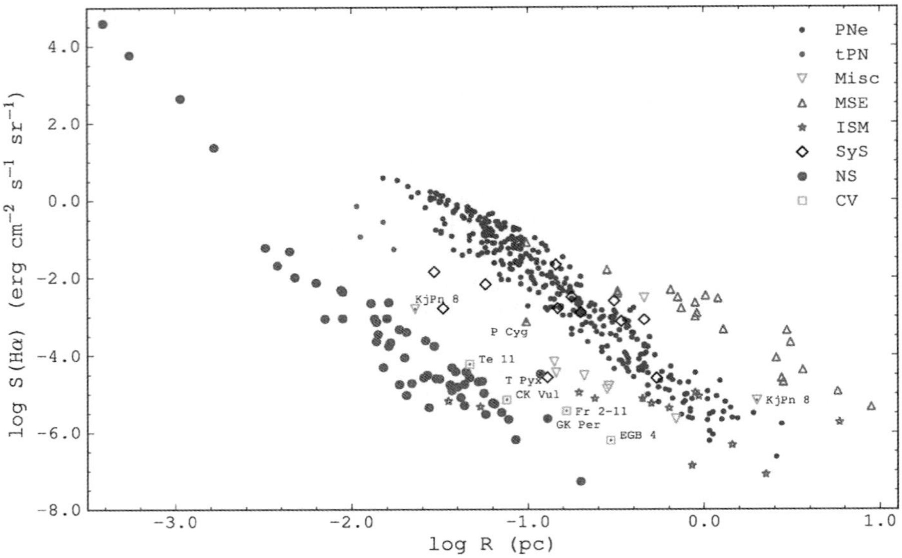

Figure 1. The Hα–radius plot as a diagnostic for ionized nebulae. Figure by David Frew.

other related areas. As photoionized regions, PN share many similar physics and diagnostic tools as HII regions and active galactic nuclei. As descendents of AGB stars and precursors of white dwarfs, research on PN rely on work on nucleosynthesis and mass loss of AGB stars and stellar atmosphere models of white dwarfs. Massive OB stars have fast stellar winds and Wolf-Rayet stars have Wolf-Rayet nebulae which have similar dynamical structures as PN. Studies of mass transfer and common envelope evolution in binary systems such as novae, cataclysmic variables and symbiotic stars have relevant lessons for PN. Some of the physical mechanisms in PN are similar to star formation regions as both have directional outflows collimated by dense equatorial tori. PN can serve as tracers of history of star formation and chemical enrichment in the Galaxy. Beyond the Galaxy, extragalatic PN can be used to map the distribution of dark matter and to establish the cosmological distance scale. Since PN are formation sites of molecules and complex organics, they play a role in the emerging new disciplines of astrochemistry and astrobiology.

It is in this context that I review the progress observed in this Symposium.

3. Surveys and catalogs

The classification and cataloging of objects is the first step to every field of research. The progress since the *Catalog of Galactic Planetary Nebulae* by Perek and Kohoutek (1967) and the *Strasbourg-ESO Catalog of Galactic Planetary Nebulae* of Acker *et al.* (1992) was reviewed by Frew (these proceedings). Narrow-band surveys have increased the number of PNs in the Galaxy to 3,300. Parker (these proceedings) describes the Hong Kong–AAO–Strasbourg Hα (HASH) database and the comprehensive functions that it offers. It is possible that many PN are not detected in optical surveys because of dust extinction and a survey in the near infrared may reveal many hidden PNs. The UWISH2 survey of the northern Galactic plane has identified 285 candidates using narrow-band imaging centered on the H_2 line (Gledhill, these proceedings). The search for

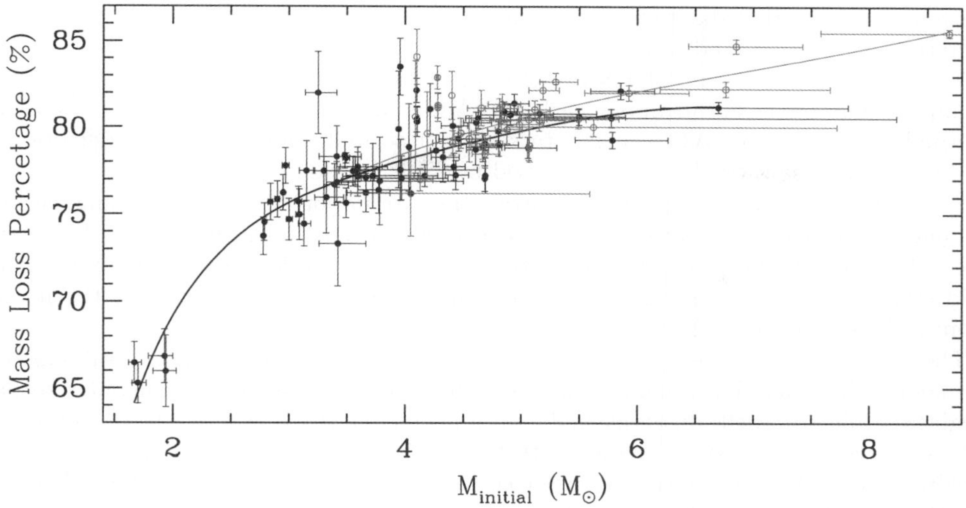

Figure 2. Percentage of total mass lost as a function of initial mass. Figure by Jeff Cummings.

proto-PN using infrared colors was reported by Szczerba (these proceedings) and Suh (these proceedings).

Since spectral and morphological properties of PN are also shared by other astrophysical objects such as symbiotic stars and compact HII regions, significant confusion problems still remain in existing catalogs. Various techniques on separating mimics from true PNs were discussed by Frew (these proceedings). A Hα surface brightness vs. radius plot was suggested to be an effective tool for this purpose (Figure 1). It is clear that we cannot embark on a meaningful discussion on the nature of PN if we do not have a consistently homogeneous group of member objects.

4. Distances

An accurate distance scale for PN is important as distances affect the luminosity determination and the distribution of PN on the Herztsprung-Russell (H-R) diagram. Statistical distances are known to have serious problems. Inaccurate distances, together with missing fluxes in the UV and infrared and poor estimates of central-star temperatures, led to the erroneous Herman-Seaton Sequence which misled a whole generation of theoreticians trying to produce evolutionary tracks of PN. Modern high-angular-resolution imaging offers hope in the use of expansion parallax (Schönberner, these proceedings). *GAIA* promises to provide accurate parallaxes to a large number of PNs, which may finally solve our long-standing problem with distances (Stanghellini, these proceedings).

5. Central stars and their evolution

The central stars of PN interact with the nebulae both radiatively and dynamically. Central stars of PN have a wide range of temperatures, emit photons of different energies, and are undergoing rapidly changing mass loss. Their time-dependent radiative and mechanical outputs determine the ionization and dynamical structures of the nebulae. The physical properties (temperature, surface gravity) of central stars can be studied by high-spectral-resolution spectroscopy and interpreted by non-LTE stellar atmosphere models. An accurate determination of these properties are essential in their accurate placement on the H-R diagram. Some central stars show X-ray emission or evidence of accretion. Some are found to be close or wide binaries.

While most central stars of PN are H-rich, the existence of H-deficient central stars requires explanation. For example, early [WC] stars are considered as products of a late thermal pulse but late [WC] central stars are not so well understood. The detection of [WN] stars and their origins were discussed by Todt (these proceedings).

PN containing binary or multiple central stars are important in determining the role of binary evolution in the formation of PN. However, correct identification of the central stars of PN is critical. A number of central stars of PN are found to be main-sequence field stars and there stars clearly are not responsible for the ionization of the nebula (Jones, these proceedings).

The existence of PN depends on the coupled evolution between the central star and the nebula. The nebular dynamical age is defined by the time that the nebular surface brightness falls below the observational limit (Jacob *et al.* 2013), whereas the evolutionary age is defined by the speed of the central star evolution. If a central star evolves too slowly, then the nebula would have dispersed before it is ionized. The consistency between the theoretical evolutionary age and the observed kinematic age is therefore a serious issue (Pereyra, these proceedings). Using different starting points, the new evolutionary tracks of Miller Bertolami (these proceedings) evolve much faster than previous tracks, resulting in much shorter lifetimes of tracks of the same mass. These new tracks have considerable effects on the initial-mass−final-mass relationship and the PN luminosity function (PNLF). These new tracks are found to be more consistent with the observed PN in the bulge, which predominantly come from an 8-12 Gyr population (Zijlstra, these proceedings). These new tracks are also found to be able to explain the rapid evolution of the central star of the Stingray Nebula (Reindl, these proceedings).

A good initial-mass−final-mass relation provides important constraints to PN evolution. Recent results from Cummings (these proceedings) show that intermediate-stars lose 60-80% of their mass during the AGB (Figure 2). This suggests that the observed ionized mass of PN ($\sim 10^{-1}$ $M_\odot$) represents only a small fraction of the total mass of the circumstellar envelope.

6. Physical structure and morphology

Planetary nebulae are interesting astrophysical objects because they contain all states of matter, from ions, neutral atoms, molecules, and solid-state dust particles. High-dynamic-range imaging of PN has revealed that PN have very complex internal structures even in the optical. Various features such as shell, rim, crown, and halo can be found in the ionized-gas component of PN (Frank *et al.* 1990). These multi-layered structures are formed as the result of interactions between consecutive stages of mass loss from the central star and their subsequent ionization by the evolving central star.

The hot bubble is formed by the shocked fast wind and the temperature structure and abundance ratios of the bubble can be studied by observing the ratios of X-ray lines (Schönberner, these proceedings). The mixing layer between the hot bubble and the optical shell has intermediate (10^5 K) temperatures and can be studied with UV spectroscopy (Guerrero, these proceedings). X-ray imaging of the hot bubble can be used to test dynamical models (Montez, these proceedings).

The observed changing abundance of molecular species with evolution provides constraints to circumstellar chemical synthesis models (Zhang, these proceedings). Molecular hydrogen is an effective tracer of the shock region in PNs (Manchado, these proceedings).

From the observed spectral energy distributions of PN, we learn that a significant fraction of the total energy output of PN is from dust emission (van de Steene, these proceedings).

Since a great deal of mass is in the neutral component, infrared mapping of the dust component is important to determine the true mass distribution of PN. We should note that the bright optical nebulae are the low-density cavities carved out by high-speed outflows and optical brightness does not translate into matter distribution (Kwok 2010). The equatorial torus, which focus the bipolar outflow, can be mapped in the mid-infrared and compared to the optical image of the lobes (Lagadec, these proceedings). The outer dust envelope can be mapped in the far infrared with space-based telescopes such as *Herschel* (van de Steene, these proceedings).

7. Complex molecules

One of the most exciting development in recent years is the detection of complex molecules and organic solids in PN. Fullerene (C_{60}) molecules are now widely detected in PN (Cami, these proceedings; Díaz-Luis, these proceedings; Otsuka, these proceedings). It is possible that hydrogenated fullerenes are also present (Zhang, these proceedings). The crucial question of the chemical structure of the carrier of the unidentified infrared emission (UIE) bands was discussed by Sloan (these proceedings). PN are the only astronomical sources where we can observe the UIE bands emerge. The short ($\sim 10^3$ yr) time scale of formation of the UIE bands provides useful clues to the pathway of synthesizing complex molecules in space. The C_{60} and UIE phenomena are linked by the presence of the 8, 12, and 17 μm plateau features, leading to the discussion of whether C_{60} is formed by a top-down or bottom-up pathway.

8. Nucleosynthesis

The elemental abundance discrepancy (AD) between determinations from optical recombination lines (ORL) and collisionally excited lines (CEL) has been with us for decades. Recently obtained high-quality spectra have been able to determine the AD factors much more accurately (Delgado Inglada, these proceedings). If the discrepancy is due to ORL and CEL being emitted in different regions, a kinematic study may reveal differences (Pena, these proceedings). Some of the most extreme discrepancy cases are found in binary systems, in particular among post-red-giant-branch objects (Wesson, these proceedings; Garcia-Rojas, these proceedings).

An accurate photoionization model is needed to properly interpret emission-line spectra. Improvements in photoionization models include taking into account stellar atmosphere models as input and 3D nebular models (Morisset, these proceedings). Incorporation of dust continuum transfer into photoionization models was discussed by Otsuka (these proceedings).

On the theoretical side, AGB nucleosynthesis models are continuously being refined. Extended nuclear networks have been developed to include reactions involving intermediate neutron processes (Doherty, these proceedings). García-Hernández (these proceedings) reminds us that O is not always a good indicator of original ISM metallicity.

One of the most interesting new results is the detection of new neutron-capture elements through lines of Rb IV, Cd IV, Se III, and Kr VI (Sterling, these proceedings; Dinerstein, these proceedings). Through nucleosynthesis models, one can use rubidium as signature of the initial mass of the star (Lugaro, these proceedings).

Pre-solar grains such as SiC represent an important link between AGB stars and the solar system. The isotopic ratios of noble gases in pre-solar grains can be actually measured in the laboratory, giving a precise test of nucleosynthesis models (Lugaro, these proceedings).

9. Binary evolution

An increasing number of AGB and PN are found to have binary central stars and there is a great deal of interest to tie the binary nature to the nebular morphology. *ALMA* offers great enhancements to sensitivity and angular resolution in molecular mapping of circumstellar envelopes. Kim (these proceedings) has found evidence for spiral structures in the circumstellar envelope of the carbon star AFGL 3068. These spirals are probably created by stellar winds of the orbiting central star, with the orbital plane of the binary system lying on the plane of the sky.

A long-term study to put limits on binaries in proto-PN through radial velocity monitoring was reported by Hrivnak (these proceedings). Since proto-PN are loosely labeled as such from objects showing infrared excesses, some proto-PN candidates were found to be circum-binary disc systems and therefore are un-related to the PN phenomenon (van Winckel, these proceedings). Dusty disks around red giant stars can also cause confusion, causing these objects wrongly labeled as post-AGB objects (Kamath, these proceedings).

Recent observations have found that planets are common around stars, so the effects of planets in AGB and post-AGB evolution is an interesting topic to investigate. Do planets play any role in morphological shaping of PN (Villaver, these proceedings; Boyle, these proceedings)?

It has been known for over 40 years that close binary systems can develop common envelopes, resulting in common-envelope ejections (Paczyński 1976). The possible relevance of common envelope ejection with PN was investigated by many authors, with numerous possible variations and scenarios possible (Iben and Tutukov 1989, de Kool 1990). In order to assess the relevance of common envelope evolution with the PN phenomenon, it would be useful to compare the fraction of common envelope objects in the PN population to the theoretical expectation from population synthesis (De Marco, these proceedings). If PN are products of binary evolution, one needs to specify which one out of several hundred possible channels (Tutukov & Yungelson 2002) that PN emerge from. Since post-ejection evolution is heavily dependent on the remnant H-envelope mass, we need to know whether the post-common-envelope-ejection object will evolve. If it does not evolve on thousand-year timescales due to too large a remnant envelope mass, do we still call it a PN?

Recognition of PN as a post-AGB phenomenon only happened in 1970, 200 years after the first discovery of PN. So we know that fundamental paradigm shift can happen. However, to challenge the current paradigm, one has to present a self-consistent alternative model which can explain evolution, morphology, abundances, kinematics, and other physical properties of PN.

10. PN as probes and tracers

The bright and narrow emission lines of PNs allow for their use as tracers of chemical and dynamical evolution of galaxies. The observed elemental abundance gradients reflect different stellar yields, infall rate laws, and star formation history (Molla, these proceedings) as well as external influence such as accretion or merger (Bresolin, these proceedings). New large-telescope observations have given accurate elemental abundances of M31, including the outer regions. The oxygen abundance from the inner regions of M31 to beyond 100 kpc is almost constant with near solar values (Fang, these proceedings).

As dynamical tracers, PN extend the coverage of rotation curves and trace stellar content in the outer regions (Coccato, these proceedings; Söldner-Rembold, these proceedings). Modern wide-field narrow-band surveys make possible the observations of PN in haloes of galaxies in the Virgo cluster. The homogeneous sample allows the determination of production rates of PN (Hartke, these proceedings). The potential effects of triaxiality was discussed by Arnaboldi (these proceedings). A key question is whether the Keplerian decline of line-of-sight velocity dispersion observed in several elliptical galaxies can be mostly attributed to a relative lack of dark matter or to the effects of radial anisotropy.

The effects of the new evolutionary tracks of Miller Bertolami on the PNLF was reviewed by Mendez (these proceedings). The possibility that PNLF can be a function of galactic location was discussed by Balick (these proceedings).

11. What is the future of PN research?

Progress in PN research has benefited greatly from technological advances such as CCD, infrared, millimeter-wave, and X-ray observational techniques. Many of the new observing facilities (*ALMA, GAIA, JWST, SOFIA*) can further extend the observational frontiers and we should try to take advantage of these facilities. It has been amply demonstrated in the past that knolwedge gained in PN research can be applied to other fields and it is important that we continue to stay relevant in the view of other disciplines.

As for the field itself, we still have many unsolved problems, in part because PN are complex objects and in part because we have so much high quality data across all wavebands to test theoretical models. We know that the ionized component does not contain most of the mass in PN, but the exact ratio of ionized to molecular mass of PN is not known. Future high-angular-resolution mappings of the ionized, molecular, and dust components can resolve this issue. Although we have detailed images of PN showing the various morphological components, the physical mechanisms for various wind components leading to the shaping of PN are yet to be identified.

In order to have a model for chemical synthesis, we need a detailed time sequence of the synthesis, including the exact relationship between the C_{60} and UIE emitters. An accurate determination of the chemical structure of organic compounds synthesized during the PN stage will be an important step in solving the mysteries of unidentified spectral phenomena such as diffuse interstellar bands, 220 nm feature, 21 and 30 μm features, and the extended red emission.

Scientific fashions come and go, but PN research has remained resilient after over 200 years. This field of research is still important as a laboratory to study physical and chemical processes and has proven useful as a diagnostic tool for other areas of research. Tremendous progress on the understanding of the nature of PN was made in the last 40 years. Whether this pace of advance can be maintained will depend on the continued injection of new talents.

Acknowledgements I wish to thank Roberto Mendez, Detlef Schönberner, Mike Barlow, and Yong Zhang for helpful comments on an earlier draft of this paper. I am grateful to Xiaowei Liu and the LOC for a very pleasant experience in Beijing. The Laboratory for Space Research was established by a special grant from the University Development Fund of the University of Hong Kong. This work is also in part supported by grants from the HKRGC (HKU 7027/11P).

References

Abell, G. O. & Goldreich, P. 1966, *PASP*, **78**, 232

Acker, A., Ochsenbein, F., Stenholm, B., Tylenda, R., Marcout, J., & Schohn, C. 1992, *Strasbourg-ESO catalog of Galactic Planetary Nebulae*, ESO publications

Corradi, R. L. M., Schönberner, D., Steffen, M., & Perinotto, M. 2003, *MNRAS*, 340, 417

de Kool, M. 1990, *ApJ*, **358**, 189

Frank, A., Balick, B., & Riley, J. 1990, *AJ*, 100, 1903

Iben, I., Jr. & Tutukov, A. V. 1989, in *IAU Symp. 131: Planetary Nebulae*, ed. Torres-Peimbert, (Kluwer), p. 505

Jacob, R., Schönberner, D., & Steffen, M. 2013, *A&A*, 558

Jacoby, G. H. 1989, *ApJ*, 339, 39

Jewitt, D. C., Danielson, G. E., & Kupferman, P. N. 1986, *ApJ*, 302, 727

Kwok, S. 2000, *The Origin and Evolution of Planetary Nebulae* (CUP)

Kwok, S. 2010, *PASA*, 27, 174

Meatheringham, S. J., Dopita, M. A., Ford, H. C., & Webster, B. L. 1988, *ApJ*, 327, 651

Paczyński, B. 1971, *Acta Astr.*, **21**, 417

Paczyński, B. 1976, in *IAU Symp. 73: Structure and Evolution of Close Binary Systems*, P. Eggleton, S. Mitton, S., & J. Whelan (eds), (Reidel:Dordrecht), p. 75

Perek, L. & Kohoutek, L. 1967, *Catalog of Galactic Planetary Nebulae*, (Academia: Praha)

Shklovsky, I. 1956a, *Astr. Zh.*, **33**, 315

Schönberner, D. 1979, *A&A*, **79**, 108

Schönberner, D., Jacob, R., Lehmann, H., *et al.* 2014, *Astron. Nachr.*, 335, 378

Tutukov, A. V. & Yungelson, L. R. 2002, *Astronomy Reports*, **46**, 667

Part 2. Poster Presentation

Planetary Nebulae: Multi-Wavelength Probes of Stellar and Galactic Evolution
Proceedings IAU Symposium No. 323, 2016
X. Liu, L. Stanghellini & A. Karakas, eds.

doi:10.1017/S1743921317003234

The Hong Kong/AAO/Strasbourg Hα (HASH) Planetary Nebula Database

Ivan S. Bojičić[1,2], **Quentin A. Parker**[1,2] and **David J. Frew**[1,2]

[1]Department of Physics, The University of Hong Kong
Hong Kong SAR, China

[2]Laboratory for Space Research, The University of Hong Kong
Hong Kong SAR, China
email: ibojicic@hku.hk

Abstract. The Hong Kong/AAO/Strasbourg Hα (HASH) planetary nebula database is an online research platform providing free and easy access to the largest and most comprehensive catalogue of known Galactic PNe and a repository of observational data (imaging and spectroscopy) for these and related astronomical objects. The main motivation for creating this system is resolving some of long standing problems in the field e.g. problems with mimics and dubious and/or misidentifications, errors in observational data and consolidation of the widely scattered data-sets. This facility allows researchers quick and easy access to the archived and new observational data and creating and sharing of non-redundant PN samples and catalogues.

Keywords. catalogs, (ISM:) planetary nebulae: general

1. Architecture

The HASH PN database (HASH) and its online interface (`http://hashpn.space`) is currently hosted and maintained on a dedicated server located on The University of Hong Kong network. The main technologies used in the web interface are PHP and JavaScript. The backend of the system (i.e. image/spectra/data parsers, image/spectra visualisation and data analysis) consist of in-house scripts written in python using several freely available scientific computing, astronomical and plotting modules (e.g. APLpy, astropy)†. The basic full system (frontend and backend) flowchart is presented in Fig. 1 and is similar to other astronomical databases (e.g. Moreau & Dubernet 2006).

The backbone of HASH is the relational database (MySQL) which provides data consistency and quick, efficient search and manipulation. Each HASH object is characterised by a unique set of parameters (i.e. unique id, coordinates of the nebular centroid and the PN G designation) and mapped to the corresponding observational data collected from the literature and astronomical data repositories.

2. HASH Content

Over 6700 Galactic objects are currently in our working database, of which ∼3500 are designated as true, likely and possible PNe (for more details and the base catalogue construction and PN identification methods see Parker *et al.* (2016), Frew & Parker, 2010 and Parker *et al.*, these proceedings). The observational data repository contains over 300k fits image cutouts from over 30 large scale imaging surveys. We mirrored the full data from some of these surveys on our server (e.g. SHASSA, NVSS, etc) thus allowing the benefit of fast data access. On the other hand, high resolution, ongoing surveys

† http://www.astropy.org/; https://aplpy.github.io/

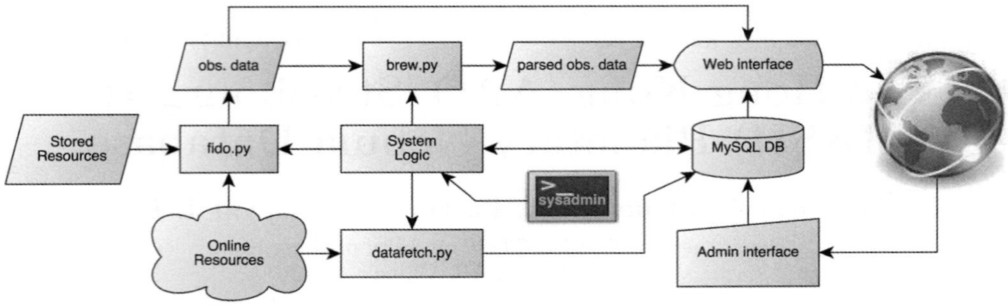

Figure 1. HASH system flowchart. After a new object is added to the database its corresponding imaging and tabulated data are first automatically collected from online and local resources using in-house written python scripts. The initial dataset is then manually vetted by the database administrators (the co-authors) applying position corrections, correcting morphological class etc, and then released to the publicly accessible web interface. We also encourage users to suggest corrections or modifications through the system itself (by adding a note on the object's individual page).

and large datasets (e.g. 2MASS, UKIDSS) are queried on demand using provided data access services or, preferably, Simple Image Access Protocol (SIAP; for details see IVOA Recommendation documents at `http://www.ivoa.net/documents/`). HASH contains over 7000 reduced spectra and over 85000 measured spectral line fluxes (relative and/or flux calibrated). One of the most important emission line data sources is the Electronic Emission-Line Catalog for Planetary Nebulae (ELCAT; Kaler *et al.*, 1997). We extended ELCAT with data for $\sim$ 500 Galactic PNe published after ELCAT was first released.

The tabulated data for objects in the database is parsed from Vizier or directly from the literature. In this version we did not include any derived parameters found in the literature (distances, physical sizes, metalicities, etc.) mainly because of inhomogeneities in methods and data.

In addition to already published observational data we provide new and more accurate informations on PN positions, angular diameters (in optical bands) and morphologies. Astrometric positions reported here are best estimates of the position of the centroid of the nebula. Positions of ionising (central) stars are catalogued separately in order to enable scrutiny and analysis of that data set separately.

3. Future work

The full catalogue of known Galactic PNe and its accompanying observational data will soon be published and integrated into the CDS database (Bojičić *et al.* in preparation). As we continue to detect new PNe and confirm additional PN candidates, together with the prospect of more accurate distance estimates (Frew *et al.* 2016, Stanghellini *et al.* these proceedings) the next version of HASH will contain a more complete sample of Galactic PNe as well as including a large range of homogeneously derived physical parameters.

References

Frew, D. J. & Parker, Q. A., 2010, *PASA*, 27,129
Frew, D. J., Parker, Q. A., & Bojičić, I. S., 2016, *MNRAS*, 455, 1459
Kaler, J. B., Shaw, R. A., & Browning, L., 1997, *PASP*, 109, 289
Moreau, N. & Dubernet, M. L., 2006, *ASP Conf.Ser.*, 351, 391
Parker, Q. A., Bojičić, I. S., & Frew, D. J., 2016, *JPhCs*, 728, 032008

Planetary Nebulae: Multi-Wavelength Probes of Stellar and Galactic Evolution
Proceedings IAU Symposium No. 323, 2016
X. Liu, L. Stanghellini & A. Karakas, eds.

doi:10.1017/S1743921317000473

Planetary Nebula Candidates Uncovered with the HASH Research Platform

Vasiliki Fragkou[1,2], Ivan Bojičić[1,2], David Frew[1,2] and Quentin Parker[1,2]

[1]The University of Hong Kong, Department of Physics, Hong Kong SAR, China
[2]The University of Hong Kong, Laboratory for Space Research, Hong Kong SAR, China
email: vfrag@hku.hk

Abstract. A detailed examination of new high quality radio catalogues (e.g. Cornish) in combination with available mid-infrared (MIR) satellite imagery (e.g. Glimpse) has allowed us to find 70 new planetary nebula (PN) candidates based on existing knowledge of their typical colors and fluxes. To further examine the nature of these sources, multiple diagnostic tools have been applied to these candidates based on published data and on available imagery in the HASH (Hong Kong/ AAO/ Strasbourg Hα planetary nebula) research platform. Some candidates have previously-missed optical counterparts allowing for spectroscopic follow-up. Indeed, the single object spectroscopically observed so far has turned out to be a bona fide PN.

Keywords. Planetary Nebulae, Cornish, Color-Color Plots

1. Introduction

Modern multi-wavelength imaging surveys can help identify planetary nebulae hidden at optical wavelengths due to extinction (Cohen *et al.* 2011). The Cornish catalogue, which is based on the 5 GHz radio continuum Cornish survey (Hoare *et al.* 2012; Purcell *et al.* 2013), covers the northern GLIPMSE region (Benjamin *et al.* 2003; Churchwell *et al.* 2009), providing us with a new tool for the detection of PN candidates.

2. Selection Method and Diagnostic Tools

After cross-correlating the Cornish (Hoare *et al.* 2012; Purcell *et al.* 2013), NVSS (Condon *et al.* 1998) and Glimpse (Benjamin *et al.* 2003; Churchwell *et al.* 2009) catalogues, we rejected objects with $S_{5\mathrm{GHz}}$ fluxes, >110 mJy or a spectral index $\alpha < -0.5$ following Anderson *et al.* (2011). We also included in our sample objects with no NVSS counterpart. After visual inspection of multi-wavelength images using the HASH database (Parker *et al.* 2016, and these proceedings), we uncovered 70 PN candidates, 21 of which have an optical detection, including our first confirmed PN (see Fig. 1).

Diagnostic tools using both emission-line and continuum fluxes from multiple wavelengths can be used to separate PNe from H II regions and non-thermal emitters (Cohen & Green 2001; Frew & Parker 2010; Parker *et al.* 2012; Frew *et al.* 2014). Previously known PNe have a median value of $F_{8\mu\mathrm{m}}/S_{843\mathrm{MHz}} = 4.7\pm1.1$, while H II regions and non-thermal emitters present MIR/radio ratios of $\approx$25 and $\approx$0.06 respectively (Cohen *et al.* 2007; Cohen *et al.* 2011). Assuming that all of our PN candidates are optically thin, we can convert their $S_{5\mathrm{GHz}}$ fluxes to $S_{843\mathrm{MHz}}$ by using the equation $S_{5\mathrm{GHz}}/S_{843\mathrm{MHz}} = (5/0.843)^{-0.1}$ (Anderson *et al.* 2011). In Figure 2 we can see the MIR/radio flux ratios of our PNe candidates overplotted on an IRAC color-color plot. The black and blue boxes indicate the areas where most known PNe and H II regions are placed, based on published data (Cohen *et al.* 2011). Data for other objects are obtained from Kurtz *et al.* (1994), Giveon *et al.* (2005) and Parker *et al.* (2012, 2016).

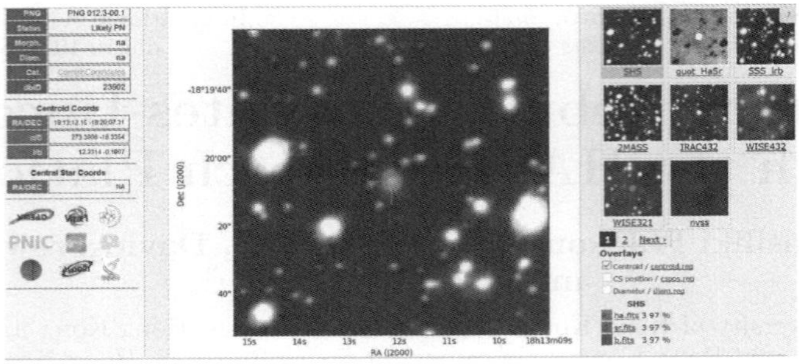

Figure 1. Example of a new PN discovered by this study, shown in HASH (Parker *et al.* 2016).

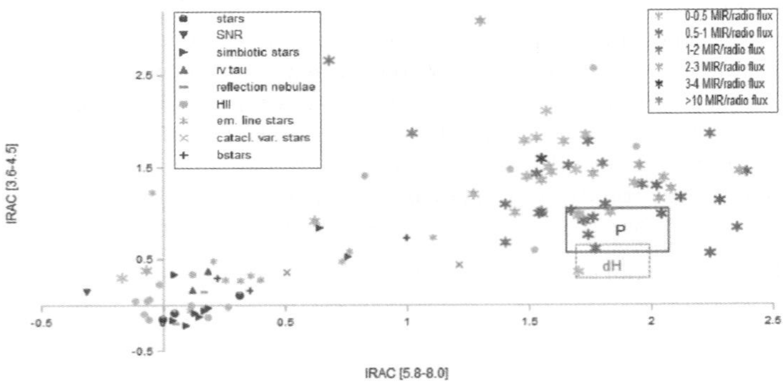

Figure 2. IRAC color-color plot of different object types.

3. Results

As shown in Fig. 2, most of our candidates are plotted on or around the box where most known PNe are located. Moreover, the MIR/radio flux ratios of most candidates agree well with previously known PNe, indicating there is a high probability of our candidates being true PNe. Follow-up studies are needed to confirm this hypothesis, especially for the 21 objects (30% of the total) that have optical counterparts allowing spectroscopic follow-up. At the time of writing, one of these has already been confirmed as a bona fide PN (Fig. 1). Future work should also investigate the sources with $S_{5\text{GHz}}$ >110 mJy.

References

Anderson, L. D., *et al.*, 2014, *ApJS*, 212, 1
Anderson, L. D., Bania, T. M., Balser, D. S., & Rood, R. T., 2011, *ApJS*, 194, 32
Benjamin, R. A., *et al.*, 2003, *PASP*, 115, 953
Churchwell, E., *et al.*, 2009, *PASP*, 121, 213
Cohen, M. & Green, A. J., 2001, *MNRAS*, 325, 531
Cohen, M., *et al.*, 2007, *MNRAS*, 374, 979
Cohen, M., *et al.*, 2011, *MNRAS*, 413, 514
Condon, J. J., *et al.*, 1998, *AJ*, 115, 1693
Frew, D. J. & Parker, Q. A., 2010, *PASA*, 27, 129
Frew, D. J., *et al.*, 2014, *MNRAS*, 440, 1345
Giveon, U., Becker, R. H., Helfand, D. J., & White, R. L., 2005, *AJ*, 130, 156
Hoare, M. G., *et al.*, 2012, *PASP*, 124, 939
Kurtz, S., Churchwell, E., & Wood, D. O. S., 1994, *ApJS*, 91, 659
Parker, Q. A., *et al.*, 2012, *MNRAS*, 427, 3016
Parker, Q. A., Bojičić, I. S., & Frew, D. J., 2016, *JPhCS*, 728, 032008
Purcell, C. R., *et al.*, 2013, *ApJS*, 205, 1

Planetary Nebulae: Multi-Wavelength Probes of Stellar and Galactic Evolution
Proceedings IAU Symposium No. 323, 2016
X. Liu, L. Stanghellini & A. Karakas, eds.

doi:10.1017/S1743921317000710

Herschel spectroscopic observations of PPNe and PNe

Pedro García-Lario[1], **J. Ramos-Medina**[2] **and C. Sánchez-Contreras**[2]

[1]European Space Astronomy Centre, ESA, E-28691 Villanueva de la Cañada, Madrid, Spain
email: pedro.garcia.lario@esa.int

[2]Centro de Astrobiología (CSIC-INTA), E-28691 Villanueva de la Cañada, Madrid, Spain

Abstract. We are building a catalogue of interactively reprocessed observations of evolved stars observed with Herschel. The catalogue will offer not only the PACS and SPIRE spectroscopic data for each observation, but also complementary information from other infrared space observatories. As a first step, we are concentrating our efforts on two main activities: 1) the interactive data-reduction of more than 500 individual spectra obtained with PACS in the 55-210 μm range, available in the Herschel Science Archive; 2) the creation of a catalogue, accesible via a web-based interface and through the Virtual Observatory. Our ultimate goal is to carry out a comprehensive and systematic study of the far infrared properties of low-and intermediate-mass evolved stars using these data and enable science based on Herschel archival data. The objects cover the whole range of possible evolutionary stages in this short-lived phase of stellar evolution, from the AGB to the PN stage, displaying a wide variety of chemical and physical properties.

---

1. Introduction

More than 200 evolved stars were observed in spectroscopic mode with Herschel (Pilbratt *et al.* 2010) either with PACS (55-210 μm) or SPIRE (200-670 μm). 71% of them correspond to low-or intermediate-mass stars, 28% are massive stars (> 8 $M_\odot$) and the 1% left are objects that are difficult to classify. Our work is initially focused on the low- and intermediate-mass evolved stars subsample for which PACS spectroscopic data is available and covers the whole range of evolutionary stages: from AGB stars (39%) to PNe (22%), including the rapidly evolving post-AGB (25%) and OH/IR stars (14%).

The automated standard data reduction pipelines used for the generation of PACS data products lack some important corrections that require in many cases an interactive analysis: spectral flat-fielding, telescope background correction, and other data reduction tasks which take into account the actual fraction of the PSF that falls out of the PACS central 'spaxel' or the slight mispointings affecting some observations are examples of the kind of corrections that can significantly improve the S/N ratio of the PACS spectra and, more important, the absolute flux calibration.

2. The THROES on-line catalogue

THROES is an on-line queriable catalogue which contains in this first release the interactively reduced PACS spectra for all evolved stars observed with Herschel. In future releases it will be further populated with the result of the interactive data reduction of the corresponding SPIRE spectroscopy observations, something on which we are currently working. For each catalogue entry, a number of columns are provided with information about the observation (Coordinates, Target Name or Observing Mode/AOT), as well as

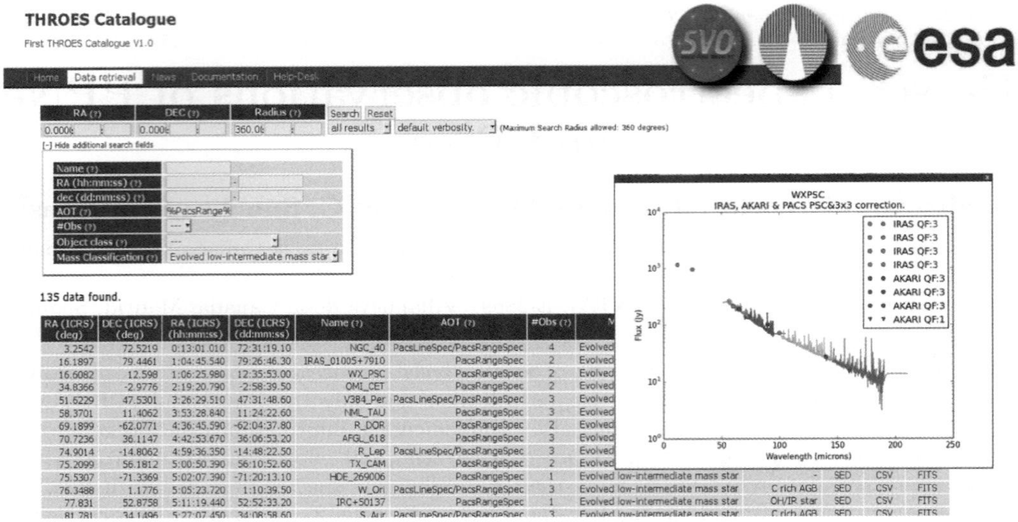

Figure 1. The THROES on-line catalogue web-based user interface.

on the object itself (low-intermediate or high mass, chemical composition - C-rich/O-rich - and evolutionary stage), including the interactively reduced PACS data, that can be downloaded in various formats or displayed for quick look, together with IRAS and AKARI photometric data (Fig. 1).

The ultimate goal of this project is enabling science exploitation of Herschel archival data. With its improved spatial resolution compared to previous space facilities covering the same spectral range (e.g LWS onboard ISO), larger FoV and better sensitivity, Herschel has the potential to lead us to a better understanding of the physical and chemical evolution experienced by evolved stars at the end of their life. Science analysis based on Herschel/PACS includes the determination of the physical and chemical properties of the molecular gas in the inner regions of their circumstellar envelopes through the analysis of the wealth of rotational molecular lines (CO, OH, H_2O, etc) displayed by many of the stars in the sample, or the analysis of dust species like forsterite (Mg_2SiO_4) at 69 μm or calcite ($CaCO_3$) at 92.6 μm that may be contained in the coolest regions of some of the envelopes. Additionally, in proto-PNe and PNe, atomic/ionized nebular emission lines are sometimes detected and can also be studied.

Reference

Pilbratt, G. L., Riedinger J. R., Passvogel T., *et al.* 2010, *A&A*, 518, L1

Planetary Nebulae: Multi-Wavelength Probes of Stellar and Galactic Evolution
Proceedings IAU Symposium No. 323, 2016
X. Liu, L. Stanghellini & A. Karakas, eds.

doi:10.1017/S1743921317001909

Emission Lines of Northern Planetary Nebulae

Nazim Aksaker[1], Sinan K. Yerli[2], Ümit Kızıloğlu[2] and Betül Atalay[3]

[1]Vocational School of Technical Sciences, University of Çukurova, Adana, Turkey
email: naksaker@cu.edu.tr

[2]Orta Doğu Teknik Üniversitesi, Department of Physics, Ankara, Turkey
email: yerli@metu.edu.tr, umk@astroa.physics.metu.edu.tr

[3]Atatürk University, Faculty of Science, Department of Physics, Erzurum, Turkey
email: betul.atalay@tubitak.gov.tr

Abstract. In this work, we present results of long slit spectrophotometric emission line flux observations of selected planetary nebulae (PNe). We have measured absolute fluxes and equivalent widths (EW) of all observable emission lines. In addition to these observations, electron temperatures (Te), densities (Ne), and chemical abundances were also calculated. The main purpose of this work is to fill the gaps in emission line flux standards for the northern hemisphere. It is expected that the measured fluxes would be used as standard data set for further photometric and spectrometric measurements of HII regions, supernova remnants etc.

Keywords. planetary nebulae: general; techniques: spectroscopic; techniques: image processing

1. Introduction

Planetary Nebulae (PNe) are the final stage of stellar evolution for progenitors have masses between 0.8 $M_\odot$ and 8 $M_\odot$. This stage comes from post-AGB or proto-planetary nebulae (PPN) after the envelope is ejected to the interstellar medium leaving hot white dwarfs at the central part of PNe. They also enrich the interstellar medium with the ejected material. Thus, obtaining chemical abundances of PNe is essential to our understanding of the galaxy and its evolution. In the past, there were several works that they have relied on standard star observations using broad band photometry of spectrophotometry, or even using photon counting detectors. Since PNe release emission lines due to hot and low density gas, spectroscopy is the key to understand the properties of them. Dopita & Hua (1995) made slitless spectrophotometric observations of southern compact PNe, using the catalogue of Acker *et al.* 1992. Later, Wright *et al.* 2005 measured the emission line fluxes for northern hemisphere with only 6 PNe targets. Aim of this work is to increase the emission line standards in the northern hemisphere.

2. Observation strategy

Our target list of PNe were chosen from the catalog of Acker *et al.* 1992. The catalog contains 1142 PNe. The following criteria were used to select suitable targets:

(*a*) First, we selected PNe located in northern hemisphere DEC (δ) > -35 degree. This reduced the total number to 838.

(*b*) Then, the catalog is filtered with RA (α) in between 7h and 12h. This reduced the total number to 805 PNe.

(*c*) From this subset, PNe with angular size $<$ 5 arcsec are selected (i.e. PNe size fitting to the available slits at the observatory). This reduced the total number to 276 PNe.

(*d*) Among this subset only 12 less studied PNe were selected.

(*e*) 5 PNe from Wright *et al.* 2005 were also added to this subset which increased the total number to 17.

3. Instrumentation

A subset of PNe was first observed with TUBITAK National Observatory's (TUG) T100 telescope to obtain visual magnitudes so that further spectroscopic exposure times could accurately be determined. TFOSC (TÜBİTAK Faint Object Spectrometer and Camera coupled with the 150 cm Russian–Turkish Telescope (RTT150)) We have used different slits for different PNe: 3.5 arcsec for point-like objects and 7.3 arcsec for wider objects.

4. Summary

In this work we have measured emission line fluxes and their equivalent widths of 17 PNe. Their physical and chemical properties were also calculated. Observed H_β fluxes of the PNe were compared with Acker *et al.* 1992. An explicit linear relation exists between Acker *et al.* 1992 and our works A similar relation was also found for Wright *et al.* 2005. The error in flux measurement of each emission line is less than 1% of the line strength. In addition to this, low SNR of the spectra on relatively bad seeing conditions produced slit losses which accumulated to no more than 10% of the signal, therefore causing a relatively low flux value. With this work the northern hemisphere coverage of PNe is extended by carrying out similar studies done by Dopita & Hua (1995) (for southern hemisphere) and Wright *et al.* 2005 (for northern hemisphere). We present new emission line fluxes which can be counted as standards for both narrow band imaging and Fabry-Perot spectroscopy. Extinction constant, electron temperature, electron density, chemical abundance and excitation class have been calculated for suitable PNe (see Aksaker *et al.* 2015 for further details).

5. Acknowledgement

We thank to TUBITAK for a partial support in using RTT150 (Russian-Turkish 1.5-m telescope in Antalya) with project number 14BT100-658 and T100 telescope with project number 10CT100-90. We also thanks to Çukurova University Project Office with grant numbers FBA-2016-6092 and FED-2016-6233.

References

Aksaker, N., Yerli, S. K., Kızıloğlu, Ü., & Atalay, B. 2015, *PASA*, 32, e003

Acker, A., Marcout, J., Ochsenbein, F., Stenholm, B., Tylenda, R., & Schohn, C. 1992, *The Strasbourg-ESO Catalogue of Galactic Planetary Nebulae. Parts I*, II. European Southern Observatory, Garching, Germany

Wright, S. A., Corradi, R. L. M., & Perinotto, M. 2005, *A&A*, 436, 967

Dopita, M. A. & Hua, C. T. 1997, *ApJS*, 108, 515

Planetary Nebulae: Multi-Wavelength Probes of Stellar and Galactic Evolution
Proceedings IAU Symposium No. 323, 2016
X. Liu, L. Stanghellini & A. Karakas, eds.

doi:10.1017/S1743921317000898

Spitzer IRS spectroscopy of planetary nebulae

G. Ramos-Larios[1], M. A. Guerrero[2], H. Mata[1], X. Fang[3], A. Nigoche-Netro[1], J. A. Toalá[4] and G. Rubio[1]

[1]Instituto de Astronomía y Meteorología, CUCEI, Universidad de Guadalajara, Av. Vallarta No. 2602, Guadalajara, Jalisco, Mexico
email: gerardo@astro.iam.udg.mx

[2]Instituto de Astrofísica de Andalucía, IAA-CSIC, C/ Glorieta de la Astronomía s/n, 18008 Granada, Spain

[3]Department of Physics, The University of Hong Kong, Pokfulam Road, Hong Kong

[4]Institute of Astronomy and Astrophysics, Academia Sinica (ASIAA), Taipei 10617, Taiwan

Abstract. We present *Spitzer Space Telescope* archival mid-infrared (mid-IR) spectroscopy of a sample of eleven planetary nebulae (PNe). The observations, acquired with the *Spitzer* Infrared Spectrograph (IRS), cover the spectral range 5.2-14.5 μm that includes the H_2 0-0 S(2) to S(7) rotational emission lines. This wavelength coverage has allowed us to derive the Boltzmann distribution and calculate the H_2 rotational excitation temperature (T_{ex}). The derived excitation temperatures have consistent values $\simeq$900$\pm$70 K for different sources despite their different structural components. We also report the detection of mid-IR ionic lines of [Ar III], [S IV], and [Ne II] in most objects, and polycyclic aromatic hydrocarbon (PAH) features in a few cases.

Keywords. Planetary nebulae: general — ISM: lines and bands — infrared: general

1. Introduction

Mid-infrared (mid-IR) spectroscopy is a powerful tool for the study of the thermal and molecular emission from PNe, since such observations are much less affected by dust extinction. Mid-IR studies of PNe have focused on imaging surveys, but mid-IR spectroscopy of PNe has only been carried out for individual objects for which chemical abundances and H_2 excitation temperatures were obtained. Currently, mid-IR studies of the excitation temperature of H_2 in PNe are scarce.

2. Results

Our analysis of archival *Spitzer* IRS SL1 and SL2 spectra of a sample of PNe has identified ionic transitions including [Ar III], [Ne II], [Ne V], and [S IV] (see example in Figure 1-left). The detection of emission lines from species of neon and argon imply that the progenitor stars of the PNe were all at least second-generation stars.

The broad PAH feature centred at 11.30 μm is detected in a number of PNe in this sample. The feature is present in the spectra of NGC 2440, NGC 3132, NGC 6445, NGC 6537, NGC 6720 and NGC 6781. There is a possible detection in NGC 2346, but it is absent in M 2-51, NGC 2818, NGC 6072 and NGC 7293.

Using the integrated flux of each H_2 line from the spectrum of each object we calculate the population levels. The population diagram with the upper temperature for

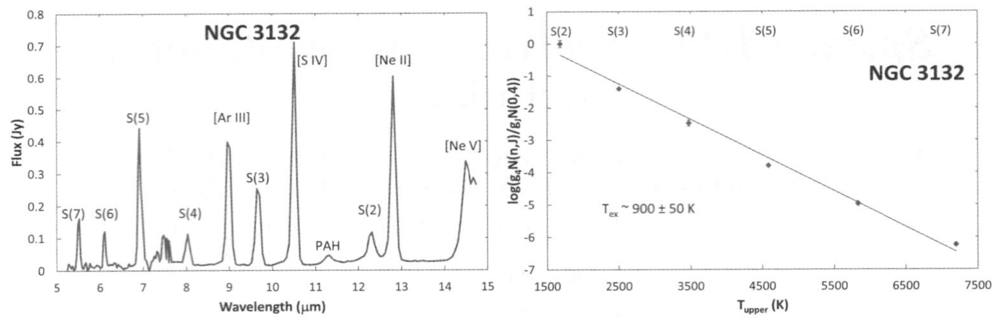

Figure 1. *(left) Spitzer* mid-IR (SL1 + SL2) spectrum of NGC 3132, with H_2 and some ionic lines identified. *(right)* The mean variation in H_2 populations for the $\nu = 0$-0 S(2)-S(7) transitions is represented in the excitation diagram for NGC 3132. The results are based on averages of line strengths for spectral selected positions (error bars are included).

the energy level of each transition is shown in (Figure 1-right). The data points in this figure show a linear trend. Since all the data points correspond to a unique transition ν, this linear trend cannot be directly interpreted as a thermal distribution of the population of the energy levels of the H_2 molecules. At any rate, the negative inverse of the slope of the linear fits can be used to measure the rotational H_2 excitation temperature, $T_{\rm ex}$(rot). These temperatures shed light on the purely rotational component of the excitation temperatures to complement the rotational-vibrational studies obtained by near-IR observations.

3. Conclusions

We present an investigation of archival *Spitzer* IRS spectra of a sample of PNe to search for mid-IR H_2 lines (see, Mata *et al.* 2016, for complete information). Among the original sample of 14 PNe with useful *Spitzer* IRS SL observations known to exhibit H_2 emission lines in the near-IR, 11 PNe show detectable mid-IR H_2 lines. This result increases the number of known PNe with mid-IR H_2 emission lines from three (NGC 6302, NGC 6781, and NGC 7293) to twelve.

The spectral analysis detects all the H_2 0-0 transitions from the S(2) to the S(7) lines. Among these lines, the H_2 0-0 S(3) λ9.66 μm line is the brightest. The analysis of the population distribution in the H_2 molecules of these PNe reveals a $T_{\rm ex}$(rot) excitation temperature of $\simeq$900 K for all of them. The conspicuous uniformity of the excitation temperature is intriguing, because the observations trace different morphological features of a non-uniform sample of PNe with different morphologies and evolutionary stages.

Ionic lines of [Ar III], [S IV], [Ne II], and [Ne V] are also detected. Their line intensities show the expected positive correlation of the [Ne V]/[Ne II] line ratio with the CSPN effective temperature. On the other hand, the anti-correlation between the [Ar III]/[Ne II] and CSPN effective temperature is not expected and may imply changes in the elemental ratios that suggest that PNe with hottest CSPNe, probably descending from more massive progenitors, are able to produce neon.

Reference

Mata, H., Ramos-Larios, G., Guerrero, M. A., *et al.* 2016, *MNRAS*, 459, 841

Planetary Nebulae: Multi-Wavelength Probes of Stellar and Galactic Evolution
Proceedings IAU Symposium No. 323, 2016
X. Liu, L. Stanghellini & A. Karakas, eds.

doi:10.1017/S1743921317002113

A New Model of NGC 6210 to Solve its Abundance Discrepancy Problem

V. Escalante[1] **and J. A. Toalá** [2]

[1]Instituto de Radioastronomía y Astrofísica
UNAM, Morelia 58341, Michoacán, México
email: v.escalante@crya.unam.mx

[2]Institute of Astronomy and Astrophysics
Academia Sinica (ASIAA), Taipei 10617, Taiwan
email: toala@asiaa.sinica.edu.tw

Abstract. We present a model that solves the abundance discrepancy problem for NGC 6210. The model proposes a high abundance of CNONe elements that lowers the temperature of the central parts of the nebula. The colder gas model reproduces the observed intensity of the strong [N II] and [O III] emission lines, and increases the predicted weak recombination lines towards their observed values. We examine how the usual nebular diagnostic line ratios depend on model abundances.

Keywords. ISM: abundances, planetary nebula: NGC 6210

The discrepancy between abundances derived from collisionally excited lines (CELs) and from optical recombination lines (ORLs) is a long standing conundrum in the theory of interstellar nebulae. Usually ORL derived abundances are larger than CEL derived abundances. A model optimized to reproduce CEL intensities calculated by Bohigas *et al.* (2015) also shows an abundance discrepancy in the sense that it underestimates ORL intensities because the predicted O^{+2} column density is too low. We used the code CLOUDY (Ferland *et al.* 2013) to construct new chemically homogeneous models of NGC 6210 with increased CNONe abundances to reproduce CELs and ORLs simultaneously.

The models propose a binary star with $T_{\rm eff} = 179300\,{\rm K}$, $L = 1635\,L_\odot$, $\log g = 7.0$, $T_{\rm eff} = 29700\,{\rm K}$, $L = 1694\,L_\odot$, $\log g = 4.5$ as suggested by the relation between binarity and the abundance discrepancy problem (Corradi *et al.* 2015). The density profile was chosen to reproduce the observed intensities of [N II], [O II], and [O III] optical CELs, the [O III] 51.80 μm, and the He I 5876 Å and He II 4686 Å lines reported by Bohigas *et al.* (2015) and Pottasch *et al.* (2009). The density profile has a tail that decreases with radius as r^{-4} to produce the extended halo of the nebula.

The O abundance in the new model, $O/H = 1.33 \times 10^{-3}$, was chosen to reproduce the O II ORLs lines to eliminate the abundance discrepancy. The model produces an inner zone with temperatures as low as 6600 K due to the higher O^{+2} concentration, where the ORL emission is produced (see Fig. 1). The temperature rises at the edges of the O^{+2} zone thus increasing the [O III] optical emission, and producing a temperature gradient with a Peimbert parameter $t^2 = 0.04$. Images obtained by García–Rojas, *et al.* (2016) of another nebula with a high abundance discrepancy, NGC 6778, show a similar concentration of ORL oxygen emission inside the CEL oxygen emission.

The model calculated by Bohigas *et al.* (2015) with a lower O/H abundance uses a density profile varying as r^2. A comparison with observations in Table 1 shows that both models reproduce the lines to comparable accuracy, except for the [O II] emission. The

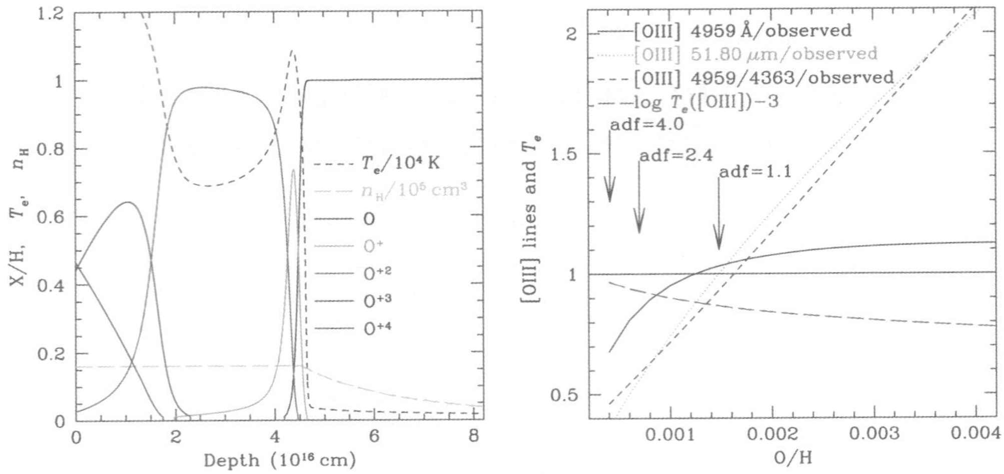

Figure 1. O ionization, electron temperature and density structure (left), and [O III] line intensities vs. O abundance (right) in our new model.

Table 1. A few comparisons of predicted and observed line intensities.

Model & Obs	O/H (10^{-4})	adf	[N II] 6584	[N II] ratio 6584/5755	[O II] 3726	[O III] 4959	[O III] ratio 4959/4363	[O III] 51.80	[O III] ratio 51.80/88.35	O II 4649 Mult V1	O II 4349 Mult V2
Obs[1]	4.10	4.0	15.68	54.4	18.33	343.40	58.3	151[4]	3.17[4]	0.446	0.113
Obs[2]	4.36	3.1	18.70	47.6	24.95	361.40	59.5			0.389	0.120
Low O model[1]	7.42	2.4	15.69	22.4	18.95	343.00	59.8	26.0	8.69	0.322	0.130
High O model[3]	13.3	1.1	15.68	43.4	48.45	332.69	61.4	53.3	8.19	0.423	0.169

[1]Bohigas *et al.* (2015), [2]Liu *et al.* (2004), [3] This work, [4]Pottasch *et al.* (2009)

[O II] emission appears asymmetrical in HST archival images (Guerrero *et al.* 2013), and therefore it is overestimated by our spherically symmetric model.

Figure 1 shows that the [O III] 4958 and 5007 Å lines vary little at high O abundances because of the decrease in electron temperature while the [O III] 51.80 μm does not depend on temperature. The abundance discrepancy factor adf=(O abundance from ORLs)/(O abundance from CELs) is given for the observed line intensities (4.0), the model abundance of Bohigas *et al.* (2015) (2.4), and the model with high O abundance (1.1).

References

Bohigas, J., Escalante, V., Rodríguez, M., & Dufour, R. J. 2015, *MNRAS*, 447, 817

Corradi, R. L. M, García–Rojas, J., Jones, D., & Rodríguez–Gil, P. 2015, *Ap. J.* 803, 99

Ferland, G. J., Porter, R. L., van Hoof, P. A. M., Williams, R. J. R., Abel, N. P., Lykins, M. L., Shaw, G., Henney, W. J., & Stancil, P. C. 2013, *RMxAA*, 49, 137

García–Rojas, J., Corradi, R. L. M., Monteiro, H., Jones, D., Rodríguez–Gil, P., & Cabrera–Lavers, A. 2016, *Ap. J. Lett*, 824, L27

Guerrero, M. A., Toalá, J. A., Medina, J. J., Luridiana, V., Miranda, L. F., Riera, A., & Velázquez, P. F. 2013, *A&A*, 557, A121

Liu, Y., Liu, X.–W., Luo, S.–G., & Barlow, M. J. 2004, *MNRAS*, 353, 1231

Pottasch, S. R., Bernard–Salas, J., & Roellig, T. L. 2009, *A&A*, 499, 249

Planetary Nebulae: Multi-Wavelength Probes of Stellar and Galactic Evolution
Proceedings IAU Symposium No. 323, 2016
X. Liu, L. Stanghellini & A. Karakas, eds.

doi:10.1017/S1743921317001685

The population of planetary nebulae near the Galactic Centre: chemical abundances

M. Mollá[1], O. Cavichia[2], R. D. D. Costa[3] and W. J. Maciel[3]

[1]CIEMAT, Avda. Complutense 40, 28040 Madrid, Spain
email: mercedes.molla@ciemat.es

[2]IFQ, Universidade Federal de Itajubá, Av. BPS, 1303, 37500-903, Itajubá-MG, Brazil
[3]IAG, Universidade de Sâo Paulo, 05508-900, Sâo Paulo-SP, Brazil

Abstract. In this work, we report physical parameters and abundances derived for a sample of 15 high extinction planetary nebulae located in the inner 2° of the Galactic bulge, based on low dispersion spectroscopy secured at the SOAR telescope using the Goodman spectrograph. The new data allow us to extend our database including older, weaker objects that are at the faint end of the planetary nebulae luminosity function. The data provide chemical compositions for PNe located in this region of the bulge to explore the chemical enrichment history of the central region of the Galactic bulge. The results show that the abundances of our sample are skewed to higher metallicities than previous data in the outer regions of the bulge. This can indicate a faster chemical enrichment taking place at the Galactic centre.

1. Introduction

Planetary nebulae (PNe) constitute an important tool to study the chemical evolution of the Milky Way and other galaxies, probing the nucleosynthesis processes, abundance gradients and the chemical enrichment of the interstellar medium. In particular, Galactic bulge PNe (GBPNe) have been extensively used in the literature to study the chemical properties of this Galactic structure. However, the presently available GBPNe chemical composition studies are strongly biased, since they were focused on brighter objects, predominantly located in Galactic regions of low interstellar reddening. The main objective of this work is to derive new physical parameters and chemical abundances of a sample of 15 high extinction planetary nebulae (PNe) located in the inner 2° of the Galactic bulge, based on low dispersion spectroscopy secured at the 4.1 m SOAR telescope.

2. Sample and Observations

The observed sample is from Jacoby and Van de Steene (2004) and the MASH catalog (Parker *et al.* 2006). They were observed with the 4.1m SOAR telescope located in Chile (optical spectra) and the 1.6 m LNA/Brazil telescope (NIR espectra). The criteria for the galactic coordinates are: $|\ell| < 2°$ and $|b| < 2°$. Also, diameters smaller than 12" and fluxes at 5 GHz lower than 100 mJy were required. The optical spectra of the GBPNe near the Galactic Centre (GC) suffer for high-level of extinction caused by the material near the Galactic plane and also in the central regions of the Galaxy. As a result, important diagnostic lines as [O III] λ4363 Å and [N II] λ5755 Å do not have enough S/N ratio to obtain the electron temperature from the temperature diagnostic diagrams. Other important temperature-sensitive lines are those from S^{+3}. The [S III] auroral line at λ6312 Å is present in most of our optical spectra. We performed near infrared (NIR) observations at the Observatório Pico dos Dias (OPD) of National Laboratory for Astrophysics (LNA, Brazil) with the 1.6 m Perkin-Elmer telescope. These observations allowed us to

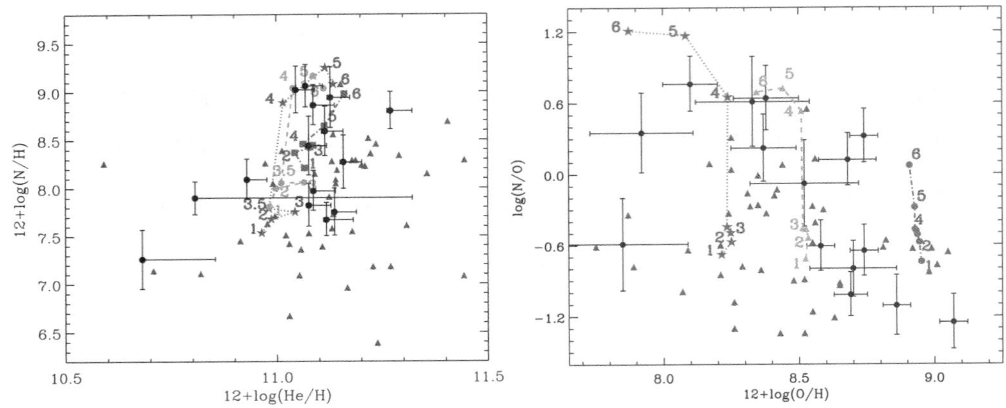

Figure 1. Abundance ratios as a function of 12 + log(He/H) and 12 + log(O/H). Filled circles with error bars are data from present work, while filled blue triangles are the PNe from CCM10 (outer bulge sample). The symbols with numbers represent the results of the models from Karakas (2010) for a given value of Z: Red stars joined by dotted line for $Z = 0.004$, yellow triangles joined by dashed line for $Z = 0.008$ and purple circles joined by dash-dotted line for $Z = 0.02$. The numbers give the initial masses of the individual models in M⊙ units.

observe the [S III] lines at $\lambda\lambda$9069 and 9532 Å, which are suitable to derive the electron temperatures taking into account the high extinction in the visible range. Te[O III] were derived from Te[S III] using the relation provided by Henry *et al.* (2004). He I emissivities are from Porter *et al.* (2013). These emissivities include collisional correction. He I abundances are calculated from the flux-weighted mean of the lines 4471, 5876 and 6678 Å. Elemental abundances for N, O, S and Ar were calculated using the ICFs from Delgado-Inglada *et al.* (2014). Only in the case of helium we opted not to use any ICF.

3. Results and Conclusions

We present the abundances for 15 PNe located near the GC, compared with the data from Cavichia *et al.* (2010) –hereafter CCM10– (outer bulge sample). The abundance ratios as a function of 12 + log(He/H) and 12 + log(O/H) are shown in Fig. 1.

The abundances of our sample are in agreement with AGB evolution models for masses < 4 M⊙. The results show that the abundances in the GBPNe near the GC are skewed to higher metallicities than previous data in the outer regions of the bulge. This can indicate a faster chemical enrichment taking place at the GC. More information can be seen at Cavichia *et al.*(2017).

References

Cavichia, O., Costa, R. D. D., & Maciel, W. J. 2010, *Rev. Mex. A&A*, 46, 159

Cavichia, O., Costa, R. D. D., Maciel, W. J., & Mollá, M. 2017, *MNRAS* (submitted)

Delgado-Inglada, G., Morisset, C., & Stasińska, G. 2014, *MNRAS*, 440, 536

Henry, R. B. C., Kwitter, K. B., & Balick, B. 2004, *ApJ*, 127, 2284

Jacoby, G. H. & Van de Steene, G. 2004, *A&A*, 419, 563

Karakas, A. I. 2010, *MNRAS*, 403, 1413

Parker, Q. A., Acker, A., Frew, D. J., *et al.* 2006, *MNRAS*, 373, 79

Porter, R. L., Ferland, G. J., Storey, P. J., & Detisch, M. J. 2013, *MNRAS*, 433, L89

Planetary Nebulae: Multi-Wavelength Probes of Stellar and Galactic Evolution
Proceedings IAU Symposium No. 323, 2016
X. Liu, L. Stanghellini & A. Karakas, eds.

doi:10.1017/S1743921317000692

Dust & Abundances of Metal-Poor Planetary Nebulae in the Galactic Anti-Center

George J. S. Pagomenos[1], **Jeronimo Bernard-Salas**[1] **and G. C. Sloan**[2,3,4]

[1]School of Physical Sciences, The Open University, Walton Hall, Milton Keynes, Bucks, MK7 6AA, UK
email: george.pagomenos@open.ac.uk

[2]Space Telescope Science Institute, 3700 San Martin Dr., Baltimore, MD 21218, USA

[3]Department of Physics and Astronomy, University of North Carolina, Chapel Hill, NC 27599-3255, USA

[4]Cornell Center for Astrophysics & Planetary Science, Cornell Univ., Ithaca, NY 14853-6801, USA

Abstract. Much of the new dust in the local ISM is produced in the last phases of stellar evolution of low- and intermediate-mass stars on the Asymptotic Giant Branch (AGB). Despite its importance, our knowledge of how dust properties depend on metallicity is limited. Studies of planetary nebulae in irregular galaxies in the Local Group (mostly focused on the LMC and SMC) have revealed a diverse spectral zoo and shown that low metallicity favours carbon-rich dust production by AGB stars. However, at ~1/3 and ~1/5 times the solar metallicity respectively, they provide two snapshots of dust composition at low metallicity, emphasising the need to investigate a region with a range of metallicity values. With its abundance gradient, the Milky Way fits this criterion and provides a good opportunity to observe the dust composition over a large metallicity range. In particular the Galactic anti-center, which is largely unexplored beyond galactocentric distances of 10 kpc, allows us to study the AGB dust *a priori* assumed to be metal-poor as well as exploring the extent of the Galactic abundance gradient. We analyse a *Spitzer* spectroscopic sample of 23 planetary nebulae towards the anti-center in order to understand how the metallicity gradient extends beyond 10 kpc from the Galactic center and to observe the dust composition in this region of our Galaxy. We find that the abundance gradients of Ne, S and Ar continue to distances of around 20 kpc (albeit with a large scatter) and the dust emission shows a carbon-rich chemistry similar to that in the Magellanic Clouds.

Keywords. planetary nebulae, dust, abundances, metallicity

---

1. Abundances

We have investigated the neon, sulfur and argon abundances, which remain unchanged throughout low- to intermediate-mass stellar evolution, within a sample of planetary nebulae (PNe) at Galactocentric distances (R_{gc}) of 11–27 kpc. It is important to investigate within this region as studies of PNe (e.g. Pottasch & Bernard-Salas 2006) and H II regions (e.g. Martín-Hernández *et al.* 2002) in the disk of the Milky Way show the presence of a metallicity gradient up to R_{gc} ~ 10 kpc. In the anti-center, abundance studies have been rare, with one finding that the gradient flattened beyond 10 kpc (Costa, Uchida & Maciel 2004) and another suggesting it may steepen (Henry *et al.* 2010).

Our abundances are plotted in Figure 1 (open squares) alongside those in the sample from the solar neighbourhood analysed by Pottasch & Bernard-Salas (2006). The anti-center abundances show no clear decrease with distance by themselves but they follow well from those of the solar neighbourhood sample up to R_{gc} = 15–20 kpc, despite a

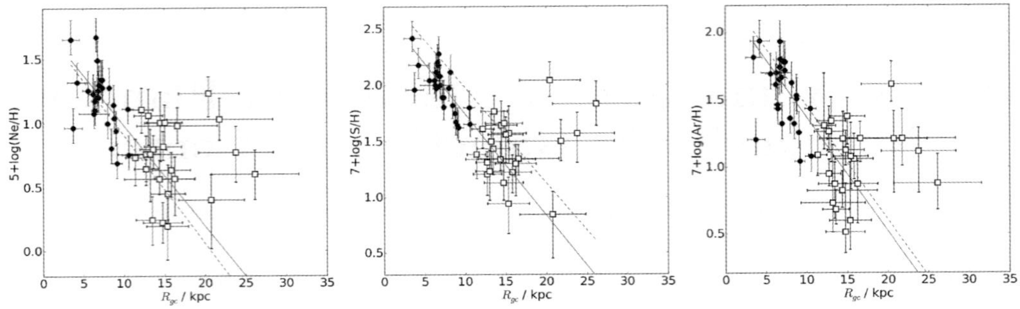

Figure 1. The abundance gradients of Ne, S and Ar in the Milky Way. Dashed lines represent the oxygen abundance gradient from within the solar neighbourhood, with a slope of −0.085 dex/kpc (Pottasch & Bernard-Salas 2006), passing through the solar elemental abundance at 8.0 kpc. Solid lines show the best fits in each plot, with slopes of −0.0788, −0.0906 and −0.0853 dex/kpc respectively. Distances are taken from Frew, Parker & Bojičić (2016). Abundance uncertainties range from 25% to 60%.

large scatter. Beyond these distances, the abundances appear to increase, potentially suggesting that these PNe are located outside the Galactic thin disk. With respect to the PNe in the solar neighbourhood, these sources are still relatively metal-poor.

2. Dust

In the solar neighbourhood, dusty PNe typically show either carbon-rich or oxygen-rich dust features (Pottasch & Bernard-Salas 2006). The central region of the Milky Way is more metal-rich, and dual-dust chemistry is relatively common (Gutenkunst *et al.* 2008). Studies show that the LMC and SMC are metal-poor and the dust in its PNe is mainly C-rich (e.g. Bernard-Salas *et al.* 2009; Sloan *et al.* 2014). Knowledge of the dust composition in the anti-center will give us insight into how dust evolves with metallicity.

From our sample of PNe in the anti-center, 17 of the 23 sources show C-rich dust emission features (e.g. PAHs), while the other six sources show no discernible dust features. Of the C-rich dust detections, there are five observations of SiC and four of C_{60} (three of which have already been detected by Otsuka *et al.* 2014), both of which are rare in the solar neighbourhood. The broad 30 μm feature that has been attributed to MgS but is also associated with aliphatic C-rich compounds has been found in six sources. We also find that O-rich dust emission and dual-dust chemistry are rare in the metal-poor anti-center with only one tentative detection of amorphous silicates. These findings agree well with those of Bernard-Salas *et al.* (2009) in the Magellanic Clouds, reinforcing the concept that low metallicity favours C-rich dust production.

References

Bernard-Salas, J., Peeters, E., Sloan, G. C. *et al.* 2009, *ApJ*, 699, 1541
Costa, R. D. D., Uchida, M. M. M., & Maciel, W. J. 2004, *A&A*, 423, 199
Frew, D. J., Parker, Q. A., & Bojičić, I. S. 2016, *MNRAS*, 455, 1459
Gutenkunst, S., Bernard-Salas, J., Pottasch *et al.* 2008, *ApJ*, 680, 1206
Henry, R. B. C, Kwitter, K. B., Jaskot, A. E. *et al.* 2010, *ApJ*, 724, 748
Martín-Hernández, N. L., Peeters, E., Morisset, C. *et al.* 2002, *A&A*, 381, 606
Otsuka, M., Kemper, F., Cami, J. *et al.* 2014, *MNRAS*, 437, 2577
Pottasch, S. R. & Bernard-Salas J. 2006, *A&A*, 457, 189
Sloan, G. C., Lagadec, E., Zijlstra, A. A. *et al.* 2014, *ApJ*, 791, 28

Planetary Nebulae: Multi-Wavelength Probes of Stellar and Galactic Evolution
Proceedings IAU Symposium No. 323, 2016
X. Liu, L. Stanghellini & A. Karakas, eds.

doi:10.1017/S1743921317001624

The systematic uncertainties introduced by atomic data in nebular abundance determinations

Leticia Juan de Dios and Mónica Rodríguez

Instituto Nacional de Astrofísica, Óptica y Electrónica (INAOE), Apdo. Postal 51 y 216, Puebla, Mexico
email: leticiajd@inaoep.mx, mrodri@inaoep.mx

Abstract. Atomic data selection is one important source of systematic uncertainty since there are important variations between the values provided by different authors. We explore the effect of using different atomic data in the determination of physical conditions and chemical abundances in a sample of Galactic planetary nebulae and H II regions. We find that the available datasets introduce significant differences in the results, especially at densities above 10^4 cm^{-3}, where O/H and N/O reach uncertainties higher than a factor of 4.

Keywords. planetary nebulae: general, ISM: abundances, atomic data, H II regions

---

The derivation of chemical abundances in ionized nebulae requires the values of transition probabilities and collision strengths for transitions involving the lower energy levels of several ions. These atomic data are calculated using different approaches that lead to variations among the existing determinations, and the uncertainties are very difficult to estimate (Aggarwal & Keenan 2013). We explore these uncertainties by calculating the impact of different combinations of atomic data in the derived electron densities, $n_{\rm e}$, electron temperatures, $T_{\rm e}$, and ionic an total abundances for all the posible combinations of atomic data. Note that a single set of atomic data, can affect all the calculations, and different combinations of atomic data will affect the results in complex ways.

To explore the effect that atomic data have in the computations of physical conditions and chemical composition in nebular objects we have selected from the literature a sample of 36 Galactic planetary nebulae (PNe) and 8 H II regions with deep observed spectra and good spectral resolution. The sample objects cover a wide range in characteristics such as density and degree of ionization so they can be considered representative of the population of ionized nebulae. We use the atomic datasets available in the software for nebular analysis PYNEB (Luridiana *et al.* 2015), for the ions O^+, O^{++}, N^+, Cl^{++}, Ar^{++}, Ar^{+3}, Ne^{++}, S^+, and S^{++} (52 atomic datasets in total). For each object we obtain the distributions of results for the values of $T_{\rm e}$[N II], $T_{\rm e}$[O III], $n_{\rm e}$[S II], $n_{\rm e}$[O II], $n_{\rm e}$[Cl III], $n_{\rm e}$[Ar IV], and the ionic and total abundances of O, N, Cl, Ar, Ne, and S, implied by the different combinations. We calculate the average density obtained from the four density diagnostics, and use $T_{\rm e}$[N II] to determine the abundances of the single ionized ions, with the exception of He^+, and $T_{\rm e}$[O III] for the remaining ions. The total abundances of O, Cl, S, Ar, and Ne are obtained using the ionization correction factors proposed by Delgado-Inglada *et al.* (2014). For N we use the traditional ICF, N/O = N^+/O^+.

We measure the total widths of the final distributions for the different parameters, which give an indication of the uncertainties introduced by atomic data, and plot them as a function of the median of the average density. The uncertainties in the average density are higher than a factor of 4 at $n_{\rm e} \geqslant 10^4$ cm^{-3}. The uncertainties in temperature

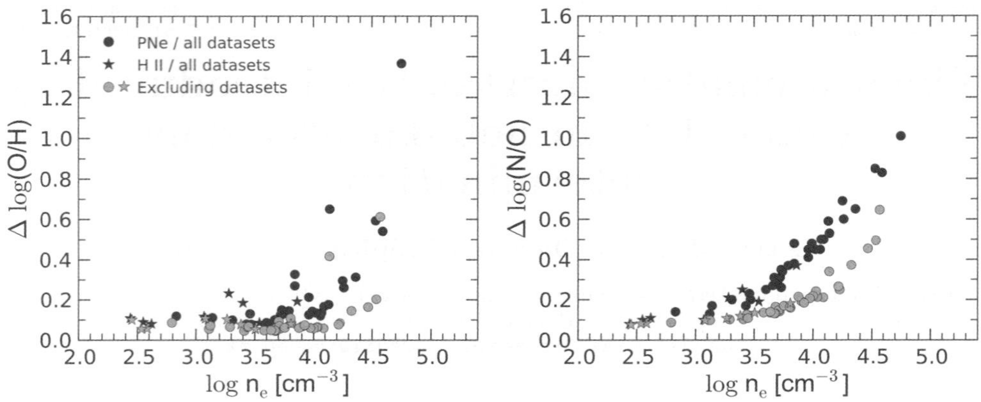

Figure 1. Total widths of the O/H and N/O abundance ratios for all the objects in the sample. The blue symbols show the results implied by all the available 52 datasets; the orange symbols show the results when we exclude the 4 datasets that lead to discordant values of density or temperature.

are below 0.05 dex at $n_e \geqslant 10^4$ cm^{-3} but reach a factor of 2 for T_e[N II] at higher densities. The spreads in the ionic abundances reach factors higher than 30 at $n_e \geqslant 10^4$ cm^{-3} for ions whose transitions have upper levels with low critical densities, like O^+, N^+, S^+. The spreads in total abundances are $\sim$ 0.2 dex at low densities, but increase to factors of 4–6 or higher at $n_e \geqslant 10^4$ cm^{-3}. The total uncertainties in the abundances of O/H and N/O are shown in Figure 1, where the dark symbols represent the results implied by all the available 52 datasets. For O/H we obtain variations of a factor of 2 at $n_e \sim 10^4$ cm^{-3} that increase to factors of 6–25 at the highest densities. Uncertainties are especially important for N/O where the differences reach a spread of one order of magnitude at $n_e \geqslant 10^4$ cm^{-3}.

We identify four datasets that lead to discordant values of density or temperature: the transition probabilities of Wiese, Fuhr, & Deters (1996) and Tayal & Zatsarinny (2010) for O^+ and S^+, respectively, and the collision strengths of Palay *et al.* (2012) and Mendoza (1983) for O^{++} and Ar^{+3}, respectively. By excluding these datasets the differences are significantly reduced, as can be seen in Figure 1, where the orange symbols show the total widths obtained excluding these atomic data. The whole analysis will be presented in Juan de Dios & Rodríguez (in prep.).

Acknowledgements. We acknowledge support from an IAU travel grant and the Mexican CONACYT grant CB-2014-240562.

References

Aggarwal, K. & Keenan, F. 2013, *ArXiv*, 1301.3002
Delgado-Inglada, G., Morisset, C., & Stasińska, G. 2014, *MNRAS*, 440, 536
Luridiana, V., Morisset, C., & Shaw, R. A. 2015, *A&A*, 30, 490
Mendoza, C. 1983, *Recent advances in atomic calculations and experiments of interest in the study of planetary nebulae*, Proc. IAU Symposium No. 103, 143
Palay, E., Nahar, S. N., Pradhan, A. K., & Eissner, W. 2012, *MNRAS*, 423, L35
Tayal, S. S. & Zatsarinny, O. 2010, *ApJS*, 188, 32
Wiese, W. L., Fuhr, J. R., & Deters, T. M. 1996, *JPCRD*, Monograph 7, 403

Planetary Nebulae: Multi-Wavelength Probes of Stellar and Galactic Evolution
Proceedings IAU Symposium No. 323, 2016
X. Liu, L. Stanghellini & A. Karakas, eds.

doi:10.1017/S1743921317002198

Mapping the Physical and Chemical Conditions of the Ring Nebula

Marcelo L. Leal-Ferreira[1], Isabel Aleman[1,2], Andrea Gaughan[3], Djazia Ladjal[4,5], Toshiya Ueta[6], Samuel Kerber[6], Blair Conn[4,5], Rhiannon Gardiner[6] and Alexander G. G. M. Tielens[1]

[1]Leiden Observatory, University of Leiden, PO Box 9513, 2300 RA, Leiden, The Netherlands

[2]IAG-USP, Universidade de São Paulo, Brazil, [3]Haverford College, USA, [4]Gemini Observatory, Chile, [5]ANU, Australia [6]University of Denver, USA, [7]Monash University, Australia

Abstract. We observed the Planetary Nebula NGC 6720 with the Gemini Telescope and the Gemini Multi-Object Spectrographs. We obtained spatial maps of 36 emission-lines in the wavelength range between 3600 Å and 9400 Å. We derived maps of c(Hβ), electronic densities, electronic temperatures, ionic and elemental abundances, and Ionization Correction Factors (ICFs) in the source and investigated the mass-loss history of the progenitor. The elemental abundance results indicate the need for ICFs based on three-dimensional photoionization models.

1. Introduction

Low- and intermediate-mass stars (0.8–8 $M_\odot$) turn into Planetary Nebulae (PNe) when the material ejected by the Asymptotic Giant Branch (AGB) projenitor star gets ionized by the remaining exposed core (central star). By investigating different spatial regions of the nebula, as in this work, we are able to access the mass-loss history of the source.

We performed slit scan observations of the PN NGC 6720 (Ring Nebula) with the 8.1 m Gemini telescope and the Gemini Multi-Object Spectrograph (GMOS). We extracted 36 emission-line maps from the data cube, as listed in Table 1. The table also shows the respective total flux of the maps, scaled to the total flux of Hβ=100.

We used the IDL tool 2D_Neb (Leal-Ferreira *et al.* 2011) to derive the maps of the extinction coefficient c(Hβ), electronic densities (N_e) and temperatures (T_e), ionic and elemental abundances, and Ionization Correction Factors (ICFs) in the Ring Nebula. The 2D_Neb follows the formalism of the well-established iraf.nebular package (De Robertis *et al.* 1987; Shaw & Dufour 1995), and it is adapted to deal with two-dimensional spatial data. It was successfully used before in the analysis of the PNe NGC 40 (Leal-Ferreira *et al.* 2011), NGC 3243 (Monteiro *et al.* 2013), and Tc 1 (Aleman *et al.* in prep).

2. Results

In the last 2 columns of Table 1, we present the median of the c(Hβ), N_e, T_e, and ionic abundance maps. Below we briefly discuss the c(Hβ) map, while the remaining maps will be presented in a forthcoming paper. We also derived elemental abundances adopting the ICF method to correct for unobserved species. We used the ICFs from Kingsburgh & Barlow (1994) and concluded that the ICFs derived from one-dimensional photoionization models are insufficient to provide a reliable total abundance result when analyzing spatially-resolved data as in the present work.

The c(Hβ) map was derived from the comparison of the observed to theoretical ratios of Hα, Hγ, and Hδ in respect to Hβ. Figure 1 shows the c(Hβ) map (left), its histogram

Table 1. Emission-lines maps of the Ring Nebula extracted from the data cube.

Specie	Wavelength (Å)	Total Flux (Hβ= 100)	Specie	Wavelength (Å)	Total Flux (Hβ= 100)	Derived Property	Median Result
[O II]	3727	570.07	[O III]	5007	1013.73	c(Hβ)	0.26
H11	3771	0.02	[N I]	5199	2.09	N$_e$ [S II] (cm^{-3})	410
H10	3798	0.40	He I	5413	0.31	T$_e$ [N II] (K)	7030
Hη	3836	0.63	[Cl III]	5517	0.04	T$_e$ [O III] (K)	10690
[Ne III]	3869	138.38	[Cl III]	5537	7.e-3	He$^+$/H$^+$	9.6E-2
Hζ	3890	19.20	[N II]	5755	2.06	He^{++}/H$^+$	1.9E-2
[Ne III]	3969	56.45	He I	5876	4.87	N^0/H$^+$	1.2E-5
Hδ	4101	25.03	[O I]	6300	28.16	N^{++}/H$^+$	3.1E-4
C II	4268	7.e-3	[S III]	6312	0.39	O^0/H$^+$	3.9E-4
Hγ	4341	45.15	[O I]	6363	9.66	O$^+$/H$^+$	1.8E-3
[O III]	4363	7.23	[N II]	6548	144.87	O^{++}/H$^+$	2.4E-4
He I	4471	3.55	Hα	6563	296.94	Ne^{++}/H$^+$	1.3E-4
He I	4542	0.02	[N II]	6584	463.12	S$^+$/H$^+$	4.7E-6
He I	4686	20.86	He I	6678	3.17	S^{++}/H$^+$	2.9E-6
[Ar IV]	4741	0.77	[S II]	6717	29.63	Cl^{++}/H$^+$	4.0E-8
Hβ	4861	100.00	[S II]	6737	28.72	Ar^{++}/H$^+$	2.9E-6
He I	4921	0.27	He I	7065	4.52	Ar^{3+}/H$^+$	5.6E-7
[O III]	4959	345.20	[Ar III]	7136	39.40		

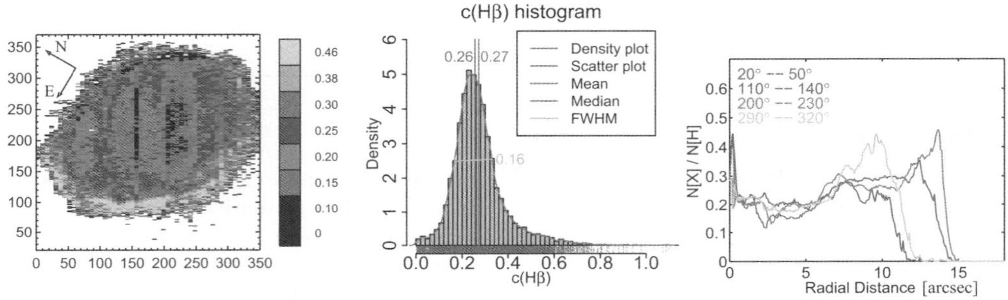

Figure 1. Extinction coefficient of NGC 6720 derived from Gemini/GMOS data. Left: Spatial distribution of c(Hβ). Center: Its histogram (the median is shown in red, mean in blue, FWHM in cyan, density plot in dark orange, and scatter plot in purple). Right: Averaged radial profiles of c(Hβ). The angles (indicated in the figure) increase counter-clockwise and 0° is to the right.

(center), and its averaged radial profile along 4 different directions (right). The spatial variation of c(Hβ) follows the structure of the nebula, a characteristic also observed in NGC 40 (Leal-Ferreira *et al.* 2011) and NGC 7009 (Walsh *et al.* 2016). The histogram helps us to show that the spatial variation of c(Hβ) is significant. If we use the full width half-maximum (FWHM) to (conservatively) determine the range where the variation is significant, we conclude that $0.18 < c(H\beta) < 0.34$. However, the radial profiles reinforce that c(Hβ) in the outer region of the source reaches values higher than 0.40 (which could be linked to a higher dust-to-gas ratio, see Walsh *et al.* (2016) for a detailed discussion).

Acknowledgments: This work was supported by CNPq, Conselho Nacional de Desenvolvimento Científico e Tecnológico - Brazil, process numbers 248503/2013-8 (MLL-F) and 157806/2015-4 (IA) and by NWO/Netherlands.

References

Aleman, I., Leal-Ferreira, M. L., *et al.* in prep., *A&A*
De Robertis, M., Dufour, J., & Hunt, R. 1987, *J. R. Astron. Soc. Canada*, 81, 195
Kingsburgh, R. & Barlow, M. 1994, *MNRAS*, 271, 257

Leal-Ferreira, M. L., Gonçalves, D. R., Monteiro, H., & Richards, J. W. 2011, *MNRAS*, 411, 1395

Monteiro, H., Gonçalves, D. R., Leal-Ferreira, M. L., & Corradi, R. L. M 2013, *A&A*, 560, 102

Shaw, R. A. & Dufour, R. J 1995, *PASP*, 107, 896

Walsh J., Monreal-Ibero A., Barlow M., Ueta T., Wesson R., & Zijlstra A. 2016 *A&A*, 588, 106

Planetary Nebulae: Multi-Wavelength Probes of Stellar and Galactic Evolution
Proceedings IAU Symposium No. 323, 2016
X. Liu, L. Stanghellini & A. Karakas, eds.

doi:10.1017/S1743921317001429

Herschel Planetary Nebula Survey (HerPLaNS): Construction of a Detailed Dusty Photoionization Model of NGC6781

Masaaki Otsuka[1], Toshiya Ueta[2], You-Hua Chu[1,3], Kentaro Asano[2] and HerPLaNS consortium

[1]ASIAA, 11F of Astronomy-Mathematics Building, AS/NTU. No.1, Sec. 4, Roosevelt Rd, Taipei 10617, Taiwan, R.O.C. email: otsuka@asiaa.sinica.edu.tw (MO)

[2]Department of Physics and Astronomy, University of Denver, 2112 E. Wesley Ave., Denver, CO 80210, USA

[3]Department of Astronomy, University of Illinois, 1002 West Green Street, Urbana, Illinois, USA

Abstract. As one of the follow-up studies of Herschel Planetary Nebula Survey (HerPlaNS; Ueta *et al.* 2014), we focus on a bipolar planetary nebula (PN) NGC6781 to characterize the dusty nebula and the central star based on our own *Herschel* data and the rich archival spectroscopic/photometric image data in the wavelengths from UV to far-IR. With CLOUDY, we constructed a comprehensive photoionization model of NGC6781 ever made including data from UV to radio. We succeeded to reproduce the observed spectral energy distribution (SED) and the atomic gas, H_2, CO, and OH molecular line fluxes. We found that about 40 % of the total dust mass would be from warm-cold dust components.

Keywords. infrared: ISM, ISM: abundances, dust, molecules, planetary nebulae: individual (NGC6781)

1. Results

We performed a comprehensive analysis of the PN NGC6781 in order to investigate the physical conditions of each of the ionized, atomic, and molecular gas components as well as the dust component in the nebula and the evolution of the central star, based on our own far-IR *Herschel* data (Ueta *et al.* 2014) and the rich archival spectroscopic/photometric data in the UV to far-R wavelength ranges.

Spitzer/IRS mid-IR spectrum shows the polycyclic aromatic hydrocarbon (PAH) 6-9 μm and 11.3 μm bands and pure rotational hydrogen molecule (H_2) lines. The excitation curve plots for the H_2 lines can be fitted by a single temperature of 1050 K. We derived the column density $N(H_2) = 9.12\times10^{18}$ cm^{-2}. Comparison of the observed H_2 line intensities relative to the H_2 v=0-0 S(2) at 9.67 μm line with the theoretical shock models by Flower & Pineau Des Forêts (2010) suggests that the H_2 lines are excited by shock interaction with the remnant asymptotic giant branch (AGB) circumstellar envelope.

We performed detailed plasma diagnostics and derived nebular abundances of the nine elements. By comparing with the AGB nucleosynthesis model of Karakas (2010), we found that the progenitor would be a 2.25-3.0 $M_{\odot}$ star. The current evolutional status and nebular elemental abundances of the H_2-rich bipolar PN NGC6720 (e.g., van Hoof *et al.* 2010) are in excellent agreement with NGC6781, suggesting that both PNe have evolved from a very similar mass star and share with their evolution.

We constructed the photoionization model with CLOUDY (Ferland *et al.* 2013) to be self-consistent with all the observations (Fig. 1). About 40 % of the total dust mass

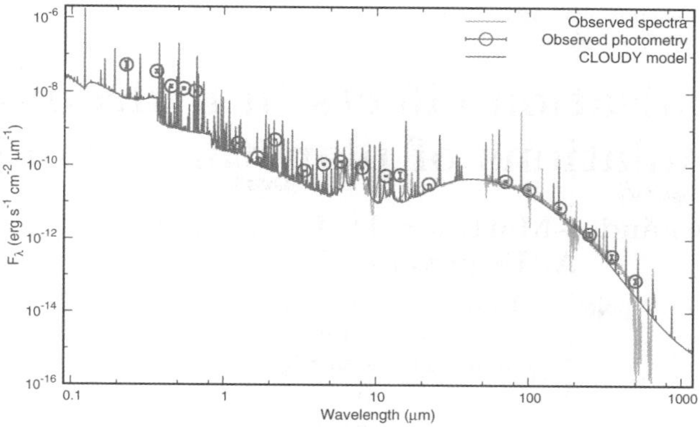

Figure 1. The full SED of the best-fit CLOUDY model of NGC6781 (red line; spectral resolution (R) = 300), compared with the observational constrains: photometry data (blue circles) and spectroscopy data (grey line).

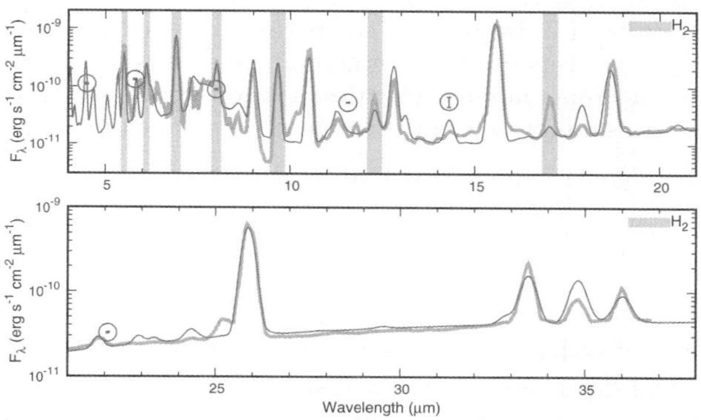

Figure 2. Comparison between the observed data (grey line and blue circles) and the best-fit CLOUDY model (red line; R = 100). The positions of the H_2 lines are highlighted.

$2.13\times10^{-3}\,M_\odot$ would be from warm-cold dust components. The derived gas mass of $0.58\,M_\odot$ corresponds to ~70 % of the total mass loss in initially $3.0\,M_\odot$ stars predicted by Karakas (2010). We found that additional heating sources are necessary to explain the observed H_2, CO, and OH line fluxes. By introducing warm regions with a constant kinematic gas temperature of ~1000 K in the photodissociation region (PDRs), we succeeded to reproduce these molecule line fluxes (Fig. 2).

To verify physical conditions of the warm regions in the PDRs introduced in our model, atomic gas and molecule emission line maps are necessary. A wide variety of molecular gas abundances are essential to further characterize the PDR conditions in NGC6781 and obtain more exact elemental abundances. A more detailed discussion will be found in Otsuka *et al.* (*in prep*).

References

Ferland, G. J., Porter, R. L., van Hoof, P. A. M., *et al.* 2013, *RMxAA*, 49, 137
Flower, D. R. & Pineau Des Forêts, G. 2010, *MNRAS*, 406, 1745
Karakas, A. I. 2010, *MNRAS*, 403, 1413
Ueta, T., Ladjal, D., Exter, K. M., *et al.* 2014, *A&A*, 565, A36
van Hoof, P. A. M., van de Steene, G. C., Barlow, M. J., *et al.* 2010, *A&A*, 518, L137

Planetary Nebulae: Multi-Wavelength Probes of Stellar and Galactic Evolution
Proceedings IAU Symposium No. 323, 2016
X. Liu, L. Stanghellini & A. Karakas, eds.

doi:10.1017/S1743921317000667

Photoionisation effects in hydrodynamical simulations of planetary nebulae

L. Hernández-Martínez, D. Estrella, P. F. Velázquez, A. Esquivel and A. C. Raga

Instituto de Ciencias Nucleares, Universidad Nacional Autónoma de México, 70-543, México, D. F. México
email: `lilihe.mtz@gmail.com`

Abstract. We explored the photoionisation effects on both the proper motion and emission of planetary nebulae NGC 6302, by means of hydrodynamical simulations. We used the GUACHO code, which includes the photoionisation due to central source (Esquivel *et al.* 2009, Esquivel & Raga 2013). We model these PNe considering an interacting stellar fast wind with and ejected toroidally shaped slow wind (Uscanga *et al.* 2014). Synthetic Hα emission maps were obtained from our numerical results in order to do a comparison between the cases with and without photoionisation. Using a wavelets fittering method on our results for the ionisation case, we do not find an increase in the proper motion velocities, however we can see an accelerated expansion in both cases. For the ionisation case the Hα emission presents an increase.

Keywords. Planetary Nebulae, Bipolarity, Hydrodinamics

1. Introduction

NGC6302 is a butterfly-shaped bipolar Planetary Nebulae (PN) with a very cumply structure and a central dense circumstellar disc. The lobes of this PN present an accelerated expansion (also known as Hubble-type expansion) maybe produced by a binary system (Szyszka *et al.* 2011). We used GUACHO to model NGC 6302 and reproduce the PN lobes. GUACHO is a massive parallel 3 dimensional hydrodynamical code with uniform mesh and that can take into account Lyα photons for the photoionisation prescriptions (Esquivel *et al.* 2009, Esquivel & Raga 2013).

2. Hydrodynamical simulations

We have performed hydrodynamical simulations with the GUACHO code and study the effects of the photoionisation on the proper movement of the clumpy structure shown in the Hα emission maps. The computational domain is 256x256x512cells (0.5x0.5x1pc) in the uniform mesh, yielding a resolution of 0.002pc. We include two winds, following the model presented in Uscanga *et al.* (2014). The physical parameters of these winds are: a) for the slow wind we used $\dot{M}_S=5\times10^{-4}$ $M_\odot$ yr^{-1}, $T_S=10^3$ K, $V_S=15$ km s^{-1} and b) for the fast wind we used $\dot{M}_F=1\times10^{-7}$ $M_\odot$ yr^{-1}, $T_F=10^6$ K, $V_F=1800$ km s^{-1}. We ran the simulations by 2000 yr, and we focus on the last 10 yr of the PN evolution in order to compare with the observations. In this simulations we imposed a turbulent density distribution (Esquivel *et al.* 2003) an ionising flux of photons of 10^{46} erg s^{-1} divided in 10^5 rays, emulating the diffuse radiation of the system. Our simulations show the density structure and the morphology like as NGC 6302 (size and structure). However, the Hα image with photons indicate us a clearly increase in the emission of the clumps. In order to study the dynamical of our numerical simulations, we processed our Hα maps (see

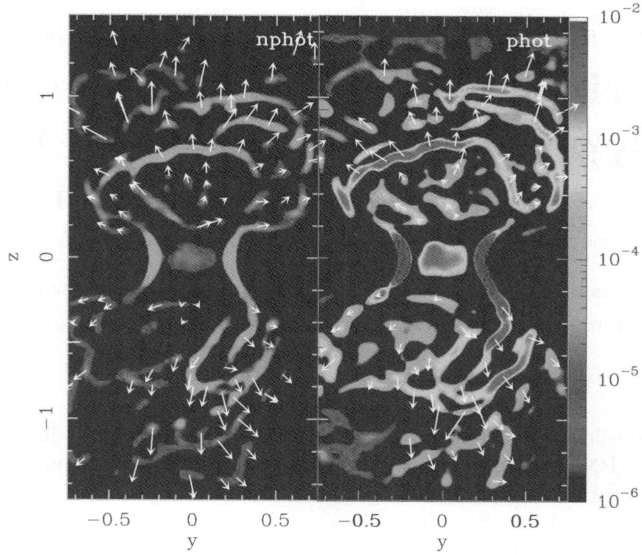

Figure 1. Synthetic Hα images of our numerical simulations, convolved with a wavelet function with sigma=5. To obtain the proper motions we followed the method described in Raga *et al.*(2016) with synthetic images that have 10yr between each other, arrows depict the proper motions.

Figure 1) with the wavelets technique presented in Raga *et al.* (2016), i.e. we convolved the synthetic images by using a Mexican-hat function with a given sigma, in our case was of 5 sigma. To obtain the proper motions we followed the method described in Raga *et al.* (2016) with synthetic images that have 10 yr between each other, arrows depict the proper motions. Then we compared with HST observations (which are going to be presented in Hernández-Martínez *et al.* in prep.)

3. Discussion and future work

We reproduced the Uscanga *et al.* 2014 model using GUACHO code which includes a central ionising source. We have compared the proper motion obtained from numerical results including or not the photoionisation of a central sourde. We can not discriminate between the two different data. So that, for our simulations the photoionisation apparently does not has an important effect in the proper motion. However, we can see a Hubble-type expansion, as it was reported by Szyszka *et al.* (2011). We also see an increase in the Hα emission in the simulation with photoionisation. These are preliminary results and we are still working on it.

References

Esquivel, A., Cantó, J., & Rodríguez-González, A. 2009, *A&A*, 507,855

Esquivel, A., Lazarian, A., Pogosyan, P., & Chou, 2003, *MNRAS*, 342, 325

Esquivel& Raga 2013, *ApJ*, 779,111

Hernández-Martínez *et al.* in prep.

Raga, A. C., Reipurth, B., Esquivel, A., & Bally, J. 2016, *AJ*, 151, 113

Szvszka, C., Zijlstra, A. A., & Walsh, J. R. 2011, *MNRAS*, 416, 715

Uscanga, L., Velazquez, P. F., Esquivel, A., Raga, A. C., & Boumis, P. 2014, *MNRAS*, 442, 3162

Planetary Nebulae: Multi-Wavelength Probes of Stellar and Galactic Evolution
Proceedings IAU Symposium No. 323, 2016
X. Liu, L. Stanghellini & A. Karakas, eds.

doi:10.1017/S1743921317001879

Characterizing global expansion evolution of the ionized matter in planetary nebulae

J. A. López[1], M. G. Richer[1], M. Pereyra[2] and M. T. García-Díaz[1]

[1]Instituto de Astronomía, Campus Ensenada, Universidad Nacional Autónoma de México, Ensenada, Baja California, C. P. 22860, México
email: jal@astrosen.unam.mx, richer@astrosen.unam.mx, tere@astrosen.unam.mx

[2]Schlumberger Foundation Fellow, University of Southampton
Southampton, SO17 1BJ, United Kingdom
email: e.m.pereyra-talamantes@soton.ac.uk

Abstract. Bulk outflow or global expansion velocities are presented for a large number of planetary nebulae (PNe) that span a wide range of evolutionary stages and different stellar populations. The sample comprises 133 PNe from the Galactic bulge, 100 mature and highly evolved PNe from the disk, 11 PNe from the Galactic halo and 15 PNe with very low central star masses and low metallicities, for a total of 259 PNe. These results reveal from a statistical perspective the kinematic evolution of the expansion velocities of PNe in relation to changing characteristics of the central star's wind and ionizing luminosity and as a function of the evolutionary rate determined by the central (CS) mass. The large number of PNe utilized in this work for each group of PNe under study and the homogeneity of the data provide for the first time a solid benchmark form observations for model predictions, as has been described by López *et al.* (2016).

Keywords. ISM: kinematics and dynamics, planetary nebulae: general, stars: evolution, low-mass

1. Introduction

We have used the SPM Kinematic Catalogue of Galactic Planetary Nebulae (López *et al.* (2012) as the single source of data for all our measurements. This catalogue contains long-slit, echelle data for nearly 700 Galactic PNe. The spectral resolution in the data ranges from 6 to 12 km s^{-1}, perfectly adequate to resolve the internal kinematics of PNe. We concentrate our measurements on the luminosity-weighted emission for the matter projected within the spectrograph slit. When line splitting is clearly present the expansion velocity is measured form the peaks separation. When the profile is unresolved we fit a gaussian and correct the observed profile for instrumental, thermal and fine structure broadening. We do not apply hydrodynamic corrections as suggested by Shoenberner *et al.* (2010) since we are not interested in measuring the velocity of the outer shock, which in any case is a pattern that cannot be measured directly. The present results have been compiled from several works that we have published on this subject since the last IAU Symposium.

Details for the bulge sample appear in Richer *et al.* (2008) and Richer, *et al.* (2010), those for the mature and highly evolved PNe in Pereyra *et al.* (2013) and for the halo and low metallicity PNe in Pereyra *et al.* (2016) and see their contribution in this volume. A detailed summary of these works has also been recently presented by López *et al.* (2016).

Figure 1 shows the overall results from these works where the approximate distribution in the H-R diagram of the different groups are shown over evolutionary tracks from Vassiliadis & Wood (1994).

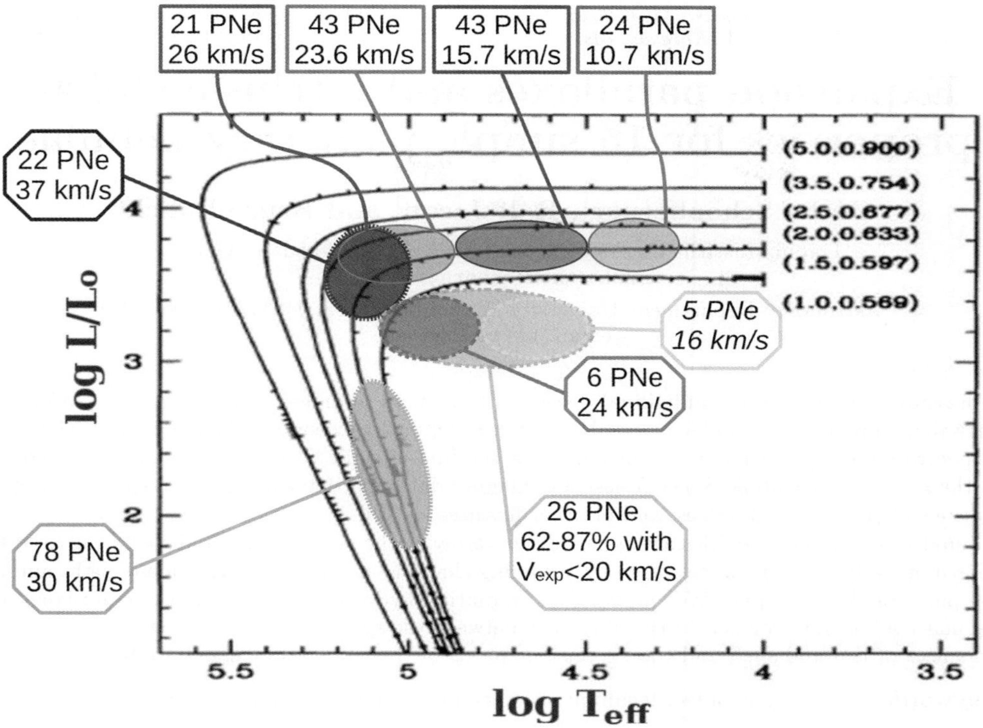

Figure 1. For each group are indicated the number of members measured and their average V_{exp}. The kinematic evolution with evolutionary stage is apparent. V_{exp} increases with T_{eff} in all cases, though within different ranges that indicate a dependance on CS masses and possibly metallicity. Labels at the top of the diagram point to the groups from the bulge. Labels on the left of the digram point to the mature and highly evolved groups. Labels within the diagram point to the halo and low metal sample.

This project has benefited from the continuous support throughout the years of DGAPA-UNAM, currently through PAPIIT project IN108416. We also thank the Observatorio Astronómico Nacional at San Pedro Mártir for generous allocations of telescope time that have allowed building the SPM Catalogue of Galactic Planetary Nebulae.

References

Lopez, J. A., Richer, M. G., & Garcia-Diaz, M. T. *et al.* 2012, *Rev. Mexicana AyA*, 67, 4679

Lopez, J. A., Richer, M. G., & Garcia-Diaz, M. T. *et al.* 2016, in: Sun Kwok & Kam-Ching Leung (eds.), *11th Pacific Rim Conference on Stellar Astrophysics: Physics and Chemistry of the Late Stages of Stellar Evolution*, Journal of Physics: Conference Series, 728, 032003

Pereyra, M., Richer, M. G., & López, J. A 2013, *ApJ*, 771, 114

Pereyra, M., López, J. A., & Richer, M. G., 2016, *AJ*, 151, 53

Richer, M. G., López, J. A., Pereyra, M., Riesgo, H., & García-Díaz, M. T. 2008, *ApJ*, 689, 203

Richer, M. G., López, J. A., García-Díaz, M. T., Clark, D. M., Pereyra & Díaz-Méndez, E. 2010, *ApJ*, 716, 857

Schoenberner, D., Jacob, R., Sandin, C., & Steffen, M. 2010, *A&A* 523 A86

Vassiliadis, E.. & Wood, P. R. 1994, *ApJS*, 92, 125

Planetary Nebulae: Multi-Wavelength Probes of Stellar and Galactic Evolution
Proceedings IAU Symposium No. 323, 2016
X. Liu, L. Stanghellini & A. Karakas, eds.

doi:10.1017/S1743921317000801

Expansion parallaxes and intrinsic stellar properties for 15 simple planetary nebulae

Detlef Schönberner[1], Ralf Jacob[1] and Bruce Balick[2]

[1]Leibniz-Institut für Astrophysik (AIP), 14482 Potsdam, Germany
email: deschoenberner@aip.de

[2]Astronomy Department University Washington, Seattle, WA 98195, USA
email: balick@uw.edu

Abstract. We determined individual distances to a small number of rather round, quite regularly shaped planetary nebulae by combining their angular expansion in the plane of the sky with a spectroscopically measured expansion along the line of sight. For this goal, we combined up to three epochs of Hubble Space Telescope imaging data and determined the respective proper motions of rim and shell edges, and of other features as well.
Ground-based radial velocities are assigned separately to the nebular rims and shells and used to determine individual distances, thereby assuming that the expansions in the line-of-sight and in the plane of sky are equal. We employed 1D radiation-hydrodynamics simulations of planetary nebulae evolution to correct for the difference between the spectroscopically measured expansion velocities of rim and shell and the expansion speeds of their respective shock fronts.

Keywords. planetary nebulae, techniques: image processing, techniques: spectroscopic

1. Context, goals, and methodology

Accurate individual distances to planetary nebulae (PNe) are vital for determining nebular luminosities, ages, and masses of their central stars. So far, reliable distances are only available for PNe in stellar populations with well-known distances (e.g. Galactic bulge, Magellanic Clouds), but only for singular cases in the Galactic disk. Statistical distances are prone to systematic errors because they rely on assumptions whose justification is not always guaranteed.

The goals are i) to obtain accurate distances for 15 simple, bright, extended H-burning disk PNe, and ii) to use these distances to derive fundamental properties of their central stars. For this purpose, the angular expansion speeds of bright rim and/or shell boundaries, $\dot{\theta}_{\rm rim/shell}$, were measured using multi-epoch HST imaging observations. The pairs of images in the lines of [O III] an [N II] span 7.6 to 14 yr for the various PNe in this survey. Two examples (IC 418, NGC 6826) are rendered in Fig. 1.

Proper motions reflect the pattern speeds of shock or ionisation fronts that propagate through the nebula. In contrast, Doppler speeds reflect bulk motions. The ratio of the two speeds must be corrected for these differences by means of 1D radiation-hydrodynamics evolution models for PNe (Marten *et al.* 1993; Mellema 2004; Schönberner *et al.* 2005; Schönberner *et al.* 2014).

Expansion distances are derived using $D_{\rm exp} = 211 \times \dot{R}_{\rm rim/shell}/\dot{\theta}_{\rm rim/shell}$, where $\dot{R}_{\rm rim/shell}$ is the pattern velocity (km s^{-1}) corresponding to $\dot{\theta}_{\rm rim/shell}$ (mas yr^{-1}) for a rim or shell. Let F be the correction factor obtained from the models such that $\dot{R}_{\rm rim/shell} = F_{\rm rim/shell} \times V^{\rm Dopp}_{\rm rim/shell}$. For the objects in our survey, we find $F_{\rm rim} = 3 \Longrightarrow 1.5$ for $V^{\rm Dopp}_{\rm rim} \simeq 5 \Longrightarrow \simeq 20$ km s^{-1}, $F_{\rm shell} = 1.25 \pm 0.05$ for all $V^{\rm Dopp}_{\rm shell}$. We measured parallaxes using the rims

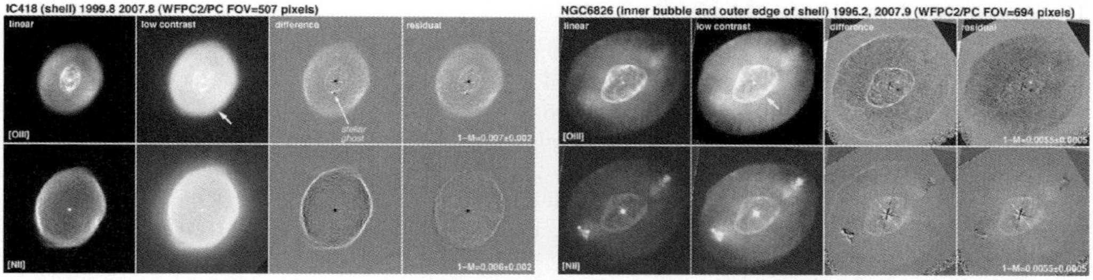

Figure 1. Examples (IC 418, NGC 6826) of the HST images, their multi-epoch changes, and post-magnification difference images. The third column shows the signal of expansion, and the fourth shows the difference after the first-epoch image was magnified by a factor M before subtraction. The angular expansion rate of a feature of angular size θ is then $\dot{\theta} = (M-1) \times \theta/\Delta t$, where Δt is the time between the epochs of the images used to measure M, ($7.6 \leqslant \Delta t \leqslant 14$ yr).

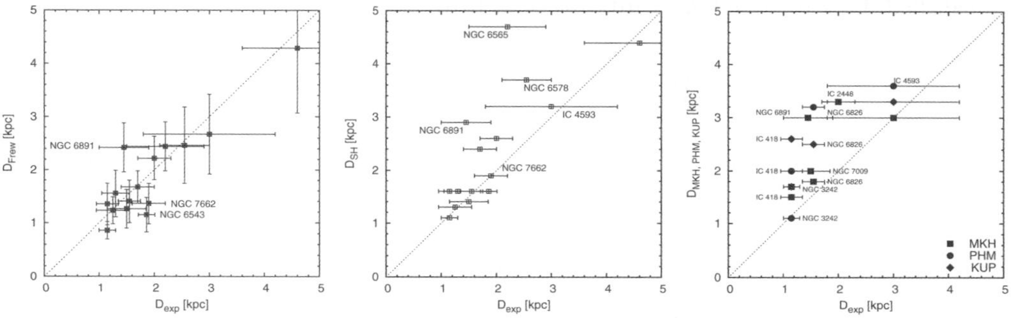

Figure 2. *Left*: distances based on the statistical calibration by Frew *et al.* (2016). *Middle*: distances based on Stanghellini & Haywood (2010). *Right*: distances derived spectroscopically.

and shells since each are independent entities with their different sources of momentum and historical motions.

2. Results and conclusion

A comprehensive presentation of the results and the conclusions drawn will be given in a forthcoming publication. Here we highlight the following (cf. Fig. 2):

1. Distances derived separately for the rims and shells agree (within the errors) after appropriate corrections for pattern speeds as explained in Sect. 1 have been applied.
2. General agreement with the statistical distances of Frew (2016) and, to a lesser degree, of Stanghellini & Haywood (2010) exists. The agreement with spectroscopic distances is poor.
3. Central-star "plateau" luminosities range from about 1500 to about 10 000 $L_{\odot}$, with a mean at about 5000 $L_{\odot}$. There is excellent agreement with the luminosities of objects from the Galactic bulge (Hultzsch *et al.* 2007) and the Magellanic Clouds (Herald & Bianchi 2004, 2007).
4. Employing the latest post-AGB evolutionary models of Miller Bertolami (2016, priv. comm.), we found a central-star mass range of about 0.53–0.60 $M_{\odot}$, with a mean of 0.55 $M_{\odot}$.

We conclude that expansion measurements of nebular edges offer, together with predictions from radiation-hydrodynamics models, a reliable method to determine distances to suited PNe.

References

Frew, D. J. Parker, Q. A., & Bojičić, I. S. 2016, *MNRAS*, 455, 1459
Herald, J. E. & Bianchi, L. 2004, *ApJ*, 611, 294
Herald, J. E. & Bianchi, L. 2007, *ApJ*, 661, 845
Hultzsch, P. J. N., Puls, J, Méndez, R. H., *et al.* 2007, *A&A*, 467, 1253
Kudritzki, R.-P., Urbaneja, M. A., & Puls, J. 2006, in Planetary Nebulae in our Galaxy and Beyond, eds. M. J. Barlow & R. H. Méndez, *IAU Symp.* 234, 119 (KUP)
Marten, H., Gesicki, K., & Szczerba, R. 1993, in Planetary Nebula, eds. R. Weinberger & A. Acker, *IAU Symp.* 155, 315
Mellema, G. 2004, *A&A*, 416, 623
Méndez, R. H., Kudritzki, R.-P., & Herrero, A. 1992, *A&A*, 260, 329 (MKH)
Miller Bertolami, M. M. 2016, *A&A*, 588, A25
Pauldrach, A. W. A., Hoffmann, T. L., & Mendéz, R. H. 2004, *A&A*, 419, 1111 (PHM)
Schönberner, D., R. Jacob & Steffen, M. 2005, *A&A*, 441, 573
Schönberner, D., Jacob, R., Lehmann, H., *et al.* 2014, *Astron. Nachr.*, 335, 378
Stanghellini, L. & Haywood, M. 2010, *ApJ*, 714, 1096

Planetary Nebulae: Multi-Wavelength Probes of Stellar and Galactic Evolution
Proceedings IAU Symposium No. 323, 2016
X. Liu, L. Stanghellini & A. Karakas, eds.

doi:10.1017/S174392131700165X

Temperature effects on the infrared spectrum of molecules in planetary nebulae

Tao Chen[1,2], Cameron Mackie[2] and Alexander G. G. M. Tielens[2]

[1]Department of Theoretical Chemistry and Biology, School of Biotechnology, Royal Institute of Technology, 10691, Stockholm, Sweden
email: chen@strw.leidenuniv.nl

[2]Leiden University, Leiden Observatory, Niels Bohrweg 2, NL-2333 CA Leiden, Netherlands

Abstract. In this work, we utilize a method based on Wang-Landau Monte Carlo sampling for studying the temperature effects of astrophysically relavant molecules. Anharmonic effects, e.g., resonances, overtones, and combination bands, are fully incoportated in this method. The calculated infrared (IR) spectra are consistent with the experimental data measured by National Institute of Standards and Technology (NIST).

Keywords. PAH, molecule, infrared, temperature, anharmonic, spectrum

1. Introduction

IR spectroscopy is a powerful tool to study molecules in space (Tielens, 2008). A key issue in such analysis is understanding the effect that temperature has on different vibrational bands, and thus interpret the IR spectra for molecules in various media. From the theoretical point of view, the effect of temperature on IR vibrational spectra depends on anharmonic effects and can be accounted for using several methods, e.g. applying empirical scaling factor on the harmonic frequencies (Merrick, 2007); classical molecular dynamics simulations (Van-Oanh, 2005), Car-Parrinello (Kumar, 2006) and Born-Oppenheimer molecular dynamics (Estácio, 2008), etc.

Alternatively, temperature effects on the IR spectra can be incoportated through inverse Laplace transformation of quantum partition functions (Romanini, 1993). In such a method, the vibrational Density of States (DoS) are estimated by direct Monte Carlo (MC) integration. However, such methods become highly inefficient for large systems due to increasing variances (Hüpper, 1999). An improved MC method, the Wang-Landau method is introduced for handling high-dimensional space problems. It has been shown that such a method could reproduce very accurate vibrational DoS for large, nonseparable systems (Basire, 2008).

We apply this Wang-Landau method on theoretical spectra at 0 Kelvin (calculated by second-order Vibrational Perturbations Theory (VPT2) (Mackie, 2015), in such spectra all anharmonic effects, i.e., resonances, overtones, and combination bands, are fully incoportated.

2. Computational details

The infrared spectra at finite temperatures are calculated in three steps: first of all, we compute the vibrational DoS by the Wang-Landau method (Basire, 2008); secondly, we

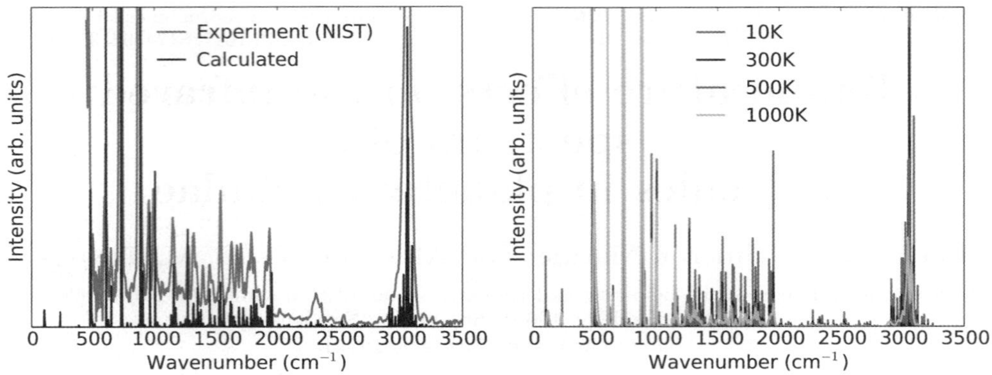

Figure 1. Gas phase IR spectra of anthracene ($C_{14}H_{10}$) molecule. The left panel compares the calculated and experimental spectra (recorded by NIST at 300K). The right figure compares the calculated spectra at different temperatures.

accumulate internal-energy dependent two-dimensional histograms of the intensities and energies of the transitions; finally, we convert the accumulated absorption histograms to microcanonical absorption spectra, which derives final the absorption intensities and energies at finite temperature through Laplace transformation. For details of the method, see Chen *et al.* (in preparation).

3. Results and Discussion

Several Polycyclic Aromatic Hydrocarbon molecules (PAHs) are tested with this method. Figure 1 shows the IR spectra of anthracene ($C_{14}H_{10}$) molecule. The left panel compares the calculated and experimental spectra (NIST at 300 K). It can be seen that the calculated spectrum is consistent with the experimental results. The right panel presents the spectra at different temperatures. Due to temperature effects, independent line shifts and broadenings can be observed clearly. Such spectra may guide infrared observations to detect PAH molecules in interstellar medium, and also aid in deeper understanding of measured spectra. For more results and discussion, see Chen *et al.* (in preparation).

References

Basire, M., *et al.*, 2008, *JCP*, 129, 081101
Estácio, S. G. & Costa Cabral, B., 2008, *J. Chem. Phys. Lett.*, 456, 170
Hüpper, B. & Pollak, E. J., 1999, *Chem. Phys.*, 110, 11176
Kumar, P. P. & Marx, D., 2006, *Phys. Chem. Chem. Phys.*, 5, 573
Mackie, C. J., *et al.*, 2015, *J. Chem. Phys.*, 143, 224314
Merrick, J. P., *et al.*, 2007, *J. Phys. Chem. A*, 111, 11683
Romanini, D. & Lehmann, K. K., 1993, *J. Chem. Phys.*, 98, 6437.
Tielens, A. G. G. M., 2008, *Annu. Rev. Astron. Astrophys.* 46, 289
Van-Oanh, N.-T., *et al.*, 2005, *Phys. Chem. Chem. Phys.*, 106, 10144

Planetary Nebulae: Multi-Wavelength Probes of Stellar and Galactic Evolution
Proceedings IAU Symposium No. 323, 2016
X. Liu, L. Stanghellini & A. Karakas, eds.

doi:10.1017/S1743921317001727

3D morphology of the PN IRAS 18333-2357

C. Muthumariappan

Indian Institute of Astrophysics, Bangalore 560 034, India. email: muthu@iiap.res.in

Abstract. We analysed the *Spitzer* 8μm and *WISE* 22μm images of the PN IRAS 18333-2357 using a 3D RT code. We describe its morphology with a H-poor disk and a spherical shell around it.

Keywords. (ISM:) planetary nebulae: individual (IRAS 18333-2357)

1. Introduction

The Halo PN IRAS 18333-2357 is a strong IR source located in the globular cluster M22. It is important in understanding the evolution of low mass stars, stellar merger at low metallicity and the dust grains ejected from metal-poor systems. The space velocity of M22 is 158 km s^{-1} and the ram pressure causes a asymmetry in the PN morphology. The nebular spectrum has forbidden lines but no H and He lines, showing It's H-poor nature with a lack of photo-ionization and presence of dust heating (Borkowski & Harrington(1991)). The nebula displays carbon rich dust chemistry (Muthumariappan *et al.* (2013)).

2. Radiative Transfer Model and Results

We used archival UV (*FUSE*), optical (Palomar), near-IR (2MASS and *DENIS*, mid- and far-IR (*WISE*, *IRAS* and *Akari*) photometric and *Spitzer*-IRS spectroscopic observations to construct its SED from far-UV to the far-IR. *Spitzer* 8μm and *WISE* 22μm images are used to trace the emission from PAH and very small grains (VSG, $\leqslant$ 200 Å) and continuum from cool dust respectively. The SED and the thermal images were modelled using a 3D Monte-Carlo RT Code Hochunk-3D to derive the nebular morphological parameters. The shell has a $1/r^n$ density function and an equatorial torus is (at 40° inclination) at the center. Thermally fluctuating grains are included in the radiative transfer. The best-fit model SED is presented in Fig. 1, and the images are shown in Fig 2.

Excess emission seen in the 3- 12μm region of SED is due to PAH/VSG which arise strongly from the disk. Disk's sky- projected sizes overlap with the optical nebula showing their co-existence, supporting nubular heating by photo-elecric emission from PAH/VSG. The central star temperature is 50 000 K and it's luminosity is 2 640 $L_{\odot}$. Nebular gas to dust mass ratio is seven. The morphology of IRAS 18333-2357, it's dust mass and it's gas to dust mass ratio (Table 1) are similar to the H-poor inner ejecta of PN A30. This similarity suggests that they possibly share similar origin: PN formed from born-again event. *WISE* 22μm image shows a faint, half moon shaped, fragmented outer envelope which is nearly perpendicular to the direction of the cluster's motion. This could be due to the PN's interaction with the Halo gas. This faint envelope could be the ejection before the Late thermal pulse and it likely to have untraced H-rich gas.

Table 1. Dust shell parameters

Parameter	Disk	Shell
Radius($''$)	inner: 2.8; outer: 6.0	inner: 13.3; outer: 25
Dust Temperature (K)	inner: 115; outer: 68	inner: 74; outer: 45
Dust mass ($M_\odot$)	1.42×10^{-3}	2.7×10^{-3}

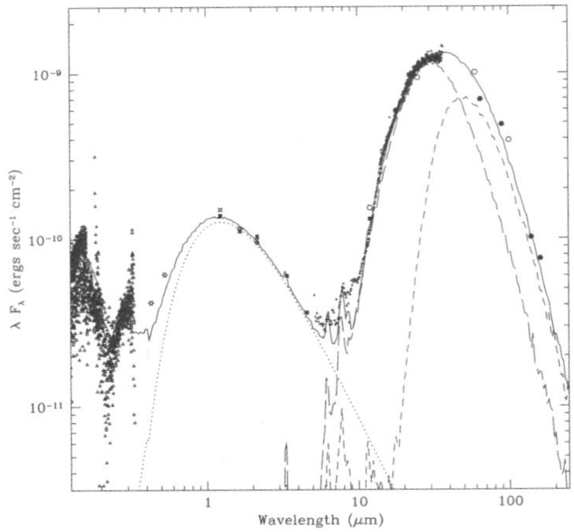

Figure 1. SED fit (solid line) to the observation. Flux from the disk (long-dashed line), from the shell (short-dashed line) and field star (dotted line) are also shown. Observations: *HST* FOS spectrum (diaginal cross), *IUE* spectrum (open triangles), *DENIS* (filled stars), *WISE* (filled squares), *Spitzer* IRS (filled triangles), *IRAS* (open hexagons) and *Akari* (filled hexagon).

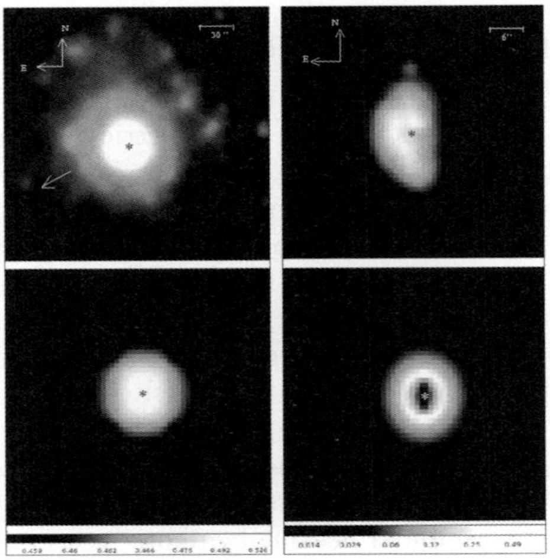

Figure 2. Top Panels: *WISE* 22 μm (left) and *Spitzer* IRAC 8 μm (right) images. Bottom Panels: model images at 22- and 8 μm. The PN motion in the sky is shown (top left picture).

References

Borkowski K. J. & Harrington J. P. 1991, *ApJ*, 379, 168

Muthumariappan C., Parthasarathy, M., & Ita Y. 2013, *MNRAS*, 435, 60

Planetary Nebulae: Multi-Wavelength Probes of Stellar and Galactic Evolution
Proceedings IAU Symposium No. 323, 2016
X. Liu, L. Stanghellini & A. Karakas, eds.

doi:10.1017/S1743921317001703

3D Hydrodynamical models of the planetary nebulae M 1-32 and M 3-15

Jackeline S. Rechy-García[1], Pablo F. Velázquez[2], Miriam Peña[1] and Alejandro C. Raga[2]

[1]Instituto de Astronomía, Universidad Nacional Autónoma de México, Apdo. Postal 70264, 04510, Ciudad de México, México.
email: `jrechy,miriam@astro.unam.mx`

[2]Instituto de Ciencias Nucleares, Universidad Nacional Autónoma de México, Apdo. Postal 70543, 04510, Ciudad de México, México.
email: `pablo,raga@nucleares.unam.mx`

Abstract. High-resolution, long-slit spectroscopic observations of two planetary nebulae, M1-32 and M3-15 are presented. The observations were obtained with the 2.1-m telescope at the OAN-SPM, and MES spectrograph. M1-32 shows wide wings on the base of its emission lines and M3-15 has two very faint high-velocity knots. To model both PNe we built a 3D model consisting in a jet interacting with an equatorially concentrated slow wind, emulating the presence of a dense torus, by using the hydrodynamical code Yguazú. The synthetic position-velocity (PV) diagrams obtained from our models reproduce well the observed PV diagrams.

Keywords. hydrodynamics, planetary nebulae: individual (M 1-32, M 3-15).

1. Introduction

More than 80% of planetary nebulae (PNe) are asymmetrical (Douchin *et al.* 2013), showing structures such as jets, knots, tori, etc. It has been proposed that this morphological variety of PNe is due to a binary-system scenario (Soker *et al.* 1992), where the progenitor star of the PN interacts with a stellar companion.

In this work, we use high resolution spectra to analyze the kinematical behavior of the gas in two PNe, M 1-32 and M 3-15, and by using a hydrodynamical model we investigate if the morphological structure of these PNe can be produced by the interaction between a jet and a dense torus. We compare the position-velocity (PV) diagrams obtained from the observations with those predicted by the hydrodynamical models.

2. Observations

High-resolution spectra for both PNe were obtained with the Manchester Echelle Spectrometer (MES; Meaburn *et al.* 2003) attached to the 2.1-m telescope of the Observatorio Astronómico Nacional at San Pedro Mártir, B.C, México (OAN-SPM). Observations were obtained for three different regions of each object. In Fig. 1 we show the PV diagrams for the slit located in the centre for both PNe. For M1-32, two bright knots (at a heliocentric systemic velocity of -86.4 km s^{-1}) and wings at high velocities (± 180 km s^{-1}, relative to the system) are visible. For M3-15, a bright condensation (at systemic velocity of 96 km s^{-1}) and jets, ending in knots at ± 90 km s^{-1} (relative to the system) are seen.

3. Simulations

We have calculated hydrodynamical models using the code Yguazú (Raga *et al.* 2000). Our model considers two components: a dense torus and a bipolar system of conical jets. In the Fig. 2 we show the synthetic PV diagrams for both PNe. The simulation for M 1-32 was carried out until an integration time of 4500 yr. In our model, the jet axis

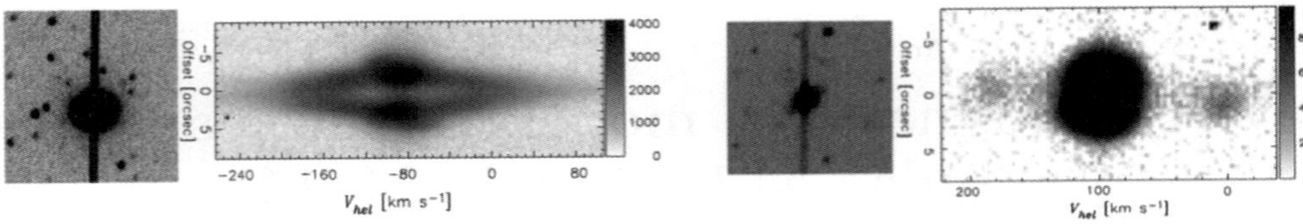

Figure 1. Slits located in the centre and PV diagrams for M 1-32 (left) and M 3-15 (right). The intense central emission corresponds to a toroidal component. The wide faint components correspond to a bipolar jet.

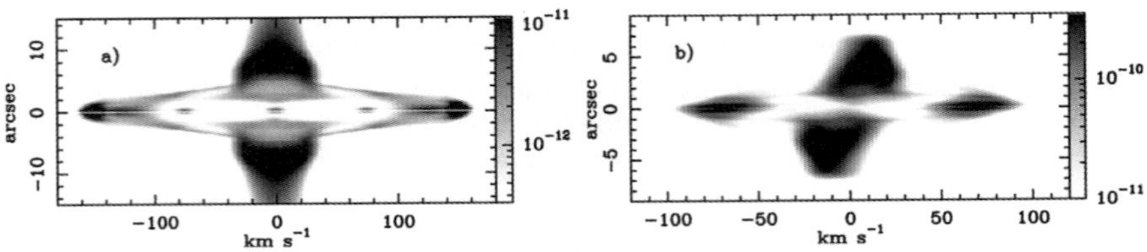

Figure 2. Synthetic position-velocity diagrams. The slit is located on the axis of the model. Left: M1-32. Right: M3-15. The logarithmic grey scale is in erg cm^{-3} sr^{-1}.

is nearly pole-on, at 0° with respect to the line of sight. We reproduce the central knots corresponding to the dense torus, and the wide wings associated with the jet material, at velocities of 160 km s^{-1}. The simulation for M 3-15 was obtained after an integration time of 3500 yr. We obtained a good agreement with observations if the jet axis was tilted by -55° with respect to the North direction, and at an inclination of 30° with respect to the line of sight.

4. Conclusions

The bipolar ejection leaves the central star, expanding through the poles of the torus at high velocities, of the order of 160 km s^{-1} and 100 km s^{-1} for M 1-32 and M 3-15, respectively. With the simulations we reproduce both the observed morphology and the PV diagrams of these nebulae, after an integration time of 4500 yr, for M1-32, and 3500 yr, for M 3-15, which are in agreement with the age of PNe of young population. In summary, we find that the spectral characteristics of M 1-32 and M 3-15 can be explained with the same physical model -a jet moving inside an AGB wind- using different parameters (physical conditions and position angles of the jet). Although the morphology of M 1-32 and M 3-15 are classified as spheroidal in the catalogs, due to their round appearance, spectroscopically both show a torus and high-velocity bipolar ejections. Therefore, we propose to call them "spectroscopic bipolars".
A full version of this work can be found in (Rechy-García *et al.* 2017).

5. Acknowledgements

This work received support from DGAPA-PAPIIT grants IN109614, IG100516 and CONACyT grant 167611. J.S.R.-G. acknowledges scholarship from CONACyT-México and the travel grant from IAU to attend the IAUS323.

References

Douchin, D., De Marco, O., Jacoby, G. H., *et al.* 2013, *Rev. Mexicana AyA*, 37, 391
Meaburn, J., López, J. A., Gutiérrez, L., *et al.* 2003, *Cent. Eur. Astrophys. Bull*, 39, 185
Raga, A. C., Navarro-González, R., & Villagrán-Muniz, M. 2000, *Rev. Mexicana AyA*, 36, 67
Rechy-García, J. S., Velázquez, P. F., Peña, M., & Raga, A. C. 2017, *MNRAS*, 464, 2318
Soker, N. & Harpaz, A. 1992, *PASP*, 104, 923

Planetary Nebulae: Multi-Wavelength Probes of Stellar and Galactic Evolution
Proceedings IAU Symposium No. 323, 2016
X. Liu, L. Stanghellini & A. Karakas, eds.

doi:10.1017/S1743921317001211

Exploring the transition into an asymmetric planetary nabulae

Mónica W. Blanco Cárdenas[1], Roberto Vázquez[1], Hans Ulrich Kaeufl[2], Marco A. Gómez-Muñoz[1], Pedro F. Guillén[1], Martín A. Guerrero[3] and Luis F. Miranda[3]

[1]Instituto de Astronomía, Universidad Nacional Autónoma de México, Km 103 Carretera Tijuana-Ensenada 22860, Ensenada, Mexico
email: mwblanco@astrosen.unam.mx

[2]European southern Observatory, Karl Schwarchild-Str 2 85748, Garching bei Muenchen, Germany

[3]Instituto de Astofísica de Andalucía, Glorieta de la Astronomía S/N 18008, Granada, Spain

Abstract. We present studies using different observational techniques, along different frequencies, aiming to resolve and investigate jets, outflows, as well as compact and innermost regions of asymmetric planetary nebulae (PNe) and objects in transition to PN. All the information gathered allow us to explore the kinematics and other important properties of the structures that play a crucial role in the shaping of complex PNe morphologies, in particular, we explore the role of disks/tori as collimating engine of extreme axisymmetric PNe.

Keywords. Stars, circumstellar mater, outflows.

1. The complex sculpting of an asymmetric planetary nebula

Several planetary nebulae (PNe) are asymmetric and high velocity jets and outflows are common features in these evolved objects. Furthermore, during the last years, steady and/or slowly rotating disks have been discovered in the cores of extreme axisymmetric PN. Infrared observations provide the opportunity to investigate the innermost dusty regions that usually are related with torii and disks. In this work, we present proto-PNe and young PNe with at least one observational proof of torus or disk, and we explore these structures using other techniques: spectro-astrometry (SA) using CRIRES-VLT (Blanco Cárdenas *et al.* 2014) and analysis of its SED using available photometry of several missions from the visible to radio frequencies (2MASS, *MSX*, *WISE*, *IRAS*, *AKARI*, etc.) and the SED building and analysis tool IRIS **(cxc.cfa.harvard.edu/iris/v2.1)**.

2. Overview and preliminary results

Proto-PN AFGL 915. This source (a.k.a The Red Rectangle) has a well-known rotating disk of few hundreds of AU (Bujarrabal *et al.* (2013)) and references there in) surrounded by a steady disk of silicates (Waters *et al.* (1998)). Blanco Cárdenas *et al.* (2014) detected a smaller spectral signature of this rotating disk (few tens of AU, ~100 mas) using the SA technique. As for the SED, three components fit well: a 7000 K Blackbody (BB), a rotating disk in mid-IR wavelengths, and a 170 K BB for the larger wavelengths.

IRAS 18454+0001: an extremely young PN? Given its compact size (~1") and the lack of an optical counterpart this source must be a bipolar PN in the making.

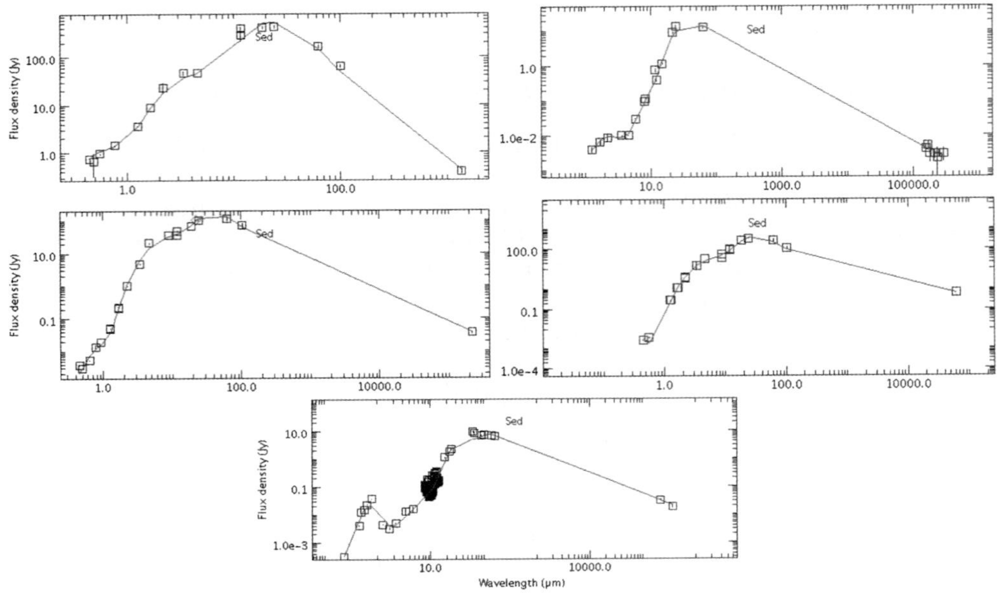

Figure 1. SEDs of the sample. From left to right and up to bottom: AFGL 915, IRAS 18454+0001, M 2-9, Mz 3, and M 1-16.

Results outstanding the equatorial estructure clearly dominated by the $\lambda 8.6\,\mu$m PAH filter of VISIR-VLT (Blanco Cárdenas *et al.* 2013), whereas the spitzer spectrum shows the presence of carbon-rich material (mid-IR PAH features) and confirms the presence of ionised material ([Ne II] $\lambda 12.8\,\mu$m). According to the SED fitting, a very hot star (83,000 K) and a disk beyond the mid-IR (larger dust grains) are needed, among other components of warmer material in the IR.

Young PNe: M 2-9, Mz 3 and M 1-16. Silicate disks were found in the central regions of M 2-9 and Mz 3 (Chesneau *et al.* 2007 and Lykou *et al.* 2011). On the other hand, CO (J=2-1) observations confirms the presence of two concentric rings in M 2-9 (Castro-Carrizo *et al.* 2012) and a torus in M 1-16 (Huggins *et al.* 2000). These three sources have fast collimated outflows (up to $\sim$300 kms^{-1}). Our CRIRES spectra does not show any evidence of disks in M 2-9 and Mz 3, only the origin of the bipolar lobes of M 2-9 is likely detected. The SEDs of M 2-9 and Mz 3 are alike. However, for M 2-9 two BB of 96,000 K and 3,200 K fit the optical data, whereas only one BB of 30,000 K fits the SED of Mz 3. A rotating disk in the mid-IR and components around 100 K fit the longer wavelengths. M 1-16 has a prominent optical component fitted by 32,000 K BB. A rotating disk fits well the data, nevertheless, this disk may have a hotter component in the near IR. Once again, colder BBs are needed to fit the data beyond mid-IR wavelengths.

References

Blanco Cárdenas, M., Kaeulf, H. U., Guerrero, M. A. *et al.* 2014, *A&A*, 566, A133
Blanco Cárdenas, Guerrero, M. A., Ramos-Larios, G *et al.* 2013, *A&A*, 551, A64
Bujarrabal, V., Castro-Carrizo, A., Alcolea, J., *et al.* 2013, *A&A*, 557, L11
Castro-Carrizo, A., Neri, R., Bujarrabal, V. *et al.* 2012, *A&A*, 545, A1
Chesneau, O., Lykou, B., Balick, B. *et al.* 2007, *A&A*, 473, L29
Huggins, P. J., Forveille, T., Bachiller, R., & Cox, P. 2000, *ApJ*, 544, 889
Lykou, B., Chesneau, O., Zijlstra, A. A. *et al.* 2011, *A&A*, 527, A105
Waters, L. B. F.. M., Waelkens, C., van Winckel, H. *et al.* 1998, *Nature*, 67, 4679

Planetary Nebulae: Multi-Wavelength Probes of Stellar and Galactic Evolution
Proceedings IAU Symposium No. 323, 2016
X. Liu, L. Stanghellini & A. Karakas, eds.

doi:10.1017/S1743921317001545

Mid-Infrared Studies of the Cool Dust Distribution in Bipolar Planetary Nebulae

Kentaro Asano[1,2], Takashi Miyata[3], Takafumi Kamizuka[3], and Toshiya Ueta[2]

[1]Department of Space Astronomy and Astrophysics Institute of Space and Astronautical Science (ISAS) Japan Aerospace Exploration Agency (JAXA), Sagamihara, Kanagawa 252-5210, Japan

[2]Department of Physics and Astronomy, University of Denver, Denver CO 80210, USA
email: Kentaro.Asano@du.edu

[3]Institute Astronomy, Graduate School of Science, The University of Tokyo, Mitaka, Tokyo 181-0015, Japan

Abstract. To understand mass loss history and mass loss and dust formation history of massive AGB stars, we carried out observations of three bipolar planetary nebulae (BPNe) in 30 micron bands from a ground-based telescope. All of our targets have a compact strong emission source in the mid-infrared (MIR) around the position of the central star of planetary nebula (CSPN). These detected emissions can be originated from cool dust. Our results show that the cool dust component is compactly distributed and much more massive than previous studies indicated. These findings suggest that they experienced a strong mass loss into the equatorial direction in past.

Keywords. Bipolar, mass losing star, planetary nebulae, dust

1. Observation and Results

We observed our target BPNe, NGC6302, Mz3, and Hb3, in the 30 μm band from the ground with MAX38 instrument at the miniTAO. MiniTAO is an infrared telescope with a 1 meter primary mirror at the summit of Co. Chajnantor in Atacama Desert, Chile. It is located in the highest astronomical observatory in the world at an altitude of 5,640 m. Thanks to the high altitude and dry weather, Atacama Desert is an extraordinary site especially for MIR observation. MAX38 is an MIR camera for MiniTAO developed by us (Asano *et al.* 2012). It is not only a rare camera covering the 25 μm in the south hemisphere, but also a unique one having a capability of imaging at the 30 μm wavelength range.

These PNe meet two criteria; (1) The declination is less than 30 degree to be accessible from the southern hemisphere, and (2) the MIR flux is bright enough for MAX38, specifically > 70 Jy at 30 μm (f30; if previously not observed, estimated by linear-interpolation with the IRAS fluxes at 25 and 60 μm. Our targets are the brightest PNe in the MIR. Therefore, they are intrinsically luminous, and hence, relatively massive. We selected bright enough PSF and flux reference stars, while three of them (VX Sgr, W Aql, and V1185 Sco) are known as variable stars. According to Schaeidt *et al.* (1996), the ISO fluxes of the reference stars themselves contains calibration errors from 10 to 30%: we adopted the 30%.

Our images are the first images of these PNe obtained in 30 μm in the world. A compact and bright MIR source is seen around the position of the CSPN in all targets. An image of NGC6302 at 31 μm is shown in Fig. 1. We performed gray-body SED fitting

Table 1. Results of cool dust component concentrated around CSPN flux, mass, luminosity.

Target	f_{18} [Jy]	f_{25} [Jy]	f_{31} [Jy]	f_{37} [Jy]	T_{cool} [K]	L_{cool} [$L_\odot$]	M_{cool} [$M_\odot$]	$\dot{M}$ [$M_\odot yr^{-1}$]
NGC6302	100	–	300	390	70	1200	2.5×10^{-2}	0.9×10^{-5}
Mz 3	45	70	100	–	120	4000	1.0×10^{-3}	1.5×10^{-6}
Hb 5	55	120	160	–	100	700	2.2×10^{-3}	4.1×10^{-6}

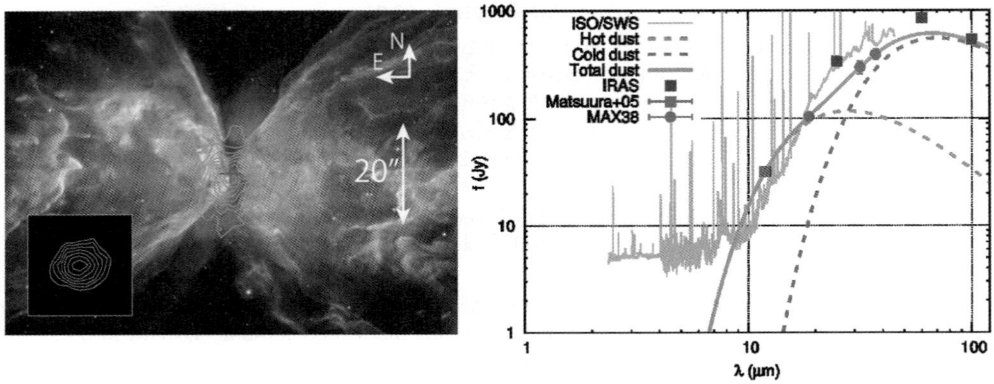

Figure 1. Left: NGC6302 31 μm overlaid image of MAX38 (green contour) on the image of HST visible. PSF image is shown in the small window of left panel. A red cross indicates position of the CSPN. Right: SED of NGC6302. Light blue solid line shows total SED of NGC6302. Green solid line, blue and red dot lines show emission from concentrated dust component.

to estimate the cold dust mass, adopting one or two component dust model with (1) cold dust component at temperature T_{cool} and (2) hot dust component at temperature T_{hot}. These results are shown in Table 1. The best fitting model of NGC 6302 is shown in the right panel of Figure 1. Our dust mass estimates revealed that there are much heavier dust around CSPN than previous studies indicated.

The structure of low temperature dust component around CSPN is naturally considered as a compact torus for the following reasons; (1) marginally or not resolved dust component with MAX38 30 μm images indicate a compact structure (< 5000 AU). (2) Bipolar shape of HST images with highly-excited ionized gas suggests a dense equatorial distribution of cold dust. (3) Torus structure of dust explains large covering factor of cold component over CSPN. We also estimated opening angles under the assumptions of optically thick dust torus and no UV absorption in polar directions.

According to Lykou *et al.* (2007), a hot and light dust disk exists around the CSPN. The presence of the cool and geometrically thick dust torus indicates that mass loss took place preferentially in the equatorial direction and that the dust distribution have changed from torus to disk. Assuming well-mixed dust and gas in the torus, the rate of mass loss can be derived from the CO expansion velocity, torus structure, and cold dust mass. Our MIR study in BPNe clearly indicates the existence of massive but compact cold dust torus around the CSPN suggestive of an enhanced mass loss in the past.

References

Asano, K., Miyata, T., Sako, S., *et al.* 2012, *ProcSPIE*, 8446, 844634
Corradi, R. L. M. & Schwarz, H. E. 1995, *AAP*, 293, 871
Lykou, F., Chesneau, O., Lagadec, E., & Zijlstra, A. 2007, *APN IV*, 97
Schaeidt, S. G., Morris, P. W., Salama, A., *et al.* 1996, *AAP*, 315, L55

Planetary Nebulae: Multi-Wavelength Probes of Stellar and Galactic Evolution
Proceedings IAU Symposium No. 323, 2016
X. Liu, L. Stanghellini & A. Karakas, eds.

doi:10.1017/S1743921317001715

Mass transfer in asymptotic-giant-branch binary systems

Zhuo Chen[1], Adam Frank[1], Eric G. Blackman[1], Jason Nordhaus[2] and Jonathan Carroll-Nellenback[1]

[1]Department of Physics and Astronomy, University of Rochester, Rochester, NY 14627
email: zchen25@ur.rochester.edu

[2]National Technical Institute for the Deaf, Rochester Institute of Technology, Rochester, NY 14623

Abstract. Binary stars can interact via mass transfer when one member (the primary) ascends onto a giant branch. The amount of gas ejected by the binary and the amount of gas accreted by the secondary over the lifetime of the primary influence the subsequent binary phenomenology. Some of the gas ejected by the binary will remain gravitationally bound and its distribution will be closely related to the formation of planetary nebulae. We investigate the nature of mass transfer in binary systems containing an AGB star by adding radiative transfer to the *AstroBEAR AMR Hydro/MHD* code.

Keywords. binaries: close, method: numerical, planetary nebulae: general, stars: AGB and post-AGB.

Introduction: Binaries are believed to be the progenitors of Type Ia supernovae, planetary nebulae and many other systems (Nordhaus & Blackman 2006; Ivanova 2013). In addition, many post-AGB stars show evidence for disks which must due to mass transfer with a companion. Our primary focus in this work is to study PNe projenitor binary systems with different orbital radii to explore changes in mass transfer.

Model: We first test our AGB-wind mass-loss model using an isolated AGB star. We model the AGB star with a pulsating ($P_{pulse} = 365d$), luminous boundary and point gravity (Bowen 1988). We assume dust forms when the equilibrium temperature is below the condensation temperature. The dust will change the opacity of the gas. We show in Fig. 1 that our AGB wind model produces a physically reasonable terminal velocity of 15 km/s and average mass loss of $2.3 \times 10^{-7}\ M_{\odot}/yr$.

We then place an accreting secondary (Krumholz *et al.* 2004) in circular orbit with the AGB star. The boundary condition of the AGB star is kept the same as in the isolated case discussed above. However, the mass of the secondary and the separation of the binary are allowed to vary. We calculate the affected mass-loss rate of the AGB star, the accretion rate of the secondary (Table 1), and examine the morphology of the outflows (Fig. 2).

Results: 1. The AGB mass-loss rate is changed by the presence of the accreting secondary.

2. The accretion rate onto the secondary can increase by up to ∼38% of the mass-loss rate. This exemplifies Wind-Roche-lobe-overflow (Podsiadlowski & Mohamed 2007).

3. Circumbinary disks and accretion disks (around the secondary) are conspicuously found in our calculations.

4. The resulting aspherical circumbinary density distribution provides insight for understanding aspects of aspherical planetary nebulae (Maercker *et al.* 2012).

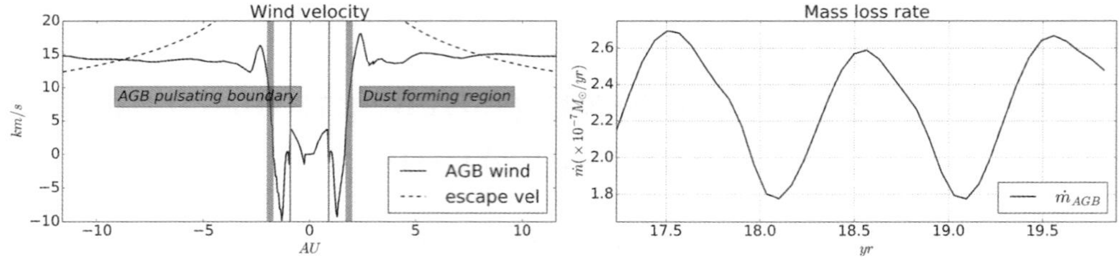

Figure 1. AGB wind's velocity profile and mass-loss rate.

model#	$m_{AGB}(M_\odot)$	$m_{sec}(M_\odot)$	$d(AU)$	$\dot{m}_{AGB}$ $(M_\odot/yr)$	$\dot{m}_{sec}$ $(M_\odot/yr)$
1	1.0	0.1	3	3.4×10^{-7}	-1.3×10^{-7}
2	1.0	0.5	4	3.1×10^{-7}	-1.0×10^{-7}
3	1.0	0.5	6	2.8×10^{-7}	-4.8×10^{-8}
4	1.0	0.5	10	2.6×10^{-7}	-5.8×10^{-9}

Table 1. The affected mass loss rate of the AGB star and the secondary.

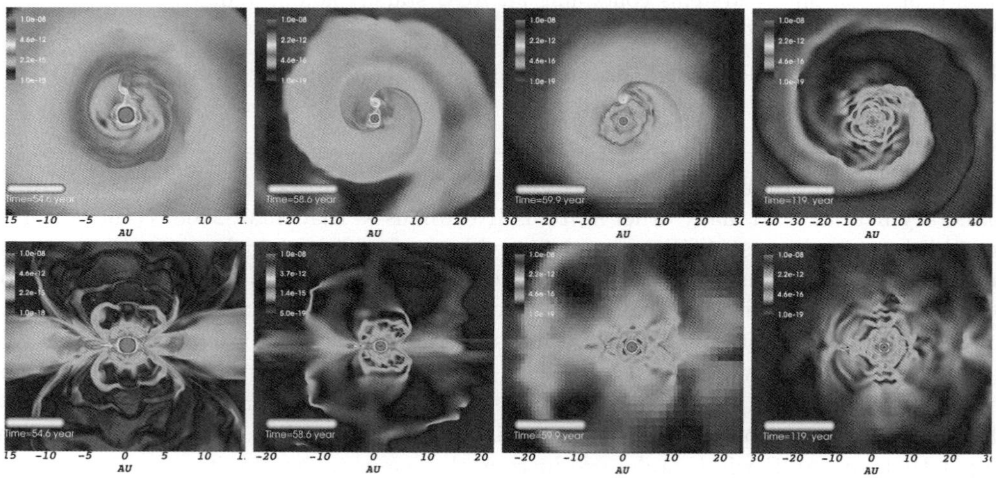

Figure 2. From left to right, the figures correspond to each of the four models listed in table 1. Figures on top show the color density plot in g/cm^3 of the middle z-plane cut. Figures on bottom show the middle x-plane cut.

References

Bowen, G. H. 1988, *ApJ*, 329, 299

Ivanova N, J.ustham S., Chen X., De Marco O., Fryer C. L., Gaburov E., Ge H., Glebbeek E., Han Z., Li X. D., & Lu G. 2013, *A&AR*, 21(1), 1-73

Krumholz, Mark R., Christopher F. McKee, & Richard I. Klein. 2004, *ApJ*, 611, 399

Maercker M., Mohamed S., Vlemmings W. H., Ramstedt S., Groenewegen M. A., Humphreys E., Kerschbaum F., Lindqvist M., Olofsson H., Paladini C., & Wittkowski M. 2012, *Nature*, 490, 232

Nordhaus, J. & Blackman, E. G. 2006, *MNRAS*, 370, 2004

Podsiadlowski, Ph., & Mohamed, S. 2007, *Baltic Astronomy*, 16, 26

Planetary Nebulae: Multi-Wavelength Probes of Stellar and Galactic Evolution
Proceedings IAU Symposium No. 323, 2016
X. Liu, L. Stanghellini & A. Karakas, eds.

doi:10.1017/S1743921317000515

AGB and post-AGB objects in the outer Galaxy

Ryszard Szczerba[1], Bosco H. K. Yung[1], Marta Sewiło[2,3], Natasza Siódmiak[1] and Agata Karska[4]

[1]Nicolaus Copernicus Astronomical Centre, PAS, ul. Rabiańska 8, 87-100 Toruń, Poland
[2]NASA Goddard Space Flight Center, 8800 Greenbelt Rd., Greenbelt, MD 20771, USA
[3]Astronomical Observatory of the Jagiellonian University, ul. Orla 171, 30-244 Kraków, Poland
[4]Centre for Astronomy, Nicolaus Copernicus University, Faculty of Physics, Astronomy and Informatics, ul. Grudziądzka 5, 87-100 Torun, Poland

Abstract. We present the results of our search for low- and intermediate mass evolved stars in the outer Galaxy using AllWISE catalogue photometry. We show that the [3.4]−[12] vs. [4.6]−[22] colour-colour diagram is most suitable for separating C-rich/O-rich AGB and post-AGB star candidates. We are able to select 2,510 AGB and 24,821 post-AGB star candidates. However, the latter are severely mixed with the known young stellar objects in this diagram.

Keywords. Astronomical data bases: miscellaneous, Stars: AGB and post-AGB, Infrared: stars

1. Introduction

We are conducting a systematic study of star formation in the outer Galaxy to uncover the population of intermediate- and low-mass young stellar objects (YSOs) and investigate the impact of the environment on the star formation process. We use the data from the "*Spitzer* Mapping of the Outer Galaxy" survey (SMOG; PI: Sean Carey) that covered $\sim$24 deg^2 region in the outer Galaxy: $l = (102°, 109°)$, $b = (-0.2°, 3.2°)$ in the IRAC 3.6–8.0 μm and MIPS 24 μm bands. This relatively unstudied region, referred by us as "L105", have different environments and star formation activities.

However, in the outer Galaxy including L105 we expect contamination from evolved stars such as asymptotic giant branch (AGB) stars, post-AGB objects and planetary nebulae (PNe). Szczerba *et al.* (2016) concentrated on identifying the location of the low- and intermediate-mass evolved stars (AGBs, post-AGBs, and PNe) on the colour-colour diagram (CCD) based on the 2MASS and *Spitzer* photometry ($K_s - [8.0]$ vs. $K_s - [24]$). This diagram allows us to separate C-rich and O-rich AGB stars quite effectively (Matsuura *et al.* 2014). Nonetheless, the total number of the SMOG sources with good photometric data in all these bands is quite limited (15,311 as compared to almost 3 millions sources detected in L105 by *Spitzer* at the shorter wavelengths). Therefore, we have found counterparts of the *Spitzer* sources in the Wide-field Infrared Survey Explorer (WISE) satellite survey at 3.4, 4.6, 12 and 22 μm and considered all CCDs based on different combinations of these bands.

2. Results

Amongst a total of 15 CCDs, the best option for separating C-rich AGB and post-AGB stars from the O-rich ones is the [3.4] − [12] vs. [4.6] − [22] CCD. In Figure 1, we show on this CCD the hydrodynamical (HD) models for gaseous dusty circumstellar shells around C-rich and O-rich stars during their final stages of AGB (red and blue dots,

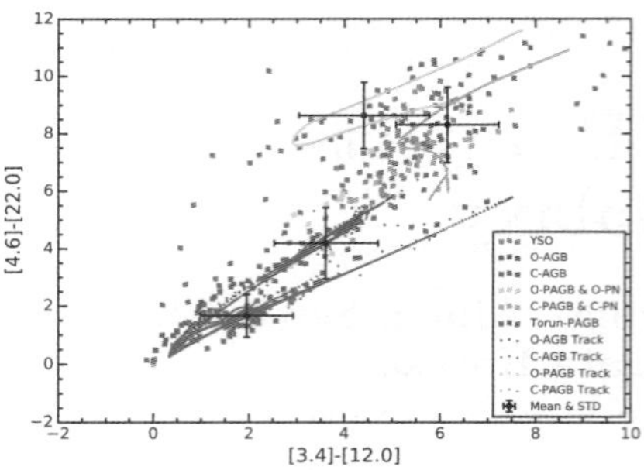

Figure 1. Comparison between HD models of AGB and post-AGB evolution with spectroscopically confirmed sources from the Magellanic Clouds and Galactic post-AGB objects. See the legend and text for details.

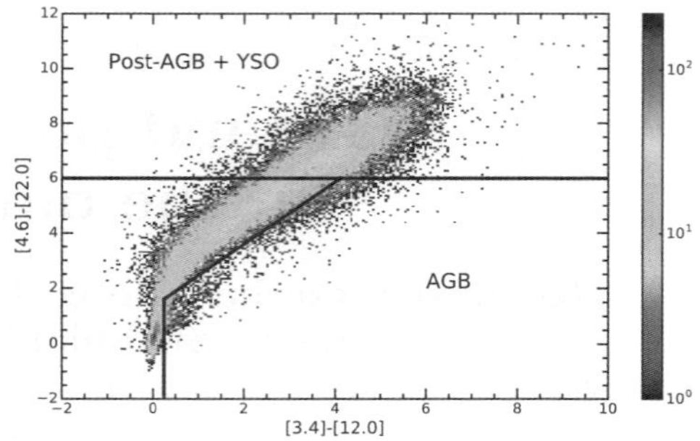

Figure 2. The Hess diagram for the SMOG sources. There are 200 × 200 bins, and the colour indicates the number of sources in each bin. The black lines represent the AGB and post-AGB star selection criteria (see text).

respectively) and post-AGB (orange and light blue dots, respectively) evolution (Steffen *et al.* 1998). The mean and standard deviation (STD) of the model distributions are also shown by the corresponding crosses. In addition, we have over-plotted known sources from the Magellanic Clouds classified by Woods *et al.* (2011) and Ruffle *et al.* (2015), as well as post-AGB stars from our Galaxy (Szczerba *et al.* 2007; 2012). From the position of the HD evolutionary tracks and the over-plotted data, we are able to distinguish typical regions for AGB and post-AGB star candidates.

Figure 2 presents the Hess diagram for 60,655 SMOG sources with AllWISE photometry with the A or B quality flag. The black lines are drawn according to the object distributions in Figure 1. Below $[4.6] - [22.0] = 6$ and to the right of the black lines we have AGB (both O- and C-rich) candidates (2,510 SMOG sources), while above $[4.6] - [22.0] = 6$ we have post-AGB candidates (24,821 SMOG sources), since the theoretical tracks and most of the Galactic post-AGB candidates are located there. However, while the AGB candidates seem relatively easier to be isolated, the post-AGB candidates are mixed with the known YSOs. The additional data analysis and follow-up spectroscopic observations are necessary to confirm the evolutionary status of the post-AGB star (or PN) candidates. The final sample of AGB/post-AGB star candidates will be used to compare the stellar evolution in the Outer and Inner Galaxy.

Acknowledgements

Authors would like to acknowledge financial support from the Polish National Science Center grants 2014/15/B/ST9/02111, 2011/01/B/ST9/02031 (R.Sz. and N.S.), and 2013/11/N/ST9/00400 (A.K.).

References

Matsuura, M., Bernard-Salas, J., & Evans, T. L., *et al.* 2014, *MNRAS*, 439 1472

Ruffle, P. M. E., Kemper, F., & Jones, O. C., *et al.* 2015, *MNRAS*, 451, 3504

Szczerba, R., Siódmiak, N., Stasińska, *et al.* 2007 *Astronomy & Astrophysics*, 378, 465

Szczerba, R., Siódmiak, N., Stasińska, G., *et al.* 2012 *Proceedings of the International Astronomical Union, IAU Symposium*, Volume 283, 506

Szczerba, R., Siódmiak, N., & Leśniewska, A., *et al.* 2016, *Journal of Physics: Conference Series*, Volume 728, article id. 042004

Steffen, M., Szczerba, R., & Schoenberner, D., 1998, *Astronomy & Astrophysics*, 337, 149

Woods, P. M., Oliveira, J. M., & Kemper, F., *et al.* 2016, *MNRAS*, 411, 1597

Planetary Nebulae: Multi-Wavelength Probes of Stellar and Galactic Evolution
Proceedings IAU Symposium No. 323, 2016
Xiaowei Liu, Letizia Stanghellini, & Amanda Karakas, eds.

doi:10.1017/S1743921317000606

A homogeneous distance catalogue for Galactic RV Tauri objects

Shane B. Vickers[1], **David J. Frew**[2,3], **Matt S. Owers**[1,4], **Quentin A. Parker**[1,2] **and Ivan S. Bojičić**[1,2]

[1]Department of Physics & Astronomy, Macquarie University, North Ryde 2109, email:shane.vickers@mq.edu.au

[2]The University of Hong Kong, Department of Physics, Hong Kong SAR, China

[3]The University of Hong Kong, Laboratory for Space Research, Hong Kong SAR, China

[4]The Australian Astronomical Observatory, North Ryde, NSW, 1670, Australia

Abstract. A subset of Post-AGB (PAGB) objects are the highly luminous RV Tauri variables that show similarities to Type-II Cepheids. By using a sample of known RV Tauri stars from the Magellanic Clouds we are able to determine period luminosity relationships (PLRs) in various bands that have been used to determine the luminosities of their Galactic counterparts. We have gathered all available photometry in order to generate an SED for each object and determine the total integrated flux. This total flux combined with a calculated or inferred intrinsic luminosity leads to a distance (Vickers *et al.* 2015). This distance catalogue has allowed us to begin to constrain the physical parameters of this poorly understood evolutionary phase and to determine links between these physical characteristics as a function of their stellar population.

Keywords. stars: AGB and Post-AGB—stars: variables: Cepheids—infrared stars

1. Introduction

The RV Tau stars are a subclass of pulsating PAGB stars named after the class prototype RV Tauri. They are luminous, variable yellow giant stars of spectral types F-K, with typical periods of 20–80 days and with alternating deep and shallow minima. The major hurdle to determining the physical characteristics of these stars is the lack of accurate distances. This is a problem that we will rectify with the release of this, the second distance catalogue that will encompass the RV Tauri stars, R Coronae Borealis stars, extreme helium stars and late thermal pulse objects. Here we will solely focus on the determination of luminosities and distances. One of our goals is to construct a volume-limited survey of PAGB stars based on our distance catalogues (Vickers *et al.* 2015, 2016), to compare with a similar survey for planetary nebulae (Frew *et al.* 2016).

2. Method

Using multi-epoch WISE photometry from the IRSA (http://irsa.ipac.caltech.edu) online database we have determined mean WISE photometry for all Galactic and extra-Galactic RV Tauri variables . The [3.4]–[4.6] WISE color is a good proxy for the presence of a dusty disk as shown in Gezer *et al.* (2015). We have built an SED for each RV Tauri in the Toruń catalogue (Szczerba *et al.* 2007, 2012). For this distance catalogue we have decided to use model atmospheres in a similar manner to Kamath, Wood & Van Winckel (2015) with the exception that we have determined interstellar extinction values as described in our first PAGB distance catalogue (Vickers *et al.* 2015).

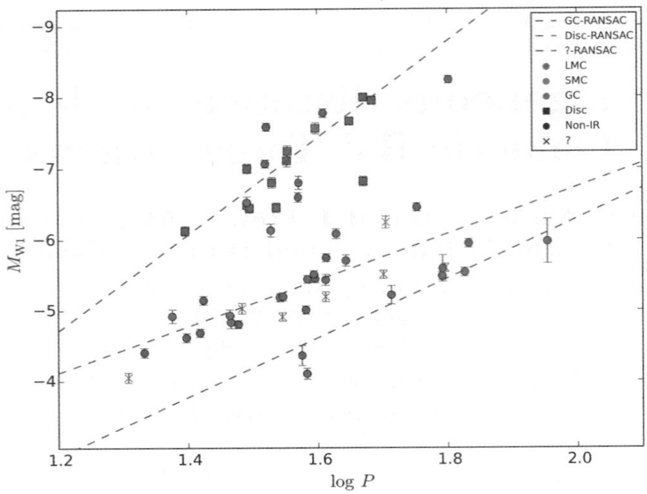

Figure 1. Period-luminosity relationship for W1 3.4 μm waveband.

3. Results

We have constructed PLRs for each of the four WISE mid-IR wavebands using all LMC and SMC objects classified as RV Tauri stars in the OGLE-III catalogue (Soszyski *et al.* 2008). Figure 1 clearly shows that the slope of the PLR for stars with a disc like SED is much steeper than those with no IR-excess. This suggests that the disc RV Tauri stars have higher intrinsic luminosities and therefore come from higher mass progenitors. The slope of the V-band PLR shows an inflection which can be attributed to the RV Tauri stars with a large IR-excess due to a circumstellar disc (Manick *et al.* 2016).

4. Conclusion and Future Work

We have constructed PLRs using the LMC and SMC RV Tauri population in the mid-IR which accentuates the bending of the PLR as mentioned in Manick *et al.* (2016). Using these PLRs we have developed a catalogue of distances for the entire Galactic RV Tauri population. The presence of a disc increases scatter in the mid-IR PLR which must be accounted for when determining distances to Galactic RV Tauri stars.

References

Frew, D. J., Parker, Q. A., & Bojičič, I. S. 2016 *MNRAS*, 455, 1459

Gezer, I., Van Winckel, H., Bozkurt, Z., De Smedt, K., Kamath, D., Hillen, M., & Manick, R. 2015, *MNRAS*, 453, 133

Kamath, D., Wood, P. R., & Van Winckel, H. 2015, *MNRAS*, 454, 1468

Manick, R., Van Winckel, H., Kamath, D., Hillen, M., & Escorza, A. *A&A* in press (arXiv:1610.00506)

Sosyński, I., Udalski, A., Szymański, M. K., Kubiak, M., Pietrzyński, G., Wyrzykowski, Ł., Szewczyk, O., Ulaczyk, K., & Poleski, R. *Acta Astr*, 58, 293

Szczerba, R., Siódmiak, N., Stasińska, G., & Borkowski, J. 2007, *A&A*, 469, 799

Szczerba, R., Siódmiak, N., Stasińska, G., Borkowski, J., García-Lario, P., Suárez, O., Hajduk, M., & García-Hernández, D. A. 2012, in Proc. IAU Symp., 283, 506

Vickers, S. B., Frew, D. J., Parker, Q. A., & Bojičič, I. S. 2015, *MNRAS*, 447, 1673

Vickers, S. B., Frew, D. J., Owers, M. S., Parker, Q. A., & Bojičič, I. S. 2016, *JPhCS*, 728, 072013

Planetary Nebulae: Multi-Wavelength Probes of Stellar and Galactic Evolution
Proceedings IAU Symposium No. 323, 2016
X. Liu, L. Stanghellini & A. Karakas, eds.

doi:10.1017/S1743921317002162

Study of protoplanetary nebulae

D. Estrella, L. Hernández-Martínez, P. F. Velázquez, A. Esquivel, and C. A. Raga

Instituto de Ciencias Nucleares, Universidad Nacional Autónoma de México
Circuito Exterior S/N – Ciudad Universitaria,
Ciudad de México, Distrito Federal – México
email: diana.trujillo@correo.nucleares.unam.mx,
liliana.hernandez@correo.nucleares.unam.mx,
pablo@nucleares.unam.mx,
esquivel@nucleares.unam.mx,
raga@nucleares.unam.mx

Abstract. We have performed 3D hydrodynamic simulations of a symmetrical jet ejection following previous works (Raga *et al.* 2009, Riera *et al.* 2014, Velázquez *et al.* 2014). The jet is emitted from a binary system in elliptical orbit, and its direction changes describing a precession cone. We have considered that the jet has a time-dependence density ejection or a time-dependence velocity ejection, in order to propose an alternative model to explain the morphology of PPNe's. Also in our description we have included the effect of the photoionization of the central source. From numerical results, synthetic Hα maps were obtained, and a proper motion study were carried out. We found that the photoionization has an important effect on the case with variation density resulting in a increse in the Hα emission.

Keywords. ISM, jets and outflows, numerical methods, planetary nebulae, radiation mechanisms

1. Simulations

1.1. *Code and initial setup*

We have performed simulations with a 3D mesh parallel HD/MHD code (Esquivel *et al.* 2009; Esquivel & Raga 2013). This code solves the gas dynamical equations and incorporates various modules, including one for radiation transfer based on a Monte Carlo ray tracing method.

We used the values of 10^{-6}M$_\odot$ yr^{-1} for the AGB mass loss rate and a wind velocity of 15km s^{-1}. For the jet, we have a velocity of 400km s^{-1} and a density of 2.15×10^{-19}g cm^{-3}. Also we imposed a precession cone angle of 15°, a density variation of 0.95%, and a velocity variation 0.5% in each model. The precession period is 62.13 yr. For the binary system, we consider a white dwarf with a mass of 0.5 $M_\odot$ and a companion star of 2.14 $M_\odot$. The total integration time of the simulations is 130 yr.

1.2. *Wavelets technique*

This method (Raga *et al.* 2016) degrades the resolution of the original images by means of a convolution with a smoothing function given by a "Mexican Hat" wavelet of half-width σ

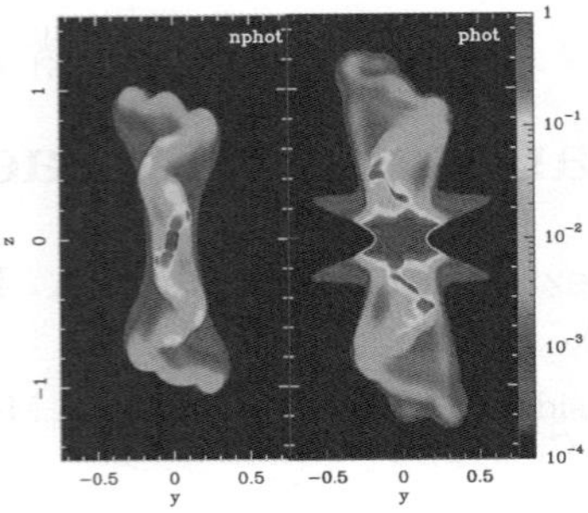

(a)

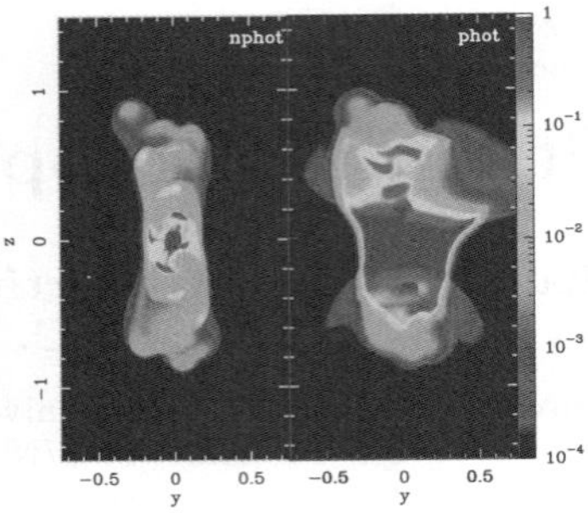

(b)

Figure 1. Synthetic Hα maps obtained for hidrodynamical simulations which consider a) a jet with a time density variation and b) a jet with time-dependent ejection velocity. Panels display the results for the simulations that include (do not include) photoionization effects. The logarithmic color scale gives de Hα emission in units of erg cm^{-2} s^{-1} sr^{-1}. Both axes are given in units of 10^{17} cm.

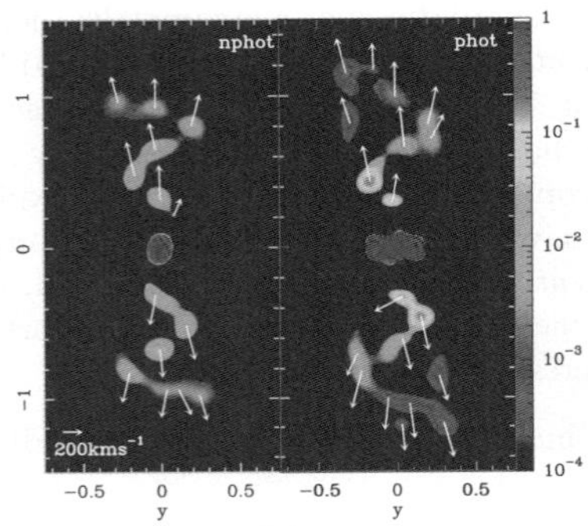

(a)

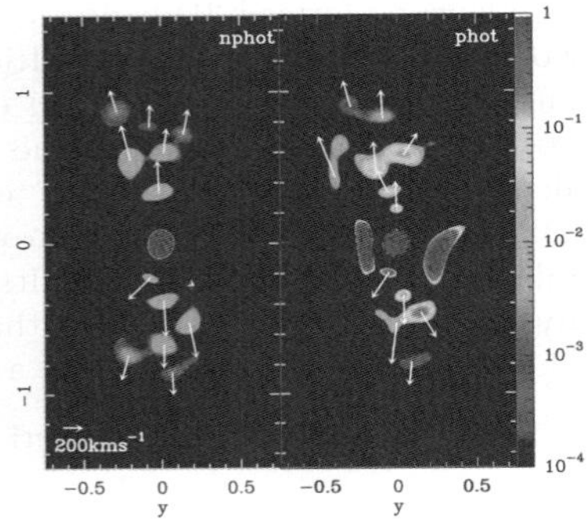

(b)

Figure 2. Proper motions for the case with a) density variation and with b) velocity variation. For both cases, we have used the wavelets technique, which allows to highlight the most intense structures obtained in the synthetic maps.

$$g_\sigma(x, y) = \frac{1}{\pi\sigma^2}\left(1 - \frac{x^2 + y^2}{\sigma^2}\right)e^{-\left(\frac{x^2+y^2}{\sigma^2}\right)}, \tag{1.1}$$

with x and y the coordinates in pixels on the plane of the images. Then the shifts in the identifiable intensity peaks are measured. For all the cases, we choose σ=10 because it shows more defined structures.

2. Results

2.1. *Density and velocity variation with and no photoionization*

We have obtained the synthetic Hα emission maps (Fig.1), where the y axis is the line of sight and the z and x axes correspond to the plane of the sky.

2.2. *Proper motions*

The proper motions are obtained using the Hα maps of two consecutive images with ten years of difference between them, for each case. In these images we apply the wavelets technique described above. Our results are shown below

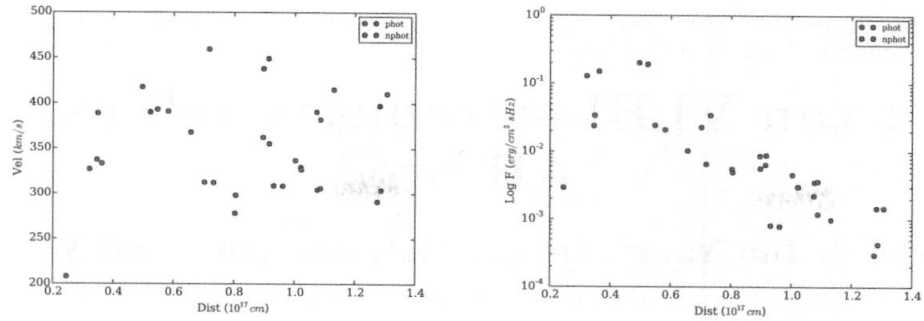

(a) (b)

Figure 3. a) Distance vs velocity plot and b) Distance vs flux plot obtained for the case of a jet with density variation, including or not the photoionization effects.

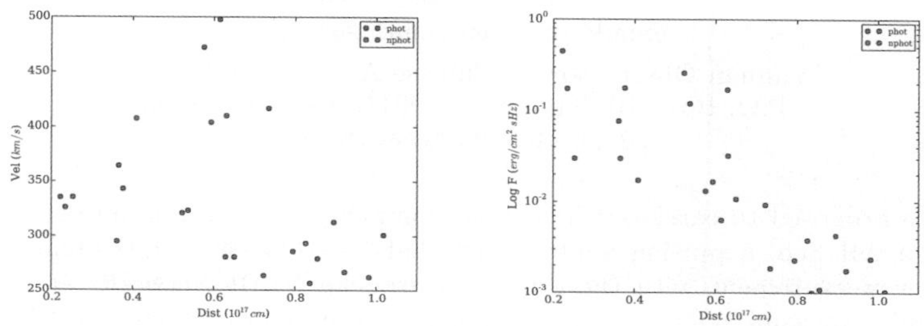

(a) (b)

Figure 4. Idem Figure 3, but for the case of a jet with a time-dependent ejection velocity.

3. Conclusions

We have made an attempt to reproduce the morphology of PPNe proposing a variation of density and velocity as a possible explanation to the observed structures in these objects. Furthermore, we have included the photoionization of the central source in the simulations. On the obtained synthetic maps, a proper motion study was performed. This study reveals that the photoionization has an important effect on both the emission and proper motion velocities.

References

Esquivel A., Raga A. C., Cantó J., Rodríguez-González A., 2009, *A&A*, 507, 855

Esquivel A., Raga A. C., 2013, *ApJ*, 779, 111

Raga, A. C., Reipurth, B., Esquivel, A., & Bally, J., 2016, *AJ*, 151, 113

Raga, A. C., Esquivel, A., Velázquez, P. F., Cantó, J., Haro-Corzo, S., Riera, A., & Rodríguez-González, A., 2009, *ApJ*, 707,1.

Riera, A., Velázquez, P. F., Raga, A. C., Estalella, R., & Castrillón, A. 2014, *A&A*, 561, A145

Riera, A., Velázquez, P. F., Steffen, W., Raga, A. C., & Cantó, J., 2012, *IAU Symposium*, Volume 283.

Sahai, R. & Trauger, J. T., 1998, *ApJ*, 116, 3.

Soker, N. & Livio, M., 1994, *ApJ*, 421, 1.

Terquem, C., Eislöffel, J., Papaloizou, J. C. B., & Nelson, R. P. 1999, *ApJL*, 512, L131

Velázquez, P. F., Raga, A, C., Cantó, J., Schneiter, E. M., & Riera, A., 2013, *MNRAS*, 428, 2.

Velázquez, P. F., Riera, A., Raga, A. C., & Toledo-Roy, J. C. 2014, *ApJ*, 794,128

Planetary Nebulae: Multi-Wavelength Probes of Stellar and Galactic Evolution
Proceedings IAU Symposium No. 323, 2016
X. Liu, L. Stanghellini & A. Karakas, eds.

doi:10.1017/S1743921317000618

What can VLBI astrometry tell us about AR Sco?

Lang Cui[1,2], **Jun Yang**[3], **Hongli Ma**[1], **Jun Liu**[1,2] **and Wen Chen**[4]

[1]Xinjiang Astronomical Observatory, Chinese Academy of Sciences,
[2]Key Laboratory of Radio Astronomy, Chinese Academy of Sciences,
150 Science 1-Street, Urumqi, 830011, China
email: cuilang@xao.ac.cn

[3]Onsala Space Observatory, Chalmers University of Technology,
SE-439 92 Onsala, Sweden
email: jun.yang@chalmers.se

[4]Yunnan Observatories, Chinese Academy of Sciences
P.O. Box 110, Kunming 650011, Yunnan, China
email: chenwen@ynao.ac.cn

Abstract. We proposed to carry out high precision very long baseline interferometry (VLBI) observations of AR Sco, a pulsing white dwarf (WD) - M dwarf (MD) binary, to provide a direct distance measurement with the e-EVN (European VLBI Network) at 5 GHz. By the proposed parallax measurement on AR Sco, not only the precise distance will be determined, but also some physical parameters, such as the luminosity, the mass and the magnetic field will be significantly tightened accordingly, even the gravitational wave amplitude from this unique binary system can be tightly constrained. In addition, the EVN observations will allow us to answer that whether there is an extended emission structure associated with AR Sco, which will help us to explain the stable continuum radio emission observed during the pulse-off state.

Keywords. VLBI, AR Sco, Binary

1. AR Sco – A radio-pulsing white dwarf binary

Recently, pulsed emission on a period of 1.97 minutes in radio, IR, optical and UV bands from a white dwarf (WD) - M dwarf (MD) binary, AR Sco, has been reported by Marsh *et al.* (2016). The brightness of the binary system also varies on a longer period of 3.56 hours due to its binary orbital motion. Marsh *et al.* (2016) argued that the electromagnetic radiation is powered by the spin-down power rather than accretion. AR Sco is the first radio 'white dwarf pulsar' detected so far. The spectral energy distribution of the pulsed emission supports a synchrotron origin. Beside the pulsed emission, AR Sco has much stronger off-pulse continuum emission, $\sim$ 10 mJy at 9 GHz (Marsh *et al.* 2016).

2. Why we need parallax measurement for AR Sco?

A model-dependent distance of 116 ± 16 pc has been reported for AR Sco and the mass of the WD is in the range of $0.81 - 1.29$ $M_{\odot}$ (Marsh *et al.* 2016). By using the observations of luminosity, the rotation frequency and the frequency derivative of the WD, Franzon & Schramm (2016) found that the WD might have a mass close to 1.29 $M_{\odot}$ and estimated its radius of $\sim$ 4080 km and the central and surface magnetic fields of $\sim 0.9 \times 10^{13}$ G and $\sim 8.95 \times 10^{11}$ G, respectively. The parameters mentioned above are important for understanding the physical processes on-going in AR Sco, unfortunately, due to the relatively large uncertainty of distance (13%), they can not be tightly constrained.

In order to better understand the unusual features and the possible scenarios presented in literatures, a high precision parallax measurement for AR Sco with VLBI phase-referencing observations is vital to improve its distance determination. The new parallax distance measurement will be expected to have an accuracy of $\sim$ 1 pc, ten times better than the present estimation (116 $\pm$ 16 pc). Once the precise distance is determined by the parallax measurement, the typical parameters, such as the physical magnitudes, the mass and the strength of magnetic field will be updated accordingly and the Period-Luminosity relationship of this periodic variable star can be revisited.

Another aspect that will benefit from the distance determination is the amplitude estimation of the gravitational wave coming from AR Sco, the pulsing WD-MD binary system. Using the model-dependent distance of $\sim$ 116 pc, the gravitational wave amplitude for AR Sco was estimated to be $\sim 10^{-(22-23)}$(Franzon & Schramm 2016), which might lead to a detectable signal by LIGO and VIRGO. By applying more precise distance, the gravitational wave amplitude from AR Sco can be tightly constrained, which would help to optimize detecting strategy for the gravitational wave detectors.

In addition, the high-resolution VLBI observations will also allow us to answer that whether there is an extended structure in this unique binary system. This may be a direct sign of the existence of the large-scale shock/interaction formed by the system. Finding them will help us to explain these stable continuum emission observed during the pulse-off status.

3. The pilot e-EVN observations

In order to significantly improve the distance determination for AR Sco, We proposed to run the astrometry observations on AR Sco with the e-EVN at 5 GHz for 4 hours per epoch in the GST time range of 14h – 18h and at four epochs with a separation of a few months. Before carrying out these observations, a short e-EVN proposal to search for nearby phase-referencing sources had been performed at 5 GHz on 2016 Sep 20. The preliminary result shows that PMN J1620-2259, 18 arcmin away from AR Sco, was detected by the e-EVN with a peak flux density of $\sim$ 10 mJy at 5 GHz, indicating that this source can be a good phase-referencing calibrator, given the recording rate of 2 Gbps for the e-EVN observations currently.

For the multi-epoch phase-referencing observations, we expect to achieve a practical image sensitivity better than 0.1 mJy per beam. With the e-EVN beam size, about 5 mas with naturally weighting, we may achieve a position precision probably better than 0.1 mas (1 sigma), which is good enough to precisely measure the large parallax ($\sim$ 9 mas) of AR Sco.

Acknowledgments

This work is supported by Xinjiang Key Laboratory of Radio Astrophysics under grant No. 2016D03020, the National Natural Science Foundation of China (NSFC) under grant No. 11503072, the Program of the Light in China's Western Region under grant No. YBXM-2014-02 and 2015-XBQN-B-01, and the NSFC under grant No. 11303079.

References

Marsh, T. R., Gänsicke, B. T., & Hümmerich, S., *et al.* 2016, *Natur.*, 537, 374
Geng, J. J., Zhang, B., & Huang, Y. F. 2016, *ApJL accepted*, arXiv:1609:02508
Franzon, B. & Schramm, S. 2016, arXiv:1609.00493

Planetary Nebulae: Multi-Wavelength Probes of Stellar and Galactic Evolution
Proceedings IAU Symposium No. 323, 2016
X. Liu, L. Stanghellini & A. Karakas, eds.

doi:10.1017/S1743921317000795

Hunting for binary central stars of planetary nebulae: exploiting archival data

Alba Aller[1], Maja Vučković[1], Jorge Lillo-Box[2], Luis F. Miranda[3], David Jones[4] and Henri M. J. Boffin[2]

[1]Instituto de Fsica y Astronomía, Facultad de Ciencias, Universidad de Valparaíso, Gran Bretaña 1111, Playa Ancha, 2360102, Valparaíso, Chile
email: alba.aller@ifa.uv.cl, maja.vuckovic@uv.cl

[2]European Southern Observatory (ESO), Alonso de Cordova 3107, Santiago, Chile; email: jlillobo@eso.org, hboffin@eso.org

[3]Instituto de Astrofísica de Andalucía - CSIC, C/ Glorieta de la Astronomía s/n E-18008 Granada, Spain; email: lfm@iaa.es

[4]Instituto de Astrofísica de Canarias, E-38205 La Laguna, Tenerife, Spain; email: djones@iac.es

Abstract. The detection of new binary central stars of planetary nebulae is crucial to definitively determine the importance of binary interactions in the nebular morphology. In this context, we are working on a project that aims to increase the low number of binary central stars detected so far. For that, we are first analyzing public archival data in order to discover potential candidates of binary central stars. These candidates will be subsequently followed-up in order to confirm and characterize them. Here we present our ongoing search and some preliminary results.

Keywords. binaries: general, planetary nebulae: general, astronomical data bases: miscellaneous

1. Overview

While binary interactions have been proposed as the main mechanism to explain the complex and amazing morphologies seen in planetary nebulae (PNe), only less than 2% of the central stars (CSs) have been found to be close binaries. This is a very small fraction among the ~3500 known PNe in the Milky Way (Parker, Bojičić & Frew 2016), specially taking into account that only about 20% of PNe are spherical! Since the sample is too big to obtain new data of all of them, archival data offer a great opportunity to search for new binary CSs candidates. For example, photometric public surveys provide light curves of a large number of objects in a long timespan which allows us to look for periodic variability in the CSs. Also, the large amount of photometric catalogues, that provide data in a very wide range of wavelengths, allow us to detect infrared excess in the spectral energy distributions (SEDs) of the stars. We are exploiting all of these archival data to search for new candidates that will be subsequently followed-up by means of multi-epoch photometric and spectroscopic observations.

2. The search for new binaries: preliminary results

In order to obtain good results, an efficient selection of the central stars is crucial. For that reason, we have carefully inspected the available photometric data of our sample of around ~3000 CSs. In the following, we briefly explain the two different approaches that have been followed to find new binary CSs candidates:

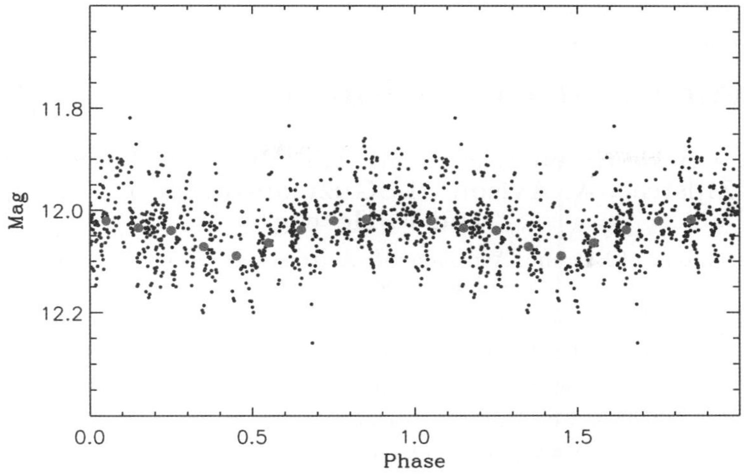

Figure 1. Phase-folded light curve of one of our candidates from the OMC-Integral archive. The red dots show the binned data with 0.1 in phase.

(*a*) Search in photometric surveys: Most of the known binary CSPNe have been detected through modulations in the light curve. With this method, Miszalski *et al.* (2009) found, by using the OGLE survey, a total of 21 binaries, more than doubling the known sample at that time. As a continuation of this working line, we have inspected the photometric surveys ASAS, Catalina Sky Surveys, OMC and SuperWasp to search for new binary CSs. We have analyzed the light curves available in these surveys for our sample looking for periodic variability. We have not only recovered all known binaries present in the surveys data but also we have found a significant number of CSs that show significant variability. Figure 1 shows an example of one of our candidates.

(*b*) Search for infrared excess: CSPNe are very hot stars and the detection of any excess flux at infrared wavelengths would indicate the presence of a cooler companion. This method has been successfully used by De Marco *et al.* (2013) and Douchin *et al.* (2015) for some individual objects. We are extending this infrared excess search to our whole sample of CSPNe in an automatic way. For that purpose, we are using VOSA (Bayo *et al.* 2008), the Virtual Observatory SED Analyzer, which provides the photometry available in the accesible catalogues and constructs the SED for the queried objects.

References

Bayo, A., Rodrigo, C., Barrado y Navascués, D., Solano, E., Gutiérrez, R., Morales-Calderón, M., & Allard, F. 2008, *A&A*, 492, 277

De Marco, O., Passy, J. C., Frew, D. J., Moe, M., & Jacoby, G. H. 2013, *MNRAS*, 428, 2118

Douchin, D., De Marco, O., Frew, D. J., Jacoby, G. H., Jasniewicz, G., Fitzgerald, M., Passy, J. C., Harmer, D., Hillwig, T., & Moe, M. 2015, *MNRAS*, 448, 3132

Miszalski, B., Acker, A., Moffat, A. F. J., Parker, Q. A., & Udalski, A. 2009, *A&A*, 496, 813

Parker, Q. A., Bojičić, I. S., & Frew, D. J. 2016, *Journal of Physics Conference Series*, 728, 3

Planetary Nebulae: Multi-Wavelength Probes of Stellar and Galactic Evolution
Proceedings IAU Symposium No. 323, 2016
X. Liu, L. Stanghellini & A. Karakas, eds.

doi:10.1017/S1743921317000588

The very fast evolution of Sakurai's object

G. C. Van de Steene[1], P. A. M. van Hoof[1], S. Kimeswenger[2], A. A. Zijlstra[3], A. Avison[3], L. Guzman-Ramirez[4], M. Hajduk[5] and F. Herwig[6]

[1]Royal Observatory of Belgium, Astronomy & Astrophysics, Ringlaan 3, Brussels, Belgium
email: g.vandesteene@oma.be

[2]Universidad Católica del Norte, Antofagasta, Chile

[3]Jodrell Bank Centre for Astrophysics, Manchester, UK

[4]Sterrewacht, Universiteit Leiden, The Netherlands

[5]Nicolaus Copernicus Astronomical Center, Torun, Poland

[6]University of Victoria, Victoria, Canada

Abstract. V4334 Sgr (a.k.a. Sakurai's object) is the central star of an old planetary nebula that underwent a very late thermal pulse a few years before its discovery in 1996. We have been monitoring the evolution of the optical emission line spectrum since 2001. The goal is to improve the evolutionary models by constraining them with the temporal evolution of the central star temperature. In addition the high resolution spectral observations obtained by X-shooter and ALMA show the temporal evolution of the different morphological components.

Keywords. planetary nebulae: individual (V4334 Sgr), evolution, Stars: AGB and post-AGB, circumstellar matter

---

1. Introduction

V4334 Sgr (a.k.a. Sakurai's object) is the central star of an old planetary nebula (PN) that underwent a very late thermal pulse (VLTP) a few years before its discovery in 1996 (Nakano *et al.* 1996). During the VLTP it ingested its remaining hydrogen rich envelope into the helium burning shell and ejected the processed material shortly afterwards to form a new, hydrogen deficient nebula inside the old PN. The star brightened considerably to become a very cool (born-again) AGB star with a spectrum resembling a carbon star. After a few years, dust formation started in the new ejecta and the central star became highly obscured, similar to R CrB stars. Emission lines were discovered: first He I 1083 nm in 1998 (Eyres *et al.* 1999), later in 2001 also optical forbidden lines from neutral and singly ionized nitrogen, oxygen and sulfur (Kerber *et al.* 2002). Sakurai baffled the scientific community with its very fast evolution, much faster than pre-discovery models predicted. The current models (Lawlor & MacDonald 2003, Herwig *et al.* 2011, Herwig *et al.* 2014, Miller Bertolami *et al.* 2006) can be improved by constraining them with the temporal evolution of the central star temperature.

2. Observations and results

2.1. *Optical monitoring*

We have been monitoring the evolution of the optical emission line spectrum since 2001 using spectra obtained with FORS at the ESO-VLT (Fig. 1). The goal of this monitoring program is to derive the stellar temperature as a function of time (van Hoof *et al.* 2007,

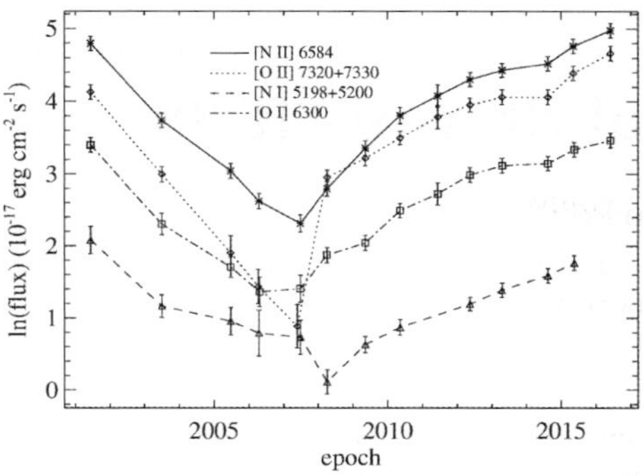

Figure 1. Evolution of emission line fluxes since 2001.

van Hoof *et al.* 2008). From 2001 through 2007 the optical spectrum showed an exponential decline in flux and the level of excitation also dropped. We see this as evidence for a shock that occurred around 1998 and cooled afterwards. A plausible explanation is that this is the fastest material ejected in the VLTP hitting older ejecta. The optical line fluxes started to increase again since 2008. The sudden jump in [O II] flux in 2008 could point to a second shock as the cause of the change in behavior. The optical spectrum shows new lines which have been emerging since 2013. Some have tentatively been identified as electronic transitions of CN. At least some of these newly emerging lines are formed very close to the central star, possibly in the disk. In the PV diagram of [N II] 658.3 nm, obtained with X-shooter, we see that the red- and blue-shifted emission come from different regions that are spatially shifted with respect to the continuum. This emission line, as well as other optical forbidden lines, originate in the bipolar lobes seen by Hinkle & Joyce (2014).

2.2. *ALMA Observations*

The continuum emission is unresolved which indicates that all the dust is in the disk. In the ALMA spectra we detect CO, CN, HC_3N, as well as ^{13}C isotopologues of these species. The CO and HC_3N (+isotopologues) emission is unresolved, coincides with the position of the central star, and most likely comes from the disk (Fig. 2). The CN and ^{13}CN emission is spatially resolved and coincides with the bipolar lobes seen by Hinkle & Joyce (2014) (Fig. 2). CN could be formed via shock-induced dissociation of HCN in the lobes.

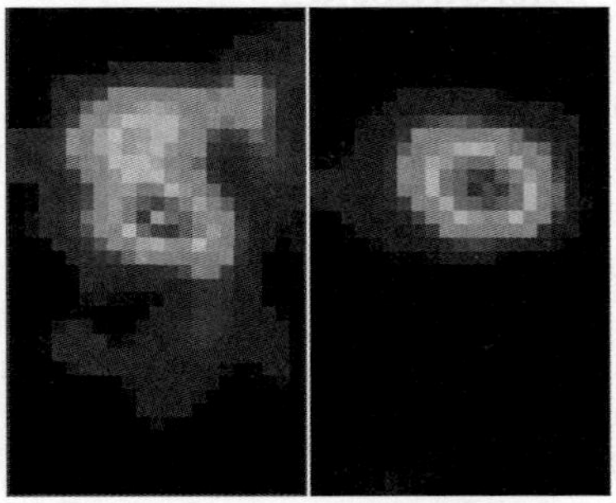

Figure 2. CN (left) and CO emission (right) detected by ALMA in 2015.

References

Chesneau, O., Clayton, G. C., Lykou, F., *et al.*, 2009, *A&A*, 493, L17
Eyres, S. P. S., Smalley, B., Geballe, T. R., *et al.*, 1999, *MNRAS*, 307, L11
Hajduk, M., Zijlstra, A. A., Herwig, F., *et al.*, 2005, *Science*, 308, 231
Herwig, F., Pignatari, M., Woodward, P. R., *et al.*, 2011, *ApJ*, 727, 89
Herwig, F., Woodward, P. R., Lin, P.-H., Knox, M., & Fryer, C., 2014, *ApJ*, 792, L3
Hinkle, K. H. & Joyce, R. R., 2014, *ApJ*, 785, 146
Kerber, F., Pirzkal, N., De Marco, O., *et al.*, 2002, *ApJ*, 581, L39
Lawlor, T. M. & MacDonald, J., 2003, *ApJ*, 583, 913
Miller Bertolami, M. M., Althaus, L. G., Serenelli, A. M., & Panei, J. A., 2006, *A&A*, 449, 313
Nakano, S., Sakurai, Y., Hazen, M., *et al.*, 1996, *IAU Circ.*, 6322, 1
van Hoof, P. A. M., Hajduk, M., Zijlstra, A. A., *et al.*, 2007, *A&A*, 471, L9
van Hoof, P. A. M., Hajduk, M., Zijlstra, A. A., *et al.*, 2008, *ASP Conf. Ser.*, vol. 391, 155

Planetary Nebulae: Multi-Wavelength Probes of Stellar and Galactic Evolution
Proceedings IAU Symposium No. 323, 2016
X. Liu, L. Stanghellini & A. Karakas, eds.

doi:10.1017/S1743921317000485

The Binary Central Star of Hf 38

Helen Barker[1]

[1]Dept. of Astronomy & Astrophysics, The University of Manchester
Oxford Road, Manchester, UK
email: helen.barker@postgrad.manchester.ac.uk

Abstract. Despite years of effort, the impact of central star binarity on planetary nebula formation and shaping remains unclear. This is hampered by the fact that detecting central star binarity is inherently difficult, and requires very precise observations. The fraction of planetary nebulae with binary central stars therefore remains elusive. This work presents initial results of central star analysis using data from the VST Hα Survey of the Southern Galactic Plane and Bulge (VPHAS+). The true central star of PN Hf 38 has been revealed, and it exhibits a 0.465$\pm$0.334 i band magnitude excess, indicative of a M0V companion.

Keywords. planetary nebulae, binarity

1. Introduction

Some planetary nebulae (PNe) are observed to have binary central stars (CS), and there is increasing evidence that the presence of a binary companion can determine the PN morphology (Hillwig *et al.* 2016). However, the total fraction of PN with binary CS remains unclear. This is due to the intrinsic difficulty in detecting binarity, even when high quality data is available.

One method of detecting CS binarity is to search for a percent-level CS i band excess, indicative of a low mass main sequence binary companion (De Marco *et al.* 2013, Douchin *et al.* 2015). VPHAS+ provides sub-arcsecond aperture photometry down to 20th magnitude in the Sloan u, g, r and i bands, and a Hα filter. This is a similar limiting brightness to the SuperCOSMOS Hα Survey (Parker *et al.* 2005) from which the Macquarie/AAO/Strasbourg Hα (MASH) catalogue (Parker *et al.* 2006, Miszalski *et al.* 2008) was created, however VPHAS+ should provide superior resolution. Once complete, there will be 1427 currently known PN within the VPHAS+ footprint. Presented here are the first results from analysis of the currently available VPHAS+ data (release 2).

2. The Central Star of Hf 38

The new, high resolution images of Hf 38 from VPHAS+ have revealed a faint blue star in the centre of the nebula. This is marked in the left panel of Figure 1. The right panel shows the Hα/r quotient image, which highlights an arc of nebulosity 0.36pc north west of the nebula, assuming a distance of 2.2kpc (Maciel 1984).

The marked star was confirmed to be the true CS by comparing the observed and expected colours. The expected CS colours were calculated by convolving the spectrum of a 100kK, log(g)=7.0, solar metallicity synthetic post-AGB atmosphere from TMAP (Werner *et al.* 2003) with the VPHAS+ bandpasses. The synthetic and observed colours were consistent within the errors.

The reddening of this PN was determined by calculating E(B-V) using three methods.

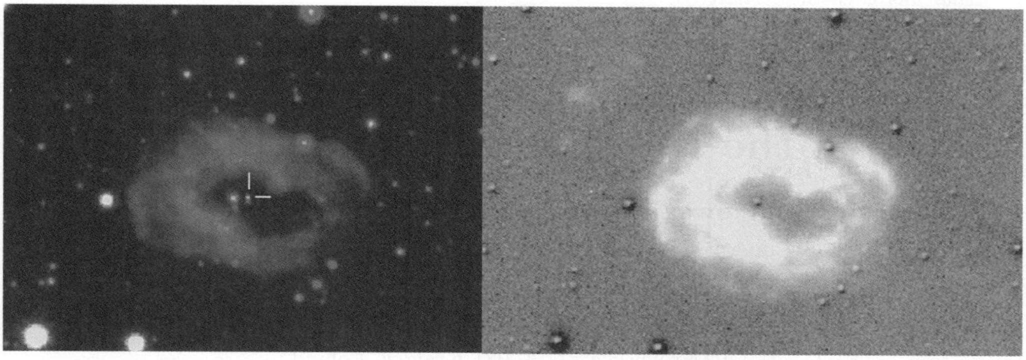

Figure 1. Images of Hf 38 from VPHAS+. On the left, r band, u band and Hα data are shown in red, green and blue respectively The right image show the Hα/r quotient image.

First, by comparing the observed to the synthetic $u-g$ colour. Second, by calculating the Hα/Hβ line ratio using the spectrum from Acker *et al.* 1992. Third, by calculating the flux ratio using literature values of Hβ (Acker *et al.* 1992) and radio flux (Murphy *et al.* 2007). These calculated values were compared to those in the literature. The calculated and literature E(B-V) values averaged to 0.90$\pm$0.08.

The observed CS colours were de-reddened using this value. The synthetic CS spectrum was scaled so the synthetic g band magnitude matched the de-reddened one. The total flux of this spectrum was then calculated. At a distance of 2.2kpc, this resulted in a CS luminosity estimate of 1070L$_{\odot}$, consistent with the expected value.

The de-reddened $g-i$ colour was then compared to the synthetic ones. While the g band will be unaffected by a low-mass main sequence companion, the companion's spectrum would peak in the i band, thereby increasing the observed $g-i$ colour. An i-band excess of 0.465$\pm$0.334 magnitudes was detected in the CS of Hf 38, consistent with a M0V binary companion.

References

Acker, A., Marcout, J., Ochsenbein, F., Stenholm, B., Tylenda, R., & Schohn, C., 1992, *The Strasbourg-ESO Catalogue of Galactic Planetary Nebulae. Parts I, II*

Amari, S., Hoppe, P., & Zinner, E., Lewis R. S.m 1995, *Meteoritics*, 30, 490

De Marco, O., Passy, J.-C., Frew, D. J., Moe, M., & Jacoby, G. H., 2013, *MNRAS*, 428, 2118

Douchin, D., De Marco, O., Frew, D. J.,Jacoby, G. H., Jasniewicz, G., Fitzgerald, M., Passy, J.-C., Harmer, D., Hillwig, T., & Moe, M., 2016, *MNRAS*, 448, 3132

Hillwig, T., Jones, D., De Marco, O., Bond, H., Margheim, S., & Frew, D., 2016, *ArXiv e-prints*: 1609.02185

&Maciel, W. J., 1984, *Astronomy & Astrophysics Supplement Series* 55, 253

Miszalski, B., Parker, Q. A., Acker, A., Birkby, J. L., Frew, D. J., & Kovacevic, A., 2008, *MNRAS*, 384, 525

Murphy, T., Mauch, T., Green, A., Hunstead, R. W., Piestrzynska, B., Kels, A. P., & Sztajer, P., 2007, *MNRAS*, 382, 382

Parker, Q. A., Phillipps, S., Pierce, M. J., Hartley, M., Hambly, N. C., Read, M. A., & MacGilliveray, H. T.. 2005, *MNRAS*, 362, 689

Parker, Q. A., Acker, A., Frew, D. J., Hartley, M., Peyaud, A. E. J., Ochsenbein, F., Phillipps, S., Russeil, D., Beaulieu, S. F., Cohen, M., Köppen, J., Miszalski, B., Morgan, D. H., Morris, R. A. H., Pierce, M. J., & Vaughan, A. E., 2006, *MNRAS*, 373, 79

Werner, K., Deetjen, J. L., Dreizler, S., Nagel T., Rauch, T., & Schuh, S. L., 2003, *Stellar Atmosphere Modeling*, 288, 31

Planetary Nebulae: Multi-Wavelength Probes of Stellar and Galactic Evolution
Proceedings IAU Symposium No. 323, 2016
X. Liu, L. Stanghellini & A. Karakas, eds.

doi:10.1017/S1743921317001892

Search for binary central stars of the Magellanic Clouds PNe

Marcin Gładkowski[1] Marcin Hajduk[2] and Igor Soszyński[3]

[1]Nicolaus Copernicus Astronomical Center, Polish Academy of Sciences,
ul. Rabiańska 8, 87-100 Toruń, Poland
email: seyfert@ncac.torun.pl

[2]National Science Centre,
ul. Królewska 57, 30-081 Kraków, Poland

[3]Obserwatorium Astronomiczne, Uniwersytet Warszawski,
Al. Ujazdowskie 4, 00-478, Warszawa, Poland

Abstract. The Optical Gravitational Experiment (OGLE) was effectively used in discovering binary central stars of planetary nebulae (CSPNe). About 50 binary CSPNe have been hitherto identified in the Galaxy, almost half of them were detected in the OGLE database. We used the OGLE data to search for binary CSPNe in the Magellanic Clouds. We also searched for PNe mimics and removed them from the PNe sample. Here, we present results of the photometric analysis for Small Magellanic Cloud (SMC) and our progress on search of binary central stars in the Large Magellanic Cloud (LMC). So far, we have discovered one binary central star of the PN beyond the Milky Way, which is located in the Small Magellanic Cloud.

Keywords. techniques: photometric, binaries: general, planetary nebulae: general, Magellanic Clouds

---

1. Introduction

The Optical Gravitational Experiment (OGLE) survey was originally dedicated to search for microlensing events. However, the massive photometric surveys provide an excellent opportunity to study many other objects and phenomena such as planetary nebulae (PNe). About 50 binary CSPNe have been hitherto identified in the Milky Way, and almost half of them were discovered using the OGLE data. The number of binary CSPNe is growing, thus it is possible to gain the knowledge about the influence of the common envelope phase of the binary evolution on the morphology of the PNe (De Marco 2009; Miszalski *et al.* 2009b; Boffin 2012). The fraction of close binary CSPNe was established as 12-21% on the basis of the OGLE photometry for the Galactic bulge (Miszalski *et al.* 2009), and is consistent with the fraction given by Bond (2000).

2. Methods of analysis

PNe for the Magellanic Clouds were selected using SIMBAD database, and then the object type and coordinate search criteria. This allowed us to search the objects. Later on, we update our list by adding the objects, which were missing in the SIMBAD database.

The objects were identified by visual comparison of the finding charts, which were available in the literature with the OGLE II-IV I-band reference images. In addition, we observe 27 PNe in the SMC with the 1.0m telescope at South African Astronomical Observatory (SAAO) with the R and H_α filters to confirm some not certain identifications.

We analysed the OGLE-II and -III I-band photometry in the case of the SMC (Hajduk, Gładkowski & Soszyński 2014), and OGLE-III-IV I-band photometry for the LMC using

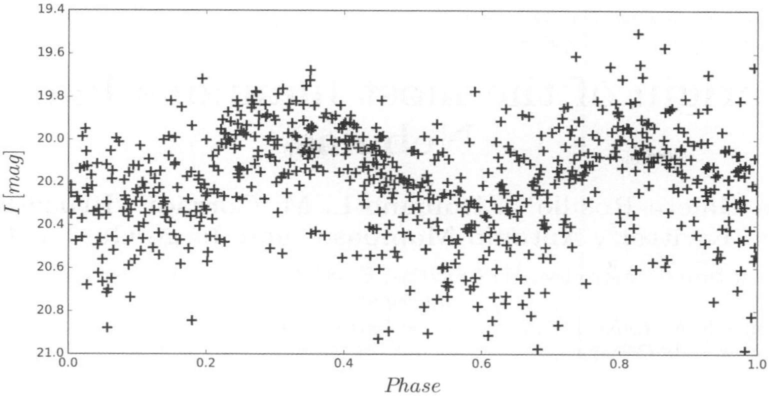

Figure 1. Phased OGLE I-band lightcurve of the Jacoby SMC 1.

the programs PERIOD04 (Lenz *et al.* 2005) and PERANSO (`www.peranso.com`). This software offers a variety of period analysis methods from which we chose the analysis of variance (ANOVA, Schwarzenberg-Czerny 1996).

3. Results

In a case of the SMC we identified 52 objects in the OGLE reference images in total out of 119 candidates. Eight object were find to be variable: six periodic and two irregular objects. We made the spectroscopic observations of five variables with the Southern African Large Telescope (SALT), and another 3 objects were observed using the Radcliffe 1.9m telescope at SAAO. Only Jacoby SMC 1 (see Fig. 1) has a binary CS with a period of 1.16 days assuming the irradiation effect as a cause of the variations, or twice that period in the case of ellipsoidal variations (Hajduk, Gładkowski & Soszyński 2014). The rest of objects turned out to be the PNe mimics.

For the LMC we have identified about 170 objects in the reference images in total of about 800 candidates so far. The work is in progress. We also found four periodic variables. Unfortunately, none of them is a close binary CSPN.

4. Conclusions

So far, we have discovered only one close binary central star of PN beyond our Galaxy - Jacoby SMC 1, which is located in the SMC. We also discovered a new symbiotic star in the SMC: [JD 2002] 11, which was study in details by Hajduk *et al.* (2015).

References

Boffin, H. M. J., Miszalski, B., Rauch, T., *et al.* 2012, *Science*, 338, 773

Bond, H. E. 2000, *ASP Conf. Ser.*, 199, 115

De Marco, O. 2009, *PASP*, 121, 316

Hajduk, M., Gładkowski, M., & Soszyński, I. 2014, *A&A*, 561, A8

Haberl, F. & Sasaki, M. 2000, *A&A*, 359, 573

Hajduk, M., Gromadzki, M., Mikołajewska, J., Miszalski, B., & Soszyński, I. 2015, *Acta Astronomica*, 65, no 2, 139-149

Lenz, P. & Breger, M. 2005, *Comm. Asteroseismol.*, 146, 53

Miszalski, B., Acker, A., Moffat, A. F. J., Parker, Q. A., & Udalski, A. 2009, *A&A*, 496, 813-825

Miszalski, B., Acker, A., Parker, Q. A., & Moffat, A. F. J. 2009, *A&A*, 505, 249

Schwarzenberg-Czerny, A. 1996, *ApJ*, 460, L107

Planetary Nebulae: Multi-Wavelength Probes of Stellar and Galactic Evolution
Proceedings IAU Symposium No. 323, 2016
X. Liu, L. Stanghellini & A. Karakas, eds.

doi:10.1017/S1743921317003465

The origin of the most luminous Planetary Nebulae

Rebeca Galera-Rosillo[1,2], Romano L. M. Corradi[3], Bruce Balick[4], Karen Kwitter[5], Antonio Mampaso[1] and Jorge García-Rojas[1]

[1]Instituto de Astrofísica de Canarias, E-38206, La Laguna, Tenerife, Spain
email: rgr@iac.es
[2]Departamento de Astrofísica, Universidad de La Laguna, E-38205, La Laguna, Tenerife, Spain
[3]Gran Telescopio de Canarias, Centro de Astrofísica de la Palma, E-38712, S/C de Tenerife, Spain
[4]Astronomy Department, University of Washington, Seattle, WA, 98195-1580, USA
[5]Department of Astronomy, Williams College, Williamstown, MA, 01267-2565, USA

Abstract. As part of a systematic effort to characterize the properties and progenitors of the most luminous planetary nebulae (PNe), we obtained a sample among the brightest PNe in two stellar systems of different metallicities: LMC ($Z/Z_{\odot}\sim0.5$) and M31 ($Z/Z_{\odot}\sim1$) by means of a combined effort with the VLT and the 10mGTC. Modelling of these data will allow us to infer the masses of the stellar progenitors, gaining insights into the controversial origin of the universal cutoff of the Planetary Nebulae Luminosity Function (PNLF).

Keywords. Planetary Nebulae: general, galaxies: abundances, galaxies: M31,LMC, ISM: evolution

1. Introduction

The Planetary Nebulae Luminosity Function (PNLF), is the number density distribution of PNe in a system over a given absolute magnitude range (Jacoby 1980). It is an important standard candle on the extragalactic distance ladder, based on the empirical evidence that PN total L in the O[III] 5007Å nebular emission line reaches a maximum value, M*, invariant with galaxy type and with only a small dependence on metallicity (Dopita *et al.* 1992, Ciardullo 2010). As PNe descend from progenitors present in any galactic subsystem, the PNLF method to determine extragalactic distances is equally effective for both spiral and elliptical galaxies, and has been applied to more than 50 galaxies.

Previous theoretical modelling suggested progenitor initial masses larger than 2 M$\odot$ (Schönberner *et al.* 2007, Méndez *et al.* 2008, Catalán *et al.* 2008) for these bright PNe, but relatively massive stars are too scarce or absent in old systems such as elliptical galaxies or haloes to produce the observed bright PNe. Alternative channels of PNe production that populate the bright end of the PNLF, as mergers, mass accretion in interacting binary stars or blue stragglers have been proposed (Ciardullo *et al.* 2005, Soker 2006). If they are able to provide the observed number of M* PNe is still controversial.

2. Observations and inmediate objectives

We present new very deep optical spectra of 8 PNe within the bright tip of the m 5007Å PNLF in the M31 spiral galaxy, from OSIRIS at GTC, and 4 spectra of PNe at the Large Magellanic Cloud, within 0.6 mag from the PNLF cutoff from FORS2 at VLT.

The first goal using these data is to obtain precise chemical abundances for a number of elements (He, N, O, S, Ne, Ar, Cl) to test the theoretical yield predictions for progenitors of different stellar masses, and determine - via Cloudy modelling of the spectra - temperatures and luminosities for the central stars which can be located on evolutionary tracks of the H-R diagram yielding mass estimates.

In the case of the VLT data, we will try besides to tackle the abundance discrepancy problem: optical recombination lines (ORLs) provide abundance values that are systematically larger than those obtained using CELs. We will determine the abundance discrepancy factor (adf) in these LMC M* PNe, which will not only quantify the uncertainty in the real nebular abundances but will also provide an independent measure of the critical C/O abundance ratio using ORLs.

References

Jacoby, G. H. 1980, *ApJs*,42, 1J

Dopita, M. A., Jacoby, G. H., & Vassiliadis, E. 1992, *Apj*, 389, 27

Ciardullo, R. 2010 *PASA*, 27, 149

Schönberner, D., Jacob, R., Steffen, M., & Sandin, C. 2007, *A&A*, 473, 467

Méndez, R. H., Teodorescu, A. M., Schönberner, D., Jacob, R., & Steffen, M. 2008, *Apj*, 681, 325-332

Catalán, S., Isern, J., García-Berro, E., & Ribas, I. 2008, *MNRAS*, 387, 1693

Ciardullo, R., Sigurdsson, S., Feldmeier, J. J., & Jacoby, G. H. 2005, *Apj*, 629, 499

Soker, N. 2006, *Apj* 640, 966

Planetary Nebulae: Multi-Wavelength Probes of Stellar and Galactic Evolution
Proceedings IAU Symposium No. 323, 2016
X. Liu, L. Stanghellini & A. Karakas, eds.

doi:10.1017/S1743921317002356

The LAMOST spectroscopic survey of planetary nebulae in M31 and M33

Maosheng Xiang[1], Xiaowei Liu[2], Meng Zhang[2], Haibo Yuan[3] and Zhiying Huo[1]

[1]National Astronomical Observatories, Chinese Academy of Science, China
email: msxiang@nao.cas.cn

[2]Department of Astronomy, Peking University, China email: x.liu@pku.edu.cn

[3]Department of Astronomy, Beijing Normal University, China

Abstract. We present LAMOST observations and radial velocity measurements of about 1500 planetary nebulae (PNe) in M31 and M33. Most of the PNe are previously known, but 36 of them are newly discovered in the outskirts of M31, and the furthest one has a projected distance larger than 50 kpc away from the galaxy center. Eighteen objects in the area of M33 are probably newly discovered PNe, and quite a few of them are associated with previously known clusters. For all the 1500 PNe, homogeneous radial velocities are measured from the LAMOST spectra, with a typical uncertainty of a few km s^{-1}.

Keywords. planetary nebulae: general; galaxies: kinematics and dynamics

1. Introduction

The Guo Shoujing telescope (LAMOST) collects low resolution (R$\sim$1800) fibre spectra of 4000 objects simultaneously in a sky area of 20 square degrees. The large number of fibres and wide field of view make it particularly suitable to search for planetary nebulae (PNe) as well as other special objects (e.g. supergiants, H II regions, globular clusters) in the M31 and M33 galaxies. PNe exhibit prominent emission lines (e.g. [O III] λ5007, Hα λ6563), thus can be easily detected and identified. Thousands of PNe and candidates in the area of M31 and M33 have been targeted by LAMOST in both the commissioning (9/2009 - 9/2011) and the survey stage (10/2011 - 6/2017). The main scientific goals include (1) Validating wavelength calibration of LAMOST spectra; (2) Searching for new PNe, especially in the out skirts of the galaxies; (3) Obtaining accurate and homogeneous measurements of radial velocities and emission line fluxes of a large sample of PNe to study stellar populations, kinematics and chemistry of M31 and M33. With the commissioning data, Yuan *et al.* (2010) discovered 36 PNe in the outskirts of M31. In this work, we present results from observations carried out in the survey stage.

2. PN identifications and radial velocity measurements

A cross-identification with the emission line object catalog of Merrett *et al.* (2006) yields 1756 observations of 1351 common objects. Comparison of [O III] λ5007 magnitudes deduced from the LAMOST spectra with those of Merrett *et al.* shows good correlation, indicating that flux calibration of LAMOST spectra is reliable. The differences in [O III] magnitudes of individual PNe have a mean and standard deviation of 0.17 and 0.58 mag, respectively. Based on those common objects, a criterion is defined in the F(5007)/F(4861) – [O III] magnitude plane to distinguish PNe and contaminations from other emission line objects, e.g. H II regions.

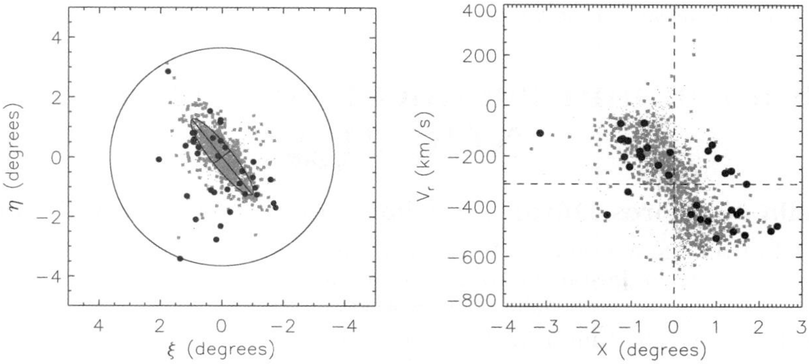

Figure 1. Left: Spatial distribution (Left) and P-V digram (Right) for newly discovered (black) and previously known (grey) PNe of M31. Previously known PNe that have LAMOST observation are shown with squares, while those do not have LAMOST observations are shown with small dots. For the left panel, the circle corresponds a projected distance of 50 kpc to the M31 center. The ellipse has a semi-major radius of 95.3 arcmin, an inclination angle of 77.°5 and a position angle of 38.°1. For the right panel, the abscissa is the projected position along the major axis of M31. The horizontal dashed line denotes a radial velocity of $-310\,\mathrm{km\,s^{-1}}$.

After inspecting literatures and the SIMBAD database, we confirm that 36 PNe are newly discovered around M31. Their spatial distribution and locations in the position-velocity (P-V) diagram are shown in Figure 1. Two of the newly discovered PNe have moderate continnum in their spectra, for which one is possibly contaminated by a nearby Wolf-Rayet star, and the other is associated with globular cluster candidate. The furtherest newly discovered PN has a projected distance larger than 50 kpc away from the M31 center, and probably associated with the giant stellar stream. In the sky area of M33, eighteen objects are probably PNe newly discovered, 5 of them are isolated with very weak continuum in their spectra, while the other 13 are associated with previously known globular clusters or Wolf-Rayet stars.

A comparison of radial velocities deduced from [O III] λ5007 lines of LAMOST spectra with those of Merrett *et al.* yields a mean difference of $-10.0\,\mathrm{km\,s^{-1}}$, and a standard deviation of $20.3\,\mathrm{km\,s^{-1}}$. While a comparison with Halliday *et al.* (2006) yields a mean difference of $-3.2\,\mathrm{km\,s^{-1}}$ and a standard deviation of $7.0\,\mathrm{km\,s^{-1}}$. For objects with [O III] magnitudes brighter than 23 mag, the standard deviation reduces to $5.0\,\mathrm{km\,s^{-1}}$. Radial velocities for ∼1500 PNe in M31 and M33 are deduced from LAMOST spectra, with a typical uncertainty of a few $\mathrm{km\,s^{-1}}$. Fluxes of prominent emission lines of those PNe are also measured.

References

Halliday C., Carter D., Bridges T. J., Jackson Z. C., Wilkinson M. I., Quinn D. P., Evans N. W., Douglas N. G., Merrett H. R., Merrifield M. R., Romanowsky A. J., Kuijken K., & Irwin M. J. 2006 *MNRAS*, 369, 97

Merrett H. R., Merrifield M. R., Douglas N. G., Kuijken K., Romanowsky A. J., Napolitano N. R., Arnaboldi M., Capaccioli M., Freeman K. C., Gerhard O., Coccato L., Carter D., Evans N. W., Wilkinson M. I., Halliday C., & Bridges T. J. 2006 *MNRAS*, 369, 120

Yuan H.-B., Liu X.-W., Huo Z.-Y., Zhang H.-H., Zhao Y.-H., Chen J.-J., Bai Z.-R., Zhang H.-T., Zhang H.-W., García-Benito R., Xiang M.-S., Yan H.-L., Ren J.-J., Sun S.-W., Zhang Y., Li Y.-P., Lu Q.-S., Wang Y., Ni J.-J., & Wang H. 2010, *Research in Astronomy & Astrophysics*, 10, 599

Planetary Nebulae: Multi-Wavelength Probes of Stellar and Galactic Evolution
Proceedings IAU Symposium No. 323, 2016
X. Liu, L. Stanghellini & A. Karakas, eds.

doi:10.1017/S1743921317000576

Kinematic and chemical analysis of PNe in NGC 3109

Sheila N. Flores-Dúran[1], Miriam Peña[1] and María T. Ruiz[2]

[1]Instituto de Astronomía, Universidad Nacional Autónoma de México, Apdo. Postal 70264, Ciudad de México, 04510, México
email: sflores@astro.unam.mx

[2]Depto. de Astronomía, Universidad de Chile, Casilla 36D, Las Condes, Santiago, Chile

Abstract. We present high resolution spectroscopy obtained with MIKE-Magellan and MES OAN-SPM of a number of planetary nebulae (PNe) and H II regions, distributed along the dwarf irregular galaxy NGC 3109 and compare their kinematical behavior with the one of H I data. We aim to determine if there is a kinematical connection among these objects. We also perform a revision of the chemical composition of PNe and H II regions in this galaxy and discuss it in comparison with stellar evolution models.

Keywords. galaxies individual: NGC 3109, galaxies: kinematics ans dynamics, ISM abundances: H II regions and Planetary Nebulae.

1. Introduction

The kinematics of Planetary Nebulae (PNe) in galaxies is a clue to understand the behavior of low-intermediate mass stars (LIMS) and their relation with other components of the galaxies.

NGC 3109 is a Magellanic-type irregular galaxy, classified as a late-type spiral SB(s)m. It is one of the most metal-poor galaxies in the vicinity of the Local Group and contains a H I rotating disk (Ott *et al.* 2012).

Peña *et al.* (2007a) reported the discovery of twenty PNe in NGC 3109 and Peña *et al.* (2007b) analyzed the chemistry of some of them finding that the oxygen abundance in PNe is systematically higher than in H II regions.

2. Overview

The heliocentric radial velocities (RV) of the objects were compared to the velocities of the H I disk at the same projected position. Physical conditions and ionic abundances of PNe and H II regions in NGC 3109, were obtained from the emission lines given by Peña *et al.* (2007b). Delgado-Inglada *et al.* (2014) ionization correction factors were used to derive the total chemical abundances.

The radial velocities (RV) were measured with a precision better than 7-8 km s^{-1}. Fig. 1 shows velocities of objects (relative to the systemic velocity of 404 km/s) vs. their position projected along the major axis of the H I disk. It can be seen that H II regions (open black squares) follow closely the rotation of the H I disk but with a large dispersion. The PNe (black dots) also rotate in the same direction that the disk. We have included RV given by Evans *et al.* (2007) for luminous blue supergiant stars (green asterisks). These relatively young objects, show a rotation similar to the H I rotation but with a large dispersion, even larger than that presented by our PN sample.

Figure 1

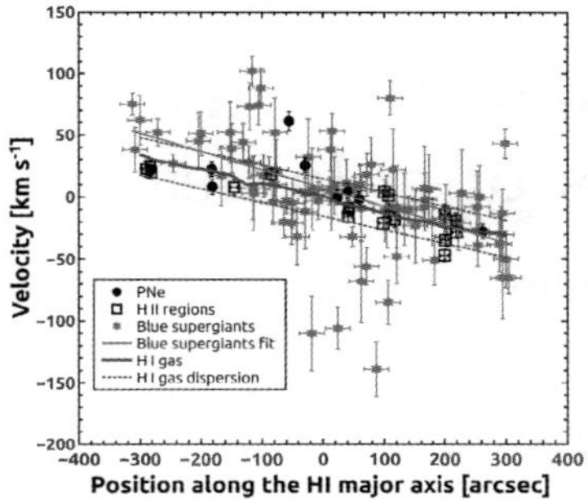

Figure 2

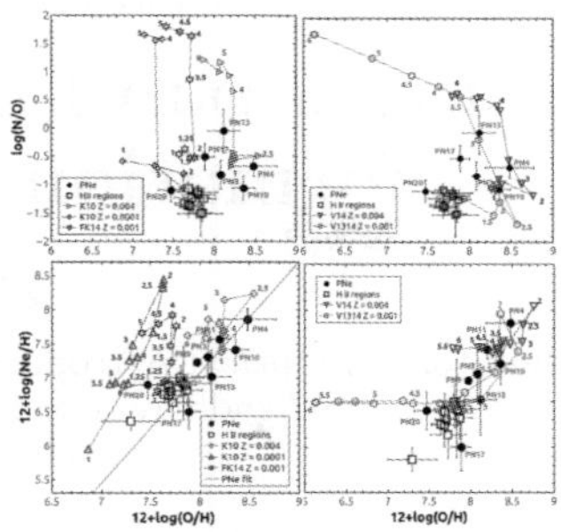

From the chemical abundance determinations we confirm that PNe are enriched in oxygen as Peña *et al.* (2007b) previously found. The oxygen abundance for H II regions is $\log O/H + 12 = 7.76 \pm 0.07$ and PNe show, in average, significantly higher oxygen abundance $\log O/H + 12 = 8.17 \pm 0.17$. Similarly to what has been suggested for some of the PNe in the Magellanic Clouds and other metal-poor galaxies, we argue that oxygen in the PNe in NGC 3109 is affected by the third dredge-up event in their progenitors. This also occurs for neon, as we notice that neon is also enriched in PNe, as predicted by stellar evolution models for a $Z = 0.004$ (Karakas 2010, K10; Ventura *et al.* (2013, 2014b, V1314) and Ventura *et al.* (2014a, V14)), see Fig. 2. We found one confirmed Peimbert Type I PN.

3. Results

We found that H II regions share the H I gas kinematics, with differences in velocities smaller than 15 km s^{-1}. On the other hand, PNe and blue supergiant stars, even when they rotate in the same direction, show a large dispersion in velocities. This could be indicating that these young-intermediate age objects (blue supergiant stars and PNe) belong to the bar that has been reported for the stellar component of NGC 3109.

The PNe in NGC 3109 are oxygen overabundant with respect to the H II regions (0.41 dex). In average Ne enrichment is also seen in agreement with predictions in K10 and Ventura *et al.* stellar evolution models. This is an important evidence supporting predictions of stellar evolution models for low-intermediate mass stars, which produce O-enriched stars a low metallicity. Such results are important to identify the best models for the chemical evolution of PNe from low metallicity stellar progenitors.

A full version of this work will be published elsewhere.

Acknowledgements

This work received financial support from DGAPA-PAPIIT (UNAM) grant IN109614.

References

Delgado-Inglada, G., Morisset, C., & Stasinska, G. 2014, *MNRAS*, 440, 536
Evans, C. J., Bresolin, F., Urbaneja, M. A., Pietrzynski, G., & Gieren, W., 2007, *ApJ*, 659, 1198
Karakas A. I. 2010, *MNRAS*, 403, 1413
Ott, J., Stilp, A. M., Warren, S. R. *et al.* 2012, *AJ*, 144, 123
Peña, M., Richer, M. G, & Stasinska, G. 2007, *A&A*, 144, 123
Peña, M., Stasinska, G., & Richer, M. G. 2007, *A&A*, 476, 745
Ventura, P., Di Criscienzo, M., Carini, R., & D'Antona, F. 2013, *MNRAS*, 431,3642
Ventura, P., Di Criscienzo, M., D'Antona, F., *et al.* 2014, *MNRAS*, 437, 3274
Ventura, P., Dell'Agli, F., Schneider, R., *et al.* 2014, *MNRAS*, 439, 977

Planetary Nebulae: Multi-Wavelength Probes of Stellar and Galactic Evolution
Proceedings IAU Symposium No. 323, 2016
X. Liu, L. Stanghellini & A. Karakas, eds.

doi:10.1017/S1743921317002009

Using *u* band emission to trace star formation rates of galaxies

Zhimin Zhou

Key Laboratory of Optical Astronomy, National Astronomical Observatories, Chinese Academy of Sciences, Beijing, 100012, China
email: zmzhou@bao.ac.cn

Abstract. Star formation rate (SFR) is one of the most important diagnostics in understanding the evolution of galaxies across cosmic times. In order to explore the possibility of using the optical u-band luminosities to estimate SFRs of galaxies, we show the correlations between u band, Hα and infrared luminosities by combing the data from the South Galactic Cap u band Sky Survey (SCUSS) with the Sloan Digital Sky Survey (SDSS) and the *Wide-field Infrared Survey Explorer* (WISE). We derive the u versus Hα relation and the u and 12 μm relations to calibrate the u-band luminosity as an SFR indicator.

Keywords. galaxies: star formation, galaxies: ISM

1. Introduction

It is well-known that star formation is the critical activity to build up the stellar mass of galaxies. The star formation rate (SFR) is one of the most important diagnostics in understanding the evolution of galaxies. The measurement of the SFRs can give clues to the physical processes in shaping the galaxies. SFRs can be traced by many indicators, including both continuum and nebular recombination emission lines (Kennicutt & Evans 2012), such as the ultraviolet (UV) continuum, Hα nebular line, infrared (IR) emission (Kennicutt 1998).

The optical u-band luminosity of a galaxy is dominated by young stars of age < 1 Gyr, and has been proved to be a reasonable star formation tracer (Hopkins *et al.* 2003). The availability of recent and upcoming deep optical surveys, such as the Canada-France-Hawaii Telescope Legacy Survey (CFHTLS), the South Galactic Cap u band Sky Survey (SCUSS; Zhou *et al.* 2016), the Large Synoptic Survey Telescope (LSST), provide an opportunity to measure SFRs of unprecedented large samples with u band data. With the advent of these deep and wide-field surveys, it becomes necessary to recalibrate the u-band SFR.

2. Data and Sample

We investigate the u band observations from SCUSS as an indicator of SFRs here in an effort to determine its feasibility and precision. SCUSS is a u band ($\lambda \sim 3538$Å) survey imaging ~5,000 deg^2 in the northern part of the Southern Galactic Cap, it provides deeper u band photometric data than SDSS. These data are used to compare with H$\alpha$and IR SFR indicators to derive an accurate $u-$SFR calibration for normal star-forming galaxies. Hαluminosities are derived from the emission line measurements of galaxy spectra from MPA-JHU spectroscopic analysis (Tremonti *et al.* 2004). IR luminosities are obtained from the ALLWISE source catalogue which provies positional and photometric information for over 747 million objects detected on the deep, coadded WISE Images.

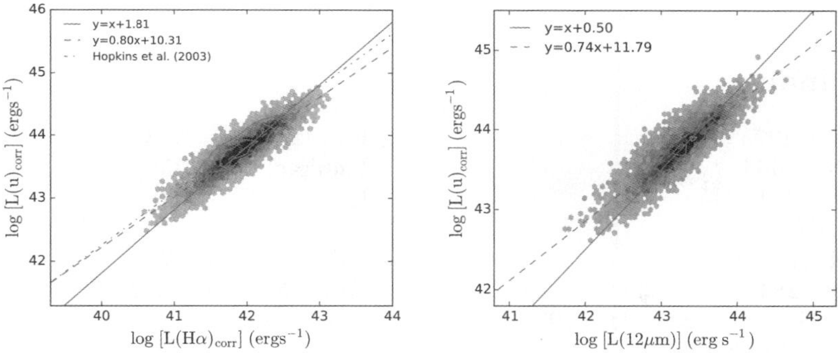

Figure 1. *Left*: Comparison of the luminosities of u band and Hα. The Hαluminosities are derived from the MPA-JHU catalogue and are corrected for the dust extinction and aperture bias. The u band luminosities are also corrected for the dust extinction. The green contours are number densities of the sample distribution. The red solid and dash lines represent the best linear and nonlinear fits in this work. The blue dash-dot line represents the relation by Hopkins *et al.* (2003) using SDSS DR1. *Right*: u-band luminosities as a function of the WISE 12 μm luminosities. The u-band luminosities are corrected for the dust extinction. The best linear and nonlinear fittings are shown with solid and dashed lines, respectively.

We match the SCUSS sources with the MPA-JHU catalogue and ALLWISE catalogue. In order to produce the accurate SFR calibration, we select the *pure* star forming galaxies with no contamination of active galactic nucleus (AGN) based on the BPT diagnostic diagram (Baldwin *et al.* 1981).

3. Result and Discussion

Figure 1 compares the luminosities of u band and Hαemission line. It can be clearly seen that the attenuation-corrected u-band luminosities correlate tightly with the Balmer decrement-corrected Hαluminosities with the Spearman rank-order correlation coefficient of 0.88. The best linear and nonlinear fits are also plotted in the figure along with the relation from Hopkins *et al.* (2003) for comparison. Our fitting results are consistent to that of Hopkins *et al.* (2003). The relation of u band and WISE 12 μm luminosities is also shown in Figure 1 along with the linear and nonlinear fits plotted. The scatter of the nonlinear fits is 0.18 dex. The systematic residuals of our calibrations are mainly affected by dust attenuation and old stellar population.

References

Baldwin J. A., Phillips M. M., Terlevich R., 1981, PASP, 93, 5
Kennicutt, R. C. 1998, ARAA, 36, 189
Kennicutt, R. C. & Evans, N. J. 2012, ARAA, 50, 531
Hopkins, A. M., Miller, C. J., Nichol, R. C., *et al.* 2003, ApJ, 599, 971
Tremonti, C. A., Heckman, T. M., Kauffmann, G., *et al.* 2004, ApJ, 613, 898
Zhou, X., Fan, X.-H., Fan, Z., *et al.* 2016, Research in Astronomy and Astrophysics, 16, 69

Author index

Object Index

Subject Index